HALOPHYTES AS A RESOURCE FOR LIVESTOCK AND FOR REHABILITATION OF DEGRADED LANDS

Tasks for vegetation science 32

SERIES EDITORS

The titles published in this series are listed at the end of this volume.

Halophytes as a resource for livestock and for rehabilitation of degraded lands

edited by

VICTOR R. SQUIRES & ALI T. AYOUB

Proceedings of the International Workshop on
Halophytes for reclamation of saline wastelands and as a resource for livestock
Problems and prospects

Nairobi, Kenya, 22–27 November 1992

organised by
United Nations Environment Programme (UNEP)
&
The University of Adelaide, Australia

sponsored by
United Nations Environment Programme (UNEP)
International Association for Ecology (INTECOL)
Australian Development Assistance Bureau (AIDAB)

SPRINGER SCIENCE+BUSINESS MEDIA, B.V.

Library of Congress Cataloging-in-Publication Data

International Workshop on Halophytes for Reclamation of Saline Wastelands and as a Resource for Livestock: Problems and Prospects (1992 : Nairobi, Kenya)
Halophytes as a resource for livestock and for rehabilitation of degraded lands : proceedings of the International Workshop on Halophytes for Reclamation of Saline Wastelands and as a Resource for Livestock: Problems and Prospects / edited by Victor Squires and Ali T. Ayoub.
p. cm. -- (Tasks for vegetation science ; 32)
Includes index.
ISBN 978-94-010-4351-9 ISBN 978-94-011-0818-8 (eBook)
DOI 10.1007/978-94-011-0818-8
1. Salt-tolerant crops--Congresses. 2. Halophytes--Utilization--Congresses. 3. Plants for land reclamation--Congresses.
I. Squires V. R. (Victor R.), 1937- . II. Ayoub, A. III. Title.
IV. Series.
SB317.S25I57 1992
636'.01'09154--dc20 93-44248

ISBN 978-94-010-4351-9

Printed on acid-free paper

Contents

PART C: REGIONAL OVERVIEWS

PART D: CASE STUDIES FROM IRRIGATED LAND

PART E: CASE STUDIES FROM RANGELAND

Foreword

Desertification (land degradation in arid, semi-arid and dry sub-humid areas resulting mainly from adverse human impacts) is the main environmental problem of dry lands, which occupy more than 40 per cent of the total global land area. The phenomenon threatens about 3.6 billion hectares and currently affects the livelihood of about 900 million people.

The world is now losing annually about 1.5 million hectares of total irrigated lands (240 million hectares) due mostly to salinization, mainly in drylands. Salt affected soils are widely distributed throughout the arid and semi-arid regions, and particularly severe in China (7 million ha), India (20 million ha), Pakistan (3.2 million ha), USA (5.2 million ha), as well as Near East, southern Europe and elsewhere.

Demands on production have increased the pressure on existing productive land and moved the limits of production onto increasingly marginal lands. Wise land-use practices have yet to be developed for such conditions. The Executive Director of UNEP reported to the Governing Council in February 1992 concerning the "Status of Desertification and Implementation of the United Nations Plan of Action to Combat Desertification (PACD)". The Report concludes that major efforts to implement the PACD had gone into supporting measures rather than concrete corrective field operations. Little evidence of progress was found in irrigated croplands, rainfed croplands or rangelands. It was recommended that every piece of land should be used in keeping with its ecological characteristics, natural capabilities and constraints.

The time for action is running short as land degradation expands, and the cost of rehabilitation measures escalates from year to year. Reserves of potentially arable land are hard to find and are unevenly distributed. Most of the available land is located within fragile and ecologically sensitive regions.

Among the objectives of the World Soils Policy adopted by UNEP Governing Council in 1982 is "to enlarge and improve the world's supply of arable agricultural land through irrigation, flood control, and reclamation". This book reflects the concern of UNEP over the vast world lands already degraded and being degraded. It is directed to government administration at various levels, environmental protection services, social and political organizations as well as research and production bodies acting in the field of environmental protection and rational management of natural resources for food production.

We acknowledge with gratitude the contributions made by the experts meeting on the subject in Nairobi, November 1992.

United Nations Environment Programme
P.O. Box 30552, Nairobi, Kenya

International Association for Ecology
Secretariat: E. Drawer
Savannah River Ecology Laboratory
Aiken, South Carolina, USA

Preface

Halophytes grow in many arid and semi arid regions around the world. Those that have greatest value for livestock and for land reclamation grow in mid-latitude regions and occur on every continent (except Antarctica) and on some important islands.

Halophytes have received much more attention lately. They have demonstrated promise in contributing to livestock production and environmental improvement, particularly in arid lands. A new and important role is in sequestration of carbon dioxide and in rehabilitation of degraded lands.

The previous volumes in this series (T:VS 2 and T:VS 27 and 28) covered such widely different fields as physiology, plant/climate interactions, soil/water relations, adaptive biology, and potential role of halophytes in irrigated agriculture and for reclamation of salt-affected lands.

The present volume is in seven parts and attempts to elucidate new aspects of the problems faced in the utilization of this special group of plants as a resource for livestock. It also explores the role of halophytes in the rehabilitation of degraded lands, especially in the world's drier regions. The emphasis here on Case Studies is very apt. The insights gained from case studies will be of immense value for designing halophyte utilization programmes that are effective in meeting the local challenges.

The drylands of the world are not only vast but quite diverse. It is in these that terrestrial halophytes are most commonly found although some maritime species have a wider geographic spread. As the areas affected by salinity increases and as populations of humans and livestock rise, greater emphasis will be placed on using plant resources which are adapted to the changing conditions. There is scope for transfer of halophyte species from one geographic region to another. Better knowledge of what species and ecotypes exist and what characteristics they have will assist in the task of species selection for any revegetation projects.

Although the definition of a halophyte will vary, emphasis in this book will be on plant species that grow naturally in saline habitats. Arid and semi-arid lands are not exclusively vegetated by halophytes but as a group they are quite often conspicuous.

There is no single taxonomic group (genus or family) which make up the bulk of halophytes. They are represented by several thousand species of trees, shrubs, forbs and grasses.

A problem for halophyte biologists working in drylands is the confused taxonomy. Similarly, little is known of the genetics, not only of the species growing there but even of the genera to which they belong.

Many grow in natural salines or in other dry environments which are rapidly deteriorating through the process of desertification. Because the pattern of distribution of halophytes has remained more or less intact it provides an immensely valuable and diverse resource with which to rehabilitate damaged habitats and provide forage for the world's burgeoning livestock populations.

A monograph on the problems and prospects for better utilization of halophytes, especially shrubs, is both timely and necessary. Timely, because of the unprecedented interest in halophytes as a rangeland resource, as a crop for direct sea water irrigation, and as a means of rehabilitating saline wastelands, and necessary because of the need to avoid expensive duplication of research effort.

The UNEP-sponsored International Workshop, held in Nairobi in November 1992, had a number of objectives, some short term and some longer term. These were to:

- strengthen and accelerate worldwide exchange of knowledge concerning halophytes and their role in the rehabilitation of saline lands and as resource for livestock husbandry;
- exchange viewpoints between individuals and organizations interested in the use, improvement and conservation of halophytes;
- deal directly with the management of both natural grazing lands (rangelands) and the animals that use these lands; emphasize those principles of greatest importance to full understanding of saline lands and their sound management and, where necessary, their rehabilitation.

In total, develop better understanding of the role of halophytes and the measures that can be taken to promote their better use so as to ensure maximum benefits, including the conservation of biodiversity.

Controversy rages over the feeding value of many halophytes since their performance in field trials in the rangeland context suggests that, at best, they are maintenance rations. Laboratory analyses on the other hand, suggest that halophyte forages are comparable with cultivated forages such as *Medicago sativa*.

Reconciliation of these opposing views might come from using halophytes in mixed rations to lot-fed animals. Such schemes are being trialled and preliminary data suggest that this may become a major role for specially-grown halophyte crops either as forage or as a by-product of oil-seeds from succulents such as *Salicornia*.

From the UNEP-sponsored Workshop, of which this book is the outcome, there emerged an awareness of the need to exploit existing knowledge as fully as possible and a challenge to the fatalism that conditions attitudes to dryland degradation.

VICTOR R. SQUIRES
University of Adelaide, Australia

ALI T. AYOUB
UNEP, Nairobi, Kenya

List of contributors

EDITORS

A.T. Ayoub, United Nations Environment Programme (UNEP), P.O. Box 47074, Nairobi Kenya (Chapter 32)

V.R. Squires, Department of Environmental Science and Rangeland Management, University of Adelaide, Roseworthy 5371, Australia (Chapter 1; Chapter 12, with E.P. Glenn, C.U. Mota, J.J. Riley, R.S. Swingle & M.C. Watson)

CONTRIBUTORS

R. Ahmad, Biosaline Research Laboratories, Department of Botany, University of Karachi, Karachi, Pakistan (Chapter 17, with M.A. Bodla, M.R. Chaudhry & S. Ismail)

F. Al-Farrajii, Department of Desertification Control, Ministry of Agricultural and Irrigation, Baghdad, Iraq (Chapter 20, with M.R. Al-Hilli)

K.H. Batanouny, Faculty of Science, Cairo University, Giza, Egypt (Chapter 11)

M.T. Bayoumi, Desert Research Center, Mataria, Cairo, Egypt (Chapter 25, with H.M. El Shaer)

M.A. Bodla, International Water and Salinity Research Institute, Lahore, Pakistan (Chapter 18, with M.S. Baig & S.R.A. Shamsi)

R.E. Brevedan, CERZOS and Departments of Agronomy and Biology, Universidad Nacional del Sur, Bahia Blanca, Argentina (Chapter 13, with O.A. Femandez & C.B. Villamil)

R. Choukr-Allah, Institut Agronomique et Veterinaire, Université Hassan II, Agadir, Morocco (Chapter 23, with G.W. Jones & L. Kenny)

H.M. El Shaer, Desert Research Center, Mataria, Cairo, Egypt (Chapter 28, with E.A. Gihad)

E.A. Gihad, Faculty of Agriculture, Cairo University, Giza, Egypt (Chapter 6, with H.M. El Shaer)

E.P. Glenn, Environmental Research Laboratory, Tucson, Arizona, USA (Chapter 12, with C.U. Mota, J.J. Riley, V.R. Squires, R.S. Swingle & M.C. Watson)

H.N. Le Houérou, CEFE/CNRS, Montpellier, France (Chapter 10)

M.A.M. Khair, Agricultural Research Corporation, Wad Medani, Sudan (Chapter 19)

A. Koocheki, College of Agriculture, Ferdowsi University of Mashad, Mashad, Iran (Chapter 21, with M.N. Mohalati)

A. Llerena, National Water Commission, SARH, Mexico City, Mexico (Chapter 30)

C.V. Malcolm, Formerly, Department of Agriculture, Perth, WA, Australia; present address: RMB, 1045 Denmark, WA 6333, Australia (Chapter 4)

S. Miyamoto, Agricultural Research Center, Texas A & M University, El Paso, Texas, USA (Chapter 5, with E.P. Glenn & N.T. Singh)

S. Nawaz, Department of Livestock Management, University of Agriculture, Faisalabad, Pakistan (Chapter 27, with S.H. Harja & R.H. Quereshi)

J.W. O'Leary, Department of Plant Sciences, The University of Arizona, Tucson, Arizona, USA (Chapter 2, with E.P. Glenn)

A.E. Osman, ICARDA, Aleppo, Syria (Chapter 22, with M.O. Shalla)

Qihong Jin, Institute of Botany, Academia Sinica, Beijing, PR China (Chapter 29)

J.J. Riley, Environmental Research Laboratory, Tucson, Arizona, USA (Chapter 26, with E.P. Glenn & C.U. Mota; Chapter 12, with E.P. Glenn, C.U. Mota, V.R. Squires, R.S. Swingle & M.C. Watson).

R.K. Ruwa, Kenya Marine & Fisheries Research Institute, Mombasa, Kenya (Chapter 14)

S.K. Saxena, Central Arid Zone Research Institute, Jodphur, Rajasthan, India (Chapters 15 and 16)

L. Stringi, Instituto di Agronomia Generale e Coltivazioni Erbacee, Universita degle Studi de Palermo, Palermo, Italy (Chapter 33, with A. Accardo, G. Amato & D. Giambalvo)

R.S. Swingle, Department of Animal Sciences, University of Arizona, Tucson, Arizona, USA (Chapter 7, with E.P. Glenn & J.J. Riley; Chapter 12, with E.P. Glenn)

I. Szabolcs, Research Institute for Soil Sciences & Agricultural Chemistry, Hungarian Academy of Sciences, Budapest, Hungary (Chapter 3)

P.N.V. Truong, Natural Resource Management Unit, Department of Primary Industries, Indooroopilly, Queensland, Australia (Chapter 31)

B.E. Warren, Department of Agriculture, WA, Sheep Industries Branch, Katanning WA 6317, Australia (Chapter 24, with T. Casson & D.H. Ryal)

A.D. Wilson, Rangeland Research and Management Consultant 'Cal Col', Deniliquin, NSW, Australia (Chapter 8)

Zhang Li-yun, Xinjiang Institute of Biology, Pedology and Desert Research, Chinese Academy of Sciences, Urumqi, Xinjiang, PR China (Chapter 9, with Xia Yang & Zou Yun)

1. Overview of problems and prospects for utilizing halophytes as a resource for livestock and for rehabilitation of degraded lands

VICTOR R. SQUIRES
Department of Environmental Science and Rangeland Management, University of Adelaide, Australia

Abstract. Halophytes are defined as plant species growing naturally in saline habitats. Their role and potential are discussed under seven main headings: land rehabilitation, as irrigated crops using saline water as feed for livestock, as medicinal plants, as fuelwoods, as shade and shelter and for sequestration of carbon dioxide. Prospects and problems are discussed.

Key words: Carbon dioxide sequestration, Deserts, Irrigation water, Saline habitats, Salt tolerance

Introduction

There are many definitions of halophytes. The principal emphasis in this paper will be on plant species growing naturally in saline habitats (Le Houérou 1994). Halophytes (halo = salt, phyte = plant) are not a single taxonomic group, but are represented by several thousand species of forbs, grasses, shrubs and trees. There are leafless succulents and leafy bushes; there are species found only in salt marshes and species that grow in the deserts. Halophytes are distributed from coastal areas to mountains and lowland deserts (Fig. 1).

Le Houérou (1994) has reviewed the sometimes confusing literature on definitions and ecophysiological and ecological classification of halophytes. He concludes that the nature and ecology of halophytes is very complex. Not surprisingly, so-called halophytes differ markedly in their degree of salt tolerance. There are some difficulties in delimiting halophytic habitats (Duncan 1974).

Halophytes occupy important niches in many (particularly arid) ecosystems. This is due to the ease with which they adapt to many diverse and generally harsh environments. Halophytes, especially xerohalophytes, are distributed in many coastal and desert regions of the world and occupy a wide range of saline to alkaline habitats (Chapman 1974).

Utilization of halophytes

The demands imposed on natural resources by population increases have caused a re-assessment of our unused or underused lands. Regions heretofore considered too dry or too salty have suddenly become an integral part of the resource. With proper management, halophytes can make a significant contribution to mankind and to world productivity.

There are several broad situations in which halophytes are used by livestock:

Firstly, as rangeland browse/grazing. This occupies the largest area particularly coastal or inland saline areas where saline soils occur (Szabolcs 1994). Rangelands which support halophytic plant communities occur in both developed and developing countries. The mode of utilization varies between the two systems. The former are predominantly fenced rangelands on commercial ranches while the latter is characterized by traditional pastoralism, often involving nomadic or transhumant grazing.

V.R. Squires & A.T. Ayoub (eds.), *Halophytes as a resource for livestock and for rehabilitation of degraded lands*, 1–6.

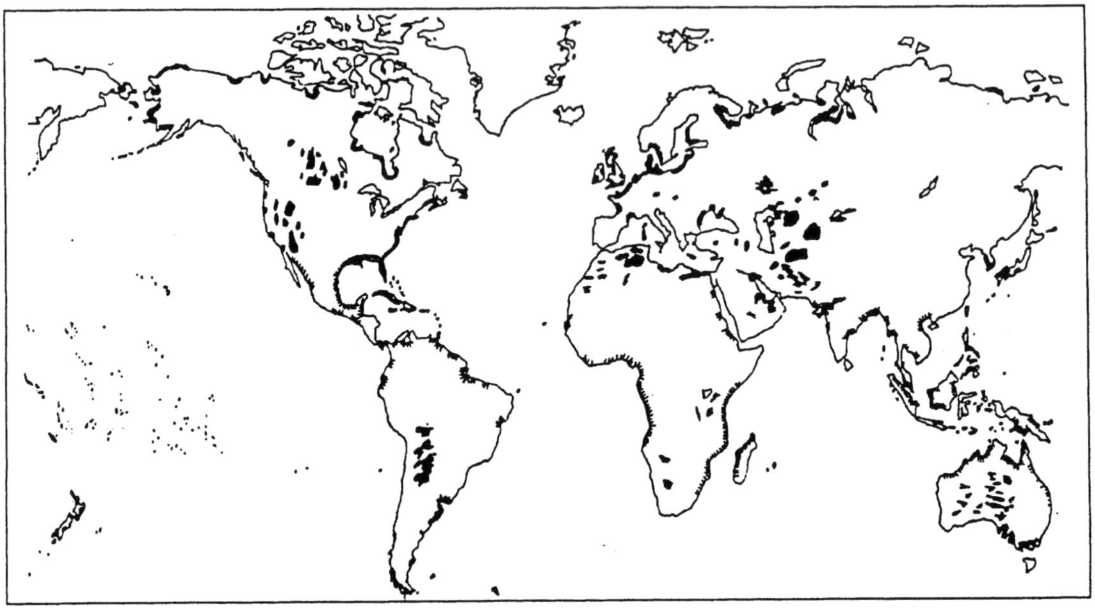

Figure 1. World distribution of salt-marshes, salt deserts (solid black) and mangrove swamps (latched). *Source*: Chapman (1974).

Secondly, halophytes are being trialled under irrigation, mainly in developed countries. The motivation may be to produce an industrial crop e.g. oil seed from *Salicornia* (Glenn et al. 1991; Riley & Abdal 1993) or to produce forage which can be fed to animals in feedlots (Watson & O'Leary 1993). This category is of more recent origin. Less is known about the potential and limitations (Swingle et al. 1994).

The way in which halophytes are assessed will very much depend on which of these systems dominates. Any evaluation must depend on viewing performance in the context (biological/economic and stage of development) in which it occurs. Much of the controversy over the role of halophytes in livestock nutrition stems from a tendency to extrapolate experience from one of the two categories to the other.

The potential role of halophytes (and some of the problems) will be considered under the following headings which apply equally to the two categories listed above:

- land rehabilitation (especially degraded drylands),
- as irrigated crops using saline water,
- as feed (fodder/forage) for livestock,
- as medicinal plants,
- as fuel wood,
- as shade and shelter,
- sequestration of carbon dioxide.

Of course there are constraints. A major constraint to the better utilization of halophytes is a lack of vision about what is possible – the extent of the plants available to perform these roles.

There are few extensive inventories of halophytes. Aronson (1989) in the most extensive world-wide study of halophytes of potential economic use published to date, found 1560 species belonging to 540 genera and 120 families. According to Le Houérou (1994) Aronson's study probably includes some 20–30% of the overall terrestrial halophytic flora of the world which could perhaps reach the figures of 5000–6000 species, i.e. 2% of the flowering plants.

Rehabilitation of degraded land

Halophytes offer advantages because of their productivity under harsh conditions, their nutritional qualities for grazing/browsing livestock, value as wildlife habitat, cover for the soil and for their general role in ecosystem functioning.

Halophyte shrubs, especially, offer tremendous potentials for mankind's benefit in making the arid and semi-arid lands of the world more productive and useful. Many of the world's rangelands, which cover about 4.5 billion ha, can support halophytes. Between 55 and 85% of these rangelands are classified as moderately to severely degraded (Dregne et al. 1991). There is a role for halophytes in the rehabilitation of these lands.

There is probably more art than science in the successful revegetation of degraded land, especially in arid environments. A basic understanding of arid land ecosystems is a must, along with some experience or training in general revegetation techniques (Malcolm 1994). There are no 'easy solutions', no one technique that will work for all sites. The factors characterizing each site (soils, available water etc.) interact in a complex harmony to dictate the ecosystem that will prevail.

In selecting species for rehabilitation, other factors to consider include availability of plant/seed material, propagation and germination problems, desired future use of the site and the ecological stability of the candidate species to ensure long term vegetative cover (Malcolm 1994).

Exploiting the desert: Saline irrigation

Some halophytes have potential as crops for direct seawater or brackish water irrigation (Glenn & Watson 1993b). Thousands of plants have been screened for their productivity and/or nutritional potential when irrigated with saline water, principally seawater (reviewed by Riley & Abdal 1993). Halophytes have demonstrated their potential as crops for forage, fuel and oilseeds (NRC 1990; Glenn et al. 1991).

The world area available for halophyte agriculture includes portions of coastal and inland salt deserts (O'Leary & Glenn 1994). The usable area for halophyte culture is approximately 1.3 million km^2, roughly comparable to the area of conventional irrigation in the world (Glenn et al. 1992b).

Halophytes can help better utilize alternative waters sources for irrigation (Glenn & Watson 1993b; Miyamoto et al. 1994). Saline water is regarded everywhere as an 'evil'. It is called marginal water, to be used only when fresh water is scarce or unavailable. However, it has been shown that saline water irrigation can actually give significant economic advantages over fresh water.

A complex research and development program has been carried out in several countries aimed at the utilization of saline water to irrigate new desert areas. This has been to offset the fact that for many years a number of arid countries have been exploiting all known freshwater supplies. The amount of freshwater available for irrigation is diminishing each year due to competition for this water by urban and industrial customers and because of deterioration in water quality.

Desert agriculture may be feasible in some cases but it is not easy. Unlike common agricultural crops where ideal seed types and cultural requirements are conventionally accepted and are the subject of continuing research, this body of research does not exist for many halophyte crops.

The question in some countries is not whether irrigated halophytes should be produced or not. They will be produced, as long as there is an economic incentive to do so. The question, rather, is: How will they be produced, with what sorts of inputs and with what consequences for long term economic and environmental costs?

Halophytes and animal nutrition

Halophytes are grazed or browsed by livestock on all six continents. The forage value of these herbaceous to woody plants has been recognized for a long time and many of the perennial shrubs are important browse species in arid and semi arid rangeland communities (Le Houérou 1992). They vary enormously in their feed value (El Shaer & Gihad). There is continuing debate about the nutritive value of the halophytes (Wilson 1994). Many halophytes contain anti-nutritional substances among their secondary metabolites (Bayoumi & El Shaer 1994).

Le Houérou (1991) indicated that a high proportion of the nitrogen in halophytes is non-protein and therefore not retained. Among this non-protein nitrogen is about 50% glycinebetaine (a quaternary ammonium compound) which is amenable to degradation by the rumen microflora when energy is available for the microorganisms. However, ruminants, and their rumen microflora, need to adapt. This may take months (Le Houérou 1991).

Under dryland conditions halophyte shrubs may be grown in plantations or in a variant of alley cropping with stands of shrubs spaced at regular intervals among dryland crops. This latter strategy has been used to good effect in Australia (Wilson 1994) and North Africa (Le Houérou 1991). Irrigated fodder crops have been used with success in the USA (Watson & O'Leary 1993), Mexico (Glenn et al. 1991) and in the Gulf states (Glenn et al. 1993b; Riley & Abdal 1993).

Halophytes as medicinal plants

Some halophytes have been used in traditional medicine for centuries (Batanouny 1994; Dakshini 1989; Paetzold 1989; Zhao & Liu 1989). There is a general lack of biological information on halophytes in relation to phytochemical, pharmacological, toxicological and medicinal importance. There is need for a systematic study of all plants used in native medicine or folk medicine on which no chemical research has yet been done. Some studies have been done on the toxicity and medicinal effects of several desert halophytes and screening is continuing (Bayoumi & El Shaer 1994).

Fuelwood and timber production

Fuelwood production under rainfed conditions can be achieved with a number of halophytes such as representatives of *Acacia*, *Eucalyptus*, *Casuarina*. *Haloxylon*, *Prosopis*. *Populus*, and *Tamarix*.

The highest production under irrigation (with moderate salinity) is from *Eucalyptus camaldulensis*, provided that the right provenance is used (Le Houérou 1994). *Casuarina* and *Populus* also contain salt tolerant representatives which have been successful under irrigation. Some species are fast growing and have very straight stems giving them potential for timber production.

Role as shade and shelter

Provision of shade and shelter (for crops, animals and humans) is a major concern in many regions, especially where salinity levels are either naturally high or have escalated through faulty irrigation and drainage practices. Several trees (e.g. *Acacia*, *Tamarix*, *Prosopis*, *Haloxylon*. *Populus*) and the tall shrub *Atriplex nummularia* have proved useful (McKell 1989).

Role of halophytes in carbon sequestration

Glenn et al. (1992b) have postulated that the growing of halophytes on deserts could play a major role in sequestering carbon. The land resources available for halophyte plantations include coastal deserts, inland saline basins, and regions of secondary salinization in arid-zone irrigation districts. Globally, these regions occupy areas of 295 million, 427 million and 65 million ha respectively. In many important respects, the case for planting halophytes in the desert compares favourably to the case that has already been made for re-afforestation (Glenn et al. 1993a).

Problems and prospects for utilizing halophytes

Halophytes occupy diverse habitats and the woody species are conspicuously present in almost all landforms in varying densities and diversities. Diversity of habitat creates a condition of high adaptive and selective potentials of plants. In shrubs, this phenomenon is marked since on the one hand they are perennial and on the other they have to occupy diverse niches in more ecologically dynamic habitats in transitional zones. Stringi et al. (1994) point to the genetic plasticity among halophytes.

In spite of their common occurrence and multiple uses shrubs, in particular, have been ignored or considered a problem and their potentials (economic, social and cultural) are yet to be evaluated. Even where grasses don't succeed, shrubs thrive successfully and sustain biota through their continuous and reliable productivities and nutritive qualities. Besides giving protection and sheltering of soil cover and their overall important role in ecosystem functioning in disturbed and degraded habitats halophytes can beautify the landscape. In the predominantly pastoral landscapes of the world's halophytes have provided fuel and fodder. They should continue to do so. But in many areas overgrazing and cutting of shrubs for fuel has led to local desertification.

For vast regions without irrigation (e.g. Australia, Central Asia) halophyte shrublands provide the foundation for livestock production (Wilson 1994; Zhang et al. 1994). The most pressing problem in most regions is the protection of halophytes. As whole they have been overused and misused, especially in harsh environments (Dahl & McKell 1987).

The introduction of sound rational and conservative management and improvement practices face severe socioeconomic constraints. The first and foremost one is the public and communal ownership or tenure of a large part of the drylands of the world. This allows unrestricted grazing for everybody. This is characterized by a lack of responsibility for the resources at the management level. Another constraint is the general neglect and indifference of politicians, decision makers, their advisors, and, chiefly, economists to the salt affected drylands.

Worldwide the areas of salt-affected soil need to be improved. Many areas need to be sown or planted to supplement the existing vegetation. Millions of hectares of soils too salty for normal crops and pastures lie idle. Salt tolerant forage shrubs may be grown on these soils. Adaptation trials are needed to determine which shrubs to use in these particular areas (Malcolm 1994).

New species should be introduced after proper screening, especially those of economic potential. In addition new technologies for improving shrublands such as inoculation with mycorrhizae, tissue culture production of elite biotypes, and genetic engineering to improve economically valuable species should be introduced and adopted as soon as possible.

Commercial development of industrial products (oil seed), biomass, fibre or other material brings a heightened awareness for the need for a better understanding of the biology and ecology of the halophytes and to better manage them for optimum and sustained production.

The immediate need is to know the plants and their full potential. Little is known about the patterns of distribution, regeneration potential, survival mechanisms and economic usage. Detailed investigation along these lines is urgently required.

The introduction of new multiple-use strategies for the management and rehabilitation of halophyte communities will require a formidable educational and research effort and great financial and moral support from local and national authorities in order to have any impact in the near future.

Acknowledgement

I am grateful to the Director, Environmental Research Laboratory, University of Arizona for access to the facilities and the support during my sabbatical from the University of Adelaide. Thanks to UNEP for the opportunity to participate in the November 1992 workshop in Nairobi.

References

Aronson, J.A. (1989). HALOPH: A data base of salt tolerant plants of the world. In: E.E. Whitehead (ed.). Office of Arid Land Studies, Univ. of Arizona, Tucson, AZ, USA, 77 pp.

Batanouny, K.A. (1994). Halophytes and halophytic plant communities in the Arab region: Their potential as a rangeland resource (this volume, Ch. 11).

Bayoumi, M.T. & El Shaer, H.H. (1994). Impacts of halophytes on animal health and nutrition (this volume, Ch. 25).

Dahl, B. & McKell, C. M. (1987). Use and abuse of China's arid lands. Rangeland J.

Chapman, V.J. (1974). Salt marshes and salt deserts of the world. In: R.J. Reimold & W.H. Queen (eds.), Ecology of halophytes. Academic Press, New York, pp. 3–19.

Dakshini, K.M.M. (1989). Shrublands of the Indian subcontinent. In: C.M. McKell (ed.), The biology and utilization of shrubs. Academic Press, San Diego, pp. 177–197.

Dregne, H., Kassas, M. & Rosanov, B. (1991). A new assessment of the world status of desertification. Desertification Control Bull. UNEP No. 20: pp. 6–18.

Duncan, W.H. (1974). Vascular halophytes of the Atlantic and Gulf coasts of North America north of Mexico. In: R.J. Reimold & W.H. Queen (eds.), Ecology of halophytes. Academic Press, New York, pp. 23–50.

El Shaer, H.M. & Gihad, E.A. (1994). Halophytes as animal feed in Egyptian deserts (this volume, Ch. 28).

Glenn, E.P., Coates, W.E., Riley, J.J., Kuehl, R.O. & Swingle, R.S. (1992a). *Salicornia bigelovii* Torr.: A seawater-irrigated forage for goats. Animal Feed Sci. Tech. 40: 21–30.

Glenn, E.P., Hodges, C.N. Lieth, H., Pielke, R. & Pitelka, L. (1992b). Growing halophytes to remove carbon from the atmosphere. Environment 34(3): 40–43.

Glenn, E.P. & O'Leary, J.W. (1985). Productivity and irrigation requirements of halophytes grown with hypersaline seawater in the Sonoran Desert. J. Arid Envir. 9: 81–91.

Glenn. E.P., O'Leary, J.W., Watson, M.C., Thompson, T.L. & Kuehl, R.O. (1991). *Salicornia bigelovii* Torr.: An oilseed halophyte for seawater irrigation. Science 251: 1065–1067.

Glenn, E.P., Pitelka, L.F. & Olsen, M.W. (1992c). The use of halophytes to sequester carbon. Water, Air & Soil Pollution 64: 251–263.

Glenn, E.P., Squires, V.R., Frye, R. & Olsen, M. (1993a). Potential for carbon sequestration in the World's Coastal Deserts, Inland Salines and Drylands. Water, Air & Soil Pollution 70: 341–355.

Glenn, E.P. & Watson, M.C. (1993b). Halophyte crops for direct salt water irrigation. In: H. Lieth & A.Al Masoom (eds.), Toward the rational use of high salinity tolerant plants. Tasks in Vegetation Science no. 27: 379–385. Kluwer Academic Publishers, Dordrecht, The Netherlands.

Glenn, E.P., Swingle, R.S., Riley, J.J., Mota, C., Watson, M.C. & Squires, V.K. (1994). North American halophytes: Potential use in animal husbandry (this volume, Ch. 12).

Le Houérou, H.N. (1991). Feeding shrubs to sheep in the Mediterranean arid zone: Intake, performance and feed value. Proc IV Intern. Rangelands Congr. Montpellier, pp. 623–627.

Le Houérou, H.N. (1992). Salt-tolerant plants for the arid regions of the Mediterranean isoclimatic zone. In: H. Lieth & A. Al Masoom (eds.), Towards the rational use of salinity tolerant plants. Tasks in Vegetation Science 27: 403–422. Kluwer Academic Publishers, Dordrecht, The Netherlands.

Le Houérou, H.N. (1994). Forage halophytes and salt-tolerant fodder crops in the Mediterranean Basin (this volume, Ch. 10).

Malcolm, C.M. (1994). Use of halophyte forages for rehabilitation of degraded lands (this volume, Ch. 4).

Miyamoto, S., Glenn, E.P. & Singh, N.T. (1994). Utilization of halophytic plants for fodder production with brackish water in subtropic deserts (this volume, Ch. 5).

McKell, C.M. (1989). The biology and utilization of shrubs. Academic Press, San Diego, 656 pp.

NRC (1990). Saline agriculture: Salt tolerant plants for developing countries. National Research Council, National Academy Press, Washington DC, USA.

O'Leary, J. (1984). The role of halophytes in irrigated agriculture. In: R.C. Staples (ed.), Salinity tolerance in plants: Strategies for crop improvement. John Wiley & Sons Inc., New York, pp. 285–300.

O'Leary, J.W. & Glenn, E.P. (1994). Global distribution and potential for halophytes (this volume, Ch. 2).

Paetzold, H. (1989). Shrublands of the USSR in Asia. In: C.M. McKell (ed.), The biology and utilization of shrubs. Academic Press, San Diego, pp. 199–223.

Riley, J.J. & Abdal, M. (1993). Preliminary evaluation of *Salicornia* production and utilization in Kuwait. In: H. Lieth & A.Al Masoom (eds.), Towards the rational use of high salinity tolerant plants. Tasks in Vegetation Science 28: 319–329. Kluwer Academic Publishers, Dordrecht, The Netherlands.

Stringi, L., Giambalvo, D., Accardo, A. & Amato, G. (1994). Characterization of progeny clones belonging to natural Sicilian *Atriplex halimus* populations (this volume, Ch. 33).

Swingle, R.S., Glenn, E.P. & Riley, J.J. (1994). Halophytes in mixed feeds for livestock (this volume, Ch. 7).

Szabolcs, I. (1994). Salt affected soils as the ecosystem for halophytes (this volume, Ch. 3).

Watson, C.M. & O'Leary, J. (1993). Performance of *Atriplex* species in the San Joaquin Valley, California, under irrigation and with mechanical harvests. Agric. Ecosystems & Envir. 43: 255–266.

Wilson, A.D. (1994). Halophytic shrubs in semi-arid regions of Australia: Value for grazing and land stabilization (this volume, Ch. 8).

Zhang, L., Xia, Y. & Zou, Y. (1994). Halophytes and halophytic communities in Inner-Asia (this volume, Ch. 9).

Zhou, S. & Liu, H. (1989). Shrublands of China. In: C.M. McKell (ed.), The biology and utilization of shrubs. Academic Press, San Diego, pp. 225–245.

2. Global distribution and potential for halophytes

JAMES W. O'LEARY[1] & EDWARD P. GLENN[2]
[1]*Department of Plant Sciences, and* [2]*Environmental Research Laboratory, The University of Arizona, Tucson, AZ, USA*

Abstract. Some 1500 species of halophytes are distributed on all the world's continents except Antarctica. Screening trials suggest that perhaps a third of these may be sufficiently salt tolerant to be grown with highly saline irrigation water, including seawater. Many species have high protein levels and other desirable nutritional characteristics, although ash (salt) content can be a problem. As much as 130×10^6 ha of salt-affected land and coastal desert may actually be usable for halophyte crops – an area twice the size of existing desert irrigation districts. Thus there is a means of providing animal feed in regions where at first sufficient land and water does not appear to be available.

Key words: Coastal deserts, Distribution, Euhalophytes, Halophytes, Irrigation, Miohalophytes, Number of species, Salt-affected soil, Taxonomy

Introduction

The use of halophytes for beneficial purposes, such as sources of animal feed, which is the focus of this book, has received increasing attention during the past two decades. As a result, reviews of progress in this area have appeared frequently during the past few years (e.g. Pasternak & San Pietro 1985; O'Leary et al. 1985; Barrett-Lennard et al. 1986; O'Leary 1988). An extensive bibliography and database of over 1500 potentially economically useful halophytes has been published (Aronson 1989), as well as a special publication of the National Research Council that describes in more detail those halophytes that have potential, especially for developing countries (BOSTID 1990). Thus, rather than discuss in detail research results that already have been adequately reviewed, we will focus on some aspects of the halophyte resource base that have not received as much attention, but that are important in the context of putting halophytes to beneficial use on a worldwide basis. That will include discussion of the number and variety of species in this group, the extent of their distribution, the variability of properties that are important relative to their potential for use as sources of animal feed, and the areas in which the potential for halophyte cultivation are greatest.

What is a halophyte?

The name 'halophyte' does not designate a member of any particular taxon or any specific geographic or physiographic area. The word literally means 'salt plant' and is used to refer to any plant that is capable of growing and reproducing in areas subject to high salinity during all or a substantial part of the time. The definition of high salinity is somewhat subjective, but most people consider a salinity equivalent to about 0.5% NaCl in the soil water as a rough approximation of the lower end of the high salinity range. This corresponds to a concentration of around 85 mM NaCl. Thus, it may appear easy to designate those plants that deserve the name halophyte. In reality, that is not the case.

V.R. Squires & A.T. Ayoub (eds.), *Halophytes as a resource for livestock and for rehabilitation of degraded lands*, 7–17.

The mere presence of a plant in an area that is considered saline is not enough. Information on the salinity of the soil in which the plant is growing is required. Unfortunately, in most cases that information was not obtained by whomever collected and identified the plant. There also is the problem of the variety of methods used to measure and quantify soil salinity (Miyamoto et al. 1994). Ecologists typically indicate the range of soil salinity with which a plant is usually associated whereas agriculturists usually report the electrical conductivity of the saturation paste extract that results in yield reduction of 50%. There are problems with each of these approaches. For example, in the former case, we might be misled in thinking that the salinity in the plant's native habitat is optimal for growth, when in fact many halophytes grow in areas where the salinity is well beyond that which results in maximum growth. On the other hand, in the latter case, the emphasis on the salinity that results in 50% reduction of growth obscures the fact that growth can be appreciable at higher salinity (Mudie 1974).

However, even when the salinity of the soil in which the collected plant is growing is known, that may not be enough (Szabolcs 1994). For example, the plant may have become established, and even accomplished most of its growth, during an earlier, less saline, episode, and the plant is now surviving, but not growing very well at the present salinity. Similarly, the vertical profile of salinity should be known, as well as the distribution of roots throughout that profile to be sure that the plant actually is experiencing the salinity indicated by the soil sample that might have been obtained. It is known, for example, that many dune species which, at first glance, might be thought to be very salt tolerant because of their location in an apparently highly saline environment have roots penetrating down into a fresh water lens suspended above the deeper, saline aquifer. This type of information is valuable to the person intending to introduce and establish that collected plant into a new area to replace or supplement the native vegetation as a way of increasing productivity, yet it rarely is available.

Growth responses to salinity

During the past decade, when we were assembling a worldwide collection of halophytes at the Environmental Research Laboratory for such purposes, we soon became aware of this limitation. Many of our accessions were obtained by collecting plants from remote sites, and the plant collectors had neither the time nor resources to obtain that kind of information. Thus, we developed a screening program that evaluated a plant's performance at several different salinity levels up to greater than seawater concentration (40 ppt) (Glenn & O'Leary 1984; Glenn 1987). When we screened 150 diverse species, from several families, all of which we were told were halophytes, we found that 57 of them (38%) survived the highest salinity treatment and, in fact, had maximum growth at the 10 ppt salinity level (170 mM). The rest of the species had maximum growth on fresh water.

Dicots and monocots responded differently to salinity. The most salt tolerant dicots are usually called euhalophytes by most halophyte specialists. As mentioned, they were stimulated by 10 ppt salinity compared to their growth on fresh water in our trials. Dicot halophytes which show their best growth on fresh water are usually called miohalophytes and few of those survived at the highest salinity level. By contrast, only one of 25 halophytic grass species were stimulated by 10 ppt salinity, so the term euhalophyte would not apply to these species, yet 15 of the 25 grasses survived on 540 mM seawater; some of them grew better than euhalophyte dicots on that salinity, so the term miohalophyte does not seem to apply, either (Glenn 1987). Perhaps halophytic grasses is the best term. To our knowledge, nobody has ever given an estimate of the proportion of the world halophyte flora consisting of each of these types. To the extent that this group of 150 species consisted of a representative sample of the world halophyte flora, it appears that only slightly more than a third would be appropriate for use in situations where the water source is highly saline, such as coastal deserts.

Maximum growth at a relatively high salinity is characteristic of the euhalophytes, and it has been demonstrated for several species of halophyte. Flowers et al. (1977) listed some in their review, and additional ones are listed in Table 1. A couple of observations are in order. The euhalophytic character apparently is limited to the dicotyledonous species, but it is not limited to only one or two families. In addition to presenting

Table 1. Salinity level for maximum growth of several halophytes

Species	Salinity level for maximum growth	References
Sarcocornia natalensis	300 mM NaCl	Naidoo & Rughunanan (1990)
Suaeda aegyptica	125 eq m^{-3} NaCl	Eshel (1985)
S. maritima	130 mM NaCl	Hagege et al. (1988)
S. maritima	200 mM NaCl	Yeo & Flowers (1980)
S australis	50–150 mM NaCl	Robinson & Downton (1985)
Atriplex nummularia	150 meq NaCl KCl, or $MgCl_2$	Greenway (1968)
A. halimus	1.8 atm. NaCl or KCl	Mozafar et al. (1970)
A. amnicola	25–50 mol m^{-3} NaCl	Aslam et al. (1986)
Salicornia bigelovii	1% NaCl	Webb (1966)
Avicennia marina	25% Seawater	Burchett et al. (1984)
Kosteletzkya virginica	85 mol m^{-3} NaCl	Blits & Gallagher (1990)
Salsola iberica (*S. kali*)	10.5 dS m^{-1} (NaCl:$CaCl_2$, 2:1)	Fowler et al. (1988)
Sesbania grandiflora	10 mmho cm^{-1} EC	Chavan & Karadge (1986)
Disphyma clavellatum	100 mol m^{-3} NaCl	Neales et al. (1983)
Crithmum maritimum	10% Seawater	Okusanya (1979)
Daucus carota spp. *gummifer*	10% Seawater	Okusanya (1979)
Ligusticum scoticum	10% Seawater	Okusanya (1979)
Lavatera arborea	5–20% Seawater	Okusanya (1979)
Inula crithmoides	10–30% Seawater	Okusanya (1979)
Spergularia rubicola	10–40% Seawater	Okusanya (1979)
Chenopodium rubrum	0.4 MPa NaCl	Warne et al. (1990)

information regarding the optimal salinity for growth of a range of halophytes, another purpose of Table 1 is to indicate the range of differences in methodology (single salts vs salt mixtures; soil vs nutrient solutions) and units used to express salinity, even in relatively recent studies. The difficulties mentioned in this section are intended to alert those involved in introducing putative halophytes into new areas that they should be as familiar with the plants as possible, and even then, preliminary testing on-site should be conducted prior to launching a large scale revegetation project or similar venture. As far as deciding whether to call a plant a halophyte or not, these difficulties may be more of a nuisance than a problem.

Classification of halophytes

Nearly all salt tolerant plants belong to the angiosperms (Flowers et al. 1986). There are two families in the ferns, the Pteridaceae and the Ophioglossaceae, that reportedly contain some salt tolerant genera, but few, if any, mosses, liverworts or even gymnosperms are salt tolerant. Thus, when we speak of halophytes, almost without exception we are referring to flowering plants. It is estimated that only about one-third of the approximately 350 families of flowering plants contain halophytes (Aronson 1989), and about half of the genera belong to 20 of those families (Flowers et al. 1986).

There is no complete catalogue of the halophyte flora of the world, for various reasons, not the least of which is the difficulty in stating for sure whether a particular species is or is not a halophyte. Surveys have been conducted (Mudie 1974; Chapman 1975; Flowers et al. 1977; Flowers 1986; Aronson 1989), but they usually have emphasized identifying halophytes with potential beneficial use, so they probably underestimate the size of this group of plants. Nevertheless, as a result of those surveys, some important bits of information are revealed. For example, even though the number of halophytes may not be large, halophytes are widely distributed among several families of flowering plants, and they occur in a wide range of habitats, from predominantly wet seacoast marshes and estuaries to inland, arid saline deserts. These two points make the potential for exploitation of this resource relatively high.

Mudie (1974) was the first to extensively catalogue halophytes based on their potential for exploitation. Her approach was to consider all species recorded from, or in the immediate vicinity of, naturally saline habitats. She compiled her list from literature on salt deserts, saline sinks, salt marshes, and mangrove swamps, and that resulted in a compilation of approximately 550 species in 220 genera in 75 families. More recently, Aronson (1989) published a compilation based not only on literature citations, but also on personal interviews with people throughout the world over several years. His list included 1,560 species in 550 genera and 117 families. Based on these studies, and the others cited above, it is clear that there is no family in which more than half of the species are halophytes. Among the dicotyledons, the Chenopodiaceae is the family most noted for its halophytic members, representing 44% of the genera in the family. After that, the Aizoaceae with 15%, the Cyperaceae with 14%, and the Caryophyllaceae with 11%, are the only other families having more than 10% halophytic genera. In the monocotyledons, there are even fewer halophytes. The Poaceae contain more halophytic genera than any other family in the group, with 45, but that represents only 7% of the genera in the family. Even though halophytic genera seem to be limited to less than half of all plant families, and even though no single family seems to be exclusively a halophyte family, that is not bad news. On the contrary, it really is good news. The fact that the limited number of halophytic species is spread among so many different families indicates that halophytism, even though it clearly is a trait controlled by several genes, is not such a complex characteristic that it only arose once during evolution. This polyphyletic origin of halophytism (Flowers et al. 1977) suggests that it is a characteristic that could be introduced into species not yet possessing it, particularly species in the same families that already contain halophytic species.

The Brassicaceae is the family to which the crop plants *Beta*, *Brassica*, and *Raphanus* belong, none of which is extremely salt tolerant, although they are more salt tolerant than many other crop plants. However, this family also contains the genus *Cakile*, a well known halophyte. *C. maritima* is a drift-line plant in Europe, and *C. edentula* occupies the same position in eastern USA. The seeds have been reported as edible (Chapman 1975), so they have received some attention as potential cultivated plants. There are 15 species of *Cakile* known, most of which have not been investigated yet. Another halophyte in this family that has received some attention is *Crambe maritima*, and there are 20 species in that genus. Another widespread genus is *Lepidium*. One species, *L. crassifolium*, is a common species of inland salt deserts in Eurasia. *L. sativum* has been used as a salad plant, so the plants must have characteristics that might make them good candidates for animal feed as well. This genus has at least 130 species, so there may be good potential for developing a good forage/fodder plant that could be grown in inland saline areas. Another member of this family that has been used as a salad plant is *Cochlearia*, of which there are about 25 species. *C. anglica* has been reported to tolerate 2% NaCl, so this genus may also have both high salt tolerance and attractive feed characteristics represented. Those features theoretically could be combined and/or improved within the genus. This is a widespread family that occurs in both northern and southern hemispheres, and it is known as a family that is partial to dry climates, as well as having members that live in very moist habitats, indicating adaptability to a wide range of sites for potential rehabilitation or development.

Some of the other dicotyledonous families with the largest numbers of halophytic genera include Asteraceae with 34 genera, Aizoaceae with 21, Apiaceae and Papilionaceae with 19, Euphorbiaceae with 15, Cyperaceae, Arecaceae, and Scrophulariaceae with 13, and Caryophyllaceae with 9 genera. As mentioned above, the dicotyledonous family with the largest number of genera is the Chenopodiaceae, which has 44. There are over 300 species in those genera, although the exact number depends on how many species of *Atriplex* are recognized.

The most comprehensive treatise on *Atriplex* is that by Osmond et al. (1980). As they point out, during the past 300 years or so, over 400 species of *Atriplex* have been described and listed in *Index Kewensis*. However, there have been many cases in which the same species was described by different authors and assigned different names. Taking into account the various synonyms in existence, they reduced that number to 245 species. That number is even high, with some Floras usually having only about half that many species in the genus. Cronquist (1981), for example, only has 150 species of *Atriplex*. The family has about 100 genera and 1500 species in all. Some of the other important halophytic genera include *Arthrocnemum* (20 species ranging from the Mediterranean to Australia), *Suaeda* (110 species), *Salsola* (about 150 species), *Anabasis* (30 species), *Allenrolfea*, and *Kochia*. Many of these genera contain both annual and perennial species.

Origin and distribution of halophytes

Since the group is defined on the basis of edaphic factors rather than botanical characteristics, it is not surprising that the distribution of halophytes is primarily determined by edaphic factors rather than climatic factors. That is not to say that climate has no influence, rather that the soil characteristics probably exert a much greater influence than climate (Szabolcs 1994).

As mentioned earlier, halophytism apparently arose several times during evolution of land plants. According to Iljin (1946) and Chapman (1960), those species probably developed primarily as seacoast plants. Migration of some of those species to inland saline areas occurred at a later time. Certain life forms appear in halophytes in higher proportions and apparently have advantages over others (Waisel 1972). For example, in the salt marshes of temperate regions, we find mostly grasses, whereas in warmer climates, and in inland saline areas, woody chamaephytes and hemicryptophytes predominate. Thus, shrubs and perennial grasses seem to be characteristic of most of the halophytic floras of the world. Few tree species are halophytic, most examples consisting of the mangroves found in the very wet, saline habitats of warm regions. An interesting comparison of the halophyte flora of Israel with the total flora of that country and of the world was made by Waisel (1972). Some of those data are presented in Table 2. The large representation among the perennial grasses and shrubs (phanerophytes, chamaephytes, and hemicryptophytes) in the halophyte flora is reflective of the overall halophyte flora mentioned above. The striking difference between the total flora for that desert region and the halophyte flora there is in the therophyte population. Those are mostly summer annuals, which often form a sizable portion of the flora of a desert region, and the large percentage of them in the total flora of Israel is not unusual. It may seem a little surprising at first that only a few of them are apparently salt tolerant, but, in general, desert ephemerals germinate and complete their life span in response to rainfall so salt tolerance is not required. It actually may be more surprising that some of them are halophytic. Presumably, those species are ones whose normal habitats are saline, so that even when rainfall occurs, the resultant soil solution is saline. For example, in areas where such soil conditions exist, such as inland salt marshes of Kansas (Ungar 1965), therophytes dominated over hemicryptophytes, the exact opposite of the situation in Israel.

Table 2. Percent distribution of Raunkiaer (1934) life forms among the halophyte and total floras of Israel and the total flora of the world

Life form	Israel		World
	Halophyte flora	Total flora	total flora
Therophytes	19.4	48.3	13.0
Cryptophytes	0	11.0	6.0
Hemicryptophytes	34.5	24.3	26.0
Chamaephytes	29.1	9.8	9.0
Phanerophytes	17.0	6.6	46.0
Total	100	100	100

Based on Table 5.3 from Waisel (1972).

In spite of the comparatively smaller number of species, the percentage of those species that are multiregional in distribution is higher in the halophytic than in the glycophytic flora (Waisel 1972). There is much less endemism, in general, among the halophytes than in the glycophytes. An area in which the halophytic flora has been more thoroughly catalogued and studied than elsewhere is the country of Israel, so it is relatively easy to make such a comparison using that information. Waisel (1972) calculated that 62% of the species in the entire flora of Israel were limited to one region with 38% occurring in two or more regions, but only 18% of the halophytic species were uniregional and 82% occurred in two or more regions.

The widespread distribution of a genus is best illustrated by *Atriplex*, which occurs on every con-

Table 3. Geographic distribution of *Atriplex* and estimated percentage of endemic species in each region

Region	Number of species	Endemic species (%)
Europe	42	7
Mediterranean	42	7
Central Asia	38	26
East Asia	11	18
Australia	59	68
South Africa	8	33
South America	73	86
Central America	51	40
North America	59	78

Based on data from Figure 2.9 and Table 2.2 in Osmond et al. (1980).

tinent but Antarctica (Table 3). Not only are there native species in each of those areas, but species native to one area have been successfully introduced and established in several of the other areas. Thus, it is not surprising that *Atriplex* has received so much attention as an introduced species for revegetation projects throughout the world. Of course, another feature of *Atriplex* that has made it attractive for use in revegetation projects (Malcolm 1994), particularly where increasing productivity of rangeland for animal feed has been an objective, is the reasonably high crude protein level in the foliage.

Potential for use as animal feed

Use of halophytes to revegetate an area is subject to the constraints posed by edaphic and climatic factors that will determine whether a particular species can be successfully established in a new area. If that species is intended to serve as a feed for animals, at least two additional constraints must be considered. One is the nutritive value of the forage, primarily the crude protein content, and the other is the ash, or salt, content of that material (Gihad & El Shaer 1994). Values for each of these parameters are available for a wide range of halophyte species, and they have been useful in guiding selection of species to be used for such purposes in many cases. However, it often is not easy to make comparisons among several candidates because the data have been obtained for each of them under different conditions. The salinity levels at which the plants were growing and the ages of the tissue may have been different, for example.

To avoid those problems when we were screening halophytes to serve as animal forages, we grew a wide range of different types of halophytes, in several families, in a common garden in a large greenhouse. All of the plants were grown in pure sand contained in a closed system through which the irrigation water could be recycled and the salinity and nutrient levels of that water maintained within a desired range. The system was divided into two units, one irrigated with artificial seawater (3.2% salinity), and the other irrigated with diluted seawater (0.5% salinity). Based on preliminary screening for salt tolerance, the plants were assigned to one of these salinity treatments. A few were grown at both salinities. After growing in the system for over a year and having been pruned several times, clippings of new growth were collected for analysis. The samples were intended to be representative of what a browsing animal might obtain. Ash content, Na and K content, and total Kjeldahl Nitrogen (TKN) content were determined (Table 4). Since it is questionable how great the error may be in converting TKN to crude protein values in halophyte tissue (O'Leary 1985), the nitrogen content is presented as TKN. There was a wide range of nitrogen contents, from as low as 15% to as high as 53%. Ash content was relatively low in the monocots, as a group, as expected, and in the dicots it ranged from 22 to 43%.

There was no close correlation between ash and nitrogen content, in general, but there are a few interesting points. For example, among the *Atriplex* species, *A. lentiformis* had one of the lowest ash contents, and it did not seem to differ much when grown on the high salinity compared to the low salinity. This species also had the highest nitrogen content. In contrast, when *A. canescens* was grown at both salinities, there were greater differences. The ash content was 50% higher when grown at high salinity, and the nitrogen level dropped substantially also. One might conclude that one species (*A. lentiformis*) would be much better suited than the other (*A. canescens*) for growing as an animal feed where the water source was extremely saline. The only other Chenopod grown at both salinities was *Rhagodia*

Table 4. Comparison of ash content, total kjeldahl nitrogen (TKN) and Na:K ratio of shoot tissue of several halophytes grown in a common garden with either diluted (05% salinity) or non-diluted (3.2% salinity) artificial seawater. Each value is the average of 3 samples

Plant	Salinity (%)	Ash (% DW)	TKN (mg g^{-1} DW)	Na:K (Molar ratio)
AIZOACEAE				
Sesuvium portulacum	0.5	33	34.6	6.4
Sesuvium sessile	0.5	28	27.2	1.5
Sessuvium sp.	0.5	35	34.4	6.1
Sessuvium verrucosum	3.2	43	22.7	3.9
AMARANTHACEAE				
Iresine rhizomatosa	0.5	29	25.1	1.5
Iresine sp.	0.5	27	28.8	1.6
APIACEAE				
Crithmum maritimum	0.5	28	19.7	2.0
ASTERACEAE				
Borrichia frutescens	0.5	26	21.7	1.3
CHENOPODIACEAE				
Atriplex atacamensis	0.5	34	20.6	2.9
A. canescens	0.5	24	33.9	0.3
A. canescens	3.2	36	25.5	4.3
A. glauca	3.2	27	42.4	6.8
A. isatidea	0.5	34	28.9	3.4
A. lentiformis	0.5	26	45.4	2.5
A. lentifomis	3.2	29	39.9	3.0
A. nummularia	3.2	31	36.4	5.0
A. quinii	0.5	43	23.7	5.6
A. sagittifolia	3.2	40	30.7	5.1
A. undulata	3.2	41	28.8	7.9
A. vestita	3.2	32	43.9	2.3
Batis maritima	3.2	38	20.2	4.7
Haloxylon recurvum	0.5	31	26.4	3.0
Maireana oppositifolia	3.2	35	34.5	6.0
M. sedifolia	0.5	26	29.4	3.5
Rhagodia gaudichaudiana	0.5	36	36.1	1.8
R. gaudichaudiana	3.2	28	36.1	3.5
Suaeda monoica	3.2	31	37.4	7.9
Suaeda cf. *californica*	0.5	28	34.7	3.4
Suaeda torreyana	0.5	31	31.5	6.6
JUNCAGINACEAE				
Triglochin striata	0.5	26	34.4	1.8
NOLANACEAE				
Nolana crassifolia	3.2	37	29.5	9.8
POACEAE				
Distichlis spicata	0.5	10	18.6	1.1
D. spicata	3.2	8	26.7	0.9
Paspalum vaginatum	0.5	17	30.0	0.9
Panicum virgatum	0.5	20	18.5	0.7
Poa languinosa	0.5	14	32.2	0.2
Puccinellia ciliata	0.5	16	30.6	0.1
Sporobolus virginicus	0.5	15	15.2	1.3
PORTULACACEAE				
Portulaca mundula	0.5	34	32.2	4.1
Portulaca sp.	3.2	39	30.4	19.0
ZYGOPHYLLACEAE				
Nitraria schuberi	3.2	22	53.5	5.6

gaudichaudiana, and there was no effect on nitrogen content and a slight effect on ash content when grown at the higher salinity. The Na:K ratios were all around 1.0 or lower for the monocots, as expected, but the values for the dicots varied considerably. The unidentified species of *Portulaca* had an unusually high value (19.0), and the *A. canescens* grown on low salinity had an unusually low value (0.3). The others had values that covered a wide range but were otherwise within the range of values characteristic of dicotyledonous halophytes.

The anomalously low Na:K ratio of *A. canescens* was investigated further because high Na levels detract from the feed value of halophyte materials, whereas high K levels are generally considered desirable. In field plantings it was found that different subspecies of *A. canescens* varied as much as 100-fold in Na:K ratios (Watson 1990; Glenn et al. 1992). Under greenhouse growth conditions, lower Na:K ratios were not associated with proportionally lower salt tolerance in *A. canescens* subspecies. This wide genetic variability at the subspecies level with respect to Na and K accumulation reinforces the view that large numbers of genes may not be involved in the physiological expressions of halophytism.

The most important conclusion to be drawn from these data is that there is great variability in these parameters among a representative sampling of the halophyte flora when grown under the same conditions, suggesting similar genetic variation among the species. This suggests that, in the short term, more extensive sampling of the halophyte flora might yield animal feed sources with both good feed value and suitable adaptability for various areas of interest, and, in the longer term, the potential for genetic improvement of those characteristics seems high.

Potential areas available for halophyte utilization

Estimates of the area of saline land in the world have been attempted by Chapman (1974), who treated only salt marshes and salt deserts, and Szabolcs (1989), who included all categories of saline soil in his estimate. By Szabolcs' more inclusive definition, there are 424 million ha of salinized soil in the world. In addition, by Meigs' (1966) estimate, there are 294 million ha of coastal desert. It is clear that there are vast areas in all parts of the world that not only are nonproductive sources of animal feed at present but also will remain so unless halophytic crops are developed for those areas (Malcolm 1994; Batanouny 1994; Le Houérou 1994).

The most likely sites are in two types of environments, desert seacoasts and inland saline, frequently arid, areas. The conditions under which halophytes could be grown there are either rainfed, or with irrigation. The source of irrigation water for the coastal areas most likely would be the adjacent seawater, whereas the source of irrigation water for the inland areas would most likely come from subsurface sources. In the latter case, the salinity of the water would usually be considerably less than that of the seawater, so the range of halophytes that might be potential candidates for inland sites should be larger than the pool from which the coastal candidates could be drawn.

The extent of the potential coastal area for utilization has been estimated. Coastal deserts border the ocean for about 30,000 km, and about half of their area lies below 33 m, which Meigs (1966) considered to be the practical limit to which irrigation water could be lifted. Based on experience with actually using off-shore ocean sources of water to irrigate halophytes, Glenn et al. (1991) decided that an upper limit of 100 m would be economically feasible. If that is the case, then almost one-half million km^2 of coastal desert is potentially usable for irrigated cultivation of halophytes. That represents approximately 17% of the total coastal desert area of the world. That area includes sites on all continents except Antarctica, and in many cases these are areas of alluvial soils that would be well-suited for irrigation with highly saline water. Included are the Ranns of Kutch, part of the deltas of the Indus, Shat-al-Arab, Nile, and Colorado Rivers.

The inland areas are more difficult to estimate, partly because the availability of a dependable underground water source is not assured in some areas. However, based on best estimates, the area of inland salt deserts available for halophytes is a little more than a half-million km^2, according to Glenn et al. (1991). That is about 15% of the estimated total saline soil area in the world. If the amount of saline land that could be used included those areas that would only have rainfed conditions, then the total area is tripled to about 1.5 million km^2 (Barrett-Lennard et al. 1986). In either case, again the area includes sites at several different spots in the world.

A third type of area that is a little more problematic is the formerly irrigated land that has now become unproductive due to salinization. The extent of that area on a worldwide basis has been estimated numerous times, and while the numbers may be undependable, they are, nevertheless, large enough in all cases to make the point that the problem is severe. It can be documented in some areas that thousands of hectares are being lost every year due to buildup of excess salinity. Thus, it would seem that this represents a potentially large area that could be used for cultivation of halophytes, made especially attractive because much of the necessary capital investment may already have been made in the form of water delivery systems, land leveling, etc. However, in many of those cases, the same conditions that have made the land no longer useful for growing conventional crops would cause similar problems when halophytes are planted there. For example, if the cause of the salinization is inadequate drainage capability, then growing halophytes won't necessarily change that. If the drainage can't be improved, salt buildup will continue, but the point at which cultivation of that land is no longer feasible is delayed for a time because of the higher salt tolerance of the halophyte. Also, if the salinization is due to poor management practices, then growing halophytes won't compensate for those poor practices (Miyamoto et al. 1994). If anything, soil/water management practices are even more important when growing halophytes. The question that needs to be answered at the outset is why the land became salinized in the first place. Some of the time, after answering such a question, the conclusion probably will be that growing halophytes is not the way to solve the problem. But sometimes the conclusion will be that growing halophytes will be beneficial.

There are basically two ways in which the salinity of the soil solution can be too high for conventional crops. One is by having reasonably good soil but water that is too saline, and the other is by having good water but soil that is highly saline. The latter situation is the one that often results from poor management practices or intractable problems such as inadequate drainage outlet in the basin in which the irrigated field sits. The former situation, on the other hand, describes the conditions under which irrigation of conventional crops probably wouldn't have been attempted in the first place, but which provides an excellent opportunity for irrigation of halophytes. This situation provides a reasonably good chance of success, with the selection of the right plant to be used there and with good soil/water management practices. The latter situation provides, in contrast, a reasonably good chance of failure, if a thorough analysis of the problem is not made before deciding that growing halophytes is the proper solution to the problem.

As an example, halophytes may have a place in irrigation districts that have developed salinity problems which preclude growing conventional crops, but for which the cost of complete reclamation by engineering methods is uneconomic (see Miyamoto et al. 1994). Since halophytes have 10- to 100-fold greater salt tolerance than conventional crops, less expensive methods of soil restoration may suffice. Land which would otherwise be abandoned may be kept in production; Dregne et al. (1991) estimated the cost of restoring salinized soil for conventional crops at up to US$ 5000 ha^{-1} whereas Le Houérou (1992) estimated the cost of establishing halophyte plantations at US$ 500–1000.

A specific example of how halophytes might fit into a reclamation program is the polder land in the Nile delta region (Boumans & Mashali 1983). This land started out saline and was reclaimed by building dikes and pumping out seawater. Salinities were reduced and in some of the region conventional crops are now grown. However, much of the land has remained too saline for conventional crops due to seepage from saline Lake Burullus, and complete reclamation will require subsurface drainage systems which may not be cost-effective. On the other hand, the reclamation program did reduce salinities even in the saltiest areas to well within the tolerance level of halophytes, so halophytes might be considered as an economic alternative to conventional crops in those areas.

Use of saline drainage water for halophytes

is another possible compatible use of halophytes within an irrigation district (Watson 1990; Watson & O'Leary 1993). For example, in the San Joaquin Valley of the United States, drainage water is contaminated with selenium and cannot be exposed for evaporation or discharged to the sea. Irrigation of salt-tolerant trees and halophytes has been recommended as an alternative disposal mechanism within the valley (San Joaquin Valley Drainage Program 1990).

Halophytes may also be valuable bio-remediation crops within irrigated fields that have high, saline water tables (see Miyamoto et al. 1994). Rows of deeply rooted, perennial halophytes interspersed in such a field may be used to scavenge subsurface saline water that would otherwise cause damage by rising into the root zone of the primary crop plants. Bioremediation is often attempted by using tree species such as *Prosopis* but often the ground water exceeds the salt tolerance level of these trees: halophyte shrubs are an alternative choice for very salty ground water conditions (see Ahmad et al. 1994).

Conclusion

The important point to be made here is that there are vast areas of potentially useful land for growth of halophytes for use as animal feeds, either under rainfed or irrigated conditions, that are distributed throughout the world. Likewise, there is a large pool of halophytes with the requisite salt tolerance for use in those areas from which suitable candidates could be selected. Thus, for most, if not all, parts of the world, there is a means for providing animal feed even though it may seem at first that sufficient land and water resources are not readily available.

References

Ahmad, R., Ismail, S., Bodla, M.A. & Chandhry, M.R. (1994). Potentials for cultivation of halophytic crops on saline wastelands and sandy deserts in Pakistan to overcome feed gap for grazing animals (this volume, Ch. 17).

Aslam, Z., Jeschke, W.D., Barrett-Lennard, E.G. Setter, T.L., Watkin, E. & Greenway, H. (1986). Effects of external NaCl on the growth of *Atriplex amnicola* and the ion relations and carbohydrate status of the leaves. Plant, Cell and Environ. 9: 571–580.

Aronson, J. (1989). HALOPH: a data base of salt tolerant plants of the world. Office of Arid Land Studies, University of Arizona, Tucson, AZ, USA, 77 pp.

Barrett-Lennard, E.G., Malcolm, C.V., Stern, W.R. & Wilkins, S.M. (eds.). (1986). Forage and fuel production from salt affected wasteland. Elsevier, New York.

Batanouny, K.H. (1994). Halophytes and halophytic plant communities in the Arab region: Their potential as a rangeland resource (this volume, Ch. 11).

Blits, K.C. & Gallagher, J.L. (1990). Salinity tolerance of *Kostelelzkya virginica*. I. Shoot growth, ion and water relations. Plant, Cell, and Environ. 13: 409–418.

BOSTID, Board on Science and Technology for International Development (1990). Saline agriculture. Salt-tolerant plants for developing countries. National Academy Press, Washington.

Boumans, J. H. &, Mashali, A.M. (1983). Seepage from Lake Burullus in the reclaimed Mansour and Zawia polder area. Agricultural Water Management 7: 411–424.

Burchett, M.D., Field, C.D. & Puikownik, A. (1984). Salinity, growth and root respiration in the grey mangrove, *Avicennia marina*. Physiol. Plant. 60: 113–118.

Chapman, V.J. (1960). Salt marshes and salt deserts of the world. Interscience, New York.

Chapman, V.J. (1974). Salt marshes and salt deserts of the world. Springer Verlag, Berlin/New York, 213 pp.

Chapman, V.J. (1975). Terrestrial halophytes as potential food plants. In: Somers, G.F. (ed.), Seed-bearing halophytes as food plants. University of Delaware, Newark, pp. 75–87.

Chavan, P.D. & Karadge, B.A. (1986). Growth, mineral nutrition, organic constituents and rate of photosynthesis in *Sesbania grandiflora* L. grown under saline conditions. Plant and Soil 93: 395–404.

Cronquist, A. (1981). An integrated system of classification of flowering plants. Columbia University Press, New York.

Dregne, H., Kassas, M. & Rosanov, B. (1991). A new assessment of the world status of desertification. Desertification Control Bulletin 20: 6–18.

El Shaer, H.M. & Gihad, E.A. (1994). Halophytes as animal feed in the Egyptian deserts (this volume, Ch. 28).

Eshel, A. (1985). Response of *Suaeda aegyptica* to KCl, NaCl, and Na_2SO_4 treatments. Physiol. Plant. 65: 308–315.

Flowers, T.J., Troke, P.F. & Yeo, A.R. (1977). The mechanism of salt tolerance in halophytes. Ann. Rev. Plant Physiol. 28: 89–121.

Flowers, T.J., Hajibaheri, M.A. & Clipson, N.J.W. (1986). Halophytes. The Quart. Rev. Biol. 61: 313–337.

Fowler, J.L., Hageman, J.H., Suzukida, M. & Assadian, H. (1988). Evaluation of the salinity tolerance of Russian-Thistle, a potential forage crop. Agron. J. 80: 250–258.

Glenn, E.P. (1987). Relationship between cation accumulation and water content of salt-tolerant grasses and a sedge. Plant, Cell & Environment 10: 205–212.

Glenn, E.P. & O'Leary, J.W. (1984). Relationship between salt accumulation and water content of dicotyledonous halophytes. Plant, Cell, and Environ. 7: 253–261.

Glenn, E.P., Kent, K.J., Thompson, T.L. & Frye, R.J. (1991). Seaweeds and halophytes to remove carbon from the atmosphere. EPRI ER/EN-7177. Electric Power Research Institute, Palo Alto, California.

Glenn, E.P., Watson, M.C. & O'Leary, J.W. (1992). Salt tolerance and osmotic adjustment of low-sodium and high-sodium subspecies of the halophyte, *Atriplex canescens*. Plant, Cell Environment 15: 711–718.

Glenn, E.P. & Watson, M.C. (1993). For direct seas water irrigation. In: H. Lieth & A.Al Massoom (eds.), Towards rational use of high salinity tolerant plants. Kluwer Academic Publishers, Dordrecht, The Netherlands, Vol. 1 pp. 379–386.

Greenway, H. (1968). Growth stimulation by high chloride concentrations in halophytes. Israel J. Bot. 17: 169–177.

Hagége, D., Kevers, C., Boucaud, J. & Gaspar, T. (1988). Activities peroxydasiques, production d'ethylene, lignification, et limitation de croissance chez *Suaeda maritima* cultiven l'absence de NaCl. Plant Physiol. Biochem. 26: 609–614.

Iljin, M M. (1946). Some results gained by studying the formation of flora in the deserts of Central Asia. Materials on the history of the flora and vegetation of the USSR, Vol. 2, pp. 196–256, (cited in Waisel, 1972).

Le Houérou, H.N. (1992). The role of saltbushes (*Atriplex* spp.) in arid-land rehabilitation in the Mediterranean Basin: A review. Agroforestry Systems 18: 107–118.

Le Houérou, H.N. (1994). Forage halophytes and salt-tolerant fodder crops in the Mediterranean Basin (this volume, Ch. 10).

Malcolm, C.V. (1994). Use of halophyte forages for rehabilitation of degraded lands (this volume, Ch. 4).

Meigs, P. (1966). Geography of coastal deserts, Paris: UNESCO, 140 pp.

Miyamoto, S., Glenn, E.P. & Singh, N.T. (1994). Utilization of halophytic plants for fodder production with brackish water in subtropic deserts (this volume, Ch. 5).

Mozafar, A., Goodin, J.R. & Oerthi, J.J. (1970). Na and K interactions in increasing the salt tolerance of *Atriplex halimus* L.: I. Yield characteristics and osmotic potential. Agron. J. 62: 478–481.

Mudie, P.J. (1974), The potential economic uses of halophytes. In: R.J. Reimold & W.H. Queen (eds.), Ecology of halophytes. Academic Press, New York, pp. 565–597.

Naidoo, G. & Rughunanan, R. (1990). Salt tolerance in the succulent, coastal halophyte, *Sarcocornia natalensis*. J. Exper. Bot. 41: 497–502.

Neales, T.F., Fraser, M.S. & Roksandic, Z. (1983). Carbon isotope composition of the halophyte *Disphyma clavellatum* (Harv.) chinnock (Aizoaceae), as affected by salinity. Aust. J. Plant Physiol. 10: 437–444.

Okusanya, O. (1979). An experimental investigation into the ecology of some maritime cliff species. III. Effect of Sea Water growth. J. of Ecology 67: 579–590.

O'Leary, J.W. (1988). Saline Environments and halophytic crops. In: E. Whitehead, C. Hutchinson, B. Timmermann & R. Varady (eds.), Arid lands today and tomorrow. Westview Press, Boulder, pp. 773–790.

O'Leary J.W., Glenn, E.P. & Watson, M.C. (1985). Agricultural production of halophytes irrigated with seawater. Plant and Soil 89: 311–321

Osmond, C.B., Bjorkman, O. & Anderson, D.J. (1980). Physiological processes in plant ecology. Toward a synthesis with *Atriplex*. Springer Verlag, Berlin.

Pasternak, D. & San Pietro, A. (1985). Biosalinity in action: Bioproduction with saline water. Martinus Nijhof Publishers, Dordrecht, The Netherlands.

Raunkiaer, C. (1934). The life forms of plants and statistical plant geography. Oxford University Press, Oxford.

Robinson, S.P. & Downton, W.J.S. (1985). Potassium, sodium, and chloride in concentrations in leaves and isolated chloroplasts of the halophyte *Suaeda australis* R. Br. Aust. J. Plant Physiol. 12: 471–479.

San Joaquin Valley Drainage Program (1990). A management plan for agricultural subsurface drainage and related problems on the Westside San Joaquin Valley. US Dept. Interior.

Szabolcs, I. (1989). Salt affected soils. CRC Press, Inc., Boca Raton, Florida, 274 pp.

Szabolcs, I. (1994). Salt affected soils as the ecosystem for halophytes (this volume, Ch. 3).

Ungar, I. (1965). An ecological study of the vegetation of the Big Salt Marsh, Stafford County, Kansas. Univ. Kansas Science Bull. 46: 1–45.

Waisel, Y. (1972). Biology of halophytes. Academic Press, New York.

Warne, P., Guy, R.D., Rollins, L. & Reid, D.M (1990). The effects of sodium sulphate and sodium chloride on growth, morphology, photosynthesis, and water use efficiency of *Chenopodium rubrum*. Can. J. Bot. 68: 999–1006.

Watson, M.C. (1990). *Atriplex* species as irrigated forage crops. Agriculture, Ecosystems and Environment 32: 107–118.

Webb, K.L. (1966). NaCl effects of growth and transpiration in *Salicormia bigelovii*, a salt-marsh halophyte. Plant and Soil 24: 261–268.

Yeo, A.R. & Flowers, T.J. (1980). Salt tolerance in the halophyte *Suaeda maritima* L. Dum.: Evaluation of the effect of salinity upon growth. J. Exp. Bot. 31: 1171–1183.

3. Salt affected soils as the ecosystem for halophytes

ISTVÁN SZABOLCS
Research Institute for Soil Science & Agricultural Chemistry, Hungarian Academy of Sciences, Budapest, Hungary

Abstract. The place and role of soil in Nature is reviewed. The relationship between the lithosphere, atmosphere, biosphere and hydrosphere, ecosystems and soils is explained. The global extent of salt-affected soils (about 10% of the earth's land surface) and the types of salinity and alkalinity are discussed. Plant responses to saline environments and the interactions between plants and soils is the principal focus. There is always a close correlation between the soils and the plants growing on them. The different types of salt affected soils are associated not only with different environmental conditions but also with the different associations of halophytes and xerophytes. The properties of different types of salt affected soils, with their environmental distribution are explained. Discussion centres on the possibilities for utilizing saline and alkaline soils to grow halophytes for fuel, browse, forage or other purposes.

Key words: Alkalinity, Electrolytes, Halophytes, Irrigation and drainage, Review paper, Salinity, Soil genesis

The place and role of the soil in nature

Soil is a substantial part of the environment. It is comprised of different substances with various properties and attributes. Soils form a special subset of any given ecosystem. They have various properties and attributes. The soils of the continents show high diversity, and their study is dealt with by several branches of soil science, e.g. classification, survey, mapping, etc. The soil, or the pedosphere, which is an environmental synonym for the soils of a given territory, has a specific place in Nature. It is a natural body, similar to rocks, waters, or biota in the sense that they, too, have their own materials, mass and energy fluxes, development, and regularities. Soils are frequently treated either as living substances or as non-biological substances. Neither of these approaches is correct. A characteristic of the soil is its complexity, the fact that it contains both living and non-living substances. Soil formation depends on both biotic and abiotic processes.

The soil, as a natural body, is inseparable from the weathering parent material on the one hand and from the biological processes, on the other. The main characteristic that distinguishes the soil from the rocks is the result of biological processes. The production of organic matter by the activities of micro-organisms, plants, invertebrates and other animals, and finally human beings transforms the rocks into soils, which are capable of supplying plants and crops with nutrients and water.

The processes of soil formation started concurrently with the appearance of life on the continents and continued during the billions of years of interaction between living substances and rocks under the influence of climatic conditions, with particular regard to the action of water, geomorphological patterns, and the time factor. As a result of their interaction specific mass and energy fluxes formed the different soil types in different environmental conditions.

With the appearance of the human race on the

V.R. Squires & A.T. Ayoub (eds.), *Halophytes as a resource for livestock and for rehabilitation of degraded lands*, 19–24.

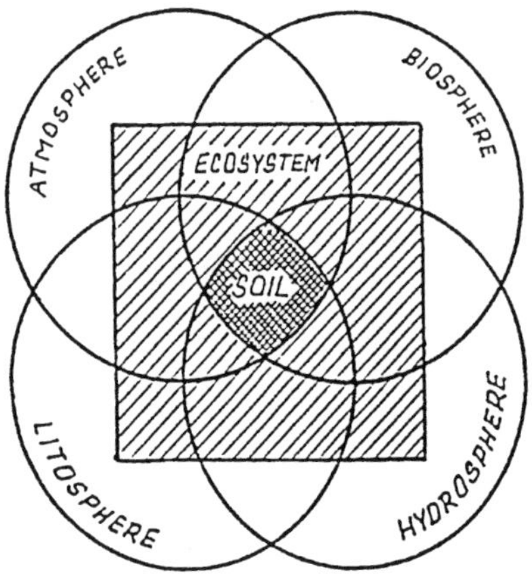

Figure 1. Schematic diagram of the interaction between the lithosphere, atmosphere, biosphere, hydrosphere, ecosystems, and soils (Szabolcs 1989).

face of earth even changes in the environment became different. Due to human activities the natural processes affected by biotic and abiotic factors accelerated and several others, which were unknown or minimal before, developed. The role of soils in Nature is complex and many-sided including biospheric, hydrospheric and lithospheric functions. Their interaction is illustrated in Figure 1.

Figure 1 clearly shows that the soil is a specific body related to ecosystems. Even the word 'soil' is very often used as a synonym of ecosystem when characterizing the given ecological conditions in a certain place. If we want to be precise we must agree that the ecosystem includes the pedon, in other words, the soils. However the soil includes different phases (solid, liquid, gaseous), living and non-living substances, plants, animals, microbes and has its own energy and material fluxes. So it can be considered an ecosystem in itself. In this respect, when speaking of soils versus their plant cover, we can consider the soils of a given place as the basis, larder, and foothold for the ecosystem developing on the spot. It is well known that close correlations exist between soil type and soil properties on the one hand and the plant cover developing on it on the other hand. It is generally accepted for instance that in savannas or in the tropical belt a well defined plant cover develops and very often the soil properties promote or limit the living conditions of certain plant species or associations.

Based upon the above considerations it can be accepted that certain soil types, when discussed as the habitat for certain plant associations, are often named as the ecosystem of the plant association concerned, as the pedon includes, apart from plants, most of the components of the ecosystem.

Evidently the soil, as a specific natural entity, is far from being identical with the vegetation and in spite of their close correlation, direct conversion between soil types and vegetation is hardly possible. Still there are soil types which more or less determine the ecological function for certain types of vegetation either by providing beneficial conditions for their development or by limiting the ecological conditions for other types of vegetation.

This is perhaps the best demonstrated in the case of salt-affected soils where high electrolyte content or extreme pH conditions limit the development of the majority of plants and serve as a habitat only for such species which can survive or tolerate the unfavourable conditions caused by the salinity and alkalinity of the soil. In such respects salt-affected soils can be considered as habitat or ecosystem for halophytes and, if we agree on this, correlations can be found between the different types of salt-affected soils and their flora and fauna as components of the ecosystem.

In order to cast light on both the theoretical and practical aspects of such considerations, it is necessary to describe briefly the properties and grouping of salt-affected soils with regard to the possibilities of the occurrence and distribution of halophytes and xerophytes developing on them.

The extent and global distribution of salt-affected soils

Nearly 10% of the total land surface is covered with different types of salt-affected soils. Table 1 shows the distribution of salt-affected soils in the world (Kovda & Szabolcs 1979) and shows that no continent on our globe is free from salt-affected soils. They are distributed not only in deserts and semi-deserts, but also frequently occur in fer-

Table 1. Salt affected soils on the continents and subcontinents

Continent	Area (ha × 10^3)
North America	15,755
Mexico and Central America	1,965
South America	129,163
Africa	80,538
South Asia	87,608
North and Central Asia	211,686
South-East Asia	19,983
Australasia	357,330
Europe	50,804
Total	*954,832*

tile alluvial plains, river valleys and coastal areas, close to densely populated areas and irrigation systems. Figure 2 shows a draft map of the distribution of salt-affected soils over the world.

Halophytes growing on different types of salt-affected soils and the possibilities of their utilization

The different groups of salt-affected soils briefly described above behave differently in relation to halophytes and sometimes xerophytes.

In Table 2 the grouping of salt-affected soils is demonstrated with the indication of their properties, environmental conditions, as well as their relation to halophytes. It is clear from Table 2 that the different types of salt-affected soils have different chemical and physical properties which in turn have different effects on the conditions of plant life. As a general statement it can be accepted that certain limiting factors exist in all types of salt-affected soils. They make life impossible for many plant species and, as a rule, only permit the development of such associations which are more or less able to tolerate the salinity and/or alkalinity of the soil. Such interrelations can be characterized for the different groups of salt-affected soils as follows:

(1) In saline soils, accounting for the overwhelming majority of all the salt-affected soils on our globe, high salt concentration and consequently high osmotic pressure constitute the limiting factors for the biota. As a rule, most of the salts are sodium salts of neutral pH (sodium chloride, sodium sulphate and sometimes sodium nitrate). Very often, when speaking of salt-affected soils, this group is considered as a subject of salinity in the top layer of the land.

The natural vegetation growing sometimes sparsely on saline soils consists of halophytes. The distribution of species depends on aridity conditions on the one hand, and on the chemical composition of salts on the other hand. The physical properties of the soil, first of all, particle size distribution, also influence the composition of halophyte species. The indicator species for salinity are very numerous and mostly perennial. There are however, a few annuals which always reveal some particular characteristics of the saline soil (humidity, chemical composition of salts, etc.).

A broad variation of different halophyte associations growing on saline soils are described in the different handbooks (e.g. International source-book 1967).

In the following only such representatives of halophyte and xerophyte vegetation will be mentioned as examples which have very direct correlations with soil conditions, i.e. the species most common in the different varieties of saline soils (*Salicornia berbera*, *Salsola soda*, *Puccinella gigantea*, *Salsola crassa*, *Suaeda microphylla*, etc).

Evidently some of them, (e.g. *Salsola soda*) grow not only on saline but also on alkali soils because it also tolerates high alkalinity. Also characteristic of both saline and alkali soils are *Statice gmelini*, *Achillea cristata*, *Aster tripolium* and others. Ligneous plants including the different types of tamarisks (*Tamarix ramosissima*, *Tamarix gallica. Tamarix balansae*) deserve attention and *Haloxylon ammodendron* on sandy soils together with *Calligonium* bushes. They are used as fuel, medicinal herbs and for technical purposes by the local population (Zhang et al. 1994; Batanouny 1994).

On saline soils with high humidity different halophyte *Carex* species can be found. In salt affected marshy or flooded areas *Bolboschoenus maritimus* and *Scirpus maritimus*, often occur together with *Carex divisa* and *Juncus maritimus*.

Figure 2. Global distribution of salt affected soils (Szabolcs 1992).

Only a few of halophyte herbs can be used as a fodder of poor value in exceptional cases.

(2) Alkali soils can be subdivided into two subgroups:
- alkali soils with a high concentration of sodium carbonate and having no typical columnar structure in the profile,
- alkali soils with comparatively low salt concentration and having a structural B horizon in the profile (solonetz type).

Consequently the halophytes and xerophytes growing on alkali soils differ in accordance with the different properties of these two subtypes. Especially in solonetz soils it is not the high salt concentration but the poor water relations and physical properties which are responsible for hindering plant life. Consequently the vegetation on these soils only partly consists of halophytes with halofrequent xerophytes accounting for a significant proportion.

In subgroup (a) *Salsola soda*, *Festuca ovina*, *Camphorosma ovata* often occur together with *Statice gmelini* and *Puccinellia limosa*. The latter also can be found in subgroup (b) in cases of high alkalinity.

Together with some halophytic species, in solonetz soils xerophytes or halofrequent xerophytes prevail; first of all different kinds of sage (*Artemisia maritima*, *Artemisia monogyna*. *Artemisia alba*, etc.), *Statice gmelini*, *Festuca ovina*, and *Festuca pseudovina*; in cases of low salt concentration *Achillea millefolium*, *Hordeum gussonearum*, *Aster tripolium* and *Poa bulbosa*, even *Poa pratensis*.

Parallel with the lowering of salt concentration the leguminae (*Melilotus indica*, *Medicago hisipida*, *Trifolium maritimum*, etc.) and herbs, typical of the non-salt affected surrounding areas, appear. Among the halophytes, particularly *Puccinellia limosa* has a high value as fodder in spite of the fact that the amount of yield is low, as a consequence of poor ecological conditions. Nevertheless in several countries e.g. in Hungary, with irrigation and the application of nitrogen fertilizers, it helps considerably in solving fodder problems in dry areas.

Experiences have been similar with a number of local xerophytes and halophytes in different countries while making efforts to relieve the fodder problem always arising in salt effected areas (Ahmad et al. 1994). Evidently in cases of low alkalinity and salinity more and more possibilities are available for producing fodder.

(3) In magnesium soils representatives of the two

Table 2. Grouping of salt affected soils and their properties in relation to halophytes

Type of salt affected soils	Electrolyte(s) causing salinity and/or alkalinity	Environment	Properties causing plant and crop stress	Indicator species of halophytes and halofrequent xerophytes	Possibilities of utilizing the natural plant cover	Methods for reclamation
Saline soils	Sodium chloride and sulphate (in extreme cases nitrate)	Arid and semi-arid	High osmotic pressure of soil solution, toxic effect of chlorides	Salicornia berbera, Salsola soda, S. crassa, Suaeda microphylla, etc. Tamarixs and Carexes	Herb and fodder of poor value Fuel	Removal of excess salt (leaching)
Alkali soils	Sodium ions capable of alkaline hydrolysis	Semi-arid, semi-humid, and humid	High (alkali) pH, poor water physical conditions	Artemisias, Statice gmelini, various Festucas and Melilotuses	Herb and fodder of poor value	Lowering or neutralizing the high pH by chemical amendments
Magnesium soils	Magnesium ions	Semi-arid and semi-humid	Toxic effect, high osmotic pressure, Ca deficiency	Spergularia salina, Suaeda maritima, etc.	See at the alkali soils	Chemical amendments, leaching
Gypsiferous soils	Calcium ions (mainly $CaSO_4$)	Semi-arid and arid	Low (acidic) pH Toxic effect	Ligneum spartum, Limonium delicatulum	See at the alkali soils	Alkaline amendments
Acid sulphate soils	Ferric and aluminium ions (mainly sulphates)	Seashores and lagoons with heavy, sulphate-containing sediments, diluvial inland slopes and depressions	High acidity and the toxic effect of aluminium	Various reeds, mangroves,white cedar, etc.	Fuel	Liming

former groups can often be found, depending on the salt concentration as well as on the alkalinity of the soil. Apart from this some specific plant associations are characteristic of magnesium soils e.g. *Spergularia salina* and *Suaeda maritima* of soils containing magnesium carbonate while *Ormenis praecox* and *Beta macrocarpa* of saline magnesium soils.

(4) Gypsiferous soils are extensive in semi-arid and arid areas and in most cases the gypsophilic xerophytes prevail among their natural plant associations. *Lygeum spartum* and *Limonium delicatulum* are specific of the calcic gypseous salt affected soils in the Middle East. On the top of micro relief *Arthrocnemum glaucum* cushions occur. *Inula crithmoides* is also frequent of occurrence on gypsiferous soils.

(5) On acid sulphate soils a diverse vegetation can be found depending on local geographical conditions. The development of the various herbaceous species is also subject to the salinity and acidity degree of the soil and they can scarcely be used as fodder. It is well-known that acid sulphate soils are well-developed in mangrove ecosystems, whose biota is determined by such circumstances. Reed vegetation in brackish swamps is also associated with acid sulphate soils, for example in West Africa where racemose is common, or in South America where *Rhizophora mangle* has been studied (Brinkmann & Pons 1973). In freshwater areas, *Imperate brasiliensis* (white cedar) is characteristic of South America, in eastern Asia Melaleuca can be observed. The latter tree has a wide ecological amplitude.

The halophytic and xerophytic vegetation, typical of the different groups of salt affected soils, has considerable significance in national parks and other environmentally protected areas. Together with native ecological factors, the preservation of native xerophytic and halophytic vegetation on salt affected soils, is a substantial part of keeping protected lands undisturbed.

Concluding remarks

The close interrelations between ecosystems and soils make possible to differentiate between the soil types associated with a certain type of vegetation. In this respect salt affected soils are closely related to halophytes.

The different groups of salt affected soils can be associated with different associations of halophytes and xerophytes. Several soil attributes are preconditions of the development of certain plant associations.

There is a high diversity of halophytes, xerophytes, and other plant associations in the different areas of the world covered by different types of salt affected soils.

In order to promote better knowledge and better utilization of the halophytic vegetation in salt affected areas further studies are necessary for revealing the complex interrelations between the soil conditions of salt affected areas and the vegetation growing on them.

Acknowledgements

It is pleasure to thank UNEP for the travel grant which allowed participation in the Nairobi workshop.

References

Ahmad, R., Ismail, S., Bodla, M.A. & Chaudry, M.R. (1994). Potentials for cultivation of halophytic crops on saline wastelands and sandy deserts in Pakistan to overcome the feed gap for grazing animals (this volume, Ch. 17).

Batanouny, K.H. (1994). Halophytes and halophytic plant communities in the Arab region: Their potential as a rangeland resource (this volume, Ch. 11).

Brinkman, H. & Pons, L.J. (1973). Recognition and prediction of acid sulphate soil conditions. Publications of the International Institute for Land Reclamation and Improvement, Wageningen 18(1): 169–203.

International source-book on irrigation and drainage of arid lands in relation to salinity and alkalinity, 1967, FAO/UNESCO, Rome/Paris.

Kovda, V. A. & Szabolcs, I. (1979). Modelling of soil salinization and alkalization. Agrokémia és Talajtan, 28, Suppl.

Szabolcs, I. (1989). Salt affected soils. CRC Press, Boca Raton, FL.

Szabolcs, I. (1992). Salinization of soil and water and its relation to desertification. Desertification Control Bulletin UNEP, No. 21, pp. 32–37.

Zhang, L, Xia, Y. & Zou, Y. (1994). Halophytes and halophytic plant communities in Inner-Asia (this volume, Ch. 9).

4. Use of halophyte forages for rehabilitation of degraded lands

C.V. MALCOLM
Formerly, Department of Agriculture, Perth, Australia

Abstract. Halophytic forage shrubs and grasses may be used to revegetate degraded saline and/or sodic environments. Testing methods have been developed for determining the best species for each site. Establishment depends on having adequate supplies of good quality seed and using specially developed niche seeding technology. Spot placements of seed are covered with mulch and the placement is sprayed to hold it in place. The use of spot treatments at a wide spacing reduces the cost to a practical level. Commercial direct seeding of forage shrubs on saline land is now an established practice involving thousands of hectares each year. The halophytic forages are sown in mixtures and used for grazing by sheep in the autumn to obviate the need for supplementary feeding. The practice is profitable for farmers/ranchers.

Key words: Establishment, Grazing, Niche seeding, Seeding, Species selection, Supplementary feeding, Utilization, Western Australia

Treating salt-affected soils

Salt-affected soils are estimated to occupy about 10% of the world's land surface (Szabolcs 1992). The severity of salting varies from mild, where the yield of sensitive crops is reduced, to severe, where only the most salt tolerant halophytes will grow. Salt-affectedness occurs on a wide range of soil types and is often complicated by interaction with factors such as waterlogging or drought.

The distribution of salt-affected soils in the world is uneven (see Szabolcs, Ch. 3, Table 1). The greatest areas occur in arid and semi-arid regions, either as a result of natural processes or, due to land use changes imposed by people. The impact of salinity problems on people is accentuated by the fact that the greatest areas of salinity occur in those regions in which there is the most critical shortage of land for agricultural production (see Table 1).

Many approaches have been developed to treat salinity problems. Efforts made to reclaim salt-affected soils for normal crops include drainage, levelling, leaching and the application of soil amendments. In some areas these measures have been highly successful but in others they have transferred the problem to other sites or have been too expensive. Some salinity problems have been treated by attacking the causes, for example, by improved irrigation management or by altering catchment use. Other workers have developed salt tolerant varieties of crop or pasture plants to improve production from salt-affected soils. Despite the application of these various measures, the extent and severity of salinity problems worldwide continue to increase. For example, in 1984 it was reported (Dudal & Purnell 1986) that 7% of the world's land surface was salt-affected and, as stated above, by 1991 the estimate had increased to 10% (Szabolcs 1994).

In nature, salt-affected soils carry a cover of vegetation. These highly salt tolerant plants are capable of growing in salt-affected sites and providing valuable products, especially forage and

V.R. Squires & A.T. Ayoub (eds.), *Halophytes as a resource for livestock and for rehabilitation of degraded lands*, 25–41.

Table 1. Regions of the world, showing areas of highly saline soils (solonchaks) and numbers of countries suffering a critical shortage of land for agriculture

Region	Number of critical countries[a]	Areas of Solonchak soils (ha × 10^3)[b]
Africa	47	43,579
S.W. Africa	42	47,233
C. America	23	242
S.E. Asia	9	nil
S. America	0	10,461

[a] From FAO, Rome (1982).
[b] These figures are from Barrett-Lennard et al. (1986) and are not exhaustive.

fuel. The use of halophytes to rehabilitate and make productive use of salt-affected land is proposed for those areas which will not respond to other measures and is discussed in this chapter.

Plant growth on salt affected soils

The most salt-tolerant plants are referred to as halophytes and they are capable of completing their life cycle and producing biomass despite the presence of salt in their root zone. Many halophytes are also tolerant of sodicity and alkalinity. Some halophytes grow on upland salt-affected soils and others grow in areas affected by groundwater within 0–3 m of the surface.

Low-lying salt-affected soils may be affected to varying degrees by inundation and surface or sub-surface waterlogging. Low soil oxygen levels, caused by excess water in soil, interact with salinity to adversely affect plant growth. Halophytes differ in their ability to withstand salt and waterlogging and as a result the species composition in salt-affected areas is zoned in response to differences in inundation or waterlogging which are often also associated with differences in salinity (Novikoff 1961). The waterlogging responsible for the zonation may be of short or long duration, unpredictable and highly variable in frequency.

Low oxygen levels caused by excess soil water cause roots to stop growing and sometimes to die. Roots lacking oxygen cannot absorb water and lose their ability to control ion (and hence salt) uptake. Potassium uptake is reduced and salt uptake and translocation to the shoots increased. Halophytes suffer similar disabilities to normal plants as a result of low root oxygen and, like non-halophytes, some have developed strategies (such as aerenchyma (root air channels) and adventitious rooting) which assist them to survive under waterlogged conditions (Galloway & Davidson 1992). Plants which lack special adaptation to low soil oxygen have their root growth restricted to those parts of the soil with adequate oxygen e.g. drained soil above the watertable or soil in raised banks.

The productivity of a plant on a salt-affected site is determined by soil depth and fertility and the salt tolerance and relative growth rate of the plant and the quantity and quality of water available in the root zone. The root zone is severely restricted in soils with a very shallow watertable or indurated layers. In addition, root growth may be affected by seasonal waterlogging of the soil above the watertable due to rainfall or run-on. The more saline and waterlogged a site, the narrower the range of plants adapted to it. The sensitivity of plants to the hydrological conditions on a salt-affected site is such that elevation changes of a few centimeters may cause a complete change in species composition of the natural vegetation.

Many halophytes on salt-affected sites are capable of surviving extremely high salinity levels if their root systems are not waterlogged. Under such conditions they make little or no growth and may shed leaves. Some halophytes absorb salt to increase their ability to survive. However the amount of salt absorbed is negligible compared to that which is excluded and high salinity levels remain in the root zone. The plants survive until rain reduces the salt levels and provides an opportunity for growth. In some cases growth is rapid, with relative growth rates approaching those of domesticated crops such as wheat (Malcolm 1986a,b,c). Many halophytes are C_4 plants (Le Houérou 1992).

Increasing the salt tolerance of crop plants

Farmers whose land has become salt-affected usually wish to have the salt removed or to be given normal crop plants of sufficient tolerance that they can continue to farm in the same way.

Hence, many scientists have endeavoured to increase the salt tolerance of crop plants.

One approach is to select from the available germ plasm in a crop species to find the most tolerant lines. Workers in California (Epstein et al. 1980) screened 6200 lines of barley for salt tolerance and managed to improve the tolerance by 23%. Unfortunately when they took the selections to the field they were unable to demonstrate differences between their selections and normal high yielding varieties.

Salt-affected fields are highly variable and the difference in salinity between the salty and non-salty patches in a field are usually much greater than the differences in tolerance between the crop varieties. The average yield taken across a field is likely to be similar, because, the loss by salt intolerant varieties on the salty patches is replaced by their higher yield on the non-salty patches. An additional problem is presented by the effects of waterlogging and poor soil structure on the saline areas. For a crop variety to really succeed on saline fields it needs to be as productive as the best varieties on non-saline areas and sufficiently salt tolerant to produce well on the saline patches. In many cases the difference in salinity between salty and non-salty patches in a field is of the order of ten-fold. No workers have been able to achieve such an increase in salt tolerance in a crop plant (Miyamoto et al. 1994).

Efforts to increase salt tolerance by genetic engineering approaches can be expected to be slow and frustrating. Salt tolerance is recognised to be a multigene property in plants and its incorporation along with all the necessary agronomic characteristics in a new crop or pasture plant represents a very complex problem.

Relative salt tolerance of crop plants and halophytes

The salt tolerances of crop plants and halophytes are compared in Fig. 1(a). The growth of most crop plants is reduced at least 50% by a chloride level of 130 mol/m^3 or less in solution culture. The most tolerant halophytes do not have their growth reduced 50% until the chloride level in the culture solution exceeds 600 mol/m^3. Halophytes are therefore adapted to growing on sites that are

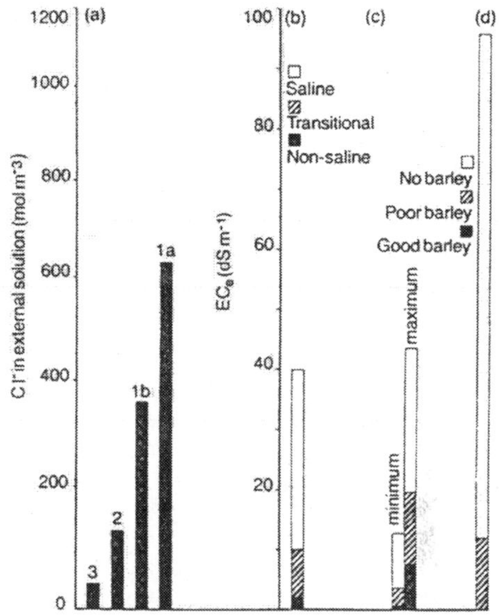

Figure 1. Comparison of salt tolerance data from glasshouse and field experiments. Scales for (a) to (d) are comparable. (a) Salinity levels associated with a 50% growth reduction (after Greenway & Munns 1980). Plants grouped according to salt tolerance: 1a and 1b, halophytes; 2, some halophytes and crops e.g. barley, cotton, tomatoes and beans; 3, very salt sensitive plants. (b) Salinity levels characteristic of non-saline, transitional and saline soils in the Rio Grande Valley of Texas (after Carter & Wiegand 1965). (c) Minimum and maximum salinity levels associated with good, poor and nil growth of barley in Saskatchewan, Canada (after Ballantyne 1962). (d) Minimum (hatcheted) and maximum (clear) levels of salinity in composite samples from salt tolerant plant test sites in south-western Australia (C.V. Malcolm, pers. comm. 1984).

of the order of four times as saline as are crop plants.

The levels of salinity and the reaction of plants to it in affected soils in widely divergent areas such as Texas, Canada and Australia are surprisingly similar. In Fig. 1(b), are data for salt-affected soils in Texas. The EC$_e$ of saline, transitional and non-saline soils was estimated to be 40, 10 and 2 dS m^{-1} respectively at 15 cm depth. These data have been graphed so that the scales for Sections (a), (b), (c) and (d) are comparable. The EC$_e$ levels in the three zones in the Texas fields correspond to recognised standards for soils too saline for any crops, suitable for moderately sensitive crops and suitable for salt sensitive crops respectively. The areas of differing salinity were so haphazardly intermixed in the fields that it was not possible for farmers to crop them separately. Levels of soil salinity in the field in Canada associated with zero, poor and good growth of barley are shown in Fig. 1(c).

The salinity levels in cropped fields in Texas and Canada may be compared with the data in Fig. 1(d) which are for sites in Western Australia. These sites had been abandoned for cropping by farmers and were used for cooperative research on the growing of halophytic forage shrubs and grasses. The halophytes grew successfully on these sites.

Comparison of the data for the various sites leads to the conclusion that cereal cropping may be conducted where the EC_e levels are about 10 dS/m or less and that halophytic shrubs and grasses may be grown on areas of higher salinity. My experience with farmers is that they will persevere with cereals on soils that are marginal for cropping. However when they discover the value of halophyte forages by trying them on their own property, they willingly turn their non-viable cereal land over to forage species on a permanent basis.

Plants for productive use of salt-affected land

A wide range of plant genera inhabit salt-affected sites and the plants adopt a variety of methods of dealing with high salt in the growth medium (trichomes, succulence, glands). These plants are referred to as halophytes and many of them are additionally adapted to surviving waterlogging (pneumatophores, aerenchyma) or drought (xerophytism, hairiness, succulence).

Halophytes are used for timber (mangroves), fuel (mangroves, *Tamarix* spp., *Atriplex* spp.) food (*Distichlis palmeri*), oil (*Salicornia bigelovii*) and forage (*Atriplex* spp., *Sporobolus* spp.). In the present chapter we are only going to consider halophytes used as forage.

The most waterlogging tolerant halophytic forages are succulent-stemmed species in genera such as *Salicornia*, *Arthrocnemum*, *Halocnemum* and *Halosarcia*. These plants often occur in areas subject to seasonal waterlogging. They are high in soluble salts and must be diluted with forage of a low salt content to prevent scouring in grazing animals. Nevertheless they are used to varying degrees by sheep, goats and camels and provide a useful supplement to dry feed. This group of plants is referred to as samphires or glassworts.

Plants which develop trichomes (or salt bladders) on their leaves are referred to as saltbushes and include the genera *Atriplex*, *Halimione*, *Rhagodia* and *Chenopodium*. These plants are less waterlogging tolerant than the samphires but some species will tolerate a few weeks of waterlogging of the whole root zone. The saltbushes have a lower salt content than the samphires and many species are readily eaten by domestic animals. A few species (e.g. *Atriplex stipitata* or bitter saltbush) are not eaten due to low palatability, possibly due to their saponin content. Major differences in palatability may even occur within a species between different ecotypes.

Other halophytic forage shrubs are in the genera *Salsola*, *Maireana*, *Eurotia*, *Suaeda*, *Kochia* and *Enchylaena*. These shrubs have succulent leaves (sometimes hairy) and most are relatively sensitive to waterlogging. Some species are readily eaten by stock and have lower salt contents than the *Atriplex* spp. Shrubs such as *Maireana brevifolia* (small leaved bluebush) should be diluted with dry feed because of their high oxalate content.

Halophytic grasses of forage value include *Sporobolus* spp., *Aeluropus* spp. and *Distichlis* spp. These species possess salt glands. Other highly salt tolerant grasses include *Puccinellia* spp. which avoid the highest salinity levels in a site by growing in winter in a mediterranean climate and being summer dormant, and *Paspalum vaginatum* (salt water couch) which is very tolerant of waterlogging. *Diplachne fusca* is tolerant of prolonged waterlogging and high pH but is less salt tolerant than the other halophytes.

It may also be beneficial to include trees in revegetation plans. Trees provide shelter and may ameliorate conditions for the growth of shrubs and grasses.

Benefits from growing halophytes

It has been established above that halophytes have the necessary salt tolerance to grow on areas where cereals fail. They also are naturally adapted to vast areas of salt-affected rangeland (Le Houérou 1992; Wilson 1994). For a farmer or grazier to choose to grow halophytes they must be able to produce forage and/or fuel in sufficient quantity and quality or provide other benefits.

The benefits from growing forage halophytes may include the following:
- improved nutrition of livestock in periods of feed shortage,
- profit from sales of forage,
- release of land previously used for growing forage, for the growth of food crops,
- reduced grazing pressure on public land,
- erosion control,
- improved use of dry residues for livestock by mixing them with green halophyte forage,
- reduced groundwater levels, due to water use by the halophyte,
- improved land value,
- improved aesthetics,
- better soil condition, and
- wildlife conservation.

Most halophytes are at present largely undomesticated. Nevertheless forage yields of up to 23.4 t/ha dry matter are reported (O'Leary 1986) from *Atriplex* spp. irrigated with highly saline water using centre-pivot overhead irrigation system on relatively fine textured soils. Yields of forage from *Atriplex* spp. grown in Western Australia on soils too saline for cereal cropping range up to 5 t ha^{-1} dry matter in areas receiving about 500 mm annual rainfall in a mediterranean type climate (Malcolm & Pol 1986). Goats fed a mixture of dry grass (on which they would normally lose weight) and green *Atriplex* spp. have gained weight (Nawaz & Hanjra 1992). Farmers have reported the wool on sheep grazing halophyte forage shrubs to be cleaner, finer and of more even fibre diameter than from sheep on dry feed plus a grain supplement. In experiments, sheep have gained body weight on a diet of *Atriplex* spp forage supplemented with dry feed (Warren & Casson 1992; Warren et al. 1994).

Ultimately the value of a forage material should be assessed in terms of animal production in a farm situation. Halophytic shrubs are best utilised in combination with dry residues especially during months of the year when there is a feed shortage (e.g. autumn to early winter in southern Australia; in Pakistan and India, probably mainly in winter). The value of stands of halophytic shrubs in Australia has been expressed in terms of the number of sheep grazing days per hectare or the saving in grain which would otherwise have had to be fed to the sheep. The sheep carrying capacity ranges from about 180 sheep grazing days (sgd) ha^{-1} for samphire on highly saline and waterlogged land to 2200 sgd ha^{-1} for stands of *Atriplex* and *Maireana* on less severe areas. These figures apply to areas receiving 300 to 400 mm annual rainfall of mainly winter incidence. Costs saved on hand feeding by grazing sheep on halophytic shrub pastures have been measured at 66 USD per ha (1990 figures, J. Hardy pers. com.). This compares favourably with the net return from growing wheat on nearby non-saline land.

Criteria for plant selection

Species which are to be used for forage production on salt-affected land must not only be salt tolerant but must satisfy agronomic criteria that fit them for practical farm or rangeland use. The species are tested to determine whether they meet these criteria by growing them in the field, subjecting them to various treatments and observing their performance. The following criteria are used in the evaluation:
- *survival duration*: a species must have the ability either to survive on a typical field site for a long time or be able to reproduce itself under an acceptable grazing system.
- *ability to flower and produce seeds*: species must be able to produce viable seeds or reproduce vegetatively.
- *production of volunteer plants*: species should be able to produce new plants in the field by natural seeding under appropriate grazing management.
- *growth habit*: plants intended for grazing by sheep should be low and spreading so that the forage is accessible to sheep; goats, cattle and camels will use taller species and also if shelter is an important aim tall bushes may be chosen.
- *feed quality*: should provide good nutrition for animals, the forage may be used in combi-

nation with other feed e.g. dry grass, stubble or hay.
- *palatability*: the forage must be sufficiently palatable for grazing animals to eat enough to satisfy their needs, this may involve combination with other feeds.
- *recovery after grazing*: plants must be capable of producing after repeated annual defoliation by grazing during the most likely period of use and a recovery period will be required.
- *effect of adverse conditions*: plants must be capable of surviving and maintaining or regaining productivity following events that characterise the field site even on an occasional basis, e.g. drought, frost, heat, insects, diseases, waterlogging, flooding.
- *additional benefits*: halophytic vegetation may offer advantages such as soil conservation, shelter, wildlife improvement, aesthetics.
- *productivity/economics*: chosen plants must be capable of producing at a profitable level. However, accurate assessments of productivity can only be made from stands planted at the optimum spacing for the species, and preferably productivity should be determined using grazing animals; these assessments are outside the realm of adaptation testing.
- *establishment ability*: plants must be able to be established readily by cheap practical field methods.
- *seed technology*: seed harvesting, storage and quality control must be possible.
- *no nuisance problems*: selected plants should not pose threats due to weediness, toxicity or other hazards.
- *salt tolerance*: the halophytic plants being considered are highly salt tolerant, the criteria listed above are those that determine the suitability of a shrub for forage production on salt-affected land.

Field testing methods for screening forage shrubs

The testing methods used for halophytic grasses are similar to those used for normal pasture species and will not be discussed here. It is necessary to consider in detail methods used for testing forage shrubs. The approach used in Western Australia to decide the best shrub species for field use has also been applied in Pakistan and is likely to have application in other areas. It is not possible to test for all of the above criteria in a single experiment. Separate experiments are required to test for site adaptation and general agronomic suitability, establishment ability and productivity.

Site adaptation and general agronomic suitability

A wide range of genera and species of reputed salt tolerance and value for grazing was collected in Western Australia in an attempt to find the best plants for on-farm use (Malcolm et al. 1984). To reduce the number of collections to a manageable size, the species were screened at two typical salt-affected sites. Some species failed to establish, others grew poorly (Malcolm & Clarke 1971) and of the shrubs under test, 25 were selected for testing at 14 sites chosen to represent the salt-affected soils of the Western Australian wheatbelt (Malcolm & Swaan 1989). Site adaptation experiments conducted in Western Australia (Malcolm & Swaan 1989) and Pakistan (Ismail et al. 1992) led to the development of a standard testing procedure referred to as ADAPT 2. The ADAPT 2 procedure has been adopted by the National Committee for Productive Use of Saline Land in Australia as a basis for adaptation testing. The ADAPT 2 programme provides a standard layout using eight single plant replicates in randomised blocks. The experiments are established on sites representing the conditions for which forage shrubs are required. Separate experiments may be conducted on zones of differing severity within a site. The zones may be identified by observing indicator species and/or testing the site for salinity. Indicator species are discussed under 'Matching Species to Sites'.

Every effort is made to minimise the soil variability within replicates. Detailed schedules are provided for characterising the site and for data to be collected on plant performance. Software is available for storing and manipulating the data. The ADAPT 2 programme allows plants suited to a site to be selected and provides information on feed quality, palatability, grazing recovery and the production of seed and volunteer seedlings. The information may be extrapolated to other sites by comparing them with the site information in the data base.

The validity of the selections made using ADAPT 2 or any other selection method is limited by the conditions experienced during the experiment. For example, one of the most important factors affecting survival on a saline site is the degree of waterlogging. Seasons giving severe waterlogging may not occur in the course of a short experiment. It is therefore desirable to arrange for species selection trials to remain in place for as long as possible. In trials in Western Australia, changes in the relative performance of species occurred after six years due to waterlogging events. Other events such as frosts or insect or disease attack are also more likely to be experienced in the course of prolonged testing.

Establishment ability

The ADAPT 2 tests are conducted by planting nursery raised plants into the field. To obtain information on field establishment ability, it is necessary to conduct tests by sowing seeds in the field. Experiments comparing the ability of species and ecotypes of *Atriplex* to establish in saline sites were conducted (H.V. Runciman, pers. com. 1986) using the Mallen Niche Seeder (see below) and by sowing weighed or counted quantities of seed by hand into measured rows (author's unpublished data). In both cases at least six replications were used to counteract the high variability encountered on saline sites. The number of germinable seeds sown must be determined and the establishment expressed as a per cent of germinable seeds sown. Experiments conducted in different years in highly divergent sites have given repeatable results, indicating that the ability to establish in the field from direct seeding is a heritable trait. As a result, seeds of two selections of *Atriplex amnicola* ('Meeberrie' and 'Rivermor') have been released in Western Australia on the basis of their superior ability to establish from seed. This ability correlates with the ability of these selections to spread in the field by natural seeding more effectively than other ecotypes under test.

The selection procedures described above have been used also to select species for rehabilitation of degraded rangeland (Ward 1992) and mine dumps in arid-areas (Fletcher et al. 1989). For both of these applications field screening for ability to establish from seeds was of special importance and screening from planted seedlings was of limited value. The ability of certain species (e.g. *A. bunburyana* and *A. vesicaria*) to establish well when direct seeded into salt-affected arid (200 mm rainfall) rangelands has been found to be repeatable from year to year and site to site.

Productivity and reaction to use

To determine the relative productivity of forage shrubs it is necessary to plant them at a specific spacing (Malcolm et al. 1988) and to compare their performance under grazed conditions (Malcolm & Pol 1986). Small shrubs such as *Atriplex vesicaria* (bladder saltbush) were found to give lower yields on a per hectare basis when planted at spacings wider than 1 × 1 m. By contrast, *Atriplex amnicola* (river saltbush) yielded the same on a per hectare basis when planted at spacings from 1 × 1 to 3 × 3 m. The costs of establishment may be greatly reduced by planting at a wider spacing. Spacing may be still further increased if shrubs are to be intersown with grasses.

Shrubs vary greatly in their ability to perform under grazed conditions. Data obtained using ADAPT 2 will indicate the relative abilities of test shrubs to recover from hand stripping and/or grazing. *Atriplex bunburyana* (silver saltbush) and *A. paludosa* (marsh saltbush) have been found not to recover as well as *A. amnicola*, *A. undulata* (wavy leaf saltbush) and *A. lentiformis* (quail brush) in hand stripping tests on individual bushes (B. H. R. Ward pers. com. 1992). Grazing experiments (Malcolm & Pol 1986) showed the ability of three *Atriplex* species to withstand heavy autumn grazing by sheep to be in descending order, *A. amnicola* > *A. undulata* > *A. paludosa*. *A. amnicola* survived heavy autumn grazing in six consecutive years without any apparent detriment to the stand. By contrast, *A. paludosa* was eliminated by subjecting it to heavy autumn grazing in three successive years.

Evaluating plant performance

The information obtained from plant experiments may be used to rank the species. The characteristics listed above under Criteria for Plant Selec-

Table 2. Relative importance of shrub performance criteria

Essential	Desirable	Optional
Survival[a]	Production of volunteers	Growth habit
Grazing recovery		Other benefits
Seed production	Seed technology	
Productivity	Feed quality	
Establishment ability	Palatability	

[a] Survival is assumed to require salt tolerance and survival of the various adverse conditions at the site.

tion are not of equal importance. In Table 2 the criteria are grouped as essential, desirable or optional.

Plants (shrubs, grasses and forbs) which do not meet the essential criteria cannot be recommended. To be useful, a plant must be able to survive the conditions on the site, including grazing, for a sufficient time to give profitable production and allow for recruitment of new plants to ensure the long term survival of the production system. The plants must be capable of producing viable seeds and of establishing by a practicable field method, preferably direct seeding.

Most plants that are well adapted to a site will produce some volunteer seedlings over time. The ability to produce volunteers has been listed as a desirable criterion in the sense of having especial ability in this area. For some applications, such as rehabilitation of rangelands, recruitment is of particular value because it permits wider spacing to be used and therefore reduces the cost of sowing. This strategy uses the ability of the plants to fill the gaps in a widely planted stand by natural colonisation. For productive use of salt-affected land in agricultural areas the level of productivity should be sufficient to pay for establishment of a complete stand initially. Under these circumstances volunteering ability is less important. It is desirable to select plants which have seeds that are readily harvested, cleaned and sown. The success of *A. undulata* in Western Australia has been partly due to the ease with which mechanical harvesting was developed by a farmer thus providing a good supply of seeds at an acceptable price.

To be useful and profitable plants must be of sufficient quality and palatability. Generally the plants worth including in test programmes will be eaten by sheep, goats or camels during periods of feed shortage and are of sufficent quality that, along with other feeds, they will benefit the animals.

The value of halophytic forages for cattle and buffaloes is not clear at this stage. Buffaloes in Pakistan readily ate left over *Atriplex* material from goat feeding trials (S. Nawaz, pers. com. 1991), and cattle on Australian rangelands are reported to consume more chenopod shrub material than sheep on mixed shrub and grass pastures due to the higher selectivity of sheep (Wilson 1994). Halophytic shrubs are normally grown where little else will grow and are therefore an improvement on the status quo. The more palatable and nutritious they are the better will be the animal response. It is often assumed that the most palatable shrubs will be the least able to recover from grazing. However, from experience in Western Australia the reverse is the case. For example, *A. bunburyana* is relatively unpalatable and susceptible to grazing and *A. amnicola* is one of the most palatable species and survives grazing extremely well.

If in addition to meeting the essential and desirable criteria, shrubs have a desirable growth habit and confer other benefits such as good shelter or erosion control, it is a bonus.

In Table 3 the performance criteria for several halophytic forage shrubs tested in salt-affected agricultural land in Western Australia are used to select the best species. By examining Table 3 it is possible to select *A. amnicola*, *A. undulata*, *A. lentiformis* and *M. brevifolia* as the only species meeting all essential criteria. Of the *Atriplex* species *A. undulata* has a slight advantage over the others because of the ease of commercial development of seed harvesting. However, *M. brevifolia* is suited especially to areas of saltland not affected by waterlogging and due to its excellent production of volunteer seedlings has an important place in saltland revegetation. In practice the four species are recommended as a mixture. It is believed the diversity of a mixture offers more ecological stability in the variable and harsh salt-affected environments. Table 3 does not account for the fact that none of the species performs adequately on highly waterlogged severely saline land. For these areas, the samphires, *Halosarcia* spp. are recommended.

Table 3. Species performance against a range of criteria

Criteria	Species[a]							
	A.am	*A.un*	*A.le*	*A.bu*	*A.pa*	*A.nu*	*A.ci*	*M.br*
Essential								
Survival	+ +[b]	+	+	+	+	5	+	+
Grazing recovery	+ +	+	+	−	−	+	?	+
Seed production	+	+	+	+	+	+	+	+
Productivity	+	+	+	?	+	?	+	+
Establishment ability	+	+	+	+	+	+	?	+
Desirable								
Volunteers	1	+	+	+	+	−	+	+
Seed technology	+	+ +	+	?	?	+	?	+
Feed quality	+	+	+	+	+	+	+	6
Palatability	+	+	+	−	+	−	+	+
Optional								
Growth habit	+ +	+ +	4	−	+ +	−	+ +	+
Shelter	2	3	+	+	−	+	−	−
Soil conservation	+	+	−	+	+	−	+ +	+

[a] *A.am* = *Atriplex amnicola*; *A.un* = *A. undulata*; *A.le* = *A. lentiformis*; *A.bu* = *A. bunburyana*; *A.pa* = *A. paludosa*; *A.nu* = *A. nummularia*; *A.ci* = *A. cinerea*; *M.br* = *Maireana brevifolia*.
[b] Ratings are: + + excellent, +good, −not satisfactory.
1. Selected lines have good volunteering ability.
2. Genotypes ranged from erect to prostrate.
3. Moderately grazed bushes develop sufficient height to protect sheep from cold winds.
4. Bushes are open and reach about 2 m in height. Farmers tend to favour the species for its showy early growth and provision of shelter.
5. Long term survival in Western Australia test sites was poor.
6. Contains 6–12% oxalate but is very high in protein and mixed with dry feed provides good nutrition.

The relative performance of a range of halophytic forage shrub species (mainly *Atriplex* spp.) on salt-affected soils in Pakistan was determined using the ADAPT 2 system (Ismail et al. 1992). The rankings are summarised in Table 4. Features of the results are the good performances by *A. lentiformis* and *A. amnicola*. Differences in species adaptation are demonstrated by the good ranking of *A. bunburyana* on the sandy Dingarh site compared to the two sites with fine textured soils. The species rankings in Pakistan are different from those in Western Australia. It must be emphasised that there is no 'wonder species' that will perform well on all sites. As in nature, different plants are adapted to different environments. Many salt-affected areas have been altered from their original condition and some of the species have been lost. Species selection trials are required to determine the best species for revegetation.

Matching species to sites

Using the information on species performance for a range of sites it is possible to develop guidelines for choosing which species to grow on a particular site. As discussed earlier, the most important factors determining which species will grow on a site are the severity of salt affectedness and the water relations of the site. The range of species can be somewhat extended by site modifications such as ridging and surface drainage to reduce the waterlogging experienced by the plant or by water harvesting measures on dry sites.

Site severity is most readily gauged by reference to indicator species (Malcolm 1986a,b,c). On salt-affected agricultural land in southern Australia, the commonest indicator species are annual salt tolerant barley grasses (*Hordeum geniculatum* or mediterranean barley grass in the west and *H. maritimum* or sea barley grass in

Table 4. Ranking of halophytic forage shrubs from tests conducted at three sites in Pakistan using the ADAPT 2 test programme

Species	Site		
	Faisalabad	Dingarh	Gundheri
Atriplex lentiformis 159	1	1	3
A. amnicola 971	2	10	2
A. lentiformis 178	3	–	–
A. amnicola 573	4	8	1
A. amnicola 949	5	5	5
A. halimus	6	4	–
A. sp (Pintharuka)	7	–	10
A. undulata 471 (1)	8	6	6
A. bunburyana (1)	9	2	12
A. bunburyana (2)	10	3	13
A. undulata 471 (2)	11	–	–
Maireana brevifolia	12	–	9
M. pyramidata	13	11	–
M. polypterygia	14	–	–
M. aphylla	15	8	–
A. cinerea	–	7	4
M. amoena	–	9	14
A. stipitata	–	–	11
A. vesicaria	–	–	8

Data from Abdullah et al. (1992); Ismail et al. (1992), and A. Asad (pers. comm.).

South Australia and the eastern states). Other genera such as *Spergularia*, *Plantago*, *Cotula* and *Pholiurus* also indicate the presence of sufficient salinity and/or waterlogging to adversely affect the growth of crop plants. Progressing from non-saline areas to saline areas the proportion of the vegetation that is composed of the salt tolerant types mentioned above increases relative to normal crop and pasture species. Also the proportion of the soil that is bare under normal farm management increases. Three generalised zones can be described to assist in matching species to sites on which they will succeed:

- *mildly salt-affected*: these areas carry a mixed but complete cover of salt tolerant annual species and have a much reduced presence of salt sensitive species; cropping will only be profitable in very favourable years.
- *moderately salt-affected*: the plant cover on these areas is patchy, interspersed with bare patches. Species composition ranges from the fresher patches which may have remnants of 'normal' pasture to highly saline bare patches which may develop a cover of annuals such as *Spergularia* spp if given special grazing protection; cropping will be patchy and unprofitable even in favourable years.
- *severely salt-affected*: under average farm management these areas are bare. In areas affected by a watertable or by seepage, the groundwater is sufficiently close to the surface in all seasons to permit the capillary fringe to intersect the surface thus enabling the development of very high levels of surface salinity due to evaporation; crops will always fail.

The three zones occur concentrically around salt-affected areas but their outlines are usually irregular making it difficult to treat them separately. All three zones are usually abandoned for cropping and regarded as salt-affected wasteland of low value and productivity.

The water relations of a salt-affected site are influenced by the rainfall, groundwater and runon. Additional water is beneficial if it can be stored in the soil profile for plant growth without causing waterlogging of the root zone. The degree, duration and season of waterlogging and/or inundation are major factors determining species survival and growth on saline sites. A simplified classification of site water relations has been combined with the above severity zones to guide species selection for saline sites. Table 5 gives a generalised guide to the selection of forage shrubs in southwestern Australia. Guidelines for species

Table 5. Guide to selection of forage shrubs for saline soils in agricultural areas in south-western Australia

Period for which the soil is saturated to the surface each year	Severity of salt-affectedness		
	Mild	Moderate	Severe
Days	Bluebush[a] Saltbushes[b]	Bluebush Saltbushes	Bluebush Saltbushes
Weeks	Saltbushes	Saltbushes	Saltbushes
Months	Saltbushes	Samphire	Samphire

[a] Small leaved bluebush (*Maireana brevifolia*).
[b] *Atriplex amnicola*, *A. undulata*, *A. lentiformis* mixture.

selection in other areas may be developed using information from a network of species adaptation experiments and the procedures described above. Data used to describe sites on which adaptation tests are conducted are shown in Table 6.

Table 6. Summary of data used to describe salt-affected sites subject to revegetation (taken from ADAPT 2 data base at Western Australian Department of Agriculture)

General
- Site name
- Location
- Elevation; Latitude; Longitude
- Topography
- Position of site in landscape
- Current vegetation

Soil
- For depths: 0–5, 5–15, 15–30, 30–60, 60–90, 90–120 cm
  - Colour, pH, EC_e and hand texture
- For depths: 0–5, 5–15, 15–30 cm
  - Total N, bicarbonate extractable P, and, on the saturation extract, Na^+, Mg^{++}, Ca^{++}, K^+ and sodium adsorption ratio
- Presence, depth and nature of any restricting layers; condition of soil surface

Water
- Depth and salinity of groundwater if within 3 m of surface
- Monthly water level measurements in test wells sunk to 0.5 and 3 m
- Amount and duration of surface ponding

Climate
- Daily records of rainfall, temperature and humidity

Revegetation of salt-affected areas

Revegetation strategies may be developed for salt-affected land by using a prescription system. An example is given in Appendix I. The role of the landholder is crucial and will be discussed later. The Prescription System leads the landholder and/or extension officer to make a series of decisions (see Appendix) to ensure that all factors are taken into account. The steps may be clarified and assisted by referring to various publications and decision aids as follows:

Step 1. Inspection and data acquisition. The National Committee on Productive Use of Saline Land in Australia has adopted the criteria in the ADAPT 2 data base for describing salt-affected areas in terms that enable species to be selected for specific sites (See Table 6). Factors such as indicator species, which are most easily determined when flowering, and, site water relations (e.g. frequency, severity and duration of inundation or waterlogging), which may only occur in wet years, must be assessed. Landholders may be assisted to evaluate ephemeral conditions and to consider how forage produced on salt-affected land fits into their production system. The purpose of revegetation is to enable the farmer to make money out of salt-affected land and to improve the environment. Economic models should be used to evaluate on-site and off-site measures (Bathgate et al. 1992). The models imitate the farm system and allow alternative management decisions to be evaluated under given market and other constraints.

Step 2. Species selection. Species adapted to sites are chosen by reference to a selection table such as Table 3. As more plant adaptation studies are conducted in more environments using systems such as ADAPT 2, it will become increasingly possible to use the data base to assist species selection in a predictive way. For some species it is important to ensure that the correct ecotype is chosen. Current research is aimed at selecting superior cultivars (Barrett-Lennard 1992). The capabilities of species are discussed in Malcolm & Swaan (1989).

Step 3. Preparation. The agronomy of salt-affected areas must be considered in a landscape and farm/rangeland context. Many salt-affected areas are improved by treating the uplands to minimise surface and subsurface flow and conduct excess water safely off the land. This requires a combination of off-site and on-site measures for recharge reduction, run-off control and drainage. Control of grazing and cutting is essential and may require fencing. Surface soil salinity can be reduced by cultivating the soil before the opening rain to encourage water penetration and salt leaching. If a site is likely to become too boggy, pre-cultivation can be omitted.

Step 4. Seed/plant preparation and sowing. It is essential to know the quality of seed to calculate the sowing rate for halophytic forages. The quality may vary greatly for non-domesticated species (Vlahos & Nicholas 1991). For many non-domesticated species standardised testing procedures are not available. For determining field sowing rates the most useful expression of seed quality is the number of germinable seeds per gram or kilogram. The Seed Testing Laboratory at the Western Australian Department of Agriculture is engaged in determining the best temperatures, germination media and reporting methods for seed quality in halophytic forages. It is desirable to have standardised methods.

Forage species may be established in the field either by direct seeding or by planting seedlings. Establishment methods will be discused in more detail later. Difficult soils such as clay and highly slaked soils are not niche seeded in southern Australian agricultural areas. The Prescription System (Appendix 1) guides decisions on seeding method or the situations where seedlings should be used. In developing countries farmers with small property sizes may find seedling planting more practical than direct seeding, however on large areas of public land direct seeding is preferable. In areas where direct seeding does not work, seedling planting is used, if the expected return justifies the cost. Grasses such as *Puccinellia ciliata* and *Lophopyrum* (*Agropyron*) *elongatum* may be sown as companion species to shrubs.

If there is likely to be restricted root growth the soil should be ripped (Barrett-Lennard 1992). Responses have been obtained to N fertiliser but not to P with shrubs and to both N and P with grasses (McCarthy 1992). The optimum spacing for shrub planting depends on the species (Malcolm et al. 1988). Good weed control is essential. Cultivation can be used to control weeds but on some sites leads to bogging problems. Knockdown herbicides such as 'Sprayseed' may be used prior to sowing and may be mixed with insecticides such as 'Le Mat' for pre-plant control of insects such as red-legged earth mites. The suitability of a range of herbicides for weed control in areas seeded to *Atriplex* spp. has been tested (Vlahos 1992) with carbetamide proving to be the most effective.

Step 5. Predator control. The pest responsible for most failures in halophyte shrub establishment by direct seeding in Australia is the red-legged earth mite. If mites are present they must be sprayed, even if seedlings have not been seen. The mites appear to attack the seedlings before emergence. Rutherglen bugs may attack perennials after annual pastures hay off and locusts will cause serious damage if present in large numbers.

Step 6. Management. Halophytic species are most valued as a feed reserve for livestock when annual plants are in short supply. This is usually in autumn in winter rainfall areas and in winter in summer rainfall areas. Halophytes recover well from utilisation at this time. In Western Australia, if 10–15% of a farm is vegetated with halophytic shrubs it is sufficient to fill the autumn feed gap. Economic models indicate a high value for this feed (Bathgate et al. 1992). Areas in excess of 10–15% of a farm are given a lesser value in economic models. Farmers with large areas of shrub pastures may be able to buy or sell livestock or forage to take advantage of the forage produced. The best value is obtained from shrub pastures if they are fed in combination with dry feed, stubble or hay (Warren & Casson 1992). It is important to have good quality water for stock grazing halophyte pastures as discussed by Wilson (1994).

Establishment of halophytes

The method of establishment used must be appropriate to the species and site. Perennial grasses with poor seed production such as *Diplachne fusca*, *Paspalum vaginatum* and *Sporobolus* spp (Truong & Roberts 1992) are planted using runners or tillers. Planting is done during the dormant period. *Puccinellia ciliata* and *Lophopyron elongatum* are planted using seeds after weed control has been achieved in the late autumn to early winter. Seeds of *P. ciliata* are sown with fertiliser on the surface of cultivated ground without soil covering, in view of their small size. Details of site selection and establishment methods are discussed by McCarthy (1992).

Halophytic forage shrubs may be established using nursery raised plants. In most situations the reliability of this method is good but the cost is a disincentive. In order to reduce the cost, mechanised seedling raising has recently been undertaken by vegetable nurserymen in South Australia, resulting in the lowering of the cost to 7 cents (US) per seedling for plants in trays of 100. The plants are about 100 mm high and have a root ball about 50 mm long. Another alternative is to use bare-rooted seedlings (no soil ball) which are produced for 6 cents (US) each. These seedlings suffer a much greater transplanting shock but are very easily transported and are raised in outdoor nursery beds. Reliability of bare-rooted stock is not as good as for seedlings with a soil ball.

A cheaper method of establishing halophytic forage shrubs is to use specially developed direct seeding machines known as niche seeders. These machines are designed to create a favourable niche for establishment. This strategy mimics the natural situation on salt-affected soils where even the most salt tolerant plants establish where seeds lodge in favourable niches. Detailed study of the problem of establishment on salt-affected soils (Malcolm 1989) led to the development of the Mallen Niche Seeder (Malcolm & Allen 1981). The seeder forms a mound using two opposed 670 mm diameter discs. A 'V' shaped wheel presses a niche on the mound and drives a mechanism which places deposits of seeds at intervals of about 2 m (adjustable) in the 'V' niche and covers them with spot placements of vermiculite mulch (about 100 mL per placement).

The seeders are now manufactured commercially and used by commercial contractors to direct-seed shrub pastures on farmers' properties. The seeder concentrates rain in the niche at the point where the seeds are placed thus encouraging leaching of salt and storage of water for the developing seedling. The geometry of the planting mound is adjusted to suit site conditions. On sites where there is a likelihood of waterlogging, the mound is raised (up to 15 cm above ground level) to reduce the waterlogging at the seeded position. On sites where there will not be a waterlogging hazard, and where rainfall is limited the niche is widened (up to 45 cm) and formed at or below ground level. For arid areas, the discs are replaced with a grader blade which creates a furrow that stores more water for the developing plant. The seeds are placed on the side of the mound, the height depending on the degree of inundation that will occur in the furrow (Ward 1992).

Research has shown that establishment can be improved by the use of black sprays (Malcolm et al. 1982). The niche seeding operation includes the option to spray the placements with material such as bitumen emulsion, latex paint, detergent or herbicide. By using these materials as spot applications the cost is greatly reduced. However, contract operators do not use the spray materials because of the cost and the work involved; they prefer to increase the rate of seed application to improve the chance of getting at least one seedling at each placement. The recommended rate of seed application is 50 germinable seeds per placement (Vlahos 1992).

Establishment costs for contract niche seeding of halophytic forage shrubs are of the order of 67–180 US$ per ha for shrubs at a spacing of 1.5–2 m. The cost for seedlings (excluding planting cost) at a comparable spacing would be 175–311 US$ ha^{-1} at 7 cents per seedling.

Participatory technology development

Problems of land degradation, including salinity, have been observed for thousands of years, but in 1977 the world community took a major initiative to combat desertification. Nearly 100 countries participated in the formulation of The Plan of Action to Combat Desertification and developed 28 recommendations for action. The Second International Rangeland Congress in 1984 provided an opportunity for scientists to evaluate progress in combatting land degradation. Various authors concluded that as a world community we had failed to arrest land degradation, in fact it was getting worse.

Reasons for failure were identified by a number of authors and it was concluded that there was a vast knowledge but that it was not being applied. The knowledge was often too specialised or impractical and ignored the needs and point of view of the land user. One of the authors (Al Sudeary 1988) identified the characteristics required for successful land rehabilitation measures. The characteristics identified were as follows:

- new measures should be economically viable,
- rural people should benefit,
- new measures should have a low cost per hectare treated,
- there should be a low recurrent cost in years following the implementation of the new measures,
- new measures should not make a high demand on the infrastructure of the country in order to be implemented;
- local initiatives, institutions and resources should be used to implement the new measures,
- the target group should participate in the solution of the problem,
- the measures used to treat the problem should not worsen the distribution of income and assets in the community,
- the solution should have people as the centrepiece.

These guidelines may be used to evaluate the methods discussed in this chapter. Halophytic forage shrubs provide an income for farmers/graziers who have been disadvantaged by having salt-affected soils; the measures are simple and cheap to apply, do not have a high infrastructure requirement or a significant recurrent cost; the measures fit readily into many farm/range land use systems and are amenable to development and use at the local level. It is concluded that the use of halophytic shrubs to make salt-affected soils productive is a strategy which should be widely evaluated for the rehabilitation of degraded lands.

It remains to suggest how the technology can best be implemented and as recommended by Al Sudeary, 'the solution should have people as the centrepiece'. There is evidence from many developing countries that farmers are more willing to adopt new measures if they have been involved in testing the technology themselves (Haverkort et al. 1991). The process is referred to as 'participatory technology development'. The author is currently engaged in running a training course in India entitled 'Community Involvement in Revegetation of Saline Land'. Farmers and their families have been involved in planting test areas of halophytic forage shrub seedlings on their land in cooperation with local scientists and non-government organisations. Simple direct seeding tests have been hand sown and shrub foliage has been fed to local goats and sheep to increase interest in the project. The simple technology has been easy to discuss and test with the farmers. What is being done in India could readily be replicated in other countries and offers a means to make productive use of the world's salt-affected land.

Acknowledgements

A travel grant was provided by the Australian International Development Assistance Bureau (AIDAB) to allow attendance at the Nairobi workshop.

Appendix: Prescription System

This Prescription System is currently under development at the Western Australian Department of Agriculture.

Prescription for establishing forage shrubs on salt-affected land

1. Off site measures
   Erosion control
   - contour cultivation
   - grade banks
   - gully repair
   Water control
   - interceptor banks
   - level banks
   - surface drainage
   Catchment management
   - cropping strategies
   - tree/shrub planting
2. Erosion and water control on site
   Surface drainage measures
   Surveyed gradelines
3. Species selection – for areas receiving less than 600 mm annual rainfall
   See Table 5.
4. Site preparation
   (a) Delineate species limits (see 3)
   (b) Arrange control of cutting or browsing
   (c) Clean up stumps, rocks, etc.
   (d) Weed control
      - graze
      - burn
      - knockdown herbicides
   (*Do not use residual herbicide*)
   (If grasses become a problem after sowing they can be controlled with grass specific herbicides such as Fusilade and Sertin).
   (e) Cultivation
      (*Avoid bogging out on wet land*)
      for weed control
      to promote salt leaching
      to develop soil tilth
      to improve root penetration
5. Seeding/planting time

Rainfall (mm/an)	Seeding	Seedlings
275–400	Late autumn/winter	Late winter
400–500	Late winter/spring[a]	Spring
500–600	Spring[a]	Spring

[a] In high rainfall areas seeding is done as soon as the land is trafficable after winter waterlogging.

6. Method of establishment
   The method of establishment is determined by soil type and economics.
   Sand to loam 10 cm or deeper
   - niche seed
   Sand to loam less than 10 cm
   - niche seed, avoid putting subsoil clay into the niche or
   - plant seedlings
   Other soils
   - plant seedlings
   In each situation the relative costs and likelihood of success will determine the method used.
7. Seeding rate and plant spacing
   Seeding rate calculation:
   For a 3 m row spacing with 1.5 m between placements (2200 ha^{-1}) and allowing 50 germinable seeds per placement.
   (SEED WITHOUT A GERMINATION TEST SHOULD NOT BE USED)

$$\text{Seeding rate (kg/ha)} = \frac{110{,}000}{\text{number of germinable seeds per kg}}.$$

   Seedling spacing:
   Atriplex amnicola 3 × 3 m
   A. cinerea 4 × 4 cm
8. Insect control
   Check regularly for insect pests.
   If any pests are observed exercise control. Do not wait for seedlings to emerge.
9. Grazing management
   Do not graze until the end of the second summer unless there has been excellent growth of the plants, e.g. >2 m diameter and flowering.

References

Abdullah, M., Akram, M., Khan, W.A. & Davidson, N.J. (1992). Selecting halophytic shrubs for the Cholistan desert. In: N. Davidson & R. Galloway (eds.). Australian Centre for International Agricultural Research Proceedings No. 42: 45–48.

Al Sudeary, A.M. (1988). Alleviation of rural poverty in arid lands. In: Whitehead, E.E., Hutchinson, C.F., Timmerman, B.N. & Varady, R.G. (eds.), Arid lands today and tomorrow. Proceedings of International Research and Development Conference, Tucson, Arizona, 20–25 October 1985, pp. 13–20.

Ballantyne, A.K. (1962). Tolerance of cereal crops to saline soils in Saskatchewan. Canad. J. Soil Sci. 42: 61–67.

Barrett-Lennard, E.G. (1992). Improving the productivity of saltland – a Western Australian experience. T.N. Herrmann (Convenor) Proc. Nat. Workshop on Productive Use of Saline Land, Adelaide, S. Australia, 22–24 Sept. 1992, pp. 67–74.

Bathgate, A.D., Young, J. & Barrett-Lennard, E.G. (1992). Economics of revegetating saltland for grazing. T.N. Herrmann (Convenor). Proc. Nat. Workshop on Productive Use of Saline Land, Adelaide, S. Australia, 22–24 Sept. 1992, pp. 87–94.

Carter, D.L. & Wiegand, C.L. (1965). Interspersed salt-affected and unaffected dryland soils of the lower Rio Grande Valley I: Chemical, physical and mineralogical characteristics. Soil Sci. 99: 256–260.

Dudal, R. & Purnell, M.F. (1986). Land resources – salt-affected soils. Reclam. and Reveg. Res. 5: 1–9.

Fletcher, M., Malcolm, C.V., Peterson, A., Jennings, B. & Burnside, D.G. (1989). Abatement of dust in the Kalgoorlie Region by Revegetation of Overburden Stockpiles. Minerals and Energy Research Institute of Western Australia Report No. 44.

Food and Agriculture Organisation (1982). Potential population supporting capabilities of land in the developing world. FAO, Rome, 130 pp.

Galloway, R. & Davidson, N.J. (1992). The interactive effect of salt and waterlogging on *Atriplex amnicola*. In N. Davidson & R. Galloway (eds.). Australian Centre for International Agricultural Research Proc. No. 42: 112–114.

Greenway, H. & Munns, R. (1980). Mechanisms of salt tolerance in non-halophytes. Ann. Rev. of Plant Physiol. 31: 149–190.

Haverkort, B., van der Kamp, J. & Waters-Bayer, A. (eds.). (1991). Joining farmer's experiments, experiences in participatory technology development. Intermediate Technology Publications, London, 269 pp.

Ismail, S., Ahmad, R. & Davidson, N.J. (1992). Design and analysis of provenance trials in Pakistan. Australian Centre for International Agricultural Research, Proceedings No. 42: 38–44.

Le Houérou, H.N. (1992). The role of saltbushes (*Atriplex* spp.) in arid land rehabilitation in the Mediterranean Basin: A review. Agrofor. Syst. 18: 107–148.

Malcolm, C.V. (1986a). Saltland management – selecting forage plants for saltland. Western Australian Department of Agriculture Farmnote No. 32/86, 2 pp.

Malcolm, C.V. (1986b). Saltland management – revegetation. Western Australian Department of Agriculture Farmnote No. 44/86, 4 pp.

Malcolm, C.V. (1986c). Production from salt-affected soils. Reclam. and Reveg. Res. 5: 343–361.

Malcolm, C.V. (1989). Forage shrub production on salt-affected soils. In: C.M. Mckell (ed.). The biology and utilization of shrubs. Academic Press, pp. 553–574.

Malcolm, C.V. & Clarke, A.J. (1971). Collection and testing of forage plants for saline and arid areas: Progress Report No. 1. Western Australian Department of Agriculture Tech. Bul. No. 8, 39 pp.

Malcolm, C.V., Clarke, A.J., D'Antuono, M.F. & Swaan, T.C. (1988). Effects of plant spacing and soil conditions on the growth of five *Atriplex* species. Agric. Ecos. and Environ. 21: 265–279.

Malcolm, C.V., Clarke, A.J. & Swaan, T.C. (1984). Plant collections for saltland revegetation and soil conservation. Western Australian Department of Agriculture Tech. Bul. No. 65, 112 pp.

Malcolm, C.V. & Pol, J.E. (1986). Grazing and management of saltland shrubs. J. Agric. W. Aust. 27(2): 59–63.

Malcolm, C.V. & Swaan, T.C. (1989). Screening shrubs for establishment and survival on salt-affected soils in south-western Australia. Western Australian Department of Agriculture Tech. Bul. No. 81, 35 pp.

McCarthy, D.G. (1992). Salt-tolerant grasses – mediterranean environment. In: T. N. Herrmann (Convenor), Proc. National Workshop on Productive Use of Saline Land, 22–24 Sept. 1992, Adelaide, S. Aust., pp. 28–35.

Miyamoto, S., Glenn, E.P. & Singh, N.T. (1994). Utilization of halophytic plants for fodder production with brackish water in subtropic deserts (this volume, Ch. 5).

Nawaz, S. & Hanjra, S.H. (1992). The productivity of goats using-different ratios of saltbush and kallar grass. In N. Davidson & R. Galloway (eds.). Australian Centre for International Agricultural Research, Proceedings No. 42: 85–87.

Novikoff, G. (1961). Contribution a l'etude des relations entre le sol et la vegetation halophile en Tunisie. Ann. Inst. Nat. Rech. Agron. de Tunisie 34: 1–339.

O'Leary, J.W. (1986). A critical analysis of the use of *Atriplex* species as crop plants for irrigation with highly saline water. In: R. Ahmad & A. San Pietro (eds.), Prospects for Biosaline Research, Dept of Botany, University of Karachi, pp. 415–432.

Rashid, A., Khattak, J.K., Khan, M.Z., Iqbal, M.J., Akbar, F. & Khan, P. (1992). Selection of halophytic forage shrubs for the Peshawar valley, Pakistan. In: N. Davidson & R. Galloway (eds.). Australian Centre for International Agricultural Research Proceedings No. 42: 56–61.

Szabolcs, I. (1992). Salt-affected soils. Proc. Internat Symp. on Strategies for Utilizing Salt Affected Lands, Bangkok, 17–25 Feb 1992, pp. 19–25.

Szabolcs, I. (1994). Salt affected soils as the ecosystem for halophytes (this volume, Ch. 3).

Truong, P.N. & Roberts, M.H. (1992). Salt tolerance of some tropical and subtropical grass species grown in Queensland. Proc. Nat. Workshop on Productive Use of Saline

Land, Adelaide, S. Australia, 22–24 Sept. 1992, pp. 36–44.
Vlahos. S. (1992). Niche seeding for the establishment of salthush on saline soils in Western Australia. In: T.N. Herrmann (Convenor). Proc. Nat. Workshop on Productive Use of Saline Land, Adelaide, S. Australia, 22–24 Sept. 1992, pp. 59–66.
Vlahos. S. & Nicholas, D. (1991). Calculating saltbush seeding rates. Western Australian Department of Agriculture Farmnote No. 81/91, 2 pp.
Ward, B.H.R. (1992). Screening species and developing direct seeding methods to regenerate arid rangeland in Western Australia. In: N. Davidson & R. Galloway (eds.). Australian Centre for International Agricultural Research Proc. No. 42, 49–52.
Warren, B. & Casson, T. (1992). Saltbush quality and sheep performance. Proc. Nat. Workshop on Productive Use of Saline Land, Adelaide, S. Australia, 22–24 Sept. 1992, pp. 79–86.
Warren, B.E., Casson, T. & Ryall, D.H. (1994). Production from grazing sheep on revegetated saltland in Western Australia (this volume, Ch. 24).
Wilson, A.D. (1994). Halophytic shrubs in semi-arid regions of Australia: Value for grazing and land stabilization (this volume, Ch. 8).

5. Utilization of halophytic plants for fodder production with brackish water in subtropic deserts*

S. MIYAMOTO,[1] E. P. GLENN[2] & N. T. SINGH[3]

[1]*Agricultural Research Center, Texas A&M University, El Paso, Texas, USA;* [2]*Environmental Research Laboratory, University of Arizona, Tucson, USA;* [3]*Indian Soil salinity Research Institute, Karnal, India*

Abstract. The arid and semi-arid regions of the subtropic zone in the Northern Hemisphere are in most parts faced with frightening rates of population growth, overgrazing, and land degradation. Among the major natural resources of arid and semi-arid deserts, brackish water (containing dissolved salts of 5 to 20 g L^{-1}) is currently the least explored resource, primarily because the salinity is too high to produce conventional crops. Irrigated production of halophytic plants (which include halophytes and salt tolerant nonhalophytes) offers potential for utilization of brackish water to supplement fodder needs. However, the knowledge required to evaluate and manage irrigated production with brackish waters has not been adequately developed. In this chapter, we examined saline tolerance characteristics of halophytic plants, ways to characterize water and soils for irrigated production, and various potential system options for irrigated production. Although there are exceptions, salt tolerance of most halophytic plants at germination does not appear to be any higher than that of conventional crops, thus requiring careful management for establishment. Many halophytic plants show optimum growth at soil solution salinity of 20 to 25 g L^{-1}, and 25 to 50% growth reductions at 30 to 40 g L^{-1}, even though they can survive soil solution salinity as high as 60 to 90 g L^{-1}. Forage yields and quality of some halophytic plants are comparable to conventional forage crops. However, excessive uptake of salts and, in some instances, toxic substances such as Se and Mo may reduce fodder values unless blended with conventional fodder. Under ordinary leaching fractions employed (less than 0.25), irrigated production of halophytic fodder plants appears to be optimum at irrigation water salinity of less than 10 g L^{-1} (13 to 15 dS m^{-1}), and would be reduced substantially when salinity of irrigation water approaches 20 g L^{-1}. Salinity of drainage water can easily reach the seawater level salinity (35 g L^{-1}) or even higher. The quantity of salts that must be leached for sustainable production is in an order of 100 tons per ha annually. Halophytic plants can be raised in conventional monoculture systems using sprinklers or surface irrigation, especially when salinity of water is too high for conventional crops (typically above 5 g L^{-1}). Multicrop culture using strip planting of halophytic crops between conventional forage crops or irrigated silvipasture systems could also be used to an advantage, especially for reducing drainage water handling. Halophytic plants grown in strips with limited irrigation can utilize saline water from the shallow water table, and can remove dissolved salts at a rate of 5 to 10 Mg ha^{-1} annually. The possibility of biofiltration of toxic trace elements from drainage water warrants extensive studies. Future prospects for halophytic fodder production seem to depend to a large extent upon the success of ongoing selection programs, enhanced breeding efforts, a considerable amount of agrotechnical evaluation, including the aspects of biofiltration and production systems, and improved understanding of the role of halophytes in saline water and soil management.

Key words: Drainage water, Fodder crops, Halophytes, Review article, Saline water, Saline water management

*Contribution from Texas Agr. Expt. Station, College Station, TX.

V.R. Squires & A.T. Ayoub (eds.), *Halophytes as a resource for livestock and for rehabilitation of degraded lands*, 43–75.

1. Introduction

The arid and semi-arid regions of the world extend 43 million km^2 (or 4.3 billion ha) or approximately one-third of the land area of the earth (Eyre 1985). In spite of limited and unreliable rainfall, and general scarcity in water resources, the population growth in the arid and semi-arid regions is occurring at an alarming rate of over 3% annually (Table 1). Approximately ninety percent of this population growth has occurred in developing countries in Asia, Africa and North and South America, and the problem of hunger has actually increased in the past several decades and is projected to continue (WFC 1991; Higgins et al. 1982).

With the exception of oil and mineral producing areas, livestock husbandry has been the key means of rural economy in these subtropical deserts. However, the continuing increase in both human and animal populations has caused the wide-spread practice of overgrazing and resulting land degradation in many of these developing countries (UNEP 1991). The problem is especially acute in the subtropic deserts of the Northern Hemisphere (e.g., Khoshoo 1986). Irrigated production of fodder crops is an effective intervention to this degradation process. It has not only increased productivity per land area, but also has provided a significant relief during the period of critical fodder needs in the event of drought. However, fresh water resources needed for irrigated production are scarce in most desert regions.

Among the major resources of the desert, brackish water containing 5 to 20 $g\,L^{-1}$ of dissolved salts is the least explored resource, mainly because salinity of such water has been viewed as too high for crop production. The availability of brackish water is well known in some areas, but not for all the regions of this desert zone. The current projection is that the quantity is large, and it takes the form of agricultural drainage water, inland saline ground water, and coastal saline water. The use of saline agricultural drainage water is especially attractive, as necessary infrastructure is available, and saline drainage water is currently considered a liability more so than an economic resource. Saline drainage water is the principal source of soil salinization, soil alkalization and waterlogging in some 43 million ha of irrigated lands of the world (UNEP 1991). This amounts to 19% of the total irrigated areas (estimated to be 228 million ha in 1990) which provides 30% of the world food supply. In arid deserts of the Middle East and Northern Africa, there are considerable known reserves of brackish inland and coastal saline waters that can be explored for irrigated production.

Table 1. The population increase in the arid zone of the world (Eyre 1985)

	Population (millions)		Increase (%)	
	1960	1985	1960–85	Annual
Asia	151.0	270.5	79	3.2
Africa	49.5	96.5	95	3.8
North America	15.5	26.0	68	2.7
South America	10.0	17.5	75	3.0
Australia	0.5	0.5	0	0

At the present time, saline water resources having salinity exceeding about 3000 $mg\,L^{-1}$ (or electrical conductivity of 4 $dS\,m^{-1}$) are rarely used for irrigation, not only because salinity often exceeds salt tolerance of most conventional crops, but also it can cause potentially adverse effects on soil resources, especially when Na is the predominant cation (e.g., Ayers & Westcot 1985; Grillot 1956). However, the use of salt-tolerant crops such as cotton, barley and date palms has pushed the limit of saline water use to the range of 5000 to 6000 $mg\,L^{-1}$ in some areas where soils are highly permeable or gypsiferous (Miyamoto et al. 1984; Frenkel & Shainberg 1975). Saline water of this level of salinity has also been shown useable even in less permeable soils when low salt water is available for conjunctive uses (e.g., Rhoades et al. 1988) or rainfall is high enough to leach some of the salts after irrigation with saline waters (Dhir 1977). Even so, the upper limit of salinity that conventional crops can tolerate is so low that much of the saline water resources with salinity greater than 5,000 or 6,000 $mg\,L^{-1}$ will remain unused, unless a quick advancement of breeding toward higher salt tolerance is achieved.

Halophytes have been viewed as having potential for irrigated production with brackish as well as seawater (e.g., Boyko 1966; Mudie 1974; O'Leary et al. 1985; Glenn & O'Leary 1985). Boyko (1966) was among the strongest advocates of producing crops in sand dunes using brackish or seawater. In spite of these earlier advocates, irrigated production of halophytes has not yet been practiced, except for experimental or pilot scales in the US, Mexico and the middle East. In addition to the general skepticism about brackish or seawater irrigation, there has been a lack of clear market identification for the products and of engineering work necessary for implementation. Using Boyko's own statement; 'Agrotechnical details for economic purposes have to be worked out as a matter of course for each species separately'

The use of halophytes for animal fodder provides a definite market or, if not, a definite application. During the past decades, a considerable body of information has accumulated in regard to halophytes or halophytic species that can be used for fodder (e.g., Mudie 1974; Gallagher 1985; Sankary 1986; Le Houérou 1986; Malcolm 1986; Jefferies & Pitman 1986; National Research Council 1990). These lists of potential halophytic fodder crops are being updated (this volume). Information on physiology of halophytes has also been accumulated (e.g., Greenway 1973; Flowers et al. 1977). What seems to be required now is to bridge these botanical, ecological and physiological interests with sound principles of soil and water management and agrotechnical knowledge.

In this chapter, we will examine, (1) saline tolerance characteristics of halophytic fodder plants, (2) ways to characterize halophytic plants, water and soils for irrigated production, and (3) various potential system options involved in irrigated production.

2. Plant characterization for irrigated production

Halophytes and salt tolerant nonhalophytes are distributed widely in saline areas ranging from salt marshes to the deserts. The use of these plants as animal fodder is not new to the indigenous population. Many species of *Atriplex*, *Salsola*, *Suaeda*, *Sporobolus*, *Distichlis*, and *Halosarcia* are known to be well-grazed by ruminants. Systematic attempts to screen these species as fodder crops are, however, comparatively new, and readers may refer to the review by Gihad & El Shaer (Ch. 6, this volume) concerning fodder values of halophyte species as well as regional reports of halophyte utilization given in this volume.

For irrigated production of halophytes, these plants must be first evaluated for agronomic performance, such as salt tolerance, water use, growth habits, fodder quality, and productivity. Unfortunately, agronomic evaluation of halophytes has been limited, and one often must infer plant performance based on a large volume of ecological observations. In this section, we will review some of the results of emerging agronomic evaluation of halophytes and salt tolerant nonhalophytes which have a potential to become a fodder crop.

2.1 *Salt tolerance during establishment*

a. *Salt effects on seed germination.* The ability of seed to germinate under high salinity is an important trait for successful establishment of halophytic plants through direct seeding. Characterization of salt effects on germination (the emergence of radicle) is complicated by interacting factors such as temperature (e.g., Okusanya 1977; Mikhiel et al. 1992), seed quality (e.g., Mohammed & Sen 1988), types of salts involved (e.g., Ryan et al. 1975), the rate of hydration, and the methods employed for germination. Thus, germination response data should be viewed as an indication of salt tolerance under a set of given conditions, rather than in any absolute term.

The methods of germination tests vary, but petri-dishes containing saline solutions seem to be used most commonly. This method is convenient, yet is subjected to salinity increases associated with evaporation and/or condensation, especially when a small quantity of saline solutions is used under fluctuating temperature regimes. Also this method is not suitable for testing seeds requiring good aeration for germination. Alternatively, a technique involving sponges

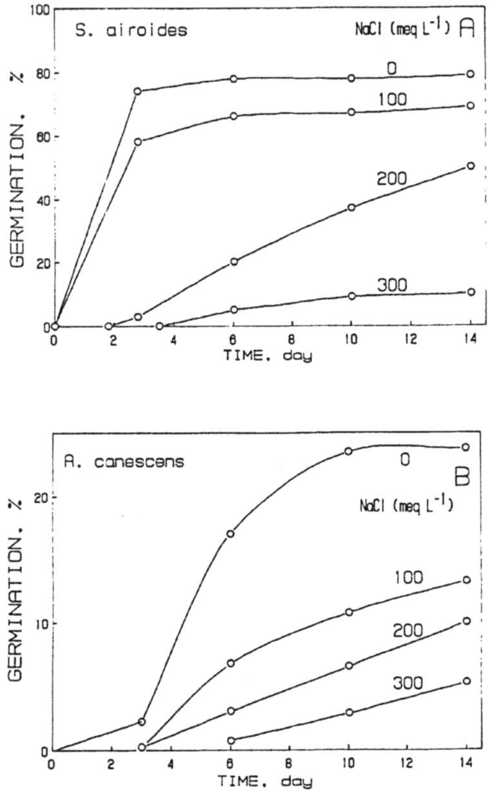

Figure 1. Seed germination of *Sporobolus airoides* (Figure 1A) and *Atriplex canescens* (Figure 1B) in NaCl solutions at 26 and 20 °C, respectively (Miyamoto 1978).

(e.g., Taylor & Lankford 1969) can be used successfully for many species. In this method, chemically inert sponges are first soaked in a desired saline solution, then excess water is drained to obtain a desirable moisture condition. Seed lots are then placed between two sheets of the sponges and incubated in an evaporation-controlled chamber or in plastic bags at a desired constant temperature.

Examples of salt effects on germination rates are shown in Figure 1, using Alkali sacaton (*Sporobolus airoides*) and fourwing saltbush (*Atriplex canescens*). Germination was determined in petridishes containing saline solutions at 26 °C for *S. aeroides* and 20 °C for *A. canescens* (Miyamoto, 1978). *Sporobolus* may represent a typical fast-germinating species, although some species, e.g., *Haloxylon* spp. can germinate in a matter of 2 hours (Sharma & Sen 1989), whereas *Atriplex* may represent a slow and poor germinating species. High salinity usually reduces the rate of germination as well as the extent of germination. Table 2 lists additional examples of seed germination under various salinity levels of the solutions consisting mostly of NaCl. Readers may also refer to Unger (1982) for additional discussions on germination of halophyte species.

Seed germination of some species is known to be enhanced by pre-soaking in water (e.g., Twitchell 1955). Presoaking of seed in a saline solution (osmotic priming) can also enhance the rate of germination (e.g., Haigh et al. 1986; Miyamoto 1989; Mohammed & Sen 1990). When the seed is pre-exposed to saline soil solutions either artificially or naturally, in-situ germination in saline soils usually occurs in a matter of one or two days after irrigation or rain. The germination threshold shown in Table 2 should be interpreted considering this process of rapid germination after the osmotic priming.

For a comparison, salinity of incubating saline solutions which causes a 50% reduction in germination of conventional crops, such as tomatoes, peppers and onions is approximately 16, 27 and 28 dS m^{-1}, respectively (e.g., Miyamoto et al. 1985; Miyamoto 1989). Saline tolerance of halophytic plants at germination is not any higher than conventional crops, although the rate of germination of some halophyte species can be higher. An exception to this rule appears to be *Saliconia* spp. which germinate readily in sea water (Joshi 1979; Stumpf et al. 1986; Glenn et al. 1991).

b. *Salt tolerance of hypocotyls.* Salt tolerance of the hypocotyl, instead of seed germination, can control seedling emergence in some plants. Guayule (*Parthenium argentatum*), a salt-tolerant shrub, is an example. The hypocotyl of this species is salt-sensitive, and can undergo mortality when the hypocotyl comes in contact with salts which are deposited at the soil surface following water evaporation (Miyamoto et al. 1985). This potential problem has not yet been examined with halophytes. However, *Atriplex* species possibly fall into this category.

Table 2. Salinity of incubating solutions which may cause a 50% reduction in seed germination of halophytic plants

Species	Solution salinity ($g\,L^{-1}$)	($dS\,m^{-1}$)	Days (No.)	Temp (°C)	References
Atriplex canescens	6 ~ 10	11 ~ 17	14	26	Miyamoto (1978)
A. lentiformis	10 ~ 11	17 ~ 19	–	5–30	Mikhiel et al. (1992)
A. linearis	10 ~ 11	17 ~ 19	–	5–30	Mikhiel et al. (1992)
Chloris gayana	5 ~ 6	9 ~ 11	–	–	Malkin & Waisel (1986)
C. virgata	10 ~ 13	12 ~ 16	8	35	Sharma & Gupta (1971)
Dactyloctenium aegyptium	15	20	8	35	Sharma & Gupta (1971)
D. sinaicum	4	6 ~ 7	8	27	Sharma & Gupta (1971)
Digitaria adscendens	4	6 ~ 7	8	30	Sharma & Gupta (1971)
Dichanthium annulatum	5	7 ~ 8	8	33	Sharma & Gupta (1971)
Echinochloa colonum	5	7 ~ 8	8	35	Sharma & Gupta (1971)
Eragrostis curvula	11	19	12	27	Ryan et al. (1975)
E. superba	11	19	12	27	Ryan et al. (1975)
Lasiurus sindicus	7 ~ 10	9 ~ 12	8	35	Sharma & Gupta (1971)
Sesuvium sesuvioides	1 ~ 2	2 ~ 4	10	28	Mohammed & Sen (1990)
Salicornia bigelovii	>40	–	–	–	Stumpf et al. (1986); Glenn et al. (1991)
S. brachiata	>36		10	20	Joshi (1979)
Salsola baryosma	5	9	10	28	Mohammed & Sen (1990)
Sporobolus airoides	11 ~ 14	19 ~ 24	14	20	Miyamoto (1978)
Suaeda depressa	15	19	–	5–15	Williams & Ungar (1972)
Trianthema triquetra	2 ~ 10	4 ~ 17	10	28	Mohammed & Sen (1990)

c. *Salt tolerance of seedlings*. Emerged seedlings are subject to high levels of salts which usually accumulate at and near the soil surface. Salt damage to the seedlings can be induced through roots or seedling leaves (Miyamoto et al. 1986). Salt damage caused by foliar salt adsorption is common after light showers splash the salts to the seedling leaves. Foliar-induced salt damage can also occur through mid-day sprinkling of water high in Na and/or Cl (e.g. Ehling & Bernstein 1959; Moore & Murphy 1979).

Saltbush (*Atriplex* spp.) is among the few halophyte species which is currently identified as being subject to foliar-induced seedling kills. Weiler (1980) reported high mortality of saltbush transplant seedlings when overhead-sprinklers were used to irrigate them with water having a conductance of 4.3 $dS\,m^{-1}$. Succulent halophytes such as *Salicornia* spp. are known to tolerate seawater level salinity, even when seedlings are sprinkler-irrigated (Glenn & O'Leary 1985).

d. *Salt and alkali tolerance of transplants*. Trees, and sometimes shrub species, are ordinarily established through transplanting seedlings. Transplant establishment is often the most critical phase of tree management (Malcolm 1994). Tomar & Yadav (1985) reported that seedling growth of two of the most salt tolerant tree species, *Prosopis juliflora* and *Acacia nilotica*, was reduced by half when irrigated with water of 5 $dS\,m^{-1}$ in a pot experiment. However, seedlings of these species have survived in saline soils having salinity of the saturation extract (EC_e) of 16 $dS\,m^{-1}$ or higher. These observations are in sharp contrast with the report of Rhodes & Felker (1988) with 'undisturbed' seedlings of *Prosopis* species and accessions. In their experiments, saline water was applied to the 'undisturbed' seedlings which had been grown with low salt water, and a significant growth reduction of *P. chilensis*, *P. articulata* and *P. juliflora* was observed when irrigation water salinity exceeded 18 $g\,L^{-1}$ (30 $dS\,m^{-1}$) or higher. Transplanted trees should be irrigated with water of low salinity, or in semi-arid regions, tree planting may coincide with the rainy season. Tree establishment by transplanting in alkali soils also met with considerable difficulties when the pH of soils approaches as high as 10. Yadav (1980), for example, reported establishment difficulties of *Prosopis* and *Acacia* species in alkali soils with a pH of around 10, although these trees can grow adequately once established. Singh et al. (1992) also reported severe growth reductions of *Eucalyptus*, *Acacia*,

Casuarina, *Delbergia* and *Pongamia* seedlings when ESP of silty clay loam was increased from 1 to 30% by adding $NaHCO_3$, and high rates of seedling mortality (>80%) of the first three species when ESP was raised to 60%. The successful establishment of trees in highly alkali soils requires a planting hole filled with amended soils (e.g., Sandhu & Abrol 1981).

Halophytes from the genera *Atriplex*, *Distichlis* and *Cressa* can be transplanted into seawater irrigated-fields with high survival rates after preconditioning them to seawater irrigation (Glenn & O'Leary 1985).

2.2 *Salt tolerance after establishment*

a. *Quantification of salt tolerance*. The quantification of salt-tolerance has been made by various means, including ecological observations, artificial salinization of soils, and artificial salinization of irrigation water. While ecological observations provide a good picture of saline adaptation of different species (e.g., Le Houérou 1986; Sankary 1986; Abbas & El-Oqlah 1992), the quantitative relation between salinity and growth rates is difficult to obtain with certainty.

Techniques involving artificial salinization of soils followed by the use of fresh or low salt water can provide a relative measure of salt tolerance of different species and accessions, especially when saline soils are to be revegetated using natural rainfall or supplemental irrigation during establishment. However, such methods are not necessarily compatible with in-situ salt tolerance involving irrigated production with saline water. The primary reason is that plant adaptation to zonal salinization provides growth considerably superior to the cases involving salinization of the entire root zone with saline water. Under the zonal salinization pattern, which is prevalent to saline soils irrigated with fresh or low salt water, plants adsorb water and certain nutrients from the lowest salt zone, and can maintain adequate growth, regardless of the salinity status of the soils (e.g., Bingham & Garber 1970; Bernstein & Francois 1973). In other words, the plant response is affected by the method and frequency of water application as much as by soil salinity levels. This behavior is especially pronounced in plant species that exhibit a strong osmotic adjustment. In addition, this system is not conducive to providing a steady salt balance or steady salt distribution profiles, which are essential for obtaining reproducible results.

Techniques involving artificial salinization of irrigation water (or irrigating solutions) can potentially offer the closest measure of salt tolerance in irrigated production with saline water, and include sand culture for screening (e.g., Glenn & O'Leary 1984) as well as outdoor lysimeters for a more realistic assessment of saline tolerance (e.g., Miyamoto et al. 1993). Plant response to salinity in solution culture may not be the same as that in the soils irrigated with the same solution, not only because of salt concentration increases associated with evaporative water losses, but also the salt concentration gradients existing between the root surface and bulk soils (Sinha & Singh 1974). In soil culture, it is essential to maintain a steady salt balance and relatively uniform salt distribution with depth. To achieve such conditions, a constant leaching fraction of greater than approximately 0.30 is usually required (Miyamoto et al. 1993). It is also important to control irrigation timing, as salinity of the solution increases with evapotranspiration.

There has also been inconsistency in expressing soil salinity levels. The use of salinity of the saturation extract (USSL 1954) has an advantage of having some quantitative relations to salinity of soil solutions of field soils. This method, although somewhat cumbersome, also provides more reliable salinity assessments than the extract of higher dilutions (such as 1:1 or 1:2 extracts), especially in soils containing gypsum. Salinity of the saturation extract, however, cannot account for the increases in soil solution salinity caused by soil water depletion. This measure was proposed originally to provide a general index of soil salinity levels that can have adverse effects on plant growth, but not specifically for saline water irrigation involving high leaching fractions.

Salinity of the soil solution is a preferred measure for expressing soil salinity in relation to saline water irrigation. It can be measured directly by some of the improved field salinity sensors or can be estimated by knowing field soil moisture regimes in addition to the saturation water content and salinity (Rhoades & Miyamoto 1990). Figure 2 shows examples of growth (ash free dry

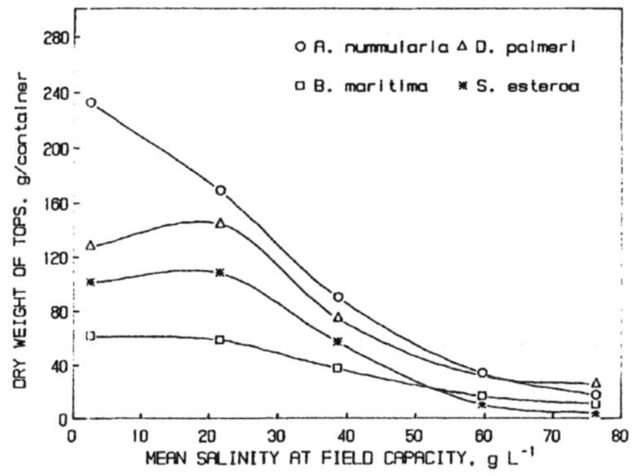

Figure 2. Dry matter (ash-free) production of four halophyte species grown for one season in lysimeters as related to the mean soil solution salinity at field capacity (Miyamoto et al., 1993).

weight) response of four halophyte species to soil salinity expressed in mean soil solution salinity at field capacity under a constant leaching fraction of 0.3 (Miyamoto et al. 1993). The composition of saline waters used for this experiment was similar to that of ocean water. The mean soil solution salinity was estimated as an arithmetic mean of salinity of irrigation and drainage waters averaged over a growing season. The actual salinity of the root zone at the mean soil water content, instead of at field capacity, is estimated to be 1.16 times the listed values, as the soil water storage was allowed to be depleted by one-third of the field capacity. Further discussion on the estimate of the mean soil solution salinity is given in Section 3.1. The data indicate that the optimum growth of those species was attained at a mean soil solution salinity of about 20 g L^{-1} or less, and that the growth declined rapidly beyond a salinity level between 20–30 g L^{-1}. These species are commonly known as true halophytes or euhalophytes, and are known for high salt tolerance among halophyte species. Under high frequency and high leaching fractions, these species provide substantial growth even at the seawater level salinity (35 g L^{-1}).

Growth response of halophytes is also known to be affected by ion compositions. Greenway (1968), for example, observed that growth of *Atriplex nummularia* was superior in NaCl-added nutrient solutions than $CaCl_2$-added nutrient solutions at the same equivalent salt concentrations. The growth inhibitory effect of $CaCl_2$ was, however, eliminated when Mg and K concentrations were increased from 5–100 meq L^{-1}, indicating a possible nutrient imbalance caused by high levels of Ca. Mozafar et al. (1970) observed that growth of *Atriplex halimus* was greatly stimulated by addition of KCl to NaCl. There are also growth differences between plants grown in NaCl or Na_2SO_4 solutions (e.g., Gale & Poljakoff-Mayber 1970; Miyamoto 1978). Air humidity can also affect growth of certain halophytes, especially at low osmotic potentials (Gale et al. 1970). Readers may refer to Flowers et al. (1977) for these physiological interactions.

b. *Salt tolerance of halophytic grasses and shrubs.* Salt tolerance of halophytic grasses and shrubs is summarized in Table 3 based on various reports. *Atriplex* species are among the shrubby fodder crops which have been studied extensively (e.g., O'Leary 1986). They are highly productive and among the most salt tolerant species as indicated in Table 3. A comparison of different *Atriplex* species in salt tolerance is not feasible from the data, because of the wide difference in the experimental methods used. Readers may refer to several ecological reports for a species comparison in salt tolerance of *Atriplex* species (e.g., Le Houérou 1986).

Several species of *Salsola* and *Suaeda* have also been used as fodder and grazed heavily in open ranges. These species are productive and as tolerant as *Atriplex* to salinity (Table 3). They also contain high levels of protein (e.g., El Shaer 1981), but tend to be high in salt content when grown in saline soils. Another productive halophyte shrub, *Halosarcia* spp. (samphire) is also used as fodder in Australia. *Kochia* is another productive shrub adapted to all types of soils, but is not as salt-tolerant as *Atriplex* or *Suaeda* (Francois 1984, Table 3).

Salicornia, a succulent annual, is among the few plants which have been grown with seawater (e.g., Glenn et al. 1982). This highly salt tolerant halophyte produces oil seed which is rich in protein. The dry straw is high in salts and low in protein, but appears to be suitable as a source of feed blend (Glenn et al. 1994; Swingle et al. 1994).

Table 3. Salinity levels that may cause 25 and 50% reductions in yields of grass and shrubs

	Reduction		Salinity in	Types of expt.	References
	25%	50%	($dS\,m^{-1}$)	or culture	
Agropyron elongatum	15	18	EC_e[a]	–	Bernstein (1964)
Allenrolfea occidentalis	50	>67	EC_w	sand	Glenn & O'Leary (1984)
Atriplex canescens	17	–	EC_e	soil	Miyamoto (1978)
A. barclayana	–	31	EC_w	sand	O'Leary (1984)
A. balimus	20	–	EC_s	sand	Mozafar et al. (1970)
A. inflata	50	–	EC_s	vermiculite	Ashby & Beadle (1957)
A. lentiformis	50	>67	EC_w	sand	Glenn & O'Leary (1984)
A. nummularia	60	–	EC_s	vermiculite	Ashby & Beadle (1957)
A. nummularia[b]	44	57	EC_s	lysimeter	Miyamoto et al. (1993)
A. patula	>18	–	EC_w	field	Hoffman & Shannon (1986)
A. patula	33	67	EC_w	sand	Glenn & O'Leary (1984)
Bassia hyssopifolia	33	50	EC_w	sand	Glenn & O'Leary (1984)
Batis maritima	50	67	EC_w	sand	Glenn & O'Leary (1984)
	48	69	EC_w	lysimeter	Miyamoto et al. (1993)
Chenopodium album	–	12	EC_w	field	Hoffman & Shannon (1986)
C. murale	33	50	EC_w	sand	Glenn & O'Leary (1984)
Chloris gayana	–	11	EC_s	sand	Malkin & Waisel (1986)
C. gayana	7	–	EC_w	field	Gausman et al. (1954)
Cynodon dactylon	–	17	EC_w	sand	Glenn (1987)
C. dactylon	5	–	EC_w	field	Gausman et al. (1954)
Diplachne fusca	10	22	EC_s	gravel	Sandhu et al. (1981)
Distichlis palmeri	16	57	EC_w	lysimeter	Miyamoto et al. (1993)
Kochia brevifolia	12	27	EC_e	field	Francois (1986)
K. prostrata	12		EC_e	field	Francois (1986)
Kosteletzkya virginica	>18		EC_w	field	Hoffman & Shannon (1986)
Hordeum vulgare	8.2	11	EC_e	soils	Bernstein (1964)
Mariana brevifolia	33	50	EC_w	sand	Glenn (1987)
Paspalum vaginatum	17	50	EC_w	sand	Glenn (1987)
Phragmites australis	17	50	EC_w	sand	Glenn (1987)
Salicornia bigelovii	30	50	EC_s	solution	Webb (1966)
S. bigelovii	50	>67	EC_w	sand	Glenn & O'Leary (1984)
Sesuvium verrucosum	33	67	EC_w	sand	Glenn & O'Leary (1984)
Spartina longispica	17	50	EC_w	sand	Glenn (1987)
Sporobolus airoides	13	–	EC_e	soil	Miyamoto (1978)
S. virginicus	17	50	EC_w	sand	Glenn (1987)
Suaeda esteroa	45	57	EC_s	lysimeter	Miyamoto et al. (1993)
S. maritima	50	–	EC_s	sand	Flowers (1972)
S. torreyana	50	>67	EC_w	sand	Glenn & O'Leary (1984)
Triticum aestivum	17		EC_w	sand	Glenn (1987)

[a] EC_e: Salinity of the saturation extract, EC_w: salinity of irrigation water. EC_S: Salinity of soil solutions.
[b] The maximum yield for this species was assumed to occur at soil salinity of approximately $10\,g\,L^{-1}$.

Distichlis spp. are also quite salt tolerant (Table 3), productive and contain low levels of salts as they have salt glands to excrete salts. *Distichlis spicata* and *D. palmeri* have yielded high biomass with seawater irrigation (Glenn & O'Leary 1985). *Distichlis* spp. are the most salt tolerant grass species and can survive wet soils. There are, however, considerable differences in salt tolerance among different accessions of salt grass (Lugg, 1983). *Distichlis spicata* was used successfully for revegetating the saline dry lake bed of Lago Texcoco in Mexico (Llerena, Ch. 30, this volume). *Sporobolus* spp. are another grass heavily grazed in most open ranges, and they are quite drought tolerant. Salt tolerance of those tested was however, much lower than *Distichlis* spp. (Table 3).

Diplachne fusca (Kallar or Karnal grass) is also

a highly productive grass. (In a recent classification, e.g., Malik et al. 1986, this grass is referred to as *Leptochloa fusca*). This grass is not only salt tolerant, but is also tolerant to alkali and waterlogging. Hussain and Hussain (1970) reported that this grass was able to grow in soils with EC_e of 22.3 dS m^{-1} and a pH of 9.8. The data of Sandhu et al. (1981), however, indicate that salinity for optimum growth is less than 10 dS m^{-1} in soil solutions. This grass has been used as animal fodder in Pakistan (e.g., Malik et al. 1986).

Chloris gayana (Rhodesgrass) has been viewed as salt-tolerant, but several salt tolerance studies suggest that a 50% reduction in growth may result at 11 dS m^{-1} in saline solutions in gravel culture (Malkin & Waisel 1986), and a 25% reduction in yield when irrigated with a solution of 7 dS m^{-1} (Gausman et al. 1954). *Chloris* species are less tolerant to salts than *Sporobolus* species, but are more tolerant to salts than *Cynodon dactylon* (Coastal Bermuda), (Table 3).

Agropyron elongatum (Tall wheat-grass) is among the few cool season grasses tolerant to both salinity and drought. They are commercially grown in the western USA both under dryland and irrigated conditions. Salt tolerance seems to vary among different accessions (e.g., Shannon 1978), but may exceed salt tolerance of *Hordeum vulgare* (Barley), which is widely viewed as among the most salt tolerant conventional small grains. Some of the recent selections of *Triticum* (wheat) are believed to be as salt tolerant as barley, if not more (Table 3).

Salt tolerant shrubs and grasses discussed above are limited to common species. Readers may refer to regional reports given in this issue for additional species which have potential as forage crops.

c. *Salt tolerance of tree species*. Many tree species are used in developing countries to provide forages during the period of dry seasons or simply to provide shade and shelters to livestock. Trees are also used to provide fuelwood and are an integral part of the rural landscape. It is difficult to obtain reliable data on salt tolerance of tree species, primarily because of the scale of the experiments required to accommodate their large sizes and the length of experiments required to obtain meaningful data. Ecological observations of natural stands (e.g., Le Houérou 1986; Abbas & El-Oqlah 1992) provide the initial screening of salt tolerant species. However, such observations leave considerable uncertainty, as soil salinity is only a component of growth limiting factors, and soil salinity levels are changeable.

Felker et al. (1981) and Rhodes & Felker (1988) have applied a mass screening technique involving sand culture of *Prosopis* seedlings. Saline solutions consisting of NaCl were applied twice a day so as to avoid salt accumulation in the growth medium (sand). Their results (Table 4) show that shoot growth of *P. juliflora*, *P. chilensis* and *P. articulata* was not significantly reduced at 18 g L^{-1} (30 dS m^{-1}) in the solutions after several months of the saline treatment. When salinity of the irrigation solution was increased to 24 g L^{-1} (39 dS m^{-1}), however, a severe growth reduction was observed, except with *P. juliflora*. Other species of *Prosopis* appear to be somewhat less tolerant to salts, as are *Casuarina* (Ng, 1987) and *Eucalyptus* species (Jain et al. 1983/1985; van der Moezel et al. 1988). Singh et al. (1991) reported saline tolerance of seedling

Table 4. Salinity levels that may cause a significant reduction (approximately 25% reduction) in shoot or trunk growth of tree seedlings

Threshold salinity		Measured in	Growth media	References
g L^{-1}	dS m^{-1}			
Sand culture with seedlings				
Prosopis juliflora, chilensis, articulata				
18	30	Solutions	Sand	Rhodes & Felker (1988)
Prosopis alba, nigra				
12	20	Solutions	Sand	Rhodes & Felker (1988)
Prosopis glandulosa, tamarugo and *velutina*				
<12	20	Solutions	Sand	Felker et al. (1981)
Casuarina equisetifolia				
7	12	Solutions	Sand	Ng (1987)
Soil culture with seedlings				
Acacia nilotica	8	Sat. Ext.	Loam	Singh et al. (1991)
Casuarina equisetifolia				
	5	Sat. Ext.	Loam	Singh et al. (1991)
Eucalyptus hybrid				
	7	Sat. Ext.	Loam	Singh et al. (1991)
Pongamia pinnata				
	5	Sat. Ext.	Loam	Singh et al. (1991)

trees of several species grown for 22 months in potted loamy soils. Their findings (Table 4) indicate that *Acacia* growth was reduced significantly at 8 dS m^{-1} in the soil saturation extract, and that of *Eucalyptus* and *Casuarina* at 7 and 5 dS m^{-1} in the saturation extract, respectively.

One of the difficulties of seedling experiments is the interpretation of the growth response relative to salt tolerance of the trees. One interpretation is that the growth reduction observed with young seedlings after successful establishment represents salt tolerance of the trees, and another interpretation is that salt tolerance may increase with increasing tree sizes. Experimental data are currently insufficient to make this judgement. If we assume that the former interpretation is correct, salt effects in observed tree responses would be arithmetically cumulative. Five-year field experiments conducted by Jain et al. (1983) seem to indicate that salt effects on tree growth is indeed cumulative, and a small reduction in initial years translates to a larger reduction over the years, as the trees in nonstressed conditions continue to grow at a faster rate, while the growth of trees under saltstress lags behind.

Jain et al. (1985) also conducted another 5 year field experiment with several tree species in highly saline soils. Their results are cited in Table 5 along with the results of their earlier experiment involving salinization of irrigation water (Jain et al. 1983). Note that salinity of the soil saturation extract reported is quite high, and fluctuated from 20 to 30 dS m^{-1} in the 27 dS m^{-1} treatment and 27 to 40 dS m^{-1} in the 33 dS m^{-1} treatment. It is quite possible that these trees have survived through high soil salinity, but the actual growth has taken place during and after the rainy season. Soil salinity values of the experiment involving salinization of irrigation water (Jain et al. 1983) were obtained in the 1:2 (soil:water) extract, and fluctuated as much as 6 fold. The lack of soil salinity control in both experiments make these results less tangible, but the relative tolerance of these species to salinity is probably a reliable indicator.

Table 5. Relative reduction in trunk diameter and mortality rates of tree species when grown in saline soils or with saline waters for 5 years (Jain et al. 1983, 1985)

	Reduction (%)	Mortality (%)
Saline water (EC_w = 9 dS m^{-1})		
Prosopis juliflora	0	0
Dishrostachys glomerata	18	0
Acacia aneura	18	20
Tamarix articulata	26	10
Leucaena leucocephala	20	40
Eucalyptus hybrid	30	60
Cassia siamea	–	100
Saline soil (EC_e = 33 dS m^{-1})		
Prosopis juliflora	20	13
Tamarix articulata	33	13
Acacia tortilis	54	56
Saline soil (EC_e = 27 dS m^{-1})		
Prosopis juliflora	17	6
Tamarix articulata	0	0
Acacia tortilis	7	0
Azadirachta indica	25	6
Eucalyptus camaldulensis	34	25
Albizzia lebb	–	81

Table 6. Soil pH and the exchangeable sodium percentages (ESP) which cause 25 and 50% reductions in yield of selected grass species (Ashok et al. 1986)

Grass species	25% reduction		50% reduction	
	pH	ESP	pH	ESP
Diplachne fusca	10.4	89	>10.4	>90
Chloris gayana	10.4	86	>10.4	>90
Cynodon dactylon	10.1	74	10.4	87
Brachiaria mutica	9.7	55	9.9	67
Panicum spp.	9.5	42	9.7	56
Pennisetum purpureum	9.2	30	9.4	40
Panicum maximum	9.0	20	9.2	27

d. *Alkali tolerance of grasses and trees.* Alkali tolerance of plants consists of their ability to tolerate toxic effects of Na, and OH ions as well as the nutritional imbalance induced by high levels of exchangeable sodium, (e.g., Na-induced K or Ca deficiency) or HCO_3 and CO_3. When soils are used as growth medium, structural deterioration is also a factor which alters plant response. Growth observations made as a function of the exchangeable Na percentages (*ESP*), thus, should not be interpreted strictly as a result of Na ion effects.

Ashok Kumar & Abrol (1986) made a series of growth response observations when grass species were grown in soils with varying exchangeable Na percentages (Table 6). Since the alteration of

ESP levels resulted in pH changes, both the pH and *ESP* values corresponding to 25 and 50% reductions are given. *Diplachne fusca* (Karnal grass or Kallar grass), *Chloris gayana* (Rhodes-grass) as well as *Cynodon dactylon* (Coastal Bermuda) appear to be highly adaptable to alkali soils. The successful adaptation of *D. fusca* to alkali soils is also reported in several field trials (e.g., Singh et al. 1991; Malik et al. 1986; Hussain & Hussain 1970). Although *Distichlis* spp. (saltgrass) are not listed in the Table, they are known to be both saline and alkali tolerant. In the successful revegetation project at the Lago Texcoco in Mexico, the pH of the soil ranged from 9.5 to 11.5 (Llerena 1994).

A large number of tree species, e.g., *Acacia*, *Prosopis*, *Casuarina* and *Sesbania*, is known to adapt to alkali soils of a pH as high as 10. There is also ample evidence to indicate that planting trees, especially *Prosopis* species helps ameliorate alkali soils. The ameliorating effects include increasing organic matter, lowering the pH of surface soils, and increasing water infiltration (e.g., Singh et al. 1990; Yadav 1978).

2.3 *Salt uptake and fodder quality*

Salinity of irrigation water and soil has a pronounced effect on salt uptake and quality of fodder. The extent of such alterations vary depending on plant species, salt levels and types, season and cultural practices, and may appear in forms of increased ash contents, elevated levels of toxic elements, and degradation of forage quality.

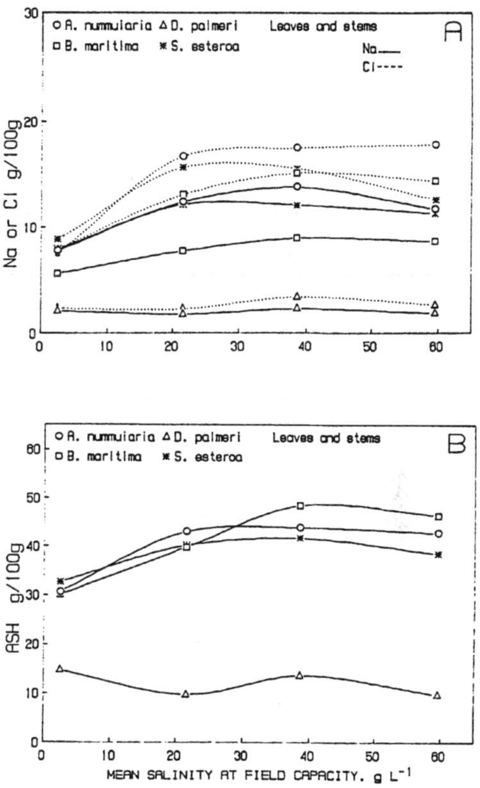

Figure 3. The concentration of Na or Cl (Figure 3A) and of ash (Figure 3B) in leaves and stems of four halophyte species grown for one season in lysimeter as related to mean soil solution salinity at field capacity (Miyamoto et al. 1993).

a. *Salt uptake and ash content.* There are two broad types of saline adaptation in halophytes. One type of plants stores salts in tissue so as to maintain the osmotic gradient across the root membrane, and another type excretes salt through salt glands to avoid toxic levels of salt accumulation. Many of the species of chenopods fall into the first type, and grass species the second type (e.g., Glenn & O'Leary 1984; Glenn 1987; Malik et al. 1986). The predominant salt elements which are taken up by halophytes are Na and Cl, and to a lesser extent K, Mg and Ca. Figure 3A shows examples of Na and Cl uptake by four halophyte species grown in lysimeters irrigated with waters having seawater compositions (Miyamoto et al. 1994). The concentrations of Na and Cl in chenopods greatly exceeded those in *Distichlis palmeri*.

Salts make up the bulk of the ash content. The ash contents vary with species, age of the plants and environmental conditions, but typically range from 20 to 40 g 100 g^{-1} of dry matter in native stands (Table 7). Under irrigated conditions, we found ash contents ranging from 30 to 50 g 100 g^{-1}, depending on salinity of the soil solutions for three species of chenopods (Figure 3B), which is greatly higher than the ash content of *D. palmeri* (Miyamoto et al. 1993). A report from Pakistan (Malik et al. 1986) shows ash contents of *Diplachne fusca* to be less than

Table 7. Ash, crude protein, crude fiber contents of some halophytes

	Ash content	Crude protein	Crude fiber	Dry matter digestability	References
Native range	(g/100 g)				
Auleropus repense	14	12	22	56	Koocheki & Mohalati (1992)
Aristida pennate	7	8	24	42	Koocheki & Mohalati (1992)
Atriplex halimus	23	13	25	66	El-Shaer & Gihad (1992)
A. nummularia	27	13	24	59	El-Shaer & Gihad (1992)
Salicornia fructiosa	14	13	19	70	El-Shaer & Gihad (1992)
Salsola rigida	11	15	16	78	Koocheki & Mohalati (1993)
S. tetrandra	36	6	36	68	El-Shaer & Gihad (1992)
Suaeda fruticosa	16	10	33	70	El-Shaer & Gihad (1992)
Kochia prostrata	6	16	18	67	Koocheki & Mohalati (1993)
Irrigated with saline water ($<20\ dS\ m^{-1}$)					
Diplachne fusca	15	12	–	–	Malik et al. (1986)
Distichlis spicata	–	12		40–50	Lugg (1983)
Irrigated with seawater					
Atriplex lentiformis	27	17	14	50	O'Leary et al. (1985)
A. barclayana	33	12	14	47	O'Leary et al. (1985)
Salicornia europaea	41	6	17	–	O'Leary et al. (1985)
Salicornia europaea (seed)	7.5	30	–	–	O'Leary et al. (1985)
Distichlis palmeri (seed)	1.3	9	–	–	Glenn et al. (1982)
Cakile maritima (seed)	5.0	21	–	–	O'Leary et al. (1985)
Cressa truxillensis (seed)	1.6	13	–	–	Glenn et al. (1982)
Conventional crop					
Alfalfa hay	–	15–22	58	64	O'Leary et al. (1985)
Soybeans (seed)	4.8	40	–	–	O'Leary et al. (1985)

$18\ g\ 100\ g^{-1}$ when irrigated with a $30\ dS\ m^{-1}$ saline solution.

High salt contents of forages may be acceptable for camels, but exceed the NRC guideline of $10\ g\ NaCl\ 100\ g^{-1}$ for sheep and goats, and must be considered among the major constraints of using chenopods as fodder. Glenn et al. (1982), however, reported that high salt contents of *Atriplex* and *Suaeda* were lowered to a range of 6 to $15\ g\ 100\ g^{-1}$ through soaking in water, then squeezing with a press to remove the water. Malik et al. (1986) reported that a large portion of Na and Cl present in Kallar grass shoots can be washed out by one minute soaking in water. Another alternative is to blend halophyte fodder with conventional fodder or halophytes with low salt uptake characteristics such as *Distichlis* or *Diplachne* spp. When fodder of high salt content is used, animal feed water must be low in salt content (Squires 1993).

b. *Nutrient contents and digestibility*. Crude protein contents vary with species, age of plants, kinds of tissue, and environmental conditions. The analyses of native stands of halophytes show that the crude protein contents of *Atriplex*, *Suaeda*, and several other species of chenopods range from 10 to $15\ g\ 100\ g^{-1}$ per ash-free dry matter (Table 7), which are not far off from the crude protein contents of alfalfa or clover. High protein contents of *Atriplex* spp. are als reported when grown with seawater (Table 7). *Kochia* also has high protein contents, whereas the protein content of grass species is somewhat lower.

The crude fiber contents also vary widely, but in general, *Salsola*, *Suaeda* and grass species have high fiber contents (Table 7). The fiber content of *Atriplex* spp. is highly variable. Blending of different forage could provide a good balance of protein and fiber content. Dry matter digestibility (in vitro) also vary considerably (Table 7). In general, grass species with high fiber contents are low in digestibility. Readers may refer to Gihad & El Shaer (1994) for detailed discussion on digestibility.

Nutritional values of many halophytes and di-

Table 8. Maximum tolerable levels of dietary minerals for domestic livestock in comparison with levels in forages (Chaney 1983)

Element	Level in plant foliage ($mg\ kg^{-1}$ dry foliage)		Maximum levels chronically tolerated[a] ($mg\ kg^{-1}$ dry diet)			
	Normal	Phytotoxic	Cattle	Sheep	Swine	chicken
As_{inorg}	0.01–1	3–10	50	50	50	50
B	7–75	75	150	(150)	(150)	(150)
Cd[c]	0.01–1	5–700	0.5	0.5	0.5	0.5
Cr[a]	0.01–1	20	(3000)	(3000)	(3000)	3000
Co	0.01–0.3	25–100	10	10	10	10
Cu	3–20	25–40	100	25	250	300
F	1–5	–	40	60	150	200
Fe^{2+}	30–300	–	1000	500	3000	1000
Mn	15–150	400–2000	1000	1000	400	2000
Mo	0.1–3.0	100	10	10	20	100
Ni	0.1–5	50–100	50	(50)	(100)	(300)
Pb[e]	2–5	–	30	30	30	30
Se	0.1–2	100	(2)	(2)	2	2
V	0.1–1	10	50	50	(10)	10
Zn	15–150	500–1500	500	300	1000	1000

[a] NRC (1980).

gestibility are close to conventional forage crops. The primary difference is the salt and ash content which is greatly higher with many (but not all) halophytes. An alternative may be to harvest seed and use it as animal feed. Seed, especially *Salicornia* seed reportedly contains high levels of protein, but is low in salts (Table 7). Glenn et al. (1991) also reported 9 and 13 g of crude protein per 100 g of seed of *Distichlis palmeri* and *Cressa truxillensis*, respectively (Table 7).

c. *Toxic metabolites*. Many plants contain metabolites which can cause toxic (or poisoning) effects to animals. These include alkaloids, nitrate, oxalate, and cyanide. Some halophyte species are known to contain these metabolites at high levels (Bayoumi & El Shaer 1994).

d. *Toxic element accumulation*. Saline water and agricultural drainage water sometimes contain elevated levels of toxic elements such as As, Cd, Cr, Co, Cu, Pb, and Hg. Some of these elements can be assimilated into plant tissue to the level that causes toxic effects to animals. The National Research Council (1980) prepared a guideline for assessing potential metal toxicity to animals based on dietary analysis (Table 8). This guideline was developed based on single element toxicological studies, and interactions among the elements are not considered. The actual toxicity can differ depending on ages and interactions, such as deficiency of certain elements due to excessive intake of other elements.

When halophytes are to be grown for animal fodder using saline water, some attention must be given to the excessive accumulation of toxic elements. At present, little information is available concerning toxic element uptake by halophytes. Based on conventional forage crop analyses, Chaney (1983) summarized normal and phytotoxic levels in forages (Table 8). According to this table, plants generally experience phytotoxicity before the concentration reaches levels toxic to animals. However, Se, Mo, Cd, Co and Mn can accumulate to the level of animal toxicity without causing phytotoxicity.

The toxicity caused by Se excess (Selenosis) in the USA is found primarily in the Northern Plains and Rocky Mountain states (Kubota & Allaway 1972). High concentration of Se is also reported in agricultural drainage water in California. (e.g., Tanji et al. 1986; Letey et al. 1987; Westcot 1988). The concentration of Se in plant tissue is highly species-dependent and usually appears in calcareous soils, rather than in acidic soils (Rosenfeld & Beath 1964). A recent greenhouse pot study indicates that *Atriplex nummularia* and *A. semibaccata* have accumulated 0.77 and 0.40 mg Se kg^{-1} of dry forage as opposed to

0.30 mg kg^{-1} by *Festuca arundinacea* (Tall fescue) when grown for one season in a soil containing 3.5 mg kg^{-1} of Se (Banuelos & Meek 1990). Considering high variability, an average Se concentration of 0.77 mg kg^{-1} in *A. nummularia* is not far off from the potential toxic level of 2 mg kg^{-1} (Table 8).

The toxicity caused by excess Mo (Molibdosis; Mo-induced Cu deficiency) is an endemic nutritional problem of ruminants primarily in the northwestern states of the USA (Kubota & Allaway 1972). Agricultural drainage water also often contains elevated levels of Mo (Westcot 1988). High soil moisture contents accentuate the plant uptake of Mo (Kubota & Allaway 1972), and legumes tend to accumulate Mo more so than grass species. The uptake of both Mo and Se are reduced with increasing SO_4 concentrations (Pratt et al. 1988). There is little information on Mo uptake by halophyte species.

The human toxicity caused by excess Cd in rice reported in Japan (e.g., Fukushima et al. 1973) has raised considerable concern. The uptake of Cd tends to increase with low levels of Ca and elevated levels of Zn. In alkali or calcareous soils of the arid region, however, the solubility of Cd is usually low enough to limit plant uptake. Likewise, Co and Mn toxicity are usually limited to acidic soils. However, high levels of Cd, Co and Mn coupled with low redox potential induced by waterlogging of organic rich soils can cause toxic metal uptake (e.g., Cajuste et al. 1991). Likewise, elevated levels of toxic metals can be found in marsh vegetation, especially when the marsh is drained. This will cause oxidation of sulfide and reduction in sediment pH. Reboredo (1991), for example, reported Cu concentrations ranging from 20–50 mg kg^{-1} in leaves when *Halimione portulacoides* (a salt marsh halophyte) was grown in an acidic sediment containing 5–50 mg kg^{-1} of Cu. This range of Cu concentration exceeds the NRC guideline if the forage is consumed by sheep (Table 8).

2.4 *Water stress, water use and production*

a. *Water stress tolerance*. The ability of plants to sustain economic growth under water stress is an important consideration from the view of irrigation and drainage management. The levels of water stress above which production decreases are well-known for most conventional crops, and range from 30–500 kPa depending on crops and growth stages (e.g., Taylor & Ashcroft 1972). In halophytic plants, growth or yield response to water stress has not yet been adequately studied, although the effect of drought stress on survival of *Atriplex* spp. and several chenopods has received considerable attention (e.g., Sharma 1982). A reason for this low interest partly stems from the fact that salt stress is likely to dominate plant response in irrigated production with high saline waters.

There are, however, many indications that water stress levels for optimum production of halophytic plants vary as much as those for conventional crops. Several ecological observations, for example, cite a wide-range of soil moisture levels that native halophytes seem to prefer. *Diplachne fusca* (Kallar or Karnal grass) and *Distichlis* spp., for example, prefer wet soils, whereas *Sporobolus* species prefer dry soils. Some halophytes such as *Salvadora* and *Atriplex* species prefer dry soils (e.g., Gupta and Saxena 1968) and other species such as *Tamarix* species grow well in wet as well as dry soils. Many *Atriplex* spp. can survive and transpire water at soil water potential well below −1.5 MPa or −15 bars (e.g., Moore et al. 1972). In a controlled greenhouse experiment using seedling trees, Tomar and Gupta (1985) further observed that *Acacia auriculiformis*, *A. catechu*, *Leucaena leucocephala* could not survive 40 days of waterlogging, whereas *Acacia nilotica*, *Cassia siamea*, *Casuarina equisetifolia*, *Prosopis juliflora* and *Terminalia arjuna* have survived the waterlogging treatment. Felker et al. (1983) observed that 3 year growth of *Prosopis* species did not decline when irrigation was initiated at 200 kPa, but was reduced by 24% when irrigated at 500 kPa (5 bars). Burman et al. (1991) observed that growth of *Azadirachta indica* was optimum near field capacity.

b. *Production and water use*. Productivity of field grown halophytes are documented by Glenn & O'Leary (1985), Watson et al. (1987), and Glenn et al. (1991). Some results obtained in a coastal desert of Mexico (Puerto Penasco) are shown in Table 9. The plots were irrigated with 10 g L^{-1}

Table 9. Production functions of some plant species

	Salinity	Yield	ET	Y/ET	References
Field experiments					
	($g\,L^{-1}$)	($kg\,ha^{-1}$)	(mm)	($kg\,m^{-3}$)	
Atriplex lentiforms	40	17940	–	–	Glenn & O'Leary (1985)
A. canescens	40	17230	–	–	
A. nummularia	40	8010	–	–	
A. barclayana	40	8630	–	–	
Batis maritima	40	17380	–	–	
Salicornia bigelovii	40	15390	–	–	
Distichlis palmeri	40	13640	–	–	
D. spicata	12	3400	–	–	Lugg (1983)
	25	5400	–	–	
	5.0	4700	–	–	
	10.0	3100	–	–	
Prosposis chilensis	–	4040	1393	2.9	Felker et al. (1983)
P. juliflora	–	1270	1443	0.09	Singh et al. (1990)
Outdoor lysimeter experiments					
	($g\,L^{-1}$)	($g\,pot^{-1}$)	(L)	($g\,L^{-1}$)	
Atriplex nummularia	10	168	65	2.6	Miyamoto et al. (1993)
Distichlis palmeri	10	144	57	2.5	
Suaeda esteroa	10	108	43	2.6	
Batis maritima	10	58	42	1.4	
Acacia auriculiformis	–	1713	1230	1.4	Chaturvedi (unpub. report)
Albizia lebbek	–	2355	1283	1.8	
Dalbergia sissoo	–	2004	1534	1.3	
Eucalyptus hybrid	–	5209	2526	2.1	
Syzygium cuminii	–	2386	1190	2.0	
Greenhouse pot experiments					
	($g\,L^{-1}$)	($g\,pot^{-1}$)	(L)	($g\,L^{-1}$)	
Atriplex triangularis	0	24.6	12.7	1.94	Philipp & Gallagher (1985)
	5.25	89.2	26.5	3.36	
	10.5	66.8	16.3	4.10	

diluted seawater at an annual rate of 16 m, most of which was probably lost as drainage. Production of many species (dry top weight) exceeded 15000 $kg\,ha^{-1}$ which is close to the yields of many conventional fodder crops. However, high ash contents must be taken into consideration when evaluating biomass production data. Lugg (1983) also reported high yields of *Distichlis spicata* when grown in a sandy loam soil with conventional practices (Table 9).

Lysimeter studies (Miyamoto et al. 1993; Chatarvidi, unpublished data) show that water use efficiency of many halophyte and tree species is in the range of 1.4–2.6 $g\,L^{-1}$. C_4 plants such as *Atriplex*, *Suaeda* and *Distichlis* appear to have higher water use efficiency than C_3 plants such as *Batis* (Table 9). Water use efficiency of some species are reported to increase somewhat with increasing salinity (e.g., Philipp & Gallagher 1985, Table 9).

Tree species grown in lysimeters (Chatarvidi, unpublished data, Table 9) or those planted densely (Felker et al. 1983) appear to yield water use efficiency comparable to that of shrubs or grasses (Table 9). However, field planted trees may experience extremely low efficiency during establishment due to excessive evaporation losses (e.g., Singh et al. 1990, Table 9). Low irrigation efficiency is certainly an overriding reality of lowering the overall water use efficiency in irrigated production of most species.

3. Water and soil characterization for brackish water irrigation

3.1 *Water quality parameters and appraisal*

a. *Salinity*. The concentration of total dissolved salts is a primary parameter of brackish water quality. It is commonly expressed in $mg\,L^{-1}$, g^{-1}, $meq\,L^{-1}$, or $mol\,L^{-1}$ or by the electrical conductivity (EC) of water. The use of the EC measure is popular among agronomists, not only because it is a convenient measure, but also relates directly to the concentration of charged ion species. For the convenience of readers, a conversion table is provided in the Appendix.

The appraisal of salinity for irrigated production depends primarily upon salt tolerance of crops and the leaching fraction that can be attained or allowed. The leaching fraction (LF) under a steady salt balance is defined as.

$$LF = D_D/D_I = C_I/C_D \tag{1}$$

where C and D represent salinity and the depth of irrigation water (I) or drainage water (D), respectively.

A first approximation of the mean salinity of the root zone upon the completion of drainage is

$$C_S = (C_I + C_D)/2. \tag{2}$$

From Eqs (1) and (2), we obtain

$$C_S = 0.50(1 + 1/LF)C_I. \tag{3}$$

The computed C_S for various combinations of C_I and LF are shown in Table 10.

In using Eqs (2) and (3), an inherent assumption is that the soil water content at the completion of drainage is at field capacity. Even though the amount of the salts left in the root zone does not change, the salt concentration of the soil solution will increase with subsequent evapotranspirational losses of water. The increase

Table 10. Projected mean salinity of soil solutions at field capacity when irrigated with water of EC_I at the specified leaching fractions (LF)

EC_I		LF = 0.5	0.4	0.3	0.2	0.1[b]
($g\,L^{-1}$)	($dS\,m^{-1}$)	($g\,L^{-1}$)				
5	8[a]	8	9	11	15	28
10	17	15	17	22	30	55
15	25	22	26	32	45	82
20	33	30	35	43	60	110
25	41	37	44	54	75	–
30	48	45	52	65	90	–
35	56	53	61	76	105	–

[a] Assume to be NaCl solutions.
[b] The soil solution salinity values at low leaching fractions is likely to be an overestimate of the effective salinity.

is assumed to be inversely related to the soil water depletion caused by evapotranspiration.

$$\bar{C}_S/C_S = FC/(FC - 0.5\,ET), \tag{4}$$

where ET is the cumulative evapotranspiration during an irrigation cycle. A factor of 0.5 is applied to ET so as to evaluate soil solution salinity at the mean soil water storage, instead of at the driest state. The soil solution salinity at the mean soil water content ($\bar{C}_S$) can be equated to salt tolerance of crops expressed in salinity of soil solutions. If salt tolerance is expressed by the salinity of the saturation extract (Ce), it must be converted before a comparison can be made.

$$\bar{C}_S = (\rho\, Wsat/\bar{\theta})Ce, \tag{5}$$

where ρ is the soil bulk density (typically 1.30 ~ 1.40 $g\,cm^{-3}$), $Wsat$ the water content of the saturated paste, and $\bar{\theta}$ the mean water content of the soils ($mL\,cm^{-3}$) used to establish the salt tolerance. When $\bar{\theta}$ is not available, it can be substituted by FC. In the case, the estimated results are equivalent to C_S instead of $\bar{C}_S$.

Many halophytes show an optimum growth at soil solution salinity of approximately 20 $g\,L^{-1}$ or 38 $dS\,m^{-1}$ (Section 2.2). If one wishes to grow such halophytes at a conventional leaching fraction of 0.2–0.3, the salinity of irrigation water suitable for attaining the optimum growth is about 10 $g\,L^{-1}$ (17 $dS\,m^{-1}$) or less (Table 10). If one can allow higher leaching fractions such as 0.4–0.5 or a greater degree of yield reductions, the permissible level of salinity in irrigation water increases to a range of 13–16 g L^{-1} (or 25–30 dS m^{-1}) or about half of sea water salinity. The attainment of desired leaching fractions is controlled by hydraulic properties of soils and irrigation management, both of which are discussed in later sections.

The above assessment of irrigation water quality for salinity control is a simplified concept, and may be conservative, especially when high frequency irrigation is used. An additional discussion on this subject is given in Miyamoto (1993).

b. *Sodicity*. Sodicity of irrigation water is usually expressed by the sodium adsorption ratio (*SAR*).

$$SAR = Na/(Ca + Mg)^{1/2}, \quad (6)$$

where the concentrations of cations are expressed in mmol L^{-1}.

The *SAR* measure is used instead of a Na to (Ca + Mg) ratio, because it relates quantitatively to the exchangeable sodium percentage (*ESP*). In water containing high levels of SO_4, however, *ESP* tends to be greater than *SAR* as SO_4 ions form ion-pairs with Ca and Mg. The relationship between *ESP* and *SAR* is also modified by types of clay, ion activity coefficients, and the presence of soluble minerals in soils such as calcite and gypsum. It is also influenced by water evaporation and/or transpiration and the presence of HCO_3 ions, all of which tend to increase *SAR* (Miyamoto 1980; Miyamoto & Pingitore 1992).

The primary concern over sodicity of irrigation water is its adverse effect on soil structural stability, water infiltration, and the leaching potential. The magnitude of sodicity effects varies not only with soil types, but also with salinity and soil management practices. Existing guidelines for appraisal of sodicity effects (e.g., Ayers & Westcot 1985) should be viewed as such (as guidelines only). In fact, a recent analysis by Pratt & Suarez (1990) indicates that there appears to be no consistent value of *SAR* or the combination of *SAR* and *EC* values that can be used universally for predicting a significant reduction in water infiltration rates.

In spite of the complexity, the general rules of sodicity effects on soil structural stability are well-known. The adverse effect of Na is, for example, most pronounced when salinity is the lowest. Under low salinity conditions, sodicity encourages slaking of weak soil aggregates or soil clods that are formed by mechanical breaking of dry soils. This effect may appear at *SAR* as low as 3 (e.g., Abu-Sharar et al. 1987). Increasing *SAR* to a range of 5 to 15 may result in dispersion of aggregates into suspended particles, commonly known as particle dispersion (e.g., Frenkel et al. 1978). The swelling of expandable clays then follows at *SAR* exceeding approximately 15 (Shainberg & Letey 1984). However, when salinity exceeds 3–5 dS m^{-1}, aggregate slaking and especially particle dispersion become minimal, because of the stabilizing effects of electrolytes. In irrigation involving brackish water having high salinity, soil structural deterioration caused by Na would be minimal, except when rainfall or application of low salt water is involved. This aspect is discussed further in Section 3.3.

c. *Toxic elements*. Boron (B) is the most common phytotoxic element that is present in brackish water. Many conventional crops are known to be affected by B when its concentration in irrigation water exceeds a few mg L^{-1} (e.g., Ayers & Westcot 1985). For halophytic plants, Boron tolerance has not yet been established, thus a meaningful appraisal of B is not feasible.

Selenium (Se) and Mo are two of the most common elements which can cause toxic effects on animals more so than phytotoxicity. Pratt (1972) recommended that the maximum concentrations of Se and Mo in irrigation water should not exceed 20 and 10 $\mu g\ L^{-1}$, respectively (Table 11). However, Pratt et al. (1988) later recommended to raise the allowable concentration of Se and Mo to 100 and 50 $\mu g\ L^{-1}$, respectively, on the basis that the presence of SO_4 reduces the uptake of Se and Mo. In addition, the uptake of Se and Mo is plant-species dependant. Blending with fodders with low Se or Mo contents offer another option for reducing or eliminating toxic effects of Se and Mo.

The recommended maximum concentration for Cd is 10 $\mu g\ L^{-1}$ (Pratt 1972). In alkali irrigation waters prevalent to the arid regions, Cd rarely exceeds 1 $\mu g\ L^{-1}$, thus Cd phytotoxicity in alkali soils is unlikely. Arsenic (As) and Cr are also phytotoxic, but again it is rare to have the concentration as high as 100 $\mu g\ L^{-1}$ in alkali

Table 11. Guidelines for the maximum concentration of trace elements in irrigation water for protection of animal health and plants (Pratt 1972; Pratt et al. 1988)

Element	($\mu g\ L^{-1}$)	
Protection of animal health		
Se	20	Protection from selenosis
	100	For water with high SO_4
Mo	10	Protection from molybdosis
	50	For water with high SO_4
Cd	10	Considering potential effects on human food chain contamination
Protection from phytotoxicity		
As	100	Phytotoxicity may occur above this concentration
Cd	10	Phytotoxicity at 100 $\mu g\ L^{-1}$ in sensitive plants in nutrient culture
Co	50	Phytotoxicity at 100 $\mu g\ L^{-1}$ in some plants in nutrient culture
Cr	100	Phytotoxicity at 500 $\mu g\ L^{-1}$ in some plants in soil culture
Cu	200	Phytotoxicity at 100–1000 $\mu g\ L^{-1}$ in plants in nutrient culture
V	100	Phytotoxicity at 500 $\mu g\ L^{-1}$ in plants in nutrient culture

water, unless industrial or mining wastewater is discharged into irrigation systems without pretreatment.

Salt marshes sometimes contain elevated levels of Cu, Cr and V, most of which are present as organo-complexes. Some of these concentrations can exceed the guidelines given in the Table. The concentrations of these elements as well as of other toxic elements in seawater free of high organic matter loading are, however, below the recommended limits for irrigation uses.

d. *Corrosion potential.* Brackish waters are known to be highly corrosive to both metals and concrete. Concrete corrosion occurs not only by low pH (below about 5.5), but also by SO_4 and Mg. Biezok (1972) indicates that concrete corrosion becomes strong when the concentration of Mg and SO_4 exceeds 300 and 600 mg L^{-1}, and becomes very strong when exceeding 1500 and 3000 mg L^{-1}, respectively.

Metal corrosion is much more complex than concrete corrosion, and readers may refer to Clarke (1980) for details.

3.2 *Soil properties for brackish water irrigation*

a. *Soil profile.* Soil profile characteristics have the dominant effect on rooting patterns, water penetration, retention and drainage. Soil horizons and their physical and chemical properties along with the rooting pattern of native vegetation must be characterized. Soil horizon properties to be determined include hardness, texture, structure, water transmissibility, water holding capacity, and salinity. Identification of petrogenetic pans is also among the priority items during the soil profile examination.

The calcic horizon ($CaCO_3$ rich layer) is probably the most common feature of upland soils of the arid and semi-arid regions. In most instances, the calcic horizon becomes restrictive to the vertical penetration of roots, especially of tree roots. Another petrogenetic soil horizon cemented with Si also occurs, especially in the watershed adjacent to sandstone or quartz formations. Such a layer will also alter the rooting pattern. When the soils containing these petrogenetic hard pans are brought under irrigation, water usually can penetrate through, but not necessarily plant roots. Depending on their properties and depths, subsoiling and/or deep holes for tree planting may be required, followed by adjustment in irrigation management.

In alluvial soils, soil profile stratification or interbedding of sand, gravel or clay layers is common. When the soil profile consists of a fine-textured layer over a coarse-textured layer such as a sand or gravel layer, water penetration ceases at the boundary until the upper layer becomes nearly saturated. Sand or gravel has minimal sorptivity. This feature must be incorporated into the decision of irrigation rates per application, or be corrected by deep plowing (e.g., Kaddah 1976). If the profile sequence is reversed (sandy soils at the surface and a fine-textured layer beneath), sub-surface drainage and salt accumulation usually become a concern. Some soils are highly compacted petrogenetically or by traffic, disking or plowing. The formation of a disk or a plow pan immediately below the disk or plow layer is common in sandy soils, including fine and medium sand. These compacted layers can restrict root developments, especially when the compacted layers are kept dry. Chiseling has been a standard measure of alleviating compaction (e.g., Tigma et al. 1991).

b. *Water intake.* One of the key factors for sustainable crop production with highly saline water is to have adequate water infiltration, penetration, and drainage. Otherwise, salts will accumulate in the soils, and irrigated production becomes not economical. Once soil profile examination and soil mapping are completed, field measurements of these parameters must be carried out for representative soil types.

Currently, various methods are used to measure water intake properties with varying complications and success. The double-ring infiltrometer (30 cm ID of inner ring and 60 cm ID outer ring) is the ASTM standard in the US, and has been used widely, presumably because of its simplicity. However, a nagging question has been the validity of this measurement in estimating infiltration rates when the entire field is irrigated under basin or border methods. This method imposes a relatively high pressure head on a very small part of the landscape, and the intended vertical flow paths tend to diverge due to the strong lateral gradient created by the surrounding drier soils. Swartzendruber & Olson (1961) studied the radial flow effect in sand, and found that an inner ring diameter of 100 cm and an outer ring diameter of 120 cm were sufficient to minimize the effect of radial flow at the wetting depth of 60 cm or less. This size requirement is greatly larger than the ASTM standard ring described earlier. Care must also be taken in ring placement, especially in soils with cracks. Otherwise, the lateral flow can induce a significant error.

If a furrow method is planned to be used for irrigation, infiltration tests should be performed using one or more furrow basins. Currently, three types of infiltration tests are used; measurements of inflow and outflow rates, measurement of inflow rates and water advance time in the furrow, and measurement of intake rates in a short furrow segment, typically 5–6 m in length. The first two methods are not convenient for initial characterization, as they require a large quantity of water for testing. The measurement of water intake into the furrow segment requires less water, and readers may refer to Walker & Willardson (1983) and Bondurant (1957) for details on these methods. In carrying out the furrow irrigation tests, it is important to settle the soils before testing. Freshly plowed and bedded fields usually yield unrealistically high initial infiltration rates.

Once infiltration data are gathered, they can be fitted to various infiltration equations. The equation proposed by Philip (1957a) has clearly defined-parameters, and for this reason, it has been used widely.

$$Z = at^{1/2} + K_S t + c \tag{7}$$

or

$$I = \mathrm{d}Z/\mathrm{d}t = 1/2at^{-1/2} + K_S, \tag{8}$$

where Z is the cumulative intake, t the time, I the intake rate; a, K_S and c are empirical coefficients. Coefficient a is commonly referred to as the sorption constant, and depends upon soil pore structure as well as soil moisture contents. In initially dry soils, the value for a typically ranges from 10–80 cm/day$^{1/2}$ (e.g., Taylor & Ashcroft 1972). Coefficient K_S is analogous to the saturated hydraulic conductivity and governs the steady infiltration rate and ranges anywhere from 1 to 500 cm day^{-1} for clay to sand. Coefficient c (which is absent in the original Philip's equation) is added here to incorporate the instantaneous water intake which occurs in dry soils, especially those which have macro-pores or cracks.

The above equation was developed for homogeneous stable porous media, and usually does not fit well to the soils which develop surface seal or those having strong stratification. In such cases, the Kostiakov equation provides better fit and readers may refer to Clemmens (1983) for other forms of infiltration equations.

When sprinkler irrigation is planned to be used, drop formers are a preferred option, but such devices are not always available. The maximum intake rate for sprinkler designs can be approximated by the steady intake rate, K_S of Eq

(8). Julander & Jackson (1983) have found a good agreement between K_S determined with a double-ring infiltrometer and the maximum steady infiltration rate determined with a drop former. The difference between the two methods is usually caused by the extent of soil surface disturbances and the radical flow influence.

c. *Water penetration and drainage*. Following the intake of water, the excess water drains out of the initial wetting depth and penetrates deeper into the profile. Thereafter, a portion of the redistributed water may escape from the root zone as drainage water. This process of redistribution is of critical importance in terms of water supply to the lower root zone as well as salt leaching. The process is influenced by various factors, including the initial soil water content, soil profile configuration, and the presence of a shallow water table.

The simplest concept to evaluate water penetration and drainage (D) is

$$p_i = (S_{oi} + p_{i-1}) - FC_i, \qquad (9)$$

where p, S_o and FC denote the percolation losses, the initial soil water content, and field capacity, respectively, and the subscript i denotes the i-th soil layer. The assumption involved in the equation is that the soil water present in the i-th layer in excess of the field capacity of the layer will drain to the next layer.

It was thought in the past that field capacity can be measured in a laboratory by subjecting soil samples to a soil water suction of 1/4 to 1/3 bar. Such a measurement, however, usually does not produce reliable enough data to be used for salinity control, especially in stratificial soils or soils with a high water table. Field capacities should be determined in-situ following deep irrigation that wets the entire projected root zone. The timing of such determination varies from one day in sandy soils to a week in clay soils. The area to be wet for this purpose must be substantially larger than the area used for a doublering infiltrometer test, since soil water flow paths diverge with increasing depths. This measurement must be taken under minimal transpiration. The field capacity typically ranges from 25–35 cm m^{-1} in alluvial soils.

The fundamental deficiency of the field capacity concept is the lack of a rate factor. If the soil moisture redistribution takes place at a rate slower than the evapotranspiration rates, drainage would not occur, even when water is applied in excess of field capacity. Even in sandy soils, no drainage may result, if irrigation water is applied at a rate less than the evapotranspiration rate.

For a quantitative evaluation of drainage potential, Gardner (1983) proposed a simple equation derived from a steady-rate flow analysis.

$$D = (1/a)\ln(K_o/ET), \qquad K_o > ET > K \qquad (10)$$

with

$$a = \left(\frac{1}{S_o - S}\right)\ln(K_o/K), \qquad (11)$$

where D is the cumulative drainage from a uniform soil layer following irrigation, K_o the hydraulic conductivity or the rate of water application at a soil water content of S_o (which is usually between FC and saturation), K the unsaturated conductivity at a soil water content of S, and a is a coefficient related to soil hydraulic properties. Equation (10) provides an estimate of drainage potential for an assumed ET when K_S and a are known, and can be used to check if water will move fast enough to satisfy Eq (9). Equation (10) is, however, not applicable when K_S is so low as to approach ET. In the case, the complete equation given by Gardner (1983) is recommended. Readers may also refer to van Genuchten (1980) and Mualem (1976) for a precise treatment of K estimates.

An alternative to the above approach is an empirical approach, such as proposed by Shaw & Thorburn (1985). In this approach, the leaching fraction observed in various soils was empirically related to certain soil properties. In theory, such an empirical relationship will provide a way to predict the leaching potential by knowing soil properties.

d. *Evaporation from the soil surface*. The evaporation from the soil surface is similar to that from free water surface when the soil surface is wet. This is then followed by a rapidly declining rate of evaporation which is controlled principally by hydraulic properties of soils. Philip (1957b) examined this process and termed them as the first stage and the second stage evaporation, respectively. During the first stage, the evaporation rate may be curtailed if there is a crop cover. Ritchie & Burnett (1971), and Villalobos & Fereres (1990) obtained essentially the same empirical relation which accounts for the effect of plant covers on the first stage evaporation

$$E = Ep\, e^{-1.4LAI}, \quad (12)$$

where Ep is the potential evaporation, and LAI the leaf area index.

The question of how long the first stage evaporation may continue has not yet been fully answered but has been determined empirically, and ranges from 0.6–1.2 cm (Ritchie 1972). This means that the first stage evaporation will terminate in a matter of a day in a hot and dry climate.

The second stage evaporation depends upon soil hydraulic properties and soil water reserves in the surface soils. When the soil water reserve is not limiting, Black et al. (1969) applied a flow equation to arrive at the following equation

$$\Sigma E = \alpha(t - t_1)^{1/2}, \quad (13)$$

where E is the cumulative evaporation, t_1 the time for the first stage evaporation, and α the characteristic coefficient which is dependent upon soil hydraulic properties, and ranges from 0.3–0.5 cm day$^{-1/2}$ (Ritchie 1972). The theoretical analysis by Black et al. (1969) shows that ΣE is proportional to the square root of diffusivity, indicating that α can be normalized by diffusivity or hydraulic conductivity. The magnitude of evaporation from the soil surface is 2 to 3 cm for the period of about 3 weeks with the largest loss from loam. The first stage evaporation accounts for 30 to 40% of such losses.

When brackish water is used for irrigation, there is a good possibility of forming a thin layer of salt crusts, which reduces water evaporation (e.g., Zawislanski et al. 1992). However, the effect of such a crust is likely to be minimal in the second stage evaporation where soil water supply rates to the surface play the dominant role of governing evaporation.

e. *Spatial variability*. Soils, especially those which have developed in river basins are spatially variable. Table 12 shows examples of spatial variation in soil properties. Bulk density and saturation water contents are among the lowest in variability with the coefficient of variation (CV) usually less than 10%. Sand, silt and clay contents and soil water storage at 10 and 1500 kPa (0.1 and 15 bar) are among the medium variability parameters with the CV ranging from 20 to 50%. Soil salinity and sodicity are also the medium variability parameters in irrigated fields consisting of the same soil type. Otherwise, the variability in salinity increases greatly. Soil hydraulic conductivity, either saturated or unsaturated, is among the most variable properties of soils with its CV often exceeding 100%.

Low to medium variability parameters usually conform to the normal frequency distribution. The sampling requirements (N) can then be computed as

$$N = (\tau/\delta)^2(CV)^2 \quad (14)$$

where τ is the normalized deviate for a given confidence level (e.g., $\tau = 1.96$, 1.64 and 1.28 at confidence levels of 5, 10 and 20% respectively), δ is the deviate range from the true mean. If one chose a 5% confidence level at 15% deviation from the true mean, the first term is 170. The sampling requirements for the CV of 0.1, 0.2, 0.3, 0.4 are 1.7, 6.8, 15, and 27, respectively. For practical soil evaluation, a lower confidence level, e.g., 20% and a larger deviate (e.g., 20%) can be used.

The methods of soil characterization outlined above are meant to be a guideline to attain some degree of standardization. Soil-related data, especially infiltration and drainage rates, are critical for assessing sustainability of irrigated production with highly saline water. These data, along with plant and water-related data provide a basis to develop appropriate irrigation scheduling. The topic of seasonal water use, irrigation scheduling and management is, however, beyond the scope of this chapter and is discussed elsewhere (Miyamoto 1993).

3.3 *Effects of brackish water and culture on soil properties*

a. *Salinity*. The use of brackish water for irrigation can increase or decrease soil salinity de-

Table 12. Examples of spatial variation in soil properties (Warrick & Nielsen 1980; Miyamoto & Cruz 1986/87)

Parameters	Mean	CV (%)	Comments
Low variation			
Bulk density ($g\,cm^{-3}$)	1.3	7	Torrif. 15 ha 5 series
	1.4	7	Torrif. 150 ha
Saturation content	40	11	Torrif. 0.2 ha
Medium variation			
Sand, silt, clay	–	28–36	Torrif. 15 ha 5 series
0.1/15 bar cont. ($g\,100\,g^{-1}$)	23/7	17–50	Torrif. 15 ha 5 series
Soil salinity ($dS\,m^{-1}$)	1.5–4.5	17–50	Torrif. Basin 1–26 ha
Soil sodicity (SAR)	2.8–17	10–30	Torrif. Basin 1–26 ha
Soil salinity ($dS\,m^{-1}$)	2	19–26	Torrif. Furrow 150 m
	4–12	17–24	Torrif. Furrow 150 m
Soil sodicity (SAR)	6–18	21–30	Torrif. Furrow 150 m
High variation			
Sat. conductivity ($cm\,hr^{-1}$)	14	190	Torrif. 15 ha 5 series
	20	110	Torrif. 150 ha 30 cm
	35	86	Torrif. 150 ha 6 depths
Unsat. conductivity ($cm\,hr^{-1}$)	0.6	280	Torrif. 150 ha 30 cm
	4.9	170	Torrif. 150 ha 6 depths

pending on the initial salinity. Salinity of irrigation water as well as the leaching fraction control the level of salt accumulation in the soils. If the assumed leaching fraction is attained, salinity of the soils should approach the projected levels (Table 10) in a matter of one growing season or less, except at low leaching fractions (<0.2).

Since soil properties are spatially variable, salt accumulation in soils will occur in spatially variable fashion, starting with less permeable portions of the field. Such sections can be diked and leached during the off-season, or may be deep-plowed or subsoiled to improve water penetration and drainage. With intensive mechanized cultivation, soils tend to lose permeability over the years, mainly due to compaction (e.g., Meek et al. 1992). This can lead to higher levels of salt accumulation in soils, unless the compacted layer is alleviated. The presence of high water tables is another common reason for salt accumulation.

b. *Sodicity*. The sodicity of the upper root zone will approach the *SAR* of irrigation water. This process usually takes a longer period than the time required for salinity equilibrium. Sodicity of the lower root zone changes more slowly than does at the surface layer, due to the translocation of cations replaced from the exchange sites of the upper root zone. Eventually, sodicity of the lower root zone exceeds sodicity of the irrigation water due to the concentration effect.

$$ESP = (1/LF)^{1/2} SAR_I, \qquad (15)$$

where *ESP* is the exchangeable sodium percentage, and SAR_I the sodium adsorption ratio of the irrigation water. In addition, Ca precipitation associated with transpirational losses of water causes *SAR* to increase (Miyamoto & Pingitore 1992).

Sodicity as well as salinity of the soil surface are usually many times greater than those within the root zone as a result of salt deposition following water evaporation. As discussed in Section 3.2, the evaporation from soil surfaces ranges from 2–3 cm. The salt deposition occurs at the surface 2–3 mm. Thus, salinity of the salt deposition sites is at least 10 times greater than those of the soil moisture present below the surface. During this process, Ca ions precipitate as $CaCO_3$ or $CaSO_4\ 2H_2O$ (gypsum), forming a white salt crust. The precipitation of Ca will increase *SAR* and *ESP* of the thin surface layer by many fold (Miyamoto & Cruz 1987).

c. *Soil structure and permeability*. The use of brackish water, as discussed briefly in Section 3.1, should not cause structural deterioration in most cases, due to stabilization effects of high electrolyte concentrations. However, the use of brackish water will have pronounced effects on structural stability of the soil surface when brought into contact with rainfall or low salt water. In addition, there is a possibility that dispersed clay particles may translocate and plug soil pores, thereby affecting the infiltration of brackish water as well. An exception to this rule would be the use of gypsiferous brackish water which forms gypsum at the soil surface.

The extent of infiltration reduction caused by surface seal is difficult to predict, as many factors affect the formation of the surface seal. These include soil texture, aggregate stability, salinity of the soil surface, and the extent of physical disturbance to the surface soils, including disking and rainfall impacts. Bare soils which are pulverized by extensive disking or tillage are most conducive to forming the surface seal. The soils with minimum tillage or under a grass cover are least susceptible to sodicity effects. Likewise, the soils which develop structural cracks are less sensitive to sodicity effects, as infiltration takes place predominantly through the cracks (e.g., Pandey 1985). Also sodicity effects on hydraulic conductivity are comparatively small in unsaturated soils (Russo & Bresler 1977). Establishment of sodicity appraisal guidelines requires field infiltration tests using local soils and prevailing management practices.

There are several remedies to Na-induced infiltration problems. Light application of gypsum prior to a rain season has been a common practice. Topdressing of soils with coarse sand or organic matter may offer an alternative in an area where gypsum can not be obtained readily. Acharya & Abrol (1976), for example, reported that placing sand on a sodic soil with *ESP* of 95% at a thickness of 1, 2 and 3 cm increased the saturated hydraulic conductivity by 48, 63 and 97 times, respectively, while no improvement in hydraulic conductivity was observed when the sand was incorporated into the soil to a depth of 7 cm. This finding may further substantiate that the flow limiting layer is at the soil surface. It is also known that vegetative covers help improve soil structural stability (e.g., Meek et al. 1992; Singh et al. 1990; Yadav 1978).

d. *Salt precipitation and pore-plugging*. A reduction in soil permeability caused by salt precipitation in water conductive pores is another potential problem of using carbonaceous or gypsiferous saline water. In irrigated alfalfa production, Frenkel et al. (1978) estimated that 10% of the effective pore space present in a root zone depth of 60–90 cm may be occupied by precipitated gypsum and calcium carbonate after about 14 years when irrigated with water containing 20 meq L^{-1} of Ca and SO_4 and 9.2 meq L^{-1} of HCO_3. The estimates made by Miyamoto (1992) for several saline waters in the Southwestern US show that the time required to occupy 10% of the pore space in a layer thickness of 30 cm ranges from 26–40 years, a range longer than the estimate by Frenkel et al. (1978). However, with gypseous water, the pores may be occupied by 10% in a matter of 8–12 years, because of the formation of gypsum. In oceanic-type brackish water, salt precipitation potential is minimal.

It is difficult at present to estimate the effect of pore space reduction on soil permeability. If the precipitation occurs in large pores, the effect can be significant. Keren et al. (1980) observed a 50% reduction in saturated hydraulic conductivity when analytical grade gypsum powder was mixed into sandy loam and loam soils at 3% by weight. The effect of pore plugging on unsaturated flow should be much smaller. However, these estimates indicate that additional research is needed to evaluate drainage impairment by precipitated salts. This problem is especially important in tree crops, as conventional subsoiling can not be used readily.

3.4 *Drainage water characterization and handling*

Many halophytes native to salt marsh can tolerate high water tables. However, drainage is still required to keep salinity in balance, especially when highly saline water is used for irrigation. For halophytes not native to salt marsh, the water table probably should be kept below at least 1 m from the soil surface as for most conventional crops. In some instances, this may require artificial drainage. If an artificial drainage system is to be installed, one must recognize that the drainage water carries large quantities of salts. When

Table 13. The US EPA criteria for protection of saline and fresh water aquatic species, and livestock

Element	Salt water aquatics		Fresh water aquatics		Livestock
	Acute	Chronic	Acute	Chronic	
As	–	–	–	–	20
Cd	43	9.3	3.9	1.1	50
Cr	1100	50.0	16	11.0	1000
Cu	2.9	2.9	18	12.0	50
Hg	2.3	0.025	2.4	0.012	1
Ni	140	71	1800.	96.0	–
Se	410	54.0	260.0	35.0	50
Ag	2.3	–	4.1	0.12	–
Zn	170	58.0	320.0	47.0	25000

brackish water containing 10 g L^{-1} of salts is used at an annual rate of 1 m, the drainage water contains approximately 100 tons of salts that need to be disposed of annually per ha. This quantity of salts is greater than the total dry matter of any crop per year. If this quantity of salts is deposited as halite in an evaporation basin, it can cover 1 ha at a thickness of 5 mm.

Another potential effect of brackish water irrigation is drainage water enrichment by nutrient elements and trace elements. Leaching of NO_3 and P is of particular concern when the drainage water is disposed into estuaries or into surface streams, as nutrient elements enhance the growth of saline algae and may cause undesirable changes in aquatic ecology. A greater concern may relate to the evaporative concentration of certain trace elements, especially Se, Cd, As, Cu, Hg, and Ag. These elements are toxic to many aquatic species as well as to livestock (Table 13), and can potentially become a factor of determining disposal options (Letey et al. 1987; Westcot 1988). At the same time, emerging research points to the fact that halophytes may take up a considerable portion of Se and possibly other elements from soil solutions (e.g., Banuelos & Meek 1990). It is entirely possible that halophyte cultivation can actually reduce the concentration of some of these elements through biofiltration. Extensive studies are needed to examine this possibility.

If drainage water from irrigated halophyte fields contain these toxic trace elements at objectional levels of concentrations, the disposal into estuaries, saline streams and wetlands may face difficulties. Evaporation ponds such as used in California would be cost prohibitive, unless markets for evaporites are developed. One potential market for evaporites is a saline solar pond. These options are discussed elsewhere (Lee 1990; Miyamoto 1992).

4. Potential system options for irrigated production

Irrigated halophyte production with brackish water may require unique production systems and cultural skills at least equal to those required for conventional crops. Some of the reasons for high management requirements are as follows:

1. The ability of many halophytes to germinate under saline conditions does not seem to be any higher than conventional crops, yet salinity of water used for irrigation would be greater by many fold.
2. The ability of halophytes to survive high salinity is remarkably high, easily exceeding the salinity of seawater. However, saline tolerance for optimum growth and perhaps for economic production is well below that of seawater for most halophytes, thus requiring proper maintenance of soil salinity.
3. The osmotic adjustment is a form of saline adaptation mechanism with many halophytes, and helps maintain growth under high salinity, yet can reduce forage quality through excessive uptake of salts, and possibly certain toxic trace elements.
4. The quantity of salts that must be leached out of the root zone from brackish water irrigation appears to be in the order of 100 tons annually, and this requires careful consideration for disposal.
5. The corrosion to concrete and metallic irrigation hardware is likely to be greatly higher with brackish water than with low salt water.

Any irrigated production system using halophytic plants and brackish water must take these constraints or characteristics into consideration. Seed treatments, special cultural techniques, the use of low salt water and/or sprinklers may help overcome the potential difficulty of halophyte establishments with brackish water. Careful soil selection, precision irrigation management, and periodic salt leaching operations should help maintain soil salinity within the range for optimum or economic production. Excessive salt and possibly toxic element uptake into plants could be dealt with through blending with fodder grown with low salt water or through a mixed planting with halophytes which excrete salts.

4.1 *Monoculture*

Production of a single crop in a given field is a standard practice in most mechanized farming. This allows standardization in many phases of production activities, thus leading to a greater degree of management control and production efficiency. Many conventional forage crops are grown in this fashion, and there seems to be no particular reason why some of the halophytic fodders can not be grown using a similar system, especially in the field abandoned solely because of high salinity of irrigation water.

The selection of irrigation systems will depend on soil properties, cropping patterns, water supply capability, and certainly financial consideration. Sprinklers are ideally suited for fodder crop production, especially in sandy soils. In the US, the center pivot system is becoming a standard, mainly because of low labor requirements, minimal requirements for land preparation, high distribution efficiency and reasonable costs. This system appears to be operating successfully in a pilot project involving *Salicornia* production using seawater in Saudi Arabia (Jim Riley, personal communication). In small tract irrigation, other forms of sprinklers are suited. Portable sprinklers are especially convenient for crop establishment in sandy fields. Sprinklers, especially those which form mists can induce foliar salt damage (e.g., Moore & Murphy 1979; Ehling & Bernstein 1959). Streak forming sprinkler nozzles or drop tubes offer an alternative to the conventional sprinkler nozzles for reducing foliar-induced salt damage (Lyle & Bordovsky 1983). When drop tube irrigation is combined with furrow diking, rainfall retention in the field can be maximized (Lyle & Dixon 1977)

Furrow irrigation as compared to either border or basin is more versatile. One of the advantages of this method is to allow the use of deep seeding and decapping techniques used for crop establishment (Miyamoto et al. 1984). In this method, the seed is planted deep, and salt crust is removed mechanically prior to hypocotyl elongation. This method provides a high degree of soil moisture conservation and salinity control during the crop establishment. Another advantage of furrow method is to be able to apply relatively shallow depths of irrigation water. The use of skip-row irrigation or surge flow techniques are especially effective in providing a shallow application of water. The skip-row irrigation is also useful for controlling salts during crop establishment, as the lateral wetting front moves across the seeded zone (Miyamoto et al. 1984). In a region with a distinct rainy season, this method is also useful for retaining rainfall in the field by means of furrow dike systems, although infiltration problems caused by the use of saline water must be dealt with.

4.2 *Multicrop culture*

Planting of two or more crops in the same field (intercropping or mixed planting) has been practiced for various reasons, including intensification of cropping, and land and water uses. This system has advantages and disadvantages, and must be targeted to achieve a particular objective. Although multicrop culture systems have not yet been used for brackish water irrigation, the following options may warrant some investigations.

a. *Strip culture*. Planting of two or occasionally three crops in alternate rows (commonly referred to as intercropping) has been practiced, primarily for furthering cropping intensity and, in some instances, to provide nitrogen by interplanting with leguminous species. Strip culture is a form of intercropping, except that the strips are wide-enough to retain a degree of independence. An example of this system is to plant strips of drought tolerant crops between irrigated cash crops requiring high moisture levels. The drought tolerant crops are ordinarily not irrigated and use lateral seepage from the adjacent irrigated strip. Under high water table situations, the drought tolerant plants may extract water from the water table. In essence, the drought tolerant crops are used to make use of 'excess' soil water which is otherwise wasted or becomes drainage water. This system is adaptable to a situation where land is plentiful, but water supply is limited and drainage water handling is costly.

A cropping system similar to the above could be developed for planting halophytes (preferably deep-rooting types) in a series of strips along with conventional crops which can tolerate moderate levels of salinity. Irrigation water is assumed to be applied to the conventional crops, and occasionally to the halophyte strips for salt leaching purposes. The main function of the halophytes planted in strips is to reduce drainage and to utilize shallow saline water if and when the water table is elevated. If the main crop happens to be a fodder crop, this system may also allow a blending of halophyte fodder with traditional fodder to attain desired fodder quality for feeding.

b. *Silvipasture*. Mixed planting of trees and pasture grass has been practiced for various reasons, including the supply of tree fodder when pasture production is low, supply of fuelwood or simply for providing shade and shelters for grazing animals. Other reasons include the optimization of water, land and nutrient uses (e.g., Shankarnarayan & Singh 1991). Deep-rooted trees may extract water which had percolated through a shallow root zone of grasses.

A production system similar to the above could be applied to surface irrigated halophytic grasses and trees. The main function of deep-rooted halophytic trees is to make use of soil water percolating out of the shallow root zone of grasses. Tree planting densities can be adjusted according to the projected area of deep percolation, e.g., high densities near unlined ditches, near water check-in points in surface irrigation methods, or in sandy portions of the spatially variable field, and low densities where minimal percolation losses are expected. There are many tree species which can develop root systems deeper than grasses, provided that the field consists of deep well-drained soils. If the soil is shallow or irrigation rates per application are low, trees are likely to compete with grass species for water. A successful experimental planting of *Prosopis juliflora* with *Diplachne fusca* is, for example, reported by Singh et al. (1990, 1991).

The above production system could be modified to strip planting under high water table conditions. Trees once established under irrigation could draw water from the water table as irrigation water supplies are curtailed. The row of trees planted in nonirrigated strips can be effective in extracting water from the table, because of the potentially large exposed leaf surfaces. An attempt to lower water tables through silviculture is reported by Heuperman et al. (1987), and Greenwood (1986). One of the constraints of this production scenario is salt tolerance of trees, which is not always high. Certain *Atriplex* species may offer an alternative (Greenway & Beresford 1980).

4.3 *Gradient culture*

The term 'gradient culture' is used to express cropping patterns which conform to salinity gradients commonly existing in natural topography. This concept is already in use in many river basin development projects where salt sensitive crops are grown upstream, and salt tolerant species downstream where salinity is typically higher. This concept may also work well in closed basin developments. Crops of low salt tolerance may be grown in high ground, and salt tolerant species in lower ground, and the drainage from the field irrigated with the highest salinity water may be disposed of to the bottom of the basin. This production system is in conformity with plant ecology, but may not be compatible with the requirements for wildlife habitats, if the drainage water contains elevated levels of toxic substances. The investigation of drainage water quality is prerequisite for a project involving surface disposal of drainage water.

The gradient culture concept can also be applied to individual farm settings, at least in theory. Drainage water from one field could be used to irrigate adjacent fields planted with crops of higher salt tolerance, and so on. The problem with this scheme is the uncertainty involved in collecting drainage water from the small targeted area. In most instances, drainage water collected into subsurface drains does not necessarily reflect the seepage from the particular field, but rather reflects the subsurface water from greater areas, especially in alluvial plains. The reliable development of gradient culture may require a scale of an irrigation district.

The production systems discussed above are some of the potential options. To make irrigated halophyte production economical and environmentally sound, innovative production systems must be developed. This requirement is especially high in existing irrigated areas suffering from the problems of high water tables and costly saline drainage water handling and disposal.

5. Future prospects

The success of irrigated production of halophytic plants for fodder will depend on continued efforts of selecting and breeding halophytic crops. There is already a considerable amount of information on halophytes which became or could become fodder crops, and additional species are likely to be added to the existing lists. However, breeding efforts certainly need to be intensified. From a broader prospective, halophyte germplasm may provide useful salt tolerant genes for genetic engineering research.

The introduction of halophytes will depend to a large extent upon socio-economic needs. Opportunities exist in the arid and semi-arid subtropics where population pressure and excessive grazing activities are causing not only the chronic shortages of fodder, arable land and forestland, but also soil erosion and related problems. Supplemental irrigation with saline agricultural drainage water or shallow tube wells in semi-arid climates may prove cost effective and more sustainable than a large irrigation scheme involving mining of deep saline water. Such attempts are already underway in several countries and seem to hold promise.

The introduction of halophytes into commercial agriculture has been viewed with a great deal of skepticism, partly due to many difficulties encountered in cultivating traditional crops with saline water and partly due to the history of chronic surplus of many commodities. However, opportunities seem to exist for halophyte cultivation in once-irrigated, but subsequently abandoned land due to salinization of irrigation water and/or soils. Such abandoned irrigated land is considerable in size and many more areas are approaching the same state (e.g., Dudal & Purnell 1986). To carry out efficient irrigated production, however, considerable amounts of agronomic evaluation must be carried out, in addition to selection and breeding efforts.

The utilization of saline agricultural drainage water for irrigated production of halophytes can offer another unique opportunity not only for producing additional fodder, but also to improve our ability to manage drainage water. Various techniques to use or reuse agricultural drainage water for irrigation of traditional crops has been developed to a great extent (e.g., Rhoades et al. 1988). However, such a scheme can not easily break the practical salinity limit of 6,000 mg L^{-1}, principally because of the limited saline tolerance of most traditional crops and the lack of incentive. This leaves a large volume of drainage water yet to be disposed of. Some of the halophytic plants can readily play a role in reuse of drainage water resulting from the cultivation of traditional crops. However, the development of innovative production systems is a prerequisite. One obvious role of halophyte cultivation is to reduce the quantity of drainage water that must be disposed of. In addition, many halophyte species have a potential to remove appreciable amounts of salts, ranging from 5–10 tons ha^{-1}. Perhaps, an even more important role of halophyte cultivation may prove to be the uptake of trace elements which are toxic to aquatic species and waterfowls, such as Hg, Ag, Cd, Cu and Se. Aquatic species can usually adapt to elevated levels of salinity, but less so to toxic chemicals. There is a strong need to study the role of halophytes in removing toxic trace elements from saline drainage water and to develop a safe and rational way of blending the halophytes with conventional fodders.

Irrigation return flow, saline streams and subsurface saline flow often constitute the largest salt input into many rivers in arid and semi-arid regions (e.g., Hedlund & Evans 1990). The diversion of such saline streams is often an effective salinity control strategy, but has met with various constraints. Irrigated production of halophytes could make the diversion strategy more attractive in some areas. Irrigated production of proven species such as *Distichlis* and *Salicornia* coupled with some breeding efforts seem to offer a realistic potential for diversion flow irrigation. Likewise, planting of halophytic trees and shrubs may help reduce saline seepage into the streams. The macro-economic gain resulting from this type of operations can be substantial with increasing economic values of fresh water streams.

Irrigated production of halophytes with coastal saline water has been advocated for some time as a means of coastal desert developments. The coastal sand dune in particular has been considered ideal by many, because of the ease of salt leaching. The development of modern irrigation equipment, especially the center pivot sprinkler has brought this potential one step closer to a reality. Another potential advantage is the relative ease of drainage handling, because of the proximity to the ocean, and low potential of encountering toxic trace elements, except for Mo, and in some instances B which may be present in coastal saline water at elevated levels. There seems to be a great deal of need to study N utilization, transformation and leaching in order to minimize the ecological disturbance of the coast through disposal of drainage water from high leaching irrigation.

While irrigated production of halophytes with brackish water can present a localized impact (either negatively or positively) on water quality, a broad picture of this endeavor may prove to be environmentally and ecologically sound. Many of the soil erosion and land degradation problems occurring in subtropic deserts are associated with overgrazing and excessive tree cuttings. Irrigated production of fodder and trees in relatively small confined areas might help minimize the ongoing large scale degradation of land and forest resources. A greater picture may relate to the concern over the CO_2 balance in the atmosphere. Improved management of forestland in humid regions can provide additional CO_2 sequestering, yet it is widely viewed that such an effort alone would be inadequate, yet alternatives appear to be costly. Some estimates indicate that an additional forest area of a few million km^2 may be required for substantial CO_2 sequestration (e.g., Sedjo 1989). Currently, desert lands carry minimal storage of organic carbon, thus placing vegetative life will greatly increase organic C storage (e.g., Glenn et al. 1992b). The area of extreme desert land is estimated to be 2.9 million km^2 along the coast, and 4.3 million km^2 in inland (UNESCO 1979), and halophyte production in these deserts may complement or even offer an alternative to reforestation for sequestering CO_2 if seawater irrigation is developed to a realm of economic feasibility.

There is increasing evidence and recognition that future soil and land management must incorporate not only the economic uses of water and land resources, but also local as well as global environmental and ecological protection. We have not yet arrived at the comprehensive understanding of the role of halophytic crops in this broadened charge of water and land resources management. However, it seems to be obvious that this challenge can be met more effectively through incorporation of halophytes in our agenda rather than excluding them.

Acknowledgement

This paper is a contribution of the Texas Agricultural Experiment Station. The acquisition of halophyte research information from India was made possible through the US-India Scientist Exchange Program administered by the National Science Foundation and the help from Dr J.C. Dagar, Senior Scientist at the Indian Soil Salinity Research Institute. Travel grants from UNEP facilitated participation of the authors in the Nairobi workshop.

Appendix: The conversion of salinity units

To convert a concentration unit of $mg\,L^{-1}$ into $mmol\,L^{-1}$ or $meq\,L^{-1}$, the concentrations must be divided by the molecular or equivalent weights of applicable salt elements or compounds shown in Table 14. To convert the concentration (C) in

Table 14. Molecular weights of common salts and the conversion coefficients

	Molecular weight	Conversion to EC (dSm^{-1})		
		a_0	a_1	a_2
NaCl	23 + 35.5 = 58.5	−0.863	0.941	0
$MgCl_2$	24.3 + (35.5)2 = 95.3	−0.863	0.931	0
$CaCl_2$	40.1 + (35.5)2 = 111.1	−0.867	0.935	0
Na_2SO_4	(23)2 + 96 = 142.0	−0.956	1.407	−0.045
$MgSO_4$	24.3 + 96 = 120.3	−0.920	0.897	−0.027
$CaSO_4$	40.1 + 96 = 136.1	−0.887	0.840	0

meq L^{-1} into an electrical conductivity (EC) unit in dS m^{-1}, the following conversion equation is provided, which is simply a numerical presentation of the information given in USDA Handbook 60 by the US Salinity Laboratory Staff (1954).

$$EC = 10^{a_0 + a_1 c + a_2 c}$$

$$c = \log_{10}(C).$$

The numerical coefficients a_0, a_1, and a_2 are shown in Table 14. When Cl is the predominant anion, a simplified relation is $EC(\text{dS m}^{-1}) = 0.10 \times (\text{meq L}^{-1})$. However, this relationship overestimates EC by 10 to 15% when the concentrations exceed approximately 100 meq L^{-1}. Another approximate conversion equation is mg $L^{-1} = 635 \times EC$ in dS m^{-1} (USSL 1954). This conversion coefficient varies considerably with ionic compositions and concentrations of irrigation water.

References

Abbas, J.A. & El-Oqlah, A.A. (1992). Distribution and communities of halophytic plants in Bahrain. J. Arid Environ. 22: 205–218.

Abu-Sharar, T.M., Bingham, J.T. & Rhodes, J.D. (1987). Stability of soil aggregates as affected by electrolyte concentration and composition. Soil Sci. Soc. Am. J. 51: 309–314.

Acharya. C.L., & Abrol, I.P. (1976). Effect of river sand on the permeability of a sodic soil. J. Indian Soc. Soil Sci. 24: 245–252.

Ashby, W.C. & Beadle, N.C.W. (1957). Studies in halophytes, III: Salinity factors in the growth of Australian saltbushes. Ecology 38: 344–352.

Ashok Kumar & Abrol, I.P. (1986). Grasses in alkali soils. CSSRI, Karnal, India, Bulletin 2.

Ayers, R.S. & Westcot, D.W. (1985). Water quality for agriculture. FAO Rome, Publication 29.

Banuelos, G.S. & Meek, D.W. 1990. Accumulation of selenium in plants grown on selenium-treated soils. J. Environ. Qual. 19: 772–777.

Bayoumi, M.T. & El Shaer, H.M. (1994). Impact of halophytes on animal health and nutrition (this volume, Ch. 25).

Bernstein, L. (1964). Salt tolerance of plants. USDA, Inf. Bull. 283.

Bernstein, L. & Francois, L.E. (1973). Leaching requirement studies: Sensitivity of alfalfa to salinity of irrigation and drainage waters. Soil Sci. Soc. Am. Proc. 37: 931–943.

Biezok, I. (1972). Concrete corrosion: Concrete protection. Hungarian Acad. of Sci. Budapest, 500 pp.

Bingham, J.T. & Garber, M.J. (1970). Zonal salinization of the root system with NaCl and boron in relation to growth and water uptake of corn plants. Soil Sci. Soc. Am. Proc. 34: 122–126.

Black, T.A., Gardner, W.R. & Thurtell, G.W. (1969). The prediction of evaporation, drainage and soil water storage for a bare soil. Soil Sci. Soc. Am. Proc. 33: 655–660.

Bondurant, J.A. (1957). Developing a furrow infiltrometer. Agr. Eng. 38: 602–604.

Boyko, H. (1966). Salinity and aridity. W. Junk Publishers, The Hague.

Burman, U., Kathju, S., Garg, B.K. & Lahiri, A.N. (1991). Water management of transplanted seedlings of *Azadirachta indica* in arid areas. Forest Eco. & Mgt. 40: 51–63.

Cajuste, L.J., Carrillo, G.R., Cota, E. & Laird, R.J. (1991). The distribution or metals from wastewater in the Mexican Valley of Mezquital. Water, Air & Soil Pollution 58: 763–771.

Chaney, R.L. 1983. Potential effects of waste constituents in the food chain. In: Parr, J.F., Marsh, P.B. & Kla, J.M. (eds.), Land Disposal of Hazardous Wastes. Noyes Data Corp., Park Ridge, NJ, pp. 152–240.

Clarke, F.E. (1980). Corrosion and encrustation in water wells: A field guide for assessment, prediction and control, FAO, Rome, Irrig. & Drainage Paper 34, 95 pp.

Clemmens, A.J. (1983). Infiltration equations for border irrigation methods. In: Adv. in infiltration. Proc. Nat. Conf. in Adv. in Infil. ASAE, pp. 266–274.

Dhir, R.P. (1977). Saline waters: Their potentiality as a source for irrigation. In: Jaiswal, P.L. (ed.), Desertification and its control. Indian Council of Agr. Res., New Delhi, pp. 130–148.

Dudal, R. & Purnell, M.F. (1986). Land Resources: Salt affected soils. Reclamation and Reveg. Res. 5: 1–9.

Ehling, C.F. & Bernstein, L. (1959). Foliar absorption of sodium and chloride as a factor in sprinkler irrigation. Am. Soc. Hort. Sci. 74: 661–670.

El Shaer, H.M. (1981). A comparative nutrition study on sheep and goats grazing southern Sinai desert range with supplements. PhD Thesis. Ani Shams University, Egypt.

El Shaer, H.M. & Gihad, E.A. (1994). Halophytes as animal feed in Egyptian deserts (this volume, Ch. 28).

Eyre, A.L. (1985). Population pressure on arid lands; Is it manageable? In: Proc. Inter. Res. & Develop. Conf. Arid lands, today and tomorrow. Tucson, AZ, Oct. 20–25. Westview Press, Boulder CO.

Felker, P., Cannell, G.H., Osborn, J.F., Clark, P.R. & Nash, P. (1983). Effects of irrigation on biomass production of 32 prosopis (Mesquite) accessions. Expt. Agri. 19: 187–198.

Felker, P., Clark, P.R., Laag, A.E. & Pratt, P.J. (1981). Salinity tolerance of the tree legumes: Mesquite (*Prosopis glandulosa* var. Torreyana, *P. velutina* and *P. articulata*), Algarrobo (*P. chilensis*), Kiawe (*P. pallida*) and Tamarugo (*P. tumaugo*) grown in sand culture on N-free media. Plant & Soil 61: 311–317.

Flowers, T.J. (1972). Salt tolerance in *Suaeda maritima* (L.) Dum. J. Expt. Bot. 23: 310–321.

Flowers, T.J., Troke, P.F. & Yeo, A.R. (1977). The mechanism of salt tolerance in halophytes. Ann. Rev. Plant Physiol. 28: 89–121.

Francois, L.E. (1986). Salinity effects on four arid zone plants (*Parthenium argentatum*, *Simmonderia chinesis*, *Kochia prostrata* and *Kochia brevifolia*). J. Arid Environ. 11: 103–109.

Frenkel, H., Goertzen, J.O. & Rhoades, J.D. (1978). Effects of clay type and content, exchangeable sodium percentage, and electrolyte concentration on clay dispersion and soil hydraulic conductivity. Soil Sci. Soc. Am. J. 42: 32–39.

Frenkel, H., Hadas, A. & Jury, W.A. (1978). The effect of salt precipitation and high sodium concentrations on soil hydraulic conductivity and water retention. Water Resour. Res. 14: 217–221.

Frenkel, H. & Shainberg, J. (1975). Irrigation with brackish water: Chemical and hydraulic changes in soils irrigated with brackish water under cotton production. In: Irrigation with brackish water. The Negev. Univ. Press, pp. 175–183.

Fukushima, M., Iahizaki, A., Sakamoto, A. & Kobayashi, M. (1973). Cadmium concentration in rice eaten by farmers in the Jinzu River Basin. Jap. J. Hyg. 28: 406–415.

Gale, J. & Poljakoff-Mayber, A. (1970) Interrelations between growth and photosynthesis of salt bush (*Atriplex halimus* L.) grown in saline media. Aust. J. Biol. Sci. 23: 937–945.

Gale, J., Naaman, R. & Poljakoff-Mayber, A. (1970). Growth of *Atriplex halimus* L. in sodium chloride salinated culture solutions as affected by the relative humidity of the air. Aust. J. Biol. Sci. 23: 947–952.

Gallagher, J.L. (1985). Halophytic crops for cultivation at seawater salinity. Plant & Soil 89: 323–336.

Gardner, W.R. (1983). Soil properties and efficient water use: An overview. In: Taylor, Jordan & Sinclair (eds.), Limitations to efficient water use in crop production. Am. Soc. Agron. Madison, WI, pp. 45–64.

Gausman, H.W., Cowley, W.R. & Barton, J.H. (1954). Reaction of some grasses to artificial salinization. Agron. J. 46: 412–414.

Gihad, E.A. & El Shaer, H.M. (1994). Utilization of halophytes by livestock on rangelands: Problems and prospects (this volume, Ch. 6).

Glenn, E.P. (1987). Relationship between cation accumulation and water content of salt tolerant grasses and a sedge. Plant, Cell & Environment 10: 205–212.

Glenn, E.P. & O'Leary, J.W. (1984). Relationship between salt accumulation and water content of dicotyledenous halophytes. Plant, Cell & Environment 7: 253–261.

Glenn, E.P. & O'Leary, J.W. (1985). Productivity and irrigation requirements of halophytes grown with seawater in the Sonoran Desert. J. Arid Environ. 9: 81–91.

Glenn, E.P., O'Leary, J.W., Watson, M.C., Thompson, T.L. & Kuehl, R.O. (1991). *Salicornia bigelovii* Torr.: An oilseed halophyte for seawater irrigation. Science 251: 1065–1067.

Glenn, E.P., Watson, M.C. & O'Leary, J.W. (1992). Salt tolerance and osmotic adjustment of low-sodium and high-sodium subspecies of the halophyte, *Atriplex canescens*. Plant, Cell & Environment 15: 711–718.

Glenn, E P., Fontes, M.R., Katzen, S. & Colvin, L.B. (1982). Nutritional value of halophytes grown on hypersaline seawater. In: San Pietro, A. (ed.), Biosaline research: A look to the future. Plenum Publishing Corp.

Glenn, E.P., Swingle, R.S., Riley, J.J., Mota, C.U., Wilson, M.C. & Squires, V.R. (1994). North American halophytes: Potential use in animal husbandry (this volume, Ch. 12).

Glenn, E., Hodges, C.N., Lieth, H., Pielke, R. & Pitelka, L. (1992). Growing halophytes to remove carbon from the atmosphere. Environment 34: 40–43.

Greenway, H. (1968). Growth stimulation by high chloride concentrations in halophytes. Israel J. Bot. 17: 169–177.

Greenway, H. (1973). Salinity, plant growth and metabolism. J. Aust. Inst. Agri. Sci. 39: 24–34.

Greenwood, E.A.N. (1986). Water use by trees and shrubs for lowering saline groundwater. Reclam. & Reveg. Res. 5: 423–434.

Greenwood, E.A.N. & Beresford, J.D. (1980). Evaporation from vegetation in landscapes developing secondary salinity using the ventilated-chamber technique II. Evaporation from *Atriplex* plantations over a shallow saline water table. J. of Hydrology 45: 313–319.

Grillot, G. (1956). The biological and agricultural problems presented by plants tolerant of saline or brackish water and the employment of such water for irrigation. In: Utilization of saline water. Unesco, Paris.

Gupta, R.K. & Saxena, S.K. (1968). Resource survey of *Salvadora oleoidis* and *S. persica* for nonedible oil in western Rajasthan. Tropical Ecology 9: 140–152.

Haigh, A.H., Barlow, E.W.R. & Milthorpe, F.L. (1986). Field emergence of tomato, carrot and onion seeds primed in an aerated salted solution. J. Am. Soc. Hort. Sci. 111: 660–665.

Hedlund, J.H. & Evans, W. (1990). Project level salinity management options. In: Tanji, K. (ed.), Agricultural Salinity Assessment and Management. ASCE.

Heuperman, A.F., Stewart, H.T.L & Wildes, R.A. (1987). The effects of Eucalyptus on water table depths in an irrigated area of Northern Victoria, Australia. In: Rana, R.S. (ed.), Afforestation of salt-affected soils. Central Soil Salinity Res. Institute, India, Vol. 3, pp. 151–161.

Higgins, G.M., Kassam, A.H., Naiken, L., Fisher, G. & Shah, M.M. (1982). Potential population support capabilities of land in the developing world. FAO, Rome, Tech. Report Project INT 175, 13 pp.

Hoffman, G.J. & Shannon, M.C. (1986). Relating plant performance and salinity. Recl. & Reveg. Res. 5: 211–225.

Hussain, M. & Hussain, A. (1970). Tolerance of *Diplachne fusca* Beauv. (Australian grass) to salt and alkali. Directorate of Land Reclamation. Lahore Research Publications, Vol. 2. No. 25.

Jain, B.L., Goyal, R.S. & Muthana, K.D. (1983). Performance of some tree species in relation to irrigation with saline waters. Annals of Arid Zone 22: 233–238.

Jain, B.L., Muthana, K.D. & Goyal, R.S. (1985). Performance of tree species in salt-affected soils in arid regions. J. Indian Soc. Soil Sci. 33: 221–224.

Jefferies, R.L. & Pitman, M.G. (1986). Perspectives of the biology of halophytes in natural habitats in relation to forage production. Reclam. & Reveg. Res. 5: 227–244.

Joshi, A.J. (1979). Physiological studies on some halophytes. PhD Thesis, Saurashtra Univ. India.

Julander, R.P. & Jackson, W. (1983). Drop formers and double ring infiltrometers: a comparison. In: Adv. in Infiltration. Proc. Nat. Conf. on Adv. in Infil. ASAE, pp. 249–253.

Kaddah, M.T. (1976). Subsoil chiseling and slip plowing on a stratified fine sandy soil. Agron. J. 68: 36–39.

Keren, R., Kriet, J.F. & Shainberg, I. (1980). Influence of size of gypsum particles on the hydraulic conductivity of soils. Soil Sci. 130: 113–17.

Koocheki, A. & Mohalati, M.N. (1994). Feed value of some halophyte range plants of arid regions of Iran (this volume, Ch. 21).

Khoshoo, T.N. (1986). Environmental priorities in India and sustainable development. Presidential Address, Proc. 73rd Indian Science Congress New Delhi.

Kubota, J. & Allaway, W.H. (1972). Geographic distribution of trace element problems. In: Micronutrients in Agriculture. Soil Sci. Soc. Am., Madison, WI, pp. 525–554.

Le Houérou, H.N. (1986). Salt tolerant plants of economic value in the Mediterranean Basin. Reclam. & Reveg. Res. 5: 319–341.

Lee, E.W. (1990). Drainage water treatment and disposal options. In: Tanji, K. (ed.), Agricultural salinity assessment and management. ASCE, No. 71.

Letey, J., Roberts, C., Penberth, M. & Vasek, C. (1987). An agricultural dilemma: Drainage water and toxic disposal in the San Joaquin Valley. Special Publication 3319. Univ. of Calif. 56 pp.

Llerena, F.A. (1994). Massive propagation of halophytes (*Distichlis spicata* and *Tamaris* spp.) on the highly saline-alkaline soils in the ex-Lake Texcoco, Mexico (this volume, Ch. 30).

Lugg, D.G. (1983). The potential of saltgrass as a forage grass irrigated with saline water. Water Resour. Res. Inst. Report 162 NMSU, Las Cruces, NM.

Lyle, W.M. & Bordovsky, J.P. (1983). LEPA irrigation system evaluation. Trans. ASAE 26: 776–781.

Lyle, W.M. & Dixon, D.R. (1977). Basin tillage for rainfall retention. Trans. ASAE 20: 1013–1021.

Malcolm, C.V. (1986). Production from salt affected soils. Reclamation and Reveg. Res. 5: 343–361.

Malcolm, C.V. (1994). Use of halophyte forages for rehabilitation of degraded lands (this volume, Ch. 4).

Malik, K.A., Aslam, Z. & Naqri, M. (1986). Kallar grass, a plant for saline land. Nuclear Institute for Agriculture and Biology. Pakistan.

Malkin, E. & Waisel, Y. (1986). Mass selection for salt resistance in Rhodes grass (*Chloris gayana*). Physiol Plant 66: 443–446.

Meek, B.D., Rechel, E.R., Carter, L.M., DeTar, W.R. & Urie, A.L. (1992). Infiltration rate of a sandy loam soil: Effects of traffic, tillage and plant roots. Soil Sci. Soc. Am. J. 56: 908–913.

Mikhiel, G.S., Meyer, S.E. & Pendleton, R.L. (1992). Variation in germination response to temperature and salinity in shrubby *Atriplex* species. J. Arid Environ. 22: 39–49.

Miyamoto, S. (1978). Tolerance of some southwestern range plants to sodium chloride and sulfate. Texas Agric. Expt. Sta. Prog. Report 3480.

Miyamoto, S. (1980). Effect of bicarbonate on sodium hazard of irrigation water: Alternative formulation. Soil Sci. Soc. Am. J. 44: 1079–1083.

Miyamoto, S. (1989). Salt effects on germination, emergence and seedling mortality of onion. Agron. J. 81: 202–207.

Miyamoto, S. (1990). A model to estimate salt accumulation and crop yields in surface-irrigated orchard. Proc. Nat. Irrig. Conf., Phoenix, AZ, 384–389, ASAE Publication 04-90.

Miyamoto, S. (1992). Potentially beneficial uses of saline water resources in the Southwestern USA. In: Leith & Massoum (eds.), Rational use of high salinity tolerant plants. Vol. 2. Kluwer Academic Publishers, Dordrecht, pp. 407–422.

Miyamoto, S. (1993). Water use and irrigation scheduling of halophytes. In: Proc. Inter. Course on Halophyte Utilization in Agriculture. Agadir, Morocco. June 1993.

Miyamoto, S. & Cruz, I. (1986). Spatial variability and soil sampling for salinity and sodicity appraisal in surface irrigated orchards. Soil Sci. Soc. Amer. J. 50: 1020–1026.

Miyamoto, S. & Cruz, I. (1987). Spatial variability of soil salinity in furrow-irrigated torrifluvents. Soil Sci. Soc. Am. J. 51: 1019–1025.

Miyamoto, S., Glenn, E.P. & Olson, M. (1994). Growth, water use and salt uptake response of four halophytes to irrigation with brackish water. J. Arid Environments (in press).

Miyamoto, S. & Pingitore, N. (1992). Predicting Ca and Mg precipitation in evaporating saline solutions. Soil Sci. Soc. Am. Proc. 56: 1767–1775.

Miyamoto, S., Piela, K. & Petticrew, J. (1985). Salt effects on germination, emergence and hypocotyl mortality of selected vegetable crops and guayule. Irrig. Sci. 6: 159–170.

Miyamoto, S., Piela, K. & Petticrew, J. (1986). Seedling mortality of several crops induced by root, stem or leaf exposure to salts. Irrig. Sci. 7: 97–106.

Miyamoto, S., Stichler, C. & Moore, J. (1984) An overview of saline water irrigation in far west Texas. In: Reppogle, J. (ed.), Proc. Irrig. & Drainage Specialty Conf., ASCE, Flagstaff, p. 222.

Mohammed, S. & Sen, D.N. (1988). A report on polymorphic seeds in halophytes, I: *Trianthema Triquetra* (L.) in Indian desert. Curr. Sci. 57: 616–617.

Mohammed, S. & Sen, D.N. (1990). Germination behavior of some halophytes in Indian desert. Ind. J. Expt. Biology. 28: 545–549.

Moore, J. & Murphy, J.M. (1979). Sprinkler irrigation with saline water. In: Miyamoto, S. (ed.), Proc. Inter. Amer. Conf. on Salinity and Water Management Techn. Texas A&M Univ. Agr. Res. Ctr. El Paso. Dec., pp. 213–221.

Moore, R.T., White, R.S. & Caldwell, M.M. (1972). Transpiration of *Atriplex confertifolia* and *Eurotia lanata* in relation to soil, plant and atmospheric moisture stress. Can. J. Bot. 50: 1411–1418.

Mozafar, A., Goodin, J.R. & Oertli, J.J. (1970). Na and K interactions in increasing salt tolerance of *Atriplex halimus*

L.; Yield characteristics and osmotic potential. Agron. J. 62: 478–481.

Mualem, Y. (1976). A new model for predicting the hydraulic conductivity of unsaturated porous media. Water Res. Res. 12: 513–522.

Mudie, P.J. (1974). The potential economic uses of halophytes. In: Reimold & Queen (eds.), Ecology of Halophytes. New York/London, Academic Press, pp. 565–597.

National Research Council (1980). Mineral tolerance of domestic animals. Nat. Acad. Sci. Washington, DC, 577 pp.

National Research Council (1990). Saline agriculture; salt-tolerant plants for developing countries. National Academy Press, Washington, DC.

Ng, B.H. (1987). The effect of salinity on growth, nodulation and nitrogen fixation of *Casuarina equisetifolia*. Plant & Soil 103: 123–125.

O'Leary, J.W. (1984). The role of halophytes in irrigated agriculture. In: Staples, R.C. (ed.), Salinity tolerance in plants: Strategies for crop improvement. John Wiley & Sons. New York, pp. 285–299.

O'Leary, J.W. (1986). A critical analysis of the use of *Atriplex* species as crop plants for irrigation with highly saline water. In: Prospects for biosaline research. Univ. of Karachi, Karachi, Pakistan.

O'Leary, J.W., Glenn, E.P. & Watson, M.C. (1985). Agricultural production of halophytes irrigated with sea water. Plant & Soil 89: 311–321.

Okusanya, O.T. (1977). The effect of seawater and temperature on the germination behaviors of *Crithmum maritimum*. Physiol. Plant. 41: 265–267.

Pandey, R.S. (1985). Effect of cracks on infiltration behavior in salt-affected soils. J. Indian Soc. Soil Sci. 33: 137–140.

Philip, J.R. (1957a). The theory of infiltration: 4. Sorptivity and algebraic infiltration equations. Soil Sci. 84: 257–264.

Philip, J.R. (1957b.) Evaporation and moisture and heat fields in the soil. J. Meteorol. 14: 354–366.

Philipp, K.R. & Gallagher, J.L. (1985). Evapotranspiration from two potential halophytic crop species. In: Adv. in evapotranspiration. Proc Nat. Conf. on Evapotrans. ASAE Dec. 16–17, pp. 259–261.

Pratt, P.F. (1972). Quality criteria for trace elements in irrigation waters. Calif. Agr. Expt., 46 pp.

Pratt, P.F., Albasel, N., Joseph, H. & Resco, C. (1988). Trace element guidelines for irrigation waters in the San Joaquin Valley. Report to the Calif. Water Res. Control Board.

Pratt, P.F. & Suarez, D.L. (1990). Irrigation water quality assessments. In: Tanji (ed.), Agricultural Salinity Assessment and Management. ASCE, pp. 220–236.

Reboredo, F. (1991). Cu and Zn uptake by *Halimione portulacoides* (L.) Aellen. A long term accumulation experiment. Bull. Environ. Contam. Toxicol. 46: 442–449.

Rhodes, D. & Felker, P. (1988). Mass screening of *Prosopis* (Mesquite) seedlings for growth at seawater salinity concentrations. Forest Ecol. and Mtg. 24: 169–176.

Rhoades, J.D., Bingham, F.T., Letey, J., Pinter, Jr., P.J., Lemert, R.D., Alves, W.J., Hoffman, G.J., Replogle, J.A., Swain, R.V. & Pacheco, P.G. (1988). Reuse of drainage water for irrigation: Results of imperial valley study. Hilgardia 56: 1–44.

Rhoades, R.J. & Miyamoto, S. (1990). Testing soils for salinity and sodicity. In: Soil testing and plant analysis. ASA Publication, Madison, WI.

Ritchie, J.T. (1972). Model for predicting evaporation from a row crop with incomplete cover. Water Res. Res, 8: 1204–1213.

Ritchie, J.T. & Burnett, E. (1971). Dryland evaporative flux in a sub-humid climate, II: Plant influences. Agron. J. 63: 56–62.

Rosenfeld, L. & Beath, O.A. (1964). Selenium. Academic Press, New York, 411 pp.

Russo, D. & Bresler, E. (1977). Effect of mixed Na/Cu solutions on the hydraulic properties of unsaturated soils. Soil Sci. Soc. Am. J. 41: 713–717.

Ryan, J., Miyamoto, S. & Stroehlein, J.L. (1975). Salt and specific ion effects on germination of four grasses. J. Range Mgt. 28: 61–64.

Sandhu, G.R., Aslam, Z., Salim, M., Sattar, A., Qureshi, R.H., Ahmad, N. & Wyn Jones, G. (1981). The effect of salinity on the yield and composition of *Diplachne fusca* (Kallar grass). Plant Cell and Environ. 4: 177–181.

Sandhu, S.S. & Abrol, I.P.: (1981). Growth responses of *Eucalyptus tereticornis* and *Acacia nilotica* to selected cultural treatments in highly sodic soil. Indian J. Agr. Sci. 51: 437–443.

Sankary, M.N. (1986). Species distribution and growth on salt affected land in Syria. Reclam. & Reveg. Res. 5: 125–141.

Sedjo, R.A. (1989). Forests: A tool to moderate global warming? Environment 31: 14–21.

Shainberg, I. & Letey, J. (1984). Response of sodic and saline conditions. Hilgardia 52: 1–57.

Shankarnarayan, K.S. & Singh, K.C. (1991). Rehabilitation of degraded pasture land in the arid region of Rajasthan. In: Abrol & Narayara (eds.), Technologies for wasteland development. ICAR.

Sharma, M.L. (1982). Aspects of salinity and water relations of Australian chenopods. In: Tasks for vegetation science, Vol. 2, pp. 155–195.

Shannon, M.G. (1978). Testing salt tolerance variability among tall wheatgrass lines. Agron. J. 70: 719–722.

Sharma, S.K. & Gupta, R.K. (1971). Effects of salts on seed germination of some desert grasses. Annals of Arid Zone 10: 33–36.

Sharma, T.P. & Sen, D.N. (1989). A new report on abnormally fast germinating seeds of *Haloxylon* spp. – An ecological adaptation to saline habitat. Curr. Sci. 58: 382–385.

Shaw, R.J. & Thorburn, P.J. (1985). Prediction of leaching fraction from soil properties, irrigation water and rainfall. Irrig. Sci. 6: 73–83.

Sinha, B.K. & Singh, N.T. (1974). Effect of transpiration rate on salt accumulation around corn roots in a saline soil. Agron. J. 66: 557–560.

Singh, G., Abrol, I.P. & Cheema, S.S. (1990). Effects of irrigation on *Prosopis juliflora* and soil properties of an alkali soil. Inter. Tree Crop. J. 6: 81–99.

Singh, G., Gill, H.S., Abrol, I.P. & Cheema, S.S. (1991). Forage yield, mineral composition, nutrient cycling and

ameliorating effects of Karnal grass grown with mesquite in a highly alkali soil. Field Crop Research 26: 45–55.
Singh, K., Yadav, J.S.P. & Singh, V. (1991). Tolerance of trees to soil salinity. J. Indian Soc. Soil Sci. 39: 549–556.
Singh, K., Yadav, J.S.P. & Singh, V. (1992). Tolerance of trees to soil sodicity. J. Indian Soc. Soil Sci. 40: 173–179.
Squires, V.R. (1993). Australian experiences with high salinity diets for sheep. In: Lieth & Masoom (eds.), Towards the rational use of high salinity tolerant plants, Vol. 1, pp. 449–457. Kluwer Academic Publishers, Dordrecht.
Stumpf, D.K., Prisco, J.T., Weeks, J., Linley, V. & O'Leary, J. (1986). Salinity and *Salicornia bigelovii*. Torr. seedling establishment. J. Exp. Bot. 37: 160–169.
Swartzendruber, D. & Olson, T.C. (1961). Model study of the double-ring infiltrometer as affected by depth of wetting and particle size. Soil Sci. 92: 219–225.
Swingle, R.S., Glenn, E.P. & Riley, J.J. (1994). Halophytes in mixed feeds for livestock (this volume, Ch. 7).
Tanji, K.A., Lauchli & Meyer, J. (1986). Selenium in the San Joaquin Valley. Environment 28: 6–11.
Taylor, R.M. & Lankford, M.K. (1969). A technique for controlled high salt germination. In: Abst. West Sec. Crop. Sci. Soc. Ann. Mtgs. Reno NV, Western Sec. Crop. Sci. Soc., p. 24.
Taylor, S.A. & Ashcroft, G.L. (1972). Physical edaphology. Freeman and Company, San Francisco.
Tigma, M., Gay, C.W. & Banner, R.E. (1991). The effects of chiseling and sulfur fertilization on infiltration, soil moisture, and vegetation parameters in the Ait Rbaa Perimeter of Central Morocco. Arid. Soil Res. & Rehab. 5: 271–288.
Tomar, O.S. & Yadav, J.S.P. (1985). Effect of saline water irrigation and fertilizer application on growth of tree seedlings. Annals of Arid Zone 24: 94–100.
Tomar, O.S. & Gupta, R.K. (1985). Performance of some forest tree species in saline soils under shallow and saline water-table conditions. Plant & Soil 87: 329–335.
Twitchell, L.T. (1955). Germination of fourwing saltbush seed as affected by soaking and chloride removal. J. Range Mgt. 8: 218–220.
UNEP (1991). Status of desertification and implementation of the UN Plan of Action to Combat Desertification. UNEP, Nairobi, Kenya.
Unesco (1979). Map of the world distribution of arid regions. Tech. Note 7, Unpub.
Unger, I.A. (1982). Germination ecology of halophytes. In: Sen & Rajpurohit (eds.), Tasks for Vegetation Science. W. Junk Publishers. The Hague, The Netherlands, Vol. 2, pp. 143–154.
US Salinity Laboratory Staff (1954). Diagnosis and improvement of saline and alkali soils. USDA, Agric. Hndbk. 60.
Van der Moezel, P.G., Watson, L.E., Pearce-Pinto, G.V.N. & Bell, D.T. (1988). The response of six *Eucalyptus* species and *Casuarina obesa* to the combined effect of salinity and waterlogging. Aust. J. Plant Physiol. 15: 465–474.
Van Genuchten, M.Th. (1980). A closed form equation for predicting the hydraulic conductivity of unsaturated soils. Soil Sci. Soc. Am. J. 44: 892–898.
Villalobos, F.J. & Fereres, E. (1990). Evaporation measurements beneath corn, cotton and sunflower canopies. Agron. J. 82: 1153–1159.
Walker, W.R. & Willardson, L.S. (1983). Infiltration measurements for simulated furrow irrigation. In: Adv. in infiltration. Proc. Nat. Conf. on Adv. in Infil. ASAE, pp. 241–248.
Warrick, A.W. & Nielsen, D.R. (1980). Spatial variability of soil physical properties in the field. In: Hillel (ed.), Appl. of soil phy. Academic Press, New York, pp. 319–344.
Watson, C., O'Leary, J. & Glenn, E.P. (1987). Evaluation of *Atriplex lentiformis* (Torr.) s. Wats. and *Atriplex nummularia* Lindl. as irrigated forage crops. J. Arid Environ. 13: 292–303.
Webb, K.L. (1966). NaCl effects on growth and transpiration in *Saliconia bigelovii*, a salt-marsh halophyte. Plant & Soil 24: 261–268.
Weiler, G.J. (1980). The use of saline water to establish fourwing saltbush and blue grama on topsoiled mine spoil. MS thesis. New Mexico State Univ., Las Cruces, NM.
Westcot, D.W. (1988). Reuse and disposal of high salinity subsurface drainage water – A review. Agr. Water Mgt. 14: 483–511.
WFC (1991). Hunger and malnutrition in the world. FAO, Rome, Italy.
Williams, M.D., & Ungar, I.R. (1972). The effect of environmental parameters on the germination, growth, and development of *Suaeda depressa* (Pursh) Wats. Am. J. Bot. 59: 912–918.
Yadav, J.S.P. (1978). Afforestation of salt-affected soils. In: Land water management in the Indus Basin (India). Proc. Nat. Symp.
Yadav, J.S.P. (1980). Salt-affected soils and their afforestation. Indian Forester. 106: 259–271.
Zawislanski, P.T., Tokunaga, T.K., Benson, S.M., Oldfather, J.M. & Narasimhan, T.N. (1992). Bare soil evaporation and solute movement in selenium contaminated soils of Kesterson Reservoir. J. Environ. Qual. 21: 447–457.

6. Utilization of halophytes by livestock on rangelands
Problems and prospects

E.A. GIHAD[1] & H.M. EL SHAER[2]
[1]*Faculty of Agriculture, Cairo University, Giza, Egypt;*
[2]*Desert Research Centre, Mataria, Cairo, Egypt*

Abstract. In arid zones and other dry lands halophytic plants often dominate. They are considered as an important source of nutrients for most classes of wild and domestic ruminants. Palatable halophytic species are generally rare but where present are a valuable feed resource, especially during drought. There are numerous halophytes which are less palatable which can be used as fodder after appropriate treatment. Feeding halophytes to ruminant livestock requires special precautions since the plants can impact on animal physiology and behavior. This paper reviews the literature on utilization of halophytes and covers the main constraints to feeding halophytic plants. Factors which affect the nutritive value include palatability and intake. Both are due to interactions between plant and animal factors. Some approaches to better utilization of halophytes as fodders and potential research topics for development of livestock production from halophytic rangelands are discussed.

Key words: Antinutritional properties, Chemical composition, Fodder, Forage, Intake, Nutritive value, Review article, Ruminants

Introduction

The rapidly growing demand for animal production, i.e. milk, meat, wool in most countries, particularly developing ones and a parallel increase in imports of livestock and their products have drawn attention to the need for intensification of efforts to develop the indigenous livestock and feedstuff resources. The shortage of animal feeds is the main constraint to increase indigenous animal productivity. Improvements in the utilization of fibrous feed resources in such countries, primarily the natural vegetation, might be a solution.

Animal husbandry, as the principal source of income for nomads, is based mostly on the natural vegetation for rearing sheep, goats, camels and some wild herbivores. Many rangelands are dominated by halophytes some of which are xerohalophytes, hygrohalophytes, or meso-euhalophytes (Batanouny 1994).

Halophytic plants are widely distributed throughout the world particularly in arid and semi-arid areas. Palatable plant species are few and are always over-grazed and disappear fast. Less palatable species are numerous and have a patchy distribution.

Unpalatable halophytes are widely distributed and represent about 60–70% of the natural rangeland vegetation in several parts in the world. Many species are highly resistant to salinity and drought – attributes which improve their survival in deserts which represents a high proportion of the developing world in the mid-latitude regions (Figure 1). Halophytes such as *Atriplex* spp., *Nitraria retusa*, *Salsola* spp. and *Suaeda* are generally considered to be extremely valuable as a fodder reserve during drought.

Despite the fact that halophytes have attracted the interest of scientists for many years knowledge about their role in animal husbandry is extremely limited.

V.R. Squires & A.T. Ayoub (eds.), *Halophytes as a resource for livestock and for rehabilitation of degraded lands*, 77–96.

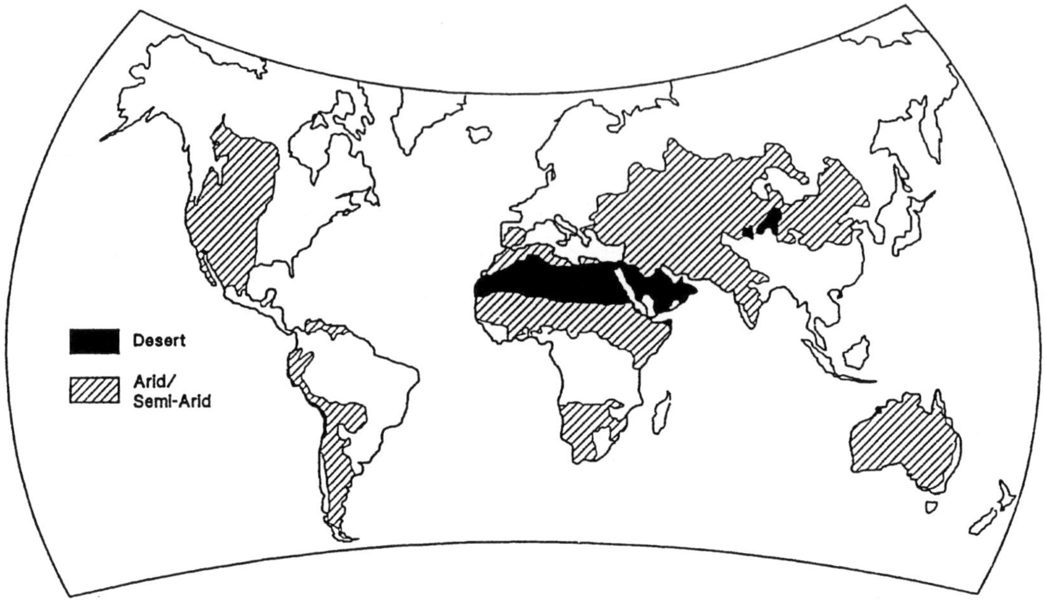

Figure 1. Sketch map of the drylands of the world (based on a map prepared by UNEP).

The present article reviews the research literature relating to the role of halophytes in feeding ruminant livestock in arid and semi-arid regions. Special emphasis is given to understanding the role of halophytes as a potential source of animal feed including consideration of problems and precautions necessary in their utilization by livestock.

Factors influencing grazing and nutritive value

Factors that affect the nutritive value of rangelands and grazing of various plant species are numerous. We focus here on plant species and varieties, stage of growth and season of use, environmental factors, range site, stocking rate and kind of grazing animals.

Plant species and ecotypes

Chemical composition, nutritive value and productivity of halophytic ranges vary greatly depending on the dominant plant species. Ecotypes within species generally differ only to a small extent in nutritive value and chemical constituents if the comparison is made at the same stage of growth and physiological development. Differences between species and genera may be larger especially in chemical composition and organic and dry matter digestibility.

Plants in the Chenopodiaceae family show an interesting geographical distribution, which is determined by the fact that they are almost all halophytic plants (Rizk 1986). The chenopod shrubs are characterized by extensive stands in which the shrubs are generally less than 1.5 meters high, they remain green during drought and may serve as a valuable forage resource either when no other feed is available, or during times of seasonal drought.

The most widespread genera of importance in terms of providing forage for livestock are *Atriplex*, *Salsola* and *Suaeda*. There are about 400 species of *Atriplex* in the world, mostly in temperate, sub-tropical and mediterranean zones between 20° and 50° of latitude north and south (Le Houérou 1991). Although the saltbushes (*Atriplex*) are characterized by their high content of NaCl, they are said to be useful as fodder plants.

Table 1. Average proximate analysis, dry matter digestibility (DMD) and nutritive values (%) of common halophytic species

Plant species		DM composition							
	DM	Ash	CP	CF	EE	NFE	DMD	TDN	DCP
Alhagi bulbosa	40	27.1	8.75	30.0	3.10	31.05	43.7	52	6.6
A. mourorum	44	25.9	9.45	29.5	4.42	30.73	46.5		
A. pseudalhagi	29	9.5	16.1	27.2	5.40	43.60		58	7.2
A. sparsifolia	32	13.7	10.5	25.1	2.35	48.35			
Arthrocnemum glaucum	22.9	31.9	3.38	12.1	1.26	51.36	50.4		
Atriplex angulata	38	17.7	15.2	12.9	3.60	50.60	62.7	54	10.2
A. barclayana		31.2	9.38				67.1		
A. canescens	32	19.6	14.2	18.3	6.1	41.80		61	10.2
A. farinosa	48.9	34.6	8.13	16.8	2.32	38.15		46.6	5.62
A. glauca	42	24.3	12.9	12.1	2.90	47.80		61	9.0
A. halimus	34.2	22.7	12.6	25.4	2.28	37.02	59	60	14.9
A. leucoclada	25.6	31.7	15.1	27.4	2.69	23.11	59.7	56	12.6
A. nummularia	24	23.2	17.2	19	4.9	35.70	57.2	51	13.4
A. semibaccata	34	17.5	14.0	20.6	3.60	44.30		59	10
A. vesicaria	32	30	12.2	13.3	4.0	40.50		51	8.4
Desmastachgo bipinnata	57	19.4	1.41	25.5	4.52	49.17		48.9	
Diplachne fusca		11.6	8.6	28.2	2.65	48.95	56.2		
Halocnemom strobiclaceum	55	27	6.41	26	1.92	35.67	63.7		
Haloxylon salicornicum	42.4	15.9	14.8	24.1	5.11	39.09	47.0	58	5.5
Juncus acutus	35	12.3	7.11	28.5	2.35	49.94	34.4		
J. rigidus	91	13	9.4	35	1.90	40.70		61	9.0
Limoniastrum monopetalum	48.6	23.6	11.5	14.6	3.49	46.81	55.5		
Limonium pruanosum	66.4	29.5	5.30	17.3	3.34	44.56	43.5	52.3	4.39
Nitraria retusa	27.4	23.8	11.4	22.6	4.15	38.05	62.8		
Phragmites spp.		8.4	10.8	41.5	1.10	38.20		55	7.1
Phragmites communis		7.5	10.0	38.1	1.71	42.69		53.4	6.9
Salicornia fructicosa	37.6	14.3	13.5	18.9	18.9	51.41	70.5		
Salsola kali	89.8	12.2	15.0	24.9	2.00	44.90	60.5		11.2
S. rigida		15.2	12.2	22.5	2.60	47.50		61	12.6
S. sieberi	61	20.9	7.7	28.9	1.4	41.10		52	4.2
S. tetrandra	51	19.7	8.79	21.2	4.79	45.62	49.2	51	8
S. vermiculata		22.1	13.1	36.1	0.60	38.10		42.0	7.7
Suaeda fruticosa	31.8	19.1	13.7	15.2	6.07	45.13	70.4		
S. maritima	28	36.3	20.7	10.3	3.01	29.60		49.0	16.3
S. pruinosa	26.6	26.4	7.79	12.9	4.93	47.98			
Tamarix aphylla	51	19.7	8.79	21.1	4.79	45.62	48.7		
T. mannifera	40	24.9	8.19	11.6	3.57	51.74	57.8	54.8	3.21
Zygophyllum album	38	12	9.4	28.2	2.9	47.50	66.2	59	5.8
Clover (Berseem) hay	92.1	14.8	13.3	30.7	2.40	39.9	53.6	52.0	11.7
Rice straw	91.7	18.6	3.4	39.5	1.40	37.1		43.0	0.80

Sources: IFI (1979); Anon. (1981, 1992); El Shaer (1981, 1986); Arieli et al. (1989); Bhattacharya (1989).

According to the chemical composition (Table 1) several species plants have a relatively high nutritive value for ruminants and may be useful as a feed supplement (Ueckert et al. 1990). The highest forage values are found during the wet season of the year when plants are green and actively growing (Chatterton et al. 1971; Kandil & El Shaer 1988, 1990). Several species of the genus *Salsola* have also received considerable attention. Most species of this genus have been successfully harvested and have value as a forage when cultivated forages are limited or unavailable. They attain relatively high levels of crude protein (CP) as shown in Table 1. The protein has an amino acid (AA) composition which was nutritionally well balanced. *Suaeda* species is considered one of the most palatable chenopod shrubs (El Shaer 1986).

In comparison with other shrub species all chenopod shrubs contain high concentrations of digestible crude protein (DCP) and Na, K, and chloride ions relative to other pasture grasses and clover hay or straw. The acceptability of most chenopod shrubs to domestic livestock is moderate to low when grasses and herbs are available but it increases as these components become scarce (Wilson 1966a,b). Under these conditions the concept of species palatability is of little value (see below).

Other forages which grow on saline soils in the desert are *Juncus* and *Diplachne* spp. The nutritive constituents (amino acids, carbohydrates and lipids) of certain *Juncus* species (family Juncaceae) have been reported (Anon. 1992). Field experiments with *J. rigida* and *J. acutus* indicate that these plants can be successfully cultivated on saline soils. Both plants have agro-industrial economic potential because of the rich chemical composition of their seeds. The two halophytes, *J. subulatus* and *Diplachne fusca* may be considered as non-conventional forage plants with nutritive values comparable to the clover hay (4th cut) based on results obtained by Anon. (1992).

Stage of growth

The stage of maturity at time of harvesting or consumption seems to be one of the most important factors affecting the feeding value. The relationship between nutritive value and seasonal growth of rangeland halophytes is derived from data of El Shaer (1981) for several halophytes grown naturally in Sinai (Figure 2). Most forages are high in nutrients and are palatable during early growth stages. The opposite is true during late stages of growth. Some plant species mature rather rapidly and as a result decrease substantially in nutritive value; other species mature rather slowly and consequently remain high in nutrient content over an extended period (Cook 1971).

Crude fiber, silica, cell wall constituents (CWC), cellulose, lignin and DM content of halophytes increase while CP, phosphorus and gross energy decrease with advancing maturity. Moreover, the process of ageing and maturation of pasture is associated also with a decline in organic matter digestibility (OMD) dry matter digestibility (DMD) and protein digestibility (Dietz 1971; Abd El Aziz 1978; El Shaer & Kandil 1990). The process of maturation is associated with the lower palatability and digestibility and consequently low feeding values. The increase in the proportions of structural to nonstructural carbohydrates accompanied by a partial lignification of the cell-wall constituents associated with ageing is likely to be the main reason for the reduction in the rate of digestibility (Balch & Campling 1962). Photosynthetic products are more rapidly converted to structural compounds thus having the effect of decreasing protein and soluble carbohydrates and increasing the structural CWC. Moreover, under poor growth condition, halophytic plants may synthesize substances, e.g. lignins, tannins, alkaloids etc. or grow thorns which make them less accessible or unpalatable (El Monayeri et al. 1986; Bayoumi & El Shaer 1994).

Season of use

Animals grazing green forages during the wet season receive adequate nutrients and forage. However, during dry seasons where forages are dry and mature, the diets are thought to be deficient in several nutrients and animals may need to be supplemented (see below). Rainfall and drought directly affect forage yields, nutrients and grazing pressure on various range sites.

Since the chemical make-up of the halophytic plant is directly related to its palatability (Malecheck & Provenza 1983) and in particular to the presence of what are termed secondary plant metabolites, rangeland plants become less palatable under drought conditions. Browse species (mainly halophytes) can furnish livestock with green foliage during drought, when there is little grass growth.

There is a general agreement that the nutritive value of range plants decrease as a result of water shortage (Hassan et al. 1980). This decrease in nutritive value is due in great part to a decrease in CP content together with an increase in the fiber component (Van Soest 1982). The increase in non-protein nitrogen (NPN) as a result of water shortage (Wilson 1966a,b) could better explain

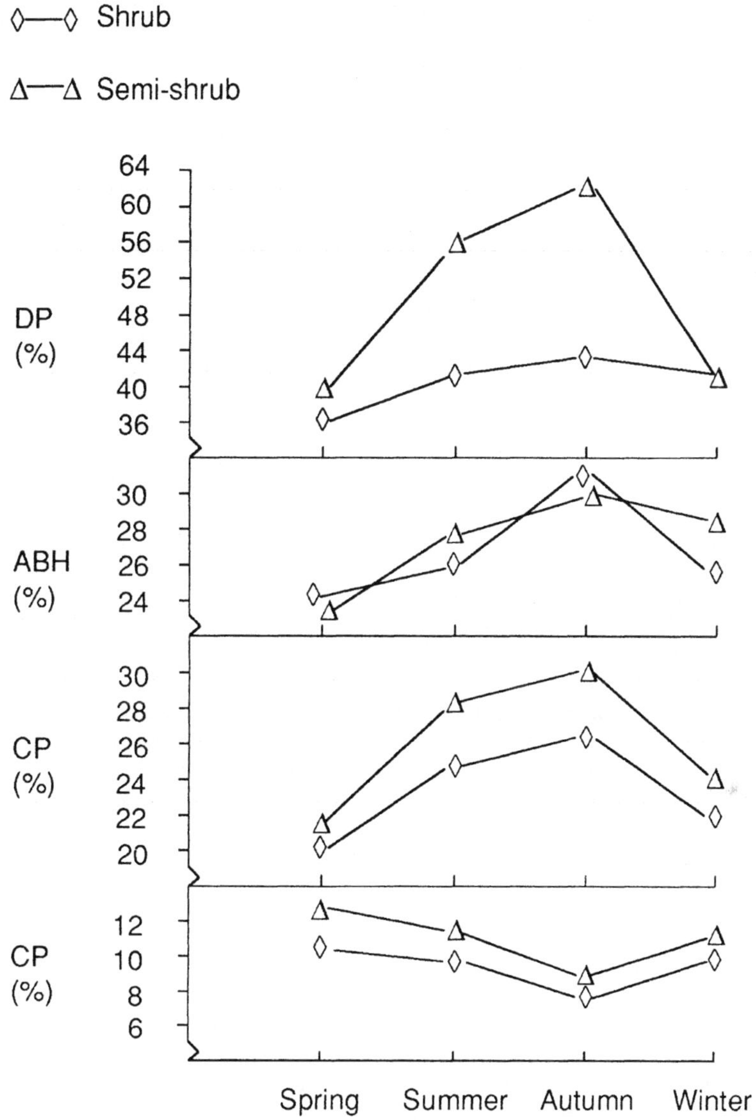

Figure 2. Seasonal variation in the chemical constituents (dry matter basis) of some halophytic plants in the Sinai peninsula. The plants were a mixture of *Atriplex halimus*, *Salsola tetrandra*, *Tamarix aphylla*, *Zygophyllum album*, *Suaeda fruticosa* and *Nitraria retusa* (Source El Shaer 1981).

the reduction in digestibility. Shrubs like *Atriplex* grown under water stress contain as high as 20% CP (Wilson 1977). Yet more than 60% of that protein is in the form of NPN which results in DMD as low as 40–45% (Benjamin et al. 1986). Water stress and extremely high temperature, as in drought seasons, are contributing factors to the low protein content of halophytes during the different stages of growth. Drought sharply reduces the forage supply from grasses and forbs and forces herbivores to subsist largely on browse. Shrubs furnish emergency forage under such conditions (Wilson & Harrington 1984).

Environmental factors

Soil and climate are the important environmental factors in determining the nutrient content of range forages under various site conditions. The effect of specific soil attributes upon the chemical composition of plants is difficult to determine because of the many interactions and confounding factors. Salinity, structure, texture, organic matter content, soil organisms and chemical composition are the most important soil factors (Szabolcs 1994). The type of the soil affects the types of rangeland and its nutritive value. Soil salinity and soil moisture are the main factors affecting halophytic rangeland plant species and their palatability to livestock (Gihad 1993).

Climatic factors, i.e. temperature, humidity, precipitation and light intensity play an important role in controlling the chemical content and nutritive value of plants as they affect assimilation, photosynthesis and metabolism.

Location

Site conditions are extremely important as they influence the growth characteristics of range plants and thus, indirectly, their nutritive value. These include soil fertility, plant development, water runoff, intensity of shade and other environmental factors. Cook (1959) reported that plants growing on favorable sites had a greater tolerance to grazing than plants growing on unfavorable sites. Therefore, poor sites frequently are expected to display poorer range condition than good range sites under the same system of management. Both carbohydrates and ash contents were significantly higher in the entire plant on favorable sites (Osman 1969; Cook 1971).

Differences in adjacent site conditions markedly affected the palatability of plants of the same species growing on both sites (Cook 1971). Degree of utilization of plants growing on poor sites was substantially greater than plants growing on adjacent favorable sites. This was believed to be a result of plant structure and to leaf:stem ratio and nutrient content which was higher compared to plant material on favorable sites. The effect of range sites on the chemical composition and nutritive value of range forage has been studied. For example, Abd El-Aziz (1982) and El Bassousy (1984) studied the effect of range site along 600 km with 50 km depth throughout the north coast of Egypt. Nine forage locations were included in these studies where range plants were mostly halophytic plants. They found that all forage locations had significant effects on the main nutrients and digestibility of range forages.

Stocking rate and kinds of grazing herbivore

Stocking rate (the amount of land allocated to each animal unit for the entire grazable period of the year) is the single most important factor influencing range usage. Heavy grazing by domestic animals leads to reduced forage production from preferred plant species. It causes the animals to alter their grazing patterns to include progressively less palatable species until the final stages of the range abuse force utilization of all species. Continuous heavy stocking is most destructive to the range. Overstocking and early grazing, together with fuel collection and excessive plowing of the rangelands for vegetable and fruit planting are the main reasons for such deterioration in much of North Africa. The long term production potential of the rangeland resources is being destroyed.

Some livestock species exercise greater dietary selection than others. For example, the camel feeds on plants or parts of plants not eaten by more conventional livestock (Wilson 1984). Its size enables it to browse higher than cattle and small ruminants, thus contributing further to reduced competition and to much enhanced complementary use in mixed species guilds. Goats place stress on the most desirable vegetation as they are more selective than other livestock and utilize greater diversity of plants than sheep or camels (Abdullahi et al. 1987; Gihad et al. 1988). Livestock, especially camels, prefer halophytes because of their high salt and water content (Wardeh et al. 1990). Therefore, stocking rangelands with more than one animal species enables more total live weight to be maintained.

Constraints to the utilization of halophytes as animal feeds

The feeding value of any rangeland is generally expressed in terms of voluntary intake and digestible nutrients per unit of feed. There are several factors which could considerably limit the feeding values of halophytic plants such as physical and chemical defenses, lignification and salinity.

Physical defenses

One common adaptation of plant against grazing animal has been called 'barbed-wire syndrome' (Grant 1984). Selectivity is also influenced by the presence of silica, fiber, spines and thorns and the ease with plants can be harvested and swallowed. Individual plant species may respond to browsing and grazing by herbivores by increasing their physical defenses. Some plant species appear to provide protection from further grazing by producing stunted branches with stiff thorns (Seif El Din & Obeid 1971).

The accumulation of silica in exposed tissues is a major physical form of defense adapted by several plant species. It accelerates tooth wear, reduces the digestibility of plant tissues and contributes to the development of esophageal cancer (McNaughton et al. 1985). Most halophytic plants grown in Egypt contain high levels (about 6%) of silica in particular during dry season (El Shaer et al. 1984). Kandil et al. (1991) found that OM intake of *Tamarix mannifera* and its utilization was poor because of its high silica content.

Chemical defenses

Chemical defenses may be the most important plant characteristic influencing food choice by many livestock species. Factors that relate to a plant's palatability and consequently nutritive value include its chemical make-up particularly the presence of so-called 'secondary metabolites' such as tannins, nitrates, nitrites, selenium, salts, oxalates some of which may be toxic (Table 2). More commonly, the toxic agents are natural organic chemical encompassing a great variety of structural types (Russell & Michael 1992; Bayoumi & El Shaer 1994). These compounds are of most interest because they are often specific to a particular plant species or genus and must therefore have been designed to serve a particular protective function. For example, compounds such as terpenes, numerous classes of phenolics, steroids, cyanogenic compounds and alkaloids.

The reduced intake of some halophytic woody plants is probably caused by relatively high concentrations of secondary metabolites in such plants in comparison to graminoids (Bryant et al. 1991). Antimicrobial substances, such as essential oils, sometimes affect rumen fermentation capabilities or the palatability of plant tissues (Oh et al. 1967, 1968). Nagy & Tengerdy (1968) have hypothesized that the absolute concentrations of volatile oils encountered in some shrubs and trees could potentially limit rumen microbial activity if these forages were the total feed of the animal. Antimicrobial substances act as repellents to limit consumption (Mueggler 1970). Nevertheless, the effect of such secondary metabolites on selectivity and utilization by rangeland livestock needs clarification, particularly with halophytic plants (Bayoumi & El Shaer 1994).

Lignification

Halophytic plants vary greatly in their fiber constituents such as neutral detergent fiber (NDF), acid detergent fiber (ADF), acid detergent lignin (ADL) cellulose and hemicellulose (Table 3).

The degree of lignification affects the utilization of range plants by livestock. The differences in voluntary intake of forage and the utilization of their nutrients may be associated mainly with the digestion of cell wall constituents (CWC). Within plant species and ecotypes, significant negative correlation usually exists between forage lignin content and either forage DM intake or

Table 2. Phytochemical screening of some halophytic plants

Plant species	P	Ox	T/S	G/C	Sa	Fl	Al	Ta	N
Aetheorhiza bulbosa	G		+	+			+	+	
Alhagi kirghisorum	S/G						+		
A. maurorum	A	+	+	+	+	+	+		+
A. pseudalhagi	A					+	+	+	
A. sparsifolia	A					+	+	+	
Arthrocnemum glaucum	C			+					+
Asparagus stipularis	C		+	+	+	+	+	+	+
Atriplex halimus	A				+	+	+	+	+
A. leucoclade	A			+	+		+	+	+
A. nummularia	A			+	+		+	+	+
Cressa cretica	Nill				+		+	+	
Cyperus capitatus			+	+		+	+	+	
Diplache fusca	A					+	+		
Frankenia pulverulanta	C	+	+	+		+	+		
Halocnemum strobiolaceum	C	+	+		+	+	+	+	+
Halopeplis perfoliata			+		+	+	+		
Haloxylon salicornicum	C	+	+		+	+	+	+	
Kochia eriophora	A		+				+	+	
Juncus acutus	A		+	+		+	+	+	+
J. arabicus	A		+	+			+	+	+
J. subulatus	A		+	+		+	+	+	
Limonium pruinosum	C	+			+		+	+	
L. monopetalum	Nill		+			+	+	+	
Lygeum spartum		+	+	+			+		
Nitraria retusa	A				+			+	+
Phragmites communis	C			+		+	+	+	
Salicornia fruticosa	C				+		+	+	
Salsola glauco	S				+	+	+		
S. kali	Nill					+	+	+	+
S. macera	G		+		+	+	+		
S. rigida	A		+			+	+	+	+
S. tetrandra	A		+		+	+	+	+	
S. vermiculata	A				+	+	+	+	
Schoenus nigricans			+	+	+	+	+	+	
Suaeda fruticosa	A		+				+	+	+
S. maritima	A		+		+		+	+	
S. monoicae	C								
S. pruinosa	C				+			+	
S. vermiculata			+				+		
Tamarix aphylla	C/G				+			+	+
T. mannifera	C				+	+		+	+
Zygophyllum album	Nill		+	+	+	+	+	+	+

P = Preferability for different animal species; S = sheep; G = goats; C = camels; A = all animals; Ox = Oxalate; T/S = Triterpenoids and/or Steroids; G/C = Glycosides and/or Carbohydrates; Sa = Saponins; Fl = Flavonoids; Al = Alkaloids; Ta = Tannins; N = Nitrates.

Sources: Massart & Janskens (1957); Rizk (1986); Anon. (1981, 1992); Wilson (1984).

nutrient utilization (Van Soest 1965; Kandil & El Shaer 1990). According to Allinson & Osbourn (1970) forage characteristics which may affect voluntary intake are:

- the content of digestible CWC,
- the content of indigestible CWC,
- the structure of CWC, and
- the content of cytoplasmic constituents.

High levels of ADL and NDF (Table 3) have been considered the most important factors limiting forage intake and digestibility (Van Soest 1965; Horn et al. 1979). Similar results on the

Table 3. Fiber constituents (%) of some common halophytes (on DM basis)

Plant species	NDF	ADF	ADL	Cel.[a]	Hemic.[b]	SC[c]
Atriplex angulata	34.5	28.7	11.3	5.8	17.4	65.5
A. halimus	66.7	40.7	9.20	26.0	31.5	33.3
A. leucoclada	36.9	29.1	11.0	3.8	18.1	63.1
A. nummularia	58.4	40.1	6.72	18.3	35.38	41.6
A. vesicaria	46.9	28.1	9.00	18.8	19.1	69.1
A. barclayana	49.5	32.1	9.78	17.4	22.4	50.5
Diplachne fusca	46.0	26.9	5.75	19.1	21.15	54.0
Halocnemum strobilaceum	70.5	40.0	7.95	30.5	32.05	29.5
Junens subulatus	42.2	25.2	4.68	17.0	20.52	57.8
J. rigidus	45.0	28.0	7.01	17.0	20.99	55.0
Nitraria retusa	38.8	29.0	8.42	9.8	20.58	66.2
Salsola tetrandra	36.0	26.0	12.1	10.0	13.9	64.0
S. vermiculata L.	39.7	30.0	12.9	9.7	17.1	60.3
Suaeda fruticosa	33.7	22.2	12.0	11.5	10.2	66.3
Tamarix aphylla	36.0	26.0	12.4	10.0	13.6	64.0
T. mannifera	40.0	28.0	11.0	12.0	17.0	60.0
Zygophyllum album	32.2	23.0	6.5	9.20	16.5	77.8
Clover (Berseem) hay	65.1	35.5	9.05	29.6	26.45	34.9
Rice straw	85.0	62.6	42.6	22.4	20.0	15.0

[a]Cellulose (Cel.) = NDF − ADF.
[b]Hemicellulose (Hemics.) = ADF − ADL.
[c]Cell soluble (CS) = 100 − NDF.
Sources: see Table 1.

saltbushes were reported (El Shaer 1981; El Shaer & Kandil 1990).

Salinity

In arid regions, natural vegetation can supply grazing animals with water specially in the green season. In the dry season, animals are obliged to depend mainly on brackish water from artesian wells. These wells vary considerably in water quality. Salt content (as total solids) may vary from <300 ppm to more than 25,000 ppm. The presence of salts in both drinking water and rangeland forage at relatively high concentration greatly affects halophyte intake by animals. Squires (1993), reviewing the Australian experience, noted that the productivity of sheep might be affected by drinking saline water. The effects were apparently correlated with the total saline load above tolerance level (Peirce 1957) and type of the salt present (Peirce 1968; Singh & Taneja 1981). The highest permitted concentrations of salt were about 2.5% NaCl for sheep and 2% for cattle (Schmidt-Nielson 1978) if the animals had a salt free diet. (Wilson 1966a) suggests 0.6% NaCl as the upper limit where a salty diet is available. Camels can tolerate much higher concentrations of salt than other livestock. Thirsty camels in drought can drink sea water of almost 3.5% salt (Salem et al. 1981). Other workers considered 1.5% saline water as the tolerable level for many animals (Peirce 1957). Moreover, Peirce (1968) indicated that the high salinity of 1.3% saline water decreased feed intake by sheep. Severe reduction in feed intake, and possibly death can occur as the salt concentration increased up to 2%. Halophytic plants, such as *Atriplex halimus* might contain as much as 3.0% NaCl on a dry weight basis. Therefore, a sheep fed 1.5 kg/day *Atriplex* will consume 300 g/day salt and require about 30 ml of water per g of salt to flush it through the kidneys. Animals can tolerate higher amounts of salt ingested through feed than via the drinking water (Gihad 1993).

Halophytic plants contain variable amounts of secondary metabolites (Table 2). Some of them are poisonous. The range of known toxic substances is very extensive and will certainly be increased as additional species are studied. The nature of the toxic principle has not always been identified (Bayoumi & El Shaer 1994). Tannins (condensed or hydrosoluble tannins, nitrates and alkaloids are common in numerous halophytic plants. Tannins are widely perceived as important because they may reduce the digestion of protein and fiber (Lindroth 1989). Tannins are bound with protein or CWC. In addition, Provenza et al. (1991) reported that herbivores may reject tannin-rich plants because they cause internal malaise. A decline in cellulose digestion as well as other changes in digestion and metabolism of lambs fed diets containing perloline were reported by Boling et al. (1975). Proper identification of the secondary metabolites that actually defend halophytes against browsing and foraging is required (Bayoumi & El Shaer 1994).

Selection and palatability

Palatability is extremely difficult to define in terms of the biological processes involved in feed selection. As commonly used, this term implies acceptability, but not necessarily desirability. Thus, a halophytic plant that is palatable may be essentially neutral with regard to preference, being neither attractive nor repellant to the taste.

The selectivity of grazing livestock may vary with animal species, available plants, stage of maturity, intensity of grazing and weather conditions. The diet selected by sheep and goats grazing halophytes has been shown to be higher in nutritive value than the plant material on offer (Wilson 1977; Kandil & El Shaer (1990). The diet ultimately selected in a particular situation is a function of many interacting and poorly understood factors (Heady 1964).

Factors that relate to a plants palatability include its chemical make-up, particularly the presence of secondary plant metabolites which are generally considered evolutionary chemical defenses of plants against herbivory (Meyer & Karazov 1991). Commonly recognized nutrients such as nitrogen (crude protein), fiber and fat also relate to plant palatability, but probably in an indirect or correlative sense.

Arnold & Hill (1972) pointed out that the receptors involved in the senses of taste and smell can detect chemical substances only at the molecular level and animals have no means of recognizing such proximate components as 'crude protein' or fiber per se, as such components are complex mixtures of many chemical compounds. Anatomical features of plants, such as thorns, awns, prickles, dense pubescence and textural features also relate to a plant's palatability but the nature of these relationships is unclear (Malecheck & Provenza 1991).

Interactions with drinking water

The high and constant water content of halophytic plants make them preferred species for animals to assure a good portion of their water requirements in arid and semi-arid areas. Water in these areas is the most limiting factor for animal production, because it is either inadequate or unsuitable for drinking. Water content of lush leaves and twigs of trees and shrubs is as high as 80% (Gauthier-Pilters & Dagg 1981). The high water content might furnish 40–50% of water requirements of animals which will be able to survive several days without drinking water (Macfarlane, 1964).

The salt content of halophytes can influence the animals water requirement because additional water is required to excrete their high salt content (about 70 ml g^{-1} of NaCl). The high salt levels of the foliage increase the water requirements of grazing animals. Salt intake approaching 200 g NaCl/sheep day have been measured for sheep grazing *Atriplex vesicaria* in summer. Sheep grazing on *Atriplex vesicaria* rangelands drank approximately 7–12 L/day, whereas, similar sheep grazing on an adjacent grassland consumed less than 3–5 L/day (Wilson 1974). The higher water intake is needed for the urinary excretion of the high salt intake. Wilson et al. 1969) found that free water intake and sodium intake from *Atriplex vesicaria* is high. Each gram of Na ingested required 74 ml free water to make full use of it. A similar figure (70.8 ml g^{-1}) was obtained by Hassan et al. (1979). Gihad (1983) reported that drinking water intake increased by 61.4% when sheep were fed *Atriplex halimus* instead of clover hay but the increase was lower in camels and higher in goats in comparison with sheep.

On arid and semi-arid rangelands the locations of water has a dominant effect on the use of vegetation (Squires 1981). In hot, dry seasons, it would be desirable to have ample fresh water at this time, but frequently fewer sources are available owing to the evaporation of shallow surface supplies and to an increase in salt content of others. Furthermore, if a sheep must drink 10 L of water daily to eliminate ingested Na, it is likely that two or three drinks daily are necessary. Therefore not only must more water be available, but also it must be available at more points. If fresh water is not readily available the grazing area around each watering point may become restricted or alternatively, the sheep may eat less saltbush. The first alternative will result in overgrazing of the areas closest to water, the second, is to the under-use of the halophyte when it is most needed, and both will lessen the chance of stock survival (Squires 1981).

There is a relationship between water consumption and moisture content of grazable halophytes and feedstuffs supplementation. Water consumption by small ruminants especially sheep increases with increasing dry matter from concentrated feedstuffs and is usually correlated with DMI (Shkolnik et al. 1980). On the other-hand, the relatively high moisture content of grazable ranges (mostly halophytes) in the wet season, decrease water intake by 60% in sheep and by 25% in goats (El Shaer et al. 1982; Benjamin et al. 1986).

Consumption by livestock

The value of a halophytic range is judged primarily upon how readily its plants are eaten by animals. Feed intake of halophytes is generally low and is influenced by all factors affecting their palatability. There are many factors affecting forage consumption such as length of the grazing day, climatic conditions, intensity of grazing and character of the plant. Forage intake is rather substantially influenced by sparse stands of palatable species because of poor condition of the range. This can be a result of dominant unpalatable species being used only slightly and the scarce palatable species used heavily (Graetz 1973). This also results from animals spending more time in search of preferred plants and in search of preferred portions of the plants in the process of grazing. The usefulness of halophytes as a sole diet without supplementation is questioned (Table 4). It must be supplemented particularly, with a readily available source of energy to better utilize it for livestock. The low intake of most halophytic plants could be principally attributed to:

- high ash content (since Na, K and NaCl levels were high (Wilson 1966a,b,c) deficiency in available carbohydrates (Hassan & Abd El Aziz 1979; Kandil & El Shaer 1988);
- higher levels of non-structural carbohydrates mainly ADL, NDF and hemicellulose limit the consumption of halophytic plants as well as the other roughages (Van Soest 1965).

However, some halophytic ranges, such as *Atriplex canescens*, *Juncus* spp., *Nitraria retusa*, *Suaeda fruticosa* and *Salsola* spp., might be valuable fodders for different livestock. Fourwing saltbush (*A. canescens*), an evergreen shrub, is recognized worldwide as a valuable browse for livestock and wildlife because of its wide range of adaptation, palatability and high nutritional value (Ueckert et al. 1990). Therefore, plantations of fourwing saltbush reserved for winter grazing (as a source of supplemental nutrients) might be valuable for range livestock and big game animals. *Juncus subulatus*, *Kochia indica* and *Diplachne fusca* may be considered as non-conventional forage producing plants comparable to berseem hay (*Trifolium alexandrinum*), they can undoubtedly contribute to overcome the problem of acute shortage in animal feeds particularly during the summer season when feedstuffs are scanty in several parts of the world (Anon. 1992).

The assumption that many *Atriplex* species have a low potential for secondary production is based on digestibility trials over short periods of time (2–5 weeks), with animals which had no prior adaptation to this feed. This is, however, contradicted by large scale field experiments with large number of animals adapted to this feed (Le Houérou 1992). The experiments showed that animals get progressively adapted to this feed over a period of 3–5 months and that the intake may show a two-fold increase with respect to the initial consumption. These results were later confirmed in Spain (Le Houérou 1991). Consumption may reach 2–2.5 kg DM/head daily with penned animals and 1.5 to 1.6 kg in field experiments with browsing animals. Le Houérou (1991) also

Table 4. Average values of halophytic forage intake (g/day/kg $W^{.75}$) with or without supplementation during the wet season

Plant species	Sheep	Goats	References
Mixture of halophytes			
Solely	31.5	44.9	El Shaer (1981)
+50% of maintenance	37.1	46.4	El Shaer (1981)
Atriplex nummularia			
Solely	34.3	–	Wilson (1977)
Solely	22.0	–	Newman (1966)
Solely	56.2	–	Weston et al. (1970)
Solely	40.0	–	Hassan & Abd El Aziz (1979)
+150 g barley	35.6	–	Hassan & Abd El Aziz (1979)
+150 g barley	59.8	58.0	Kandil & El Shaer (1988)
+250 g barley	91.7	79.0	Shehata et al. (1987)
Atriplex halimus (natural)			
Solely fresh	19.4	5.8	El Shaer et al. (1980)
Solely dried	13.1	21.0	El Shaer et al. (1980)
Solely fresh	13.4	14.8	Kandil & El Shaer (1990)
+100 g barley	26.4	27.6	Kandil & El Shaer (1990)
+150 g	33.4	29.5	El Shaer & Kandil (1980)
Atriplex halimus (cultivated)			
+150 g barley	41.1	38.2	El Shaer & Kandil (1990)
Halocnemum strobiolacum			
Fresh solely	13.2	14.2	El Shaer et al. (1990)
Dried solely	9.45	8.23	El Shaer et al. (1990)
Silage	59.5	59.0	El Shaer et al. (1990)
Silage	66.9	68.4	El Shaer et al. (1991)
Tamarix aphylla			
Fresh solely	9.69	8.72	El Shaer et al. (1991)
Dried solely	Nill	Nill	El Shaer et al. (1991)
Silage	43.3	49.2	El Shaer et al. (1991)
Zygophyllum album			
Fresh solely	2.12	2.94	El Shaer et al. (1990)
Dried solely	Nill	Nill	El Shaer et al. (1990)
Silage	32.7	28.2	El Shaer et al. (1990)
Salsola kali	27.3	–	Newman (1966)
Tamarix mannifera	40.2	–	Kandil et al. (1991)
Grasses	30.3	–	Newman (1966)
Forbs	36.6	–	Newman (1966)

reported that *Atriplex* forage is an excellent corrective to poor rangeland, to stubble and to straw. Adding an *Atriplex* complement to free ranging animals allowed for daily bodyweight gains of 100 g (or slightly more in sheep in already good condition).

Ensiling broiler litter with the unpalatable or less palatable halophytes has been a way to increase the consumption (Table 4), digestibility and nitrogen utilization by sheep and goats (El Shaer et al. 1990, 1991). Silage may be a potential source of supplementary feeding where unpalatable and/or less-palatable halophytes are abundant and easily harvested.

Nutrient content and utilization

One of the first requirements to be met by range animals is energy because they frequently travel long distances to acquire feed and water. In addition, they must maintain body temperature during the winter without the aid of shelters. When energy-supplying carbohydrates and fats are inadequate, the animal will use protein for energy. This will further aggravate any protein deficiency already present in the basal diet.

Recent research shows that saltbushes definitely possess a rather low energy value of around 0.5–0.7 MJ ME kg^{-1} DM (= 2.5–3.5 MJ NE) just enough to cover the maintenance needs of sheep if they consume 1.2–1.5 kg DM/day (Le Houérou 1991) but it is not enough to cover production requirement for growth, pregnancy or lactation. When small ruminants are grazing on *Atriplex nummularia* or *A. halimus* as a sole feed, they do not satisfy their production requirements of energy. Sheep and goats may not obtain even their maintenance energy requirements as animals tend to lose weight (El Shaer & Kandil 1990). The increase in structural carbohydrates (i.e. NDF, ADF and ADL) with seasonal progression indicates a concomitant decrease in digestibility as these non-digestible portions are negatively correlated with both DMI, DMD and protein digestibility of halophytic plants (Kandil & El Shaer 1988, 1990). It apparently acts as a barrier to micro-organisms attempting to attack the cellulose in the cell wall and various intercellular nutrients. Mertens (1977) concluded that changes in the composition of the CWC involving lignin and possibly silica limit the potential extent of digestion whereas the rate of digestion is limited less by chemical entities than by the crystalline or physical nature of the fiber. As a result of changes in these properties during maturation, the time needed to chew food particles before they can pass out of the rumen may increase two- or three-fold and intake falls.

The term crude fat (usually expressed as ether extract or EE) includes a wide variety of substances soluble in certain organic solvents and insoluble in water. The true lipids included are: simple lipids, true fats and oils, compound lipids and derived lipids such as saturated and unsaturated fatty acids. Most of halophytes lipids and fats are digestible in varying degrees by ruminants. Fats serve as an important form of storage food for shrubs and as a respiratory substrate in buds and twigs. Most of halophytic plants like *A. halimus* and *A. nummularia* contain moderate amounts of ether extract (Table 1). The high EE content of such halophytic plants could be attributed to the high ratio of waxes and essential oils to true fats present in such plants (Wardeh et al. 1990). This fact should be noted when calculating energy content of rations for animals under range condition. Fats and related substances in halophytes tend to decrease with seasonal progression.

Crude protein (CP) content is significantly correlated with digestible protein content in most plants so that the CP level can give a reasonably reliable indication of nutritive value. The situation with halophytes is less clear as a variable part of the nitrogen is non-protein nitrogen (NPN). It is more useful to quote N^+ not CP levels (O'Leary & Glenn 1994). It is well known that the amount of N compounds present in halophytic plants varies greatly with the age or stage of development and kind of tissue and the season of growth. The nitrogen content of halophytes generally might have a wide range. It is as low as 0.2% in coarse dry materials and high as 4.8% in fresh green growth of preferred plant species such as *Suaeda fruticosa*, *Nitraria retusa*, *Juncus subulatus* (Wardeh et al. 1990; Anon. 1992).

Digestibility of N is 65% in some *Atriplex* spp. but only 55% of the digestible N is retained (Le Houérou 1991). The 45% digestible N which is not retained is NPN, about 50% of it is glycinebetaine. The N retention of saltbushes by ruminants may be considerably increased when additional energy is present in the diet and the micro-organisms of the gut have had time to adjust (El Shaer 1991; Kandil & El Shaer 1980; Le Houérou 1991).

Several studies in Egypt on *Atriplex* spp. showed that sheep and goats fed halophytic plants as a sole ration were in negative N balance (Hassan et al. 1878; El Shaer 1981; Kandil & El Shaer 1988; El Shaer et al. 1890). Shortage of readily available carbohydrates is a problem in such halophytes as *A. nummularia*, *A. halimus*, *H. strobolicum*, *Tamarix mannifera*, and *T. aphylla*. This, together with the rapid hydrolysis of its CP in the rumen, led to accumulation of ammonia which is inefficiently utilized by the rumen micro-organisms giving rise to blood urea-N and consequently increased urinary nitrogen excretion. Such trends have confirmed the earlier ones obtained by the Australian workers (Wilson 1977; Wilson et al. 1969; Weston et al. 1970). High protein content of halophytes would satisfy most of the protein requirements of grazing animals to perform their physical functions, meanwhile, those with a low protein content fail to meet even maintenance requirements.

Attention has been drawn to the high digestible CP content of the leaves of the saltbushes.

The protein value of saltbushes may be appreciably lower than the CP content would suggest (Hassan et al. 1879; Kandil & El Shaer 1988). In all of the saltbushes studied, there was considerable degradation of dietary CP to ammonia in the rumen. The quantity of CP in forms other than ammonia digested in the intestines presumably are the principal source of amino acids (AAs) for the sheep tissues and equivalent to only 60% of the total DCP (Hassan et al. 1979).

The protein from *Atriplex nummularia*, *A. repens* and *A. halimus* have a fairly well-balanced amino acid (AA) composition. The percentage of the essential AAs were at or above the level used by FAO. Moreover, the protein content of several species of *Salsola* had an AA composition which was nutritionally balanced. Wild halophytic plants have been successfully harvested and used as forages in mixed rations particularly when cultivated forages are limited or are unavailable (Swingle et al. 1994).

Minerals

Most halophytic plants contain a high ash content particularly during droughts. The ash content of may be as high as 18–30% of the total DM; about 50% of it is as dissolved salts, mostly NaCl. Sheep consuming 1.5 kg DM daily would ingest on average about 360 g of minerals or 180–200 g of soluble salts. Such animals must drink 5–6 L of water daily, and 12 L in hot weather in order to eliminate these salts (Wilson 1986b; Le Houérou 1991).

The average mineral composition of numerous halophytic plants collected from several sources are summarized in Table 5. Generally, calcium (Ca) content of all halophytes seems to be well above the dietary requirements of 0.18–0.27% recommended by the NRC (1975). Phosphorous levels of halophytes vary greatly and several species appear to meet the NRC (1975) recommendations of 0.18–0.43%. Such changes in P content may be attributed to changes in both stem:leaf ratio and in chemical content of the plant parts themselves (Cook 1971). The minimum sodium (Na) level needed by ruminants is 0.06% as has been established by NRC. The Na content of halophytes at different stages of growth is more than sufficient. The high Na content of these plants may not be harmful because animals can tolerate large amounts of salt if adequate amounts of water are available (Kearl 1982).

The high content of Ca, Na and Mg in halophytic plants may be attributed to the following:
- halophytes originated in salt marsh habitats,
- the three mineral elements are considered essential nutrients for halophytic plants, and
- the high osmotic pressure is an adaptive mechanism through which halophytes cope with saline conditions.

Information about minor elements found in halophytic plants is rare. El Shaer (1986) reported Cu, Zn, Mn and Fe levels in some halophytes of the Sinai desert in Egypt (Table 6). All halophytes studied had adequate levels of Zn Fe, Mn and Cu although levels of Fe and Mn were well above the recommendations of NRC (1975), and Kearl (1982).

Supplementary feeding on halophytic rangelands

In most of the rangelands of the world there is a wide seasonal variation in both the quantity and quality of the forage available but livestock numbers remain more or less stable. Supplemental feeding is a widely accepted defence against scarcity of forage under arid and semi-arid conditions. Proper feeding practices are closely related to rangeland management. It includes establishing policies for proper and wise use of supplementary feeding during dry seasons or other times when the animal's feed requirements exceed availability of nutrients. These are expected to differ between the different environments and according to the type of animal, the production process and whether or not the cyclic pattern of nutrient availability and animal needs for certain physiological processes (pregnancy and lactation, growth) coincide.

The lack of sufficient energy in halophytes is the most serious deficiency for grazing animals kept for production of milk, fiber or work (Crampton 1957) and the most expensive to counteract. Several factors affect and limit the intake of halophytes. The lower consumption of the many halophytic ranges is one of the main limitations to their utilization as a forage which consequently impairs the performance and productivity of grazing or browsing livestock. This can be overcome when the salty rangelands have

Table 5 Average values of some major elements in halophytic ranges (on DM basis, %)

Plant species	Na	K	Ca	P	Cl	Mg	S
Atriplex angulata	3.3	1.8	–	–	–	–	–
A. canescens	3.90	2.08	1.43	0.16	–	–	–
A. glauca	6.30	1.60	2.14	0.13	–	0.56	0.41
A. halimus	4.41	1.33	1.51	0.29	–	0.32	0.17
A. leucoclada	4.1	3.0	1.36	0.49	8.4	0.35	–
A. nummularia	4.15	3.19	1.26	0.11	7.8	0.30	0.20
A. semibaccata	3.32	1.35	1.29	0.11	10.0	–	–
A. vesicaria	5.87	4.05	1.56	0.15	9.1	0.32	–
A. farinosa	1.69	1.11	2.32	–	3.19	0.23	–
Juncus spp.	–	2.35	0.30	0.25		–	–
Kochia pyramidata	6.6	1.5	–	–	–	3.4	
Nitraria retusa	5.35	0.66	1.96	0.22	–	0.36	0.14
Phragmites communis	–	0.70	0.49	0.20	–	–	–
Salsola kali	5.71	0.63	2.33	0.37	–	–	–
S. rigida	1.30	1.14	8.40	0.44	–	0.22	–
S. sieberi	0.22	1.50	2.03	0.04	–	–	–
S. tetrandra	5.65	1.45	3.98	0.16	–	0.59	0.12
Suaeda maritima	2.74	1.12	2.02	0.28	–	–	–
S. fruticosa	4.06	1.28	0.11	0.41	–	0.30	0.20
Tamarix aphylla	2.75	0.78	3.73	0.16	–	0.43	0.12
Clover (Berseem) hay	0.88	2.22	0.41	0.07	–	–	–

Sources: IFI (1979); El Shaer (1986); Leigh (1972).

Table 6. Average values of trace elements in some halophytic plants (on DM basis, as ppm)

Plant species	Zn	Cu	Fe	Mn
Atriplex halimus	65	11	503	51
Nitraria retusa	32	11	578	62
Salsola tetrandra	44	9	664	97
Suaeda fruticosa	55	13	674	88
Tamarix aphylla	38	13	874	60
Zygophyllum album	41	7.8	393	52
Clover (Berseem) hay	19	10	–	36

Source: El Shaer (1986).

a good green cover of herbaceous plants (Wilson & Graetz 1980) or when supplemented with a source of readily available carbohydrate such as barley or molasses (Table 7). Saltbush intake by sheep or goats was affected ($p < 0.05$) by the level of energy supplement. Animals which were given barley grain tended to consume more saltbush than those fed on forage alone (El Shaer 1991). Sheep and goats raised on the salty rangelands (mainly *Atriplex halimus*, *A nummularia*, *Nitraria retusa*, *Salsola tetrandra*, *Tamarix aphylla* and *Suaeda fruticosa*) lost weight all the year round. Weight loss was least in spring and the highest in summer. Barley (energy) supplements improved the growth of both sheep and goats, decreased water intake and rumen ammonia-N and increased the rumen volatile fatty acids (Hassan et al. 1982; Kandil & El Shaer 1988, 1989). Moreover, the amount of feedstuffs fed as a supplement to grazing/browsing ruminants, has an effect on their productive and reproductive performance and on the amount of range plants consumed. Barley supplements were fed to grazing sheep and goats on a mixture of halophytic plants in Sinai, Egypt. Ewes grazing on halophytic rangelands in Sinai were fed barley as an energy supplement at one of three levels (50, 100 or 150% of the maintenance requirement). High levels of supplements given to ewes and does in late pregnancy or early lactation led to a significant ($p < 0.05$) improvement in milk production, birth weights of lambs and kids as compared to either the medium or low level of supplements. Reproductive efficiency, as judged by lambing percentage of ewes and does grazing natural halophyte-dominated rangelands without supplements in Israel, was enhanced by the supplements. Benjamin et al. 1986 showed that when a supplement of a high energy feed was given, in

Table 7. Performance of ewes and does grazed mixture of halophytes with barley supplementation

Criteria measured	Ewes[a]			Does[a]		
	L	M	H	L	M	H8
Milk yielding (kg milk/kg $W^{.75}$)						
First 10 wks	1.08	1.89	2.15	3.04	4.33	7.01
Last 6 wks	0.31	0.73	0.78	1.88	1.90	2.77
Total	1.39	2.56	2.94	4.32	6.23	9.80
Live weight of offsprings						
Birth weight, kg	3.00	3.56	4.05	1.42	1.62	1.88
At 10 wks, kg	8.43	12.1	15.1	5.83	6.28	6.31
At 16 wks, kg	11.2	17.4	20.4	8.30	9.0	9.20
ADG of offsprings (g/day)						
First 10 wks	77.6	12.2	15.8	6.3	66.6	63.3
Last 6 wks	66.0	12.6	12.6	69.5	64.8	68.8
Overall	73.2	12.4	14.6	61.4	65.9	65.4

[a]L, M and H represent the levels of barley supplementation as 50, 100 and 150% of the maintenance requirements according to ARC (1965). ADG = average daily gain.
Source: El Shaer (1981).

addition to grazing, lambing percentage was increased by 15% units. Lambs per lambing was 1.33 and 1.53 for ewes grazing *Atriplex* spp. only and those grazing *Atriplex* plus a barley supplement, respectively.

Le Houérou (1991) reported that *Atriplex nummularia* and *A. canescens* intakes over a 9 month period were able to maintain body weight of mature sheep indefinitely without supplements. When these shrubs were supplemented sheep had body weight gains of 80 g day^{-1}. Similar findings have been reported in Spain with *A. halimus* and *A. nummularia*. (Le Houérou 1991) and in Egypt with wild and cultivated *A. halimus* and *A. nummularia* (Kandil & El Shaer 1988; El Shaer & Kandil 1990). The type of supplementary feeding of lactating ewes and does is one of the principal factors affecting body weight changes of dams and their milk yield and composition in addition to daily gain of the offspring (El Shaer et al. 1986).

Encouraging results have been obtained in Egypt from using agro-industrial by-products such as grape pulp and date seeds as energy supplements for sheep and goats on halophytic rangelands. Ground date seeds and olive pulp could be used economically in supplementing sheep and goats grazing a mixture of halophytic rangelands. This would reduce dependence on the highly expensive conventional feedstuffs (Abou El Nasr 1985; El Shaer et al. 1986; Khamis et al. 1989).

All over the world livestock production on arid and semi-arid halophytic rangelands is practiced in the traditional way where supplementary feeding is used without knowledge of either the animal's needs or of the contribution of nutrients from the rangelands at different seasons of the year. Feedstuff supplement programs, unless carefully designed and practiced can cause damage to the rangeland productivity.

Development of livestock production on halophytic rangelands

Development of livestock husbandry on rangelands can be achieved through appropriate range and grazing management which allow a sustainable use pattern and prevent further deterioration. It must be based on control of animal numbers, better grazing systems, use of sown pastures and where appropriate, integration with irrigated areas.

Control of animal numbers

The grazing zones could be divided into grazing districts as the establishment of such districts is a major factor for the development and proper utilization of the rangeland resource. One of the main functions of the grazing districts is to control and regulate animal numbers in order to ensure an optimum sustained level of forage production. It is also necessary to control long distance movement within the region to reduce losses due to such migration of flocks and herds.

The hema grazing systems

Many range sites have been protected to demonstrate the importance of range management in improving forage production and for seed collection. Natural regeneration of rangelands is being achieved through re-introduction of the 'hema' system (Draz 1978). The Hema system was once widely used in many countries of the Near and Middle East. Hema cooperatives are now being formed to regulate grazing in place of the destructive open range systems which currently prevail. Hema, as a range reserve, could be managed for use by individuals, families, tribes, villagers or even government agencies. This system is classified according to the type of protection offered (Draz 1983):
- animal grazing totally prohibited,
- grazing and/or cutting permitted but restricted to certain seasons of the year, and
- grazing allowed all year round but the kinds and numbers of animals are specified.

Plantation of fodder shrubs

Rehabilitation of rangelands with suitable shrubs, perennial grasses and tress can considerably add to feed resources and if properly used can arrest soil erosion (Malcolm 1994). Suitable plant types are usually locally-adapted ecotypes. Among the forages *Atriplex* spp., *Juncus* spp, *Tamarix aphylla*. *Salsola* spp. and *Diplachne fusca* have shown promise (Osman 1994; Le Houérou 1991; Anon. 1992). The nutritional quality of cultivated *A. halimus*, a species improved by domestication, has been studied by El Shaer & Kandil (1990).

Draz (1978) reported that the Australian oldman saltbush (*A. nummularia*) was highly adaptable to local dryland conditions in the marginal areas of the Syrian steppe. Its massive perennial growth provides enough green feed reserve which can be used in the late summer and early autumn – time of great nutritional stress in the semi arid Mediterranean climate region (Le Houérou 1994). The shrub is highly drought resistant and salt tolerant and provides good soil protection and helps with regeneration of natural plant cover. Because of this *A. nummularia* has been planted in many countries as a fodder reserve.

Suggested topics for further research

Further development of the range livestock industry in the semiarid regions where halophytic vegetation dominated would be:
- integrated studies on animal pasture relationships. It would cover such aspects as range nutrient cycling as affected by grazing strategy and by supplementary feeding; grazing management strategies in relationship to carrying capacity and the determination of carrying capacities in different environments under different conditions;
- supplementary feeding of sheep and goats grazing on halophytic rangelands. It would focus on evaluating the conventional and unconventional systems of supplementary feeding; prospects for adopting intermittent feeding and watering systems; explore the possibility of exploiting the compensatory growth phenomenon; and critically evaluate the economics of different supplementary feeding systems;
- assess the impact of salinity of feed and water; assess the tolerance of different animal species and breeds;
- assess the role and significance of trace elements – deficiency and excess;
- assess the impact of chemical metabolites as toxic materials and/or nutritional factors using halophytic plants as livestock feeds.

Acknowledgements

We are grateful to UNEP and to the Australian Development Assistance Bureau (AIDAB) for travel grants to allow attendance at the Workshop and to Dr Squires for editing the manuscript.

References

Abd El Aziz, H.M. (1978). The chemical content and nutritive value of some forage plants of north western coast of the UAR. Ph.D. Thesis Fac. Agric. Ain Shams Univ., Egypt.

Abd El Aziz, H.M. (1982). A study of the nutritive value of some range plants in the north western coastal desert. PhD Thesis, Fac. Agric. Ain Shams Univ. Egypt.

Abdullah, A.N., Hassan, N.I., Bayoumi, M.A. & El Shorbagy, M.A. (1987). Forage preference of sheep, goats and camels in the Mediterranean type arid region of Wadil-Azieb, Syria. Int. Conf. on Anim. Prod. in Arid Zones, ICAPAZ, 7–12 Sept. Damascus, pp. 964–988.

Abou El Nasr, H.M. (1985). A study on the possibility of using desert agricultural by-products in feeding livestock. PhD Thesis Fac. Agric. Cairo Univ. Egypt.

Allison, D.W. & Osbourn, D.F. (1970). The cellulose-lignin complex in forages and its relationship to forage nutritive value. J. Agric.Sci. Camb. 74: 23.

Anon. (1981). Agricultural and water investigations of Sinai, performed by the Desert Research Institute, Mataria, Cairo, Dames & Moore under their Phase 1 Contract, Oct. 1981.

Anon. (1992). Revegetation of salt affected lands of deltaic Mediterranean coast. Egypt Proj. 90009 Phase II, Final Report Jan 1992–June 1992, Supreme Council of Universities.

A.R.C. (1965). Nutrient requirements of livestock. No. 2 Ruminants. Agric. Res. Counc., London, 264 pp.

Arieli, A., Naim, E., Benjamin, R.W. & Pasternak, D. (1989). The effect of feeding saltbush and sodium chloride on energy metabolism of sheep. Anim. Prod. 49: 451.

Arnold, G.W. & Hill, J.L. (1972). Chemical factors affecting selection of foods by ruminants. In: J.B. Harborne (ed.), Phytochemical ecology, Academic Press, London, p. 71.

Balch, C.C. & Campling, R.C. (1962). Regulation of voluntary food intake in ruminants. Nutr. Abstr. & Rev. 23: 669.

Batanouny, K.H. (1993). Adaptation of plants to saline conditions in arid regions. In: H. Lieth & A. Al Masoom (eds.), Towards rational use of high salinity tolerant plants. Vol. 1, pp. 387–401.

Batanouny, K.H. (1994). Halophytes and halophytic plant communities in the Arab region: Their potential as a rangeland resource (this volume, Ch. 11).

Bayoumi, M.T. & El Shaer, H.H. (1994). Impacts of halophytes on animal health and nutrition (this volume, Ch. 25).

Benjamin, R.W., Barkai, D., Lasvie, Y. & Forti, M. (1986). Evaluation of introduced species in agro-pastoral systems. In: Fodder production and its utilization by small ruminants in arid regions, CALAR Project (FOPAR) APRI, Dokki, Cairo, Egypt, pp. 28.

Boling, J.A., Buch, L.P., Buckner, R.C., Pendlum, L.C., Burrus, P.B., Yates, S.G., Rogovin, S.P. & Tooky, H.L. (1975). Nutrient digestibility and metabolism of lambs fed added perloline. J. Anim. Sci. 40: 972–976.

Bryant, J.P., Provenza, F.D., Pastor, J., Reichardt, P.B.,Clausen, T.P. & du Toit, J.T. (1991). Interaction between woody plants and browsing mammals mediated by secondary metabolites. Ann. Rev. Ecol. Syst.

Chatterton, N.J., Goodin, J.R., Mc Kell, C.M., Parker, R.V. & Rible, J.M. (1971). Monthly variation in the chemical composition of desert saltbush J. Range Mange. 12: 289.

Cook, C.W. (1971). Effects of season and intensity of use on desert vegetation. Utah Agric. Exp. Sta. Bull. 483.

Crampton, E.W. (1957). Interrelationship between digestible nutrient and energy intake, voluntary dry matter intake and overall feeding value of forages. J. Anim. Sci. 16: 464.

Devendra, C. (1977). Studies in the intake and digestibility of two varieties (Serdang and Colonino) of guinea grass (*Panicum maximum*) by goats and sheep. 1. Long grass. MARDI Res. Bull. 5: 91.

Dietz, D.R. (1971). Nutritive value of shrubs. In: C.M. McKell, J.P. Blaisdell & J.R. Goodin (eds.), Wildland shrubs – their biology and utilization. USDA, Forest Service, INT-1-1972, pp. 289–302.

Draz, O. (1978). Revival of the hema system of range reserves as a basis or Syrian range development program. Proc. Ist Int. Rangeland Congr. Denver, p. 100.

El Bassousy, A.A. (1984). A study of the nutritive value of some range plants from Sallom to Mersa Muttroh. PhD Thesis, Fac. Agric. Ain Shams Univ. El-Khema, Egypt.

El-Monayeri, O.M., Khafagi, O.A., Ahmed, A.M. & El Tantawy, H. (1986). Contribution to the chemical composition of plants belonging to various ecological groups in the Red Seas area. Desert Inst. Bull. Egypt, 36: 405.

El Shaer, H.M. (1986). A comparative nutrition study on sheep, and goats grazing southern Sinai desert range with supplements. PhD thesis Fac. Agric, Ain Shams Univ. Egypt.

El Shaer, H.M. (1986). Seasonal variation of the mineral composition of the natural pastures of southern Sinai grazed by sheep. Proc. 2nd Egyptian-British Conf. Anim and Poultry Prod. Aug 26–28, Bangor.

El Shaer, H.M., Hassan, N.I., El Ashry, A.M. & El-Serafy, M.A. (1984). Nutritional studies on pastures indigenous to southern Sinai. Effect of level of supplement on some rumen and blood metabolites in sheep and goats. Proc. 6th Int. Conf. on Animal and Poultry Prod. Sept. 27–30, 1982, Zagazig, Egypt.

El Shaer, H.M. & Kandil, H.M. (1990). Comparative study on the nutritional value of wild and cultivated *A. halimus* by sheep and goats. Comm.Sci. and Dev. Res. 29: 181.

El Shaer, H.M., Salem, O.A., Khamis, H.S., Shalaby, A.S. & Fardi, M.F.A. (1990). Nutritional comparison studies of goats and sheep fed broiler litter ensiled with desert shrubs in Sinai. Proc. Int. Goat Prod. Symp. Oct. 22–26, Tallahassee, FL, p.70.

El Shaer, H.M., Kandil, H.M. & Khamis, H.S. (1991). Salt marsh plants ensiled with dried broiler litter as a feedstuff for sheep and goats. J. Agric Sci Mansoura Univ. 16: 1524.

Gauthier-Pilters, H. & Dagg, A.I. (1981). The camel: its evolution, ecology, behaviour and relationshlp to man. The Univ. Chicago Press, Chicago, London.

Gihad, E.A. (1993). Utilization of high salinity tolerant plants and saline water by desert animals. In: H. Lieth & A. Al Massoom (eds.), Towards rational use of high salinity tolerant plants. Kluwer Academic Publishers, Dordrecht, The Netherlands. Vol. 1, pp. 443–447.

Gihad, E.A. El-Gallad, T.T., Sooud, A.E., Abou El-Nasr, M.M. & Farid, M.F. (1988). Feed and water intake, digestibility and nitrogen utilization by camels compared to sheep and goats fed low protein desert by-products. Options Mediteran. Mediterraenes-Serie Seminaire no. 2, p. 75.

Graetz, R.D. (1973). Biological characteristics of Australian Acacia and chenopodiaceous shrublands relevant to their pastoral use. Proc 3rd Workshop US/Australia Rangelands Panel, Tucson, AZ, March 26–April 5, p. 33.

Hassan, N.I. & Abdel Aziz H.M. (1979). Effect of barley supplementation on the nutritive value of saltbush (*Atriplex nummularia*). World Rev. Anim. Prod. 15: 47.

Hassan, N.I., Abdel Aziz, H.M. & el-Tabbkh, A.E. (1979). Evaluation of some forages introduced to newly-reclaimed areas in Egypt. World Rev. Anim. Prod. 15: 2.

Hassan, N.I. El-Serafy, A.M. & Farid, M.F.A. (1980). Studies on pasture indigenous to South Sinai 1. Chemical composition and *in vitro* digestibility. Proc. 2nd Ann. Meeting of the ASAS, Cornell Univ. Ithaca, 26–30 July.

Hassan, N.I., El Shaer, H.M. & Kandil, H.M. (1982). Performance of lambs and kids on salty pastures supplemented with different levels of concentrates. World Rev. Anim. Prod 18: 37.

Heady, H.F. 1964. Palatability of herbage and animal preference. J. Range Manage. 17: 76.

Horn, F.P., Telford, J.P., McGroskey, J.E., Stephens, D.F., Whiteman, J.V. & Robert T. (1979). Relationship of animal performance and dry matter intake of chemical constituents of grazed forage. J. Anim. Sci. 49: 1051.

I.F.I. (1979). Arab and Middle East Tables of feed composition. Int. Feedstuffs Inst. Utah Agric. Exp. Sta. Logan Utah Res. Rept. 30.

Kandil, H.M. & El Shaer, H.M. (1988). The utilization of A. nummularia by goats and sheep in Sinai. Proc. Int. Symp. on constraints & possibilities of Ruminant Prod. in the dry tropics. November 5–7 1988, Cairo, Egypt.

Kandil, H.M. & El Shaer, H.M. (1990). Comparison between goats and sheep in utilization of high fibrous shrubs with energy feeds. Proc Int. Goat Prod. Symp. Oct, 22–26, Tallahassee, FL.

Kandil, H.M., El Shaer, H.M., Khamis, H.S. & Ahmed, A.M. (1991). Nutritional value of hyperarid forage species for sheep in Upper Egypt. J. Agric. Sci, Mansoura Univ. 16: 518.

Kearl, L.C. (1982). Nutrient requirements of ruminants in developing countries, Int. feedstuffs Inst. Utah Agric. Exp. Sta. Logan, Utah.

Khamis. H.S., El Shaer, H.M., Farid, M.F.A., Salem, O.A. & Shalaby, A.S. (1989). Utilization of date seeds and olive pulp as supplementary feed for lactating ewes in Sinai. Proc 3rd Egyptian British Conf. in Animal & Poultry Prod. Oct 7–10, Alexandria, Egypt.

Le Houérou, H.N. (1991). The role of saltbushes (*Atriplex* spp) in the development and rehabilitation of arid and semi-arid zones in the Mediterranean Basin. Int. Conf. on the Agric. Manage. of salt-affected Areas, Agidir, Morocco, April 26–May 3.

Le Houérou, H.N. (1994). Forage halophytes and salt-tolerant fodder crops in the Mediterranean Basin (this volume, Ch. 10).

Leigh, J.H. (1972). Saltbush and other chenopod shrubs. In: N. Hall (ed.), The use of trees and shrubs in the dry country of Australia. Aust. Govt. Publ. Serv. Canberra.

Lindroth, R.L. (1989). Mammalian herbivore-plant interactions. In: Plant animal interactions, McGraw Hill, New York, pp. 163–204.

Macfarlane, W. V. (1964). Terrestrial animals in dry heat: Ungulates. In: Dill (ed.), Handbook of Physiology. Section 4, Amer. Physiol. Soc. Washington, pp. 5090–539.

Malcolm, C.V. (1994). Use of halophyte forages for rehabilitation of degraded lands (this volume, Ch. 4).

Malecheck, J.C. & Provenza, F.D. (1983). Feeding behavior and nutrition of goats on rangelands. World Anim. Rev. 47: 38.

McNaughton, S.J., Tarrants, J.L., McNaughton, M.M. & Davis, R.H. (1985). Silica as a defense against herbivory and a growth promoter in African grasses. Ecology 66: 628.

Mertens, D.R. (1977). Dietary fiber components: relationships to the rate and extent of ruminant digestion. Feed Processes 36: 187.

Meyer, M.W. & Karazov, W.H. (1991). Deserts. In: T. Psalo & C. Rabbins (eds.), Plant defenses against mammalian herbivory. CRC Press, Boca Raton.

Muegller, W.F. (1970). Objectionable characteristics of range plants. Range and wildlife habitat evaluation A research symp. USDA Misc. Publ. 1147, 220 p.

Nagy, J.G. & Tengerdy, R.P. (1968). Antibacterial action of essential oils of *Artemisia* as an ecological factor II. Antibacterial action of volatile oils of *Artemisia tridentata* (Big sagebrush) on bacteria from the rumen of mule deer. Appl. Microbiolo. 16: 441.

Newman, D.M.R. (1969). The chemical composition digestibility and intake of some native pasture species in central Australia. Aust. J. Exp. Agric. Anim. Husb. 9: 599.

N.R.C. (1975). Nutrient requirements of domestic animals. No. 5, Nutrient requirements of sheep, 5th ed. National Academy of Science, National Res. Coun., Washington.

Oh, H.K., Jones, M.B. & Longhurst, W.M. (1967). Comparison of rumen microbial inhibition resulting from various essential oils isolated from relatively unpalatable plant species. Appl. Microbiol. 16: 39.

Oh, H.K., Jones, M.B. & Longhurst, W.M. (1968). Comparison of rumen microbial inhibition resulting from various essential oils isolated from relatively unpalatable plant species. Appl. Microbiol. 17.

O'Leary, J.W. & Glenn, E.P. (1994). Global distribution and potential for halophytes (this volume, Ch. 2).

Osman, A.E. (1969). Variation in chemical content of native forage plants as influenced by different sites at Sida-Barrani. MSc Thesis Fac. Agric. Ain Shams Univ., Egypt.

Osman, A.E. & Shalla, M.A. (1994). Use of shrubs in pasture environment under Mediterranean environment in northern Syria (this volume, Ch. 22).

Peirce, A.W. (1957). Studies on salt tolerance of sheep. I. The tolerance of sheep to mixtures of sodium chloride in the drinking water. Aust. J. Agric Res. 10: 725.

Peirce, A.W. (1968). Studies on salt tolerance of sheep. VII. The tolerance of ewes and their lambs in pens for drinking water of the types obtained from underground sources in Australia. Aust. J. Agric. Res. 19: 577.

Provenza, F.D., Burritt, E.A., Clausen, T.P., Bryant, J.P. & Eichardt. P.B.P. (1991). Conditioned flavor aversion: A mechanism for goats to avold condensed tannins in blackbrush, Amer. Natur. 136: 810.

Rizk, A.M. (1986). The phytochemistry of the flora of Qatar. Sci. & Appl. Res. Center, Univ. Qatar. 33 pp.

Rozenthal, G.A. & Jansen, D.H.J, eds., (1979). Herbivores: Their interaction with secondary plant metabolites. Academic Pres, New York.

Russell, J.M. & Michael, H.R. (1992). Plant toxins and palatability to herbivores. J. Range Manage. 45: 13.

Salem, M.S. Ashour, G. & Gihad, E.A. (1991). Important physiological and nutritional aspects of camel diets. 3rd Conf. Anim. Nutr. Ministry Agric. Sakha, Egypt. Nov. 1991, p. 177.

Seif El Din, A. & Obeid, M. (1971). Ecological studies on the vegetation of Sudan. IV. The effect of simulated grazing on the growth of *Acacia senegal* (L.) Willd. seedlings. J. Appl. Ecol. 8: 211.

Shehata, E., Swidan, F., Ramah, A. & Heider, A. (1977). Utilization of *Atriplex nummularia* by desert sheep and goats. Proc 2nd Int. Conf. on Desert Development. Cairo, Egypt, p.25.

Shkolnik, A., Maltz, B. & Gordin, S. (1980). Desert conditions and goat milk production. J. Dairy Sci. 623: 1744.

Schmidt-Neilsen, K. (1978). Desert animals: Physiological problems of heat and water, 3rd ed. Dover Publ, New York.

Singh, N. & Taneja, G.C. (1981). Variability for salt tolerance of in-bred and cross-bred sheep of Rajasthan desert. Indian J. Anim. Sci. 51: 377.

Squires, V.R. (1981). Livestock management in the Arid Zone. Inkata Press, Melbourne, 271 pp.

Squires, V.R. (1993). Australian experiences with high salinity diets and for sheep. In: H. Lieth & A. Al Masoom (eds.), Towards the rational use of high salinity tolerant plants. Vol. 1: 449–457.

Swingle, R.S., Glenn, E.P., Riley, J.J. (1994). Halophytes in mixed feeds for livestock (this volume, Ch. 7).

Szabolcs, I. (1994). Salt affected soils as the ecosystem for halophytes (this volume, Ch. 3).

Ueckert, D.N., Lawrence, B.K., Petersen, J.L. & Huston, J.E. (1990). Performance of yearling Angora does grazing fourwing saltbush during winter. Texas Agric. Exp. Sta. Prog. Rept. 4785, pp. 47–48.

Van Soest, P.J. (1965). Symposium on factors influencing the voluntary intake in relation to chemical composition and digestibility. J. Anim. Sci. 24: 834.

Van Soest, P.J. (1982). Nutritional ecology of the ruminant. O & B Books, Corvallis.

Wardeh, M.F., Dauze, M. & Ould-Mustafa, M.M. (1990). The nutritive value of plant species eaten by camels. Proc Int. Conf Camel Prod. & Impr. Tobruk, Libya, December 10, 1990.

Weston, R.H., Hogan, J.P. & Hemsley, J.A. (1970). Some aspects of the digestion of *Atriplex nummularia* (saltbush) by sheep. Proc. Aust. Soc. Anim. Prod. 8: 517.

Wilson, A.D. (1966a). The intake and excretion of sodium by sheep fed on species of *Atriplex* and *Kochia* (bluebush). Aust. J. Agric. Res. 17: 155–163.

Wilson, A.D. (1966b). The tolerance of sheep to sodium chloride in drinking water. Aust. J. Agric. Res. 17: 503–514.

Wilson, A.D. (1966c).The value of *Atriplex* (saltbush) and *Kochia* (bluebush) species as food for sheep. Aust. J. Agric. Res. 17: 147.

Wilson, A.D. (1974). Water consumption and water turnover of sheep grazing semi-arid pasture communities in New South Wales. Aust. J. Agric. Res. 25: 339.

Wilson, A.D. (1975). Water requirements and productivity of animal populations. In: R.L. Reid (ed.), Proc. III World Conf. Anim. Prod. Sydney Univ. Press, p. 268.

Wilson, A.D. (1977). The digestibility and voluntary intake of the leaves of trees and shrubs by sheep and goats. Aust. J. Agric. Res. 28: 501.

Wilson, A.D. & Graetz, R.D. (1980). Cattle and sheep production on an *Atriplex vesicaria* (saltbush) community. Aust. J. Agric. Res. 31: 369.

Wilson, A.D. & Harrington, G.N. (1984). Nutritive value of Australian plants. In: H.N. Le Houérou (ed.), Browse in Africa: The current state of knowledge. Proc Int. Symp. Addis Ababa, April 8–12, 1990.

Wilson, A.D., Leigh, J.H., Hindley, N.L. & Mulham, W.E. (1975). Comparison of the diets of goats and sheep on a *Casuarina cristata-Heterodenrum oleifolium* woodland community in western New South Wales. Aust. J. Exp. Agric. Anim. Husb. 15: 45.

Wilson, A.D., Leigh, J.H. & Mulham, W.E. (1969). A study of Merino sheep grazing a bladder saltbush (*Atriplex vesicaria*) cotton bush (*Kochia aphylla*) community on the Riverine Plain Aust. J. Agric. Res. 20: 1123.

Wilson, R.T. (1984). The camel. Longmans, London.

7. Halophytes in mixed feeds for livestock

R. SPENCER SWINGLE,[1] EDWARD P. GLENN[2] & JAMES J. RILEY[2]
[1]*Department of Animal Sciences, University of Arizona, Tucson, AZ, USA*;
[2]*Environmental Research Laboratory, Tucson, AZ, USA*

Abstract. Halophyte forages, grains, oil seeds and various byproducts could become available in larger quantities for use as livestock feeds if efforts to develop irrigated halophyte crops continue to be successful. Incorporation into mixed diets would minimize potential adverse effects of the high salt content, low energy concentration, and content of antinutritional factors in halophyte ingredients on ruminant livestock production, and would likely yield higher economic returns than would be possible from direct grazing of halophyte resources. Animal feeding studies demonstrate that organic constituents of halophytes are highly digestible, and that diets containing moderate levels of halophyte ingredients are readily consumed by livestock. Salts contributed by halophyte ingredients in mixed diets would probably not have detrimental effects on animal health and weight gains if fresh drinking water is provided, but dilution of organic constituents by the high mineral content would reduce energy concentration of diets and, consequently, feed conversion efficiency. In regions where other livestock feeds cannot be grown, or wherever there is agronomic justification for their production, halophytes could be effectively used in livestock production.

Key words: *Atriplex*, Feed ingredients, Livestock, Nutritive value, Poultry, Ruminant, *Salicornia*

Introduction

In some areas of the world, native and introduced halophytes are important forage resources, especially for sustaining grazing livestock when other feeds are scarce (Le Houérou 1992). However, little, if any, attention was given to the possibility that halophyte ingredients could be useful in mixed feeds for livestock before the concept of growing halophytes as irrigated crops was proposed It has been clearly established that irrigated production of halophytes is technically feasible and could make available substantial quantities of several harvested products (forages, grains, oil seeds) and/or byproducts (straw, seed meals. seed hulls) with potential for use as livestock feeds (Glenn et al. 1991). Demonstration that these materials could be used successfully in practical livestock production systems would be an important boost to, if not a requisite for, continued development of halophyte-based irrigated agriculture.

The purposes of this paper are to examine the potential advantages of utilizing halophytes as components of mixed feeds for ruminant livestock, to review animal feeding studies in which halophyte blends have been evaluated, and to examine non-nutritional considerations of halophyte use in animal feeding.

Advantages to use of halophytes in mixed feeds

The general nutritional characteristics of halophytes have been well defined (Gihad & El Shaer 1994). Although nutrient composition varies by species, plant part, stage of maturity and is affected by many variables, halophytes usually contain

V.R. Squires & A.T. Ayoub (eds.), *Halophytes as a resource for livestock and for rehabilitation of degraded lands*, 97–100.

sufficient crude protein and/or carbohydrate components to have significant nutritional potential. However, the typically high content of minerals (much of which is sodium chloride), low content of available energy (at least partly due to dilution of organic constituents by the high mineral content) and possible presence of antinutritional factors including tannins, alkaloids, oxalates and saponins are factors which compromise the usefulness of halophytes in animal production (Bayoumi & El Shaer 1994).

Incorporation of halophytes into blended diets would be advantageous because a level of inclusion could be selected to minimize adverse effects of the factors listed above on animal growth performance. Also, mixtures of halophytes and higher energy ingredients, such as legumes, cereals, molasses or fat, would be more likely (than halophytes grazed or fed alone) to support levels of animal productivity required to justify the expense of growing halophytes under irrigated conditions.

Halophyte ingredients in ruminant diets

The surprisingly large number of studies in which harvested halophytes have been fed in combination with other dietary ingredients are reviewed in detail elsewhere (Glenn et al. 1994). In general, results show organic constituents of halophyte forages are highly digestible by goats (Wiley 1982) and sheep (De la Llata 1991), and that mixed diets containing halophyte forages are acceptable to sheep (De la Llata 1991; Madrid, unpub. data) goats (Wiley 1982; Glenn et al. 1992) and cattle (Madrid, unpubl. data; Watson, unpubl. data). Halophyte forages have usually been included as 15 to 35% of diet dry matter, but De la Llata (1991) reported that a diet with 70% *Atriplex* forage was readily consumed by lambs.

Information on effects of including halophyte ingredients in mixed diets on the economically important characteristics of weight gain and efficiency of feed conversion by livestock is lacking. Limited studies conducted with goats in the Arabian Gulf (Glenn et al. 1992; Riley et al. 1994), with lambs and cattle in Mexico (Madrid, unpubl. data), show that diets containing *Atriplex* or *Salicornia* forage can support positive weight gains. Other studies have shown that *Salicornia* seed meal can be used in poultry diets (Glenn et al. 1991). However, because *Salicornia* meal contains saponins, it might be more useful in diets for ruminants (Glenn et al. 1994). A growth trial with lambs has been conducted at Arizona (Swingle et al. unpubl. data) to compare *Atriplex* forage and *Salicornia* straw with bermudagrass hay (at 30% of the diet), and to compare *Salicornia* meal (10% of the diet) with cottonseed meal (Table 1).

Ultimately, the role of halophyte ingredients, especially halophyte forages, in diets for productive livestock will be constrained by the high content of minerals. For halophytes used in blended diets, the indirect effect of ash to dilute energy concentration will probably be more important than the direct effects of salt ingestion on dry matter intake or physiological status of animals.

It is hypothesized that minerals provided by moderate levels of halophytes in blended diets could be considered 'physiologically inert' if sufficient fresh drinking water can be provided to allow excretion of excess salts. The effect of halophyte intake on weight gain would depend on the ability of animals to adjust feed intake to compensate for the lower dietary energy density. In high-roughage diets, where intake is limited by gastrointestinal fill, energy intake and weight gains would likely be lower on halophyte diets than could be achieved using conventional feedstuffs. However, in higher-concentrate diets, where intake is not regulated by gastrointestinal fill, halophyte diets might be consumed at a level that would allow energy intakes and gains equivalent to those possible on diets containing only conventional ingredients.

As a result of the lower energy density in halophyte ingredients, weight gains will necessarily be reduced if dry matter intake is the same or lower than can be achieved on conventional diets; alternatively, intake will have to be higher in order to achieve weight gains possible on conventional diets. Either scenario would increase feed requirements per unit gain (reduce gain per unit feed).

The reduced feed conversion efficiency will

Table 1. Growth performance of lambs fed diets containing halophyte ingredients (University of Arizona, 17 February through 11 May 1993 – 84 days)

Item	Diet				SEM
	Control	*Atriplex* forage	*Salicornia* straw	*Salicornia* meal	
Lambs (n)	6	6	6	6	
Live weight (kg)					
Initial	30.5	30.5	30.1	29.9	0.53
Final	50.5	50.0	51.1	52.1	1.34
Daily dry matter intake					
Weight (kg)[a,b]	1.30	1.44	1.67	1.33	0.070
% of Body wt[a,b]	3.20	3.57	4.11	3.28	0.132
Daily gain (kg)	0.24	0.23	0.25	0.27	0.015
Efficiency					
Gain: DMI[a]	0.19	0.16	0.15	0.20	0.005
DMI: gain[a]	5.39	6.32	6.74	5.05	0.204

[a]Diets with bermudagrass hay (Control and *Salicornia meal*) differ from diets with halophyte forages (*Atriplex* forage and *Salicornia* straw), $p < 0.05$.
[b]*Atriplex* forage diet differs from *Salicornia* straw diet, $p < 0.05$.
Source: Swingle et al. (unpublished data).

likely be a constant 'penalty' for using halophyte ingredients in livestock feeding systems as attempts to improve nutritional value of halophyte forages by water leaching to reduce the mineral content have not been encouraging (Wiley 1982; Glenn et al. 1992). However, the significance of the feed conversion penalty will vary. For example, if irrigated production of halophytes could support significant animal production, thus providing high quality products for human consumption, in locations where conventional feedstuffs are not available, the magnitude of the feed : gain ratio would be inconsequential. In areas where conventional feeds are abundant, reduced feed efficiency would be a significant deterrent to use of halophyte ingredients. In these cases halophyte ingredients would find application only if their cost per unit energy or cost per unit product is competitive with that of conventional feeds.

Other (non-nutritional) considerations influencing utility of halophytes as harvested feed ingredients

Forage processing. Halophyte forages must be chopped or ground to facilitate blending with other feedstuffs. In studies at the University of Arizona, *Atriplex* and *Salicornia* forages are processed using a conventional rotary-type hay mill (United Farm Tools, Turlock, CA) equipped with a 15-cm screen and powered by a 100 horsepower electric motor. It is likely that other types of equipment (e.g., hammermills, tub grinders) could also be satisfactory for processing halophytes. Relative cost for processing halophyte and conventional forages, and maintaining processing equipment, are not known, but are likely to be higher for harvested *Atriplex* forage, in comparison with grass or legume forages, because of the presence of large diameter, woody stems.

Fate of sodium and other salts. Excretion of excess salts via the urine could create an environmental quality concern in locations where halophyte diets are fed to high concentrations of livestock over extended periods of time.

Diet sorting and disposal of orts. Livestock are capable of sorting coarser halophyte forage particles (especially *Atriplex* stems) from mixed diets. Selective refusal of these coarse particles complicates evaluation of halophyte ingredients and, under practical conditions, disposal of orts would increase costs and labor requirements. Sorting could be lessened by processing forages to progressively smaller mean particle sizes, or by limit-feeding animals. However, these remedies would increase processing costs, decrease digestible energy intake, and reduce performance potential.

Conclusions

Based on observations and research to-date, it is concluded that halophytes could be effective components in mixed diets for livestock. The principal advantages would be that adverse effects due to the high mineral content of halophyte tissues could be minimized, that animal performance and economic returns would likely be higher than could be achieved by direct grazing of halophytes, and that non-forage fractions and halophyte byproducts could be utilized effectively. The principal disadvantage to use of halophyte ingredients will be reduced feed conversion efficiency, due to the dilution effect of minerals on energy density. Longer-term, more extensive animal feeding studies will be necessary to fully establish the nutritive value of halophytes in comparison to feedstuffs currently used in practical animal production systems.

Acknowledgements

Thanks are due to UNEP who provided a travel grant to one of us (EPG) to attend the workshop in Nairobi.

References

Bayoumi, M.T. & El Shaer, H.M. (1994). Impacts of halophytes on animal health and nutrition (this volume, Ch. 25).

De la Llata, M.M. (1991). Nutritive value of *Atriplex deserticola* and *Salicornia* forage for ruminants. MS Thesis, University of Arizona, Tucson, AZ.

Gihad, E.A. & El Shaer, H.M. (1994). Utilization of halophytes by livestock on rangelands: problems and prospects (this volume, Ch. 6).

Glenn, E.P., O'Leary, J.W., Watson, M.C., Thompson, T.L. & Kuehl, R.O. (1991). *Salicornia bigelovi* Torr.: An oilseed halophyte for seawater irrigation. Science 251: 1065–1067.

Glenn, E.P., Coates, W.E., Riley, J.J., Kuehl, R.O. & Swingle, R.S. (1992). *Salicornia bigelovi* Torr.: A seawater irrigated forage for goats. Anim. Feed Sci. Technol. 40: 21–30.

Glenn, E.P., Swingle, R.S., Riley, J.J., Mota, C.U., Watson, M.C. & Squires, V.R. (1994). North American halophytes: Potential use in animal husbandry (this volume, Ch. 12).

Le Houérou, H.N. (1992). The role of saltbushes (*Atriplex* spp.) in the Mediterranean basin: A review. Agrofores. Systems 18: 107–148.

Riley, J.J., Glenn, E.P. & Mota, C.U. (1994). Small ruminant feeding trials on the Arabian Peninsula with *Salicornia bigelovi* Torr. (this volume, Ch. 26).

Wiley, S.T. (1982). Nutritive value for goats of *Atriplex* species grown with hypersaline water. MS Thesis, University of Arizona, Tucson, AZ.

8. Halophytic shrubs in semi-arid regions of Australia
Value for grazing and land stabilization

A.D. WILSON
Rangeland Research and Management Consultant, 'Cal Col', Deniliquin, NSW, Australia

Abstract. This paper presents an overview of the grazing ecology and utilization of halophytes and halophytic plant communities in Australia. Halophyte species are dominant over large areas of the arid and semi-arid winter rainfall regions. They support an extensive Merino sheep industry. Most halophyte shrubs are high in protein content and of moderate digestibility. However, food intakes are low when fed as a sole ration, so that their value is limited to providing a green maintenance forage during the dry season when alternative forage is both scarce and of low quality. Livestock performance is improved when the shrubs are fed as mixtures with grasses. This takes advantage of the high protein and avoids the problems of anti-nutritional compounds. In natural communities the chenopod species are low in the preference hierarchy and are only eaten in substantial quantities during periods of low rainfall when other forage species are scarce. They require periodic rests from grazing, either through their natural low dietary preference during the growing season, or through the removal of stock. It is concluded that there is considerable potential for reestablishment of chenopod shrubs in areas that have been overgrazed in the past. A number of species also have potential for planting as drought forage reserves and others are proving valuable for the reclamation of saline land.

Key words: *Atriplex* spp., Chenopods, Nutritive value

Introduction

Halophytic and xeromorphic plant communities play a significant part in providing stability to soils and forage for grazing livestock across southern Australia. The forage contribution is central to the value of many arid plant communities and forms the backbone of the pastoral industry in the southern inland. The carrying capacity of these lands is low, but forage quality remains high, in part due to the contribution from halophyte chenopod shrubs. This gene pool, together with introduced species, is now being utilized for establishment on land were salt has been brought to the surface by rising water tables. This has arisen from both irrigation and the clearing of trees from semiarid landscapes where salt was previously stored at depth.

The contribution to land stability is no less important. Halophytes provide perennial cover, both reducing erosion and improving the fertility of the soil surface. Fertility is improved through the trapping of nutrients under their canopy and by providing protection to other herbaceous species that grow beneath or between the shrub canopies (D.J. Tongway, pers. commun.).

This paper presents an overview of the halophytic species and plant communities in Australia and their pastoral significance. It also includes an overview of the use of halophytes in areas of induced salinity, although this topic is dealt with in more detail in a companion paper (Malcolm 1994).

V.R. Squires & A.T. Ayoub (eds.), *Halophytes as a resource for livestock and for rehabilitation of degraded lands*, 101–113.

Geography

Natural halophytic plant communities are common around the littoral margins of the continent and within inland basins with terminal drainage. Examples are the mangroves of the northern coasts and tussocky or rhizomatous grasses such as *Sporobolus virginicus* (Carnahan 1955). The shrub *Nitraria billardieri*, which is found on saline soils, both inland and littoral, has been used for stabilization of disturbed land such as mine dumps (Noble & Whalley 1978).

Natural communities and species that are useful for livestock mainly occur in the low rainfall inland regions of southern Australia (Lat. 24° S–34° S). This region receives 150–400 mm of rainfall annually, with 30–50% falling in summer, with a high variability. The winter component is more reliable and hence there is a predominance of growth during the cool season (Oxley 1979).

These shrubs occur over a wide range of soil types, from heavy grey clay to sandy calcareous and rocky soils, with each species having a distinct soil preference. In some regions, such as the Nullabor and Riverine Plains, chenopod shrubs are the dominant species, whilst in others they form an understory within open woodlands. They also occur along saline drainage lines and on skeletal rocky upland soils. In total they comprise some 500,000 km^2 of the Australian landscape (Graetz & Wilson 1984).

Introduced halophytes that have found a useful place for colonizing saline land in agricultural districts include tall wheat-grass (*Agropyron elongatum*), puccinellia (*Puccinellia ciliata*), and *Atriplex* spp. The grass *Vetiveria vaginatum* has also been found useful for soils of low salinity in subtropical climates (Truong 1994). Further detail on these species is presented by Malcolm (1994).

Growth and ecology

Species

The species of principal interest in Australia are from the family Chenopodiaceae. There are some 200 species within the genera *Arthrocnemum*, *Atriplex*, *Chenopodium*, *Enchylaena*, *Maireana*, *Pachycornia*, *Rhagodia*, *Sclerolaena* etc. Some species also have a number of distinct ecotypes (Parr-Smith & Calder 1979; Parr-Smith 1982), although the level of ecotypic variation is poorly understood. In natural communities the species are specific to certain soil types, with a dominance of *Atriplex* spp on clay soils and *Maireana* spp on sandy loam, calcareous and texture contrast soils. The larger species are deep rooted and some (e.g. *A. amnicola*) withstand periods of waterlogging (C.V. Malcolm, pers. comm.). The major perennial species of value as forage plants are listed in Table 1.

The family Chenopodiaceae includes species with both C3 and C4 metabolic pathways (Chapman & Jacobs 1979). Growth in well watered situations can continue up to 45 °C in C4 species such as *Atriplex nummularia* and they are capable of growth rates similar to cultivated crop species. Growth optima are nearer 25 °C for C3 species (Jones et al. 1969). However, growth in natural communities is usually limited in all species by water deficit, so that production is usually moderate and the main growing seasons are spring and autumn. Insect attack is uncommon, but areas of *A. vesicaria* are killed at times from attack by an unknown caterpillar (Clift et al. 1987).

Salt tolerance

Many of these species are salt accumulators, which grow more vigorously at moderate levels of salinity (100 mM NaCl) than at low levels of salinity (1–10 mM NaCl). They tolerate high levels of salinity, although growth rate is reduced (Ashby & Beadle 1957; Sharma 1982). *Atriplex* species secrete salt into bladder cells attached to the surface of the leaves. As a consequence mineral levels of the leaves (including the bladders) are high, with important consequences for herbivores grazing on them (see below). Sodium levels may reach 8% of the leaf dry weight and ash levels 35% of the dry weight (Wilson 1966). Atriplex species maintain a gradient between leaf salinity and culture solution salinity of about −12 bars. They are also very drought tolerant and have an ability to extract 40% more water from the soil than the conventional wilting point of −15 bars (Sharma 1976). Leaf salt levels are thus at a maximum under drought or high levels of soil salinity, and the salinity levels of surface soil may increase beneath the leaf canopy as a consequence of leaf drop (Sharma & Tongway 1973).

Table 1. The main perennial chenopod species that are grazed by livestock in Australia

Species	Form	Palatability*	Natural habitat
Atriplex amnicola	shrub	moderate	river flood plains
A. bunburyana	shrub	low	flood plains, sand dunes
A. cinerea	shrub	moderate	coastal dunes
A. eardleyae	subshrub	moderate	range of soils
A. lentiformis	shrub	moderate	loam soils
A. nummularia	shrub	low	clay soils
A. paludosa	shrub	moderate	clay and duplex soils
A. rhagodioides	shrub	moderate	saltlake margins
A. semibaccata	prostrate	moderate	loam to clay soils
A. stipitata	slender shrub	low	mallee soils
A. undulata	shrub	moderate	Argentinian rangeland
A. vesicaria	low shrub	moderate	range of soils
Chenopodium auricomum	shrub	moderate	clay soils, lowlying
C. nitrariaceum	shrub	moderate	clay soils, lowlying
Enchylaena tomentosa	shrub	moderate	range of soils
Frankenia sp.	low shrub	low	salt pans, clay flats
Halosarcia pergranulata	subshrub	very low	saline, waterlogged
Maireana aphylla	shrub	low	clay plains
M. astrotricha	shrub	moderate	calcareous sandy loams
M. brevifolia	shrub	moderate	range of soils, salin
M. convexa	sparse shrub	moderate	flood plains, stony
M. excavata	forb	moderate	clay loam soils
M. georgei	low shrub	moderate	sands to loam soils
M. glomerifolia	subshrub	moderate	stony plains
M. pentagona	forb	moderate	red and brown clays
M. planifolia	low shrub	moderate	alluvial
M. platicarpa	low shrub	moderate	alluvial plain
M. polypterygia	low shrub	low	alluvial plain
M. pyramidata	shrub	very low	calcareous sandy loam
M. sedifolia	shrub	low	calcareous sandy loam
M. suedifolia	slender shrub	moderate	sand dunes, alluvial
M. villosa	subshrub	moderate	gravelly skeletal soil
Rhagodia eremaea	low shrub	moderate	sandy loam soils
R. nutans	trailing forb	moderate	range of soils
R. spinescens	shrub	moderate	range of soils
Sclerolaena brachyptera	subshrub	moderate	range of soils
S. diacantha	subshrub	moderate	sandy to clay loam soils
S. lanicuspis	subshrub	moderate	range of soils

Sources: Wilcox & Morrissey (undated); Cunningham et al. (1981), Mitchell & Wilcox (1988).

Where shrubs are growing above a saline watertable, increases in soil salinity may occur within the root zone (Malcolm et al. 1988). This indicates that there are long-term implications for the use of chenopod shrubs on saline land. Salt will be redistributed and long-term yields may differ to those recorded in the short-term.

Growth and establishment

In natural situations chenopod species present a reasonably constant biomass, at least in contrast to herbaceous plants. They are thus a relatively more important component of pastures during dry times than during moist periods. However, leaf fall can occur during droughts. As an example, the proportion of leaf of *A. vesicaria*, varies from a high of 54% during moist conditions to a low of 22% during dry conditions (Graetz & Wilson 1984).

The temperature optimum for germination of *Maireana* and other species is 20–25 °C (Williams 1979). There is also a requirement for a long period of high moisture, which acts to remove inhibiting salts from the surrounding bracts or membranes (Burbidge 1946). Natural germination is therefore restricted to periods of high rainfall (e.g. 50 mm spread over a number of days (Williams 1979). The juxtaposition of favourable temperature and moisture conditions is rare and natural establishment of even those species that germinate freely, such as *A. vesicaria*, occurs only once every 3–5 years (Crisp 1978). In *Maireana* species, establishment is even less frequent because the seed rapidly loses viability. A sequence of favourable rainfall events is required, initially to provide viable seed and subsequently to germinate and establish that seed (Williams 1979). The species *M. sedifolia* may establish only once every 50 years under natural conditions, but has a longevity in excess of 150 years (Crisp 1978). The inverse relationship between frequency of establishment and longevity of the chenopod species reflects differences in basic life strategy (Graetz and Wilson 1984). Turnover rates of 50% per annum for *A. vesicaria*, 3–8% for *M. pyramidata* and 4% for *M. astrotricha* have been recorded (Eldridge et al. 1990). The ease of establishment and longevity are major factors to consider in choosing species for sowing.

Germination and establishment can be improved under cultivated situations by using methods that reduce the salinity of the soil surface, reduce moisture loss and increase soil temperature. Successful techniques include the use of raised beds, covering the seed with a vermiculite mulch and/or a latex paint. Germination can be increased 2–3 fold by these treatments (Malcolm & Swaan 1985; Malcolm 1994). The same treatments may be used to improve establishment in arid areas. In addition, the use of gypsum on saline or clay soils and the placement of seed beside furrows that collect water, may increase germination and establishment 3-fold (Ward 1992).

Grazing ecology

In natural communities, these chenopods occur as mixed stands, sometimes with several chenopod shrubs or subshrubs, and always with numerous cool season herbaceous annuals or biennials growing on the intervening soil after rain. These communities may have a chenopod foliar cover of only 10–15%, but this is highly significant for protecting the soil from erosion and supporting livestock during the dry season. The shrubs are important for providing stability to the vegetation, soil and animal production (Graetz & Wilson 1984). In these mixed communities the shrubs are usually the least palatable of the species, at least while they are green. A typical heirarchy of selectivity by sheep would be annual herbs > perennial grasses > dry herbage > chenopod shrub. Where shrub stands are mixed there is also a heirarchy of selectivity between shrubs, such as *Atriplex vesicaria* > *M. astrotricha* > *M. sedifolia* > *M. pyramidata* (Graetz & Wilson 1984). Cattle have a similar preference heirarchy, but less ability to select small green plants from between bushes and therefore eat much more of the larger shrubs than sheep (Graetz & Wilson 1980). A consequence of selection for grasses is that chenopod shrubs are rested from grazing after rain. It appears that these rests from grazing are essential for shrub survival, because continuous grazing leads to the early demise of shrub populations. This has occurred over wide areas of their natural distribution (e.g., Lay 1979; Stanley 1983).

On the other hand the shrubs also benefit from regular grazing. For instance it has been observed that the shrub *A. vesicaria* is more vigorous and has a greater cover of edible foliage when grazed intermittently than when left ungrazed (Andrew & Lange 1984). The requirement is for intermittent defoliation, but not continuous grazing. The best result is achieved through resting over the growing season, which also has the advantage of resting young seedlings that are more susceptible to defoliation than older plants (e.g., Milthorpe 1978).

The impact of grazing is seen not only as a loss of shrubs, but as shifts in composition from the more to the less palatable chenopods. Thus so called 'piospheres' form around sources of drinking water, with the species that are least palatable and most tolerant of grazing dominating in the innermost circle near water and the more palatable species relegated to the outer circles some 1.5–2 km out from water. A typical sequence formed in the South Australian arid lands shows *Maireana brevifolia* near water, *M. georgei* and *M. pyramidata* at intermediate distances and *Atriplex vesicaria* further out (Barker & Lange 1969). Many smaller shorter-lived chenopods also occur in this sequence. In planted situations in Western Australia, *M. brevifolia* has been shown to be more palatable than *A. vesicaria*, but better able to withstand heavy grazing (C.V. Malcolm, pers. comm.).

Management

The general conclusion on management of the most defoliation susceptible species, such as *A. vesicaria*, is that grazing lands must be lightly stocked to ensure that heavy continuous grazing does not occur and that the land must be subdivided by fencing to prevent large numbers of stock congregating on one area. The successful prescription detailed by Lange et al. (1984) is that sheep numbers be kept low ($<0.15\,ha^{-1}$ for a 200 mm rainfall) and that there be sufficient subdivision so that there is limit of 250–350 sheep in any paddock and on any single water point. Where the chenopod shrubs form only part of the landscape (e.g. the alluvial soils in an *Acacia aneura* arid woodland), they must be fenced separately and used as specialist forage during the dry autumn and winter period, otherwise they will be overgrazed (Morrissey 1984). Grazing distribution can also be improved by placing water supplies away from preferred grazing areas.

The more defoliation tolerant species require rests from grazing to allow regrowth and seeding. This may be achieved naturally through the grazing of more palatable grasses during the wet season, or through reservation from grazing until the dry season.

It is apparent that the Australian chenopod halophytes have particular requirements for establishment and management if they are to be retained under grazing for any length of time. In natural situations this is achieved by rigorous control of livestock numbers and the natural rests from grazing that occur after rainfall when other species are preferentially grazed. Grazing regimes must also allow for intermittent seeding and reestablishment. For long-lived species re-establishment can be infrequent, but can only be achieved through careful control of grazing during periods of favorable rainfall. Without these controls on grazing there is a change in botanical composition towards the less palatable and shorter lived species that are of less value for maintaining livestock during the dry season. In cultivated plantations, species can be chosen for a greater tolerance to defoliation and for longevity, but they still require a form of rotational grazing so that long periods of rest are allowed during the main growing season. Differences in species tolerance to grazing can be considerable, with species such as *A. amnicola*, *A. brevifolia* and *A. nummularia* recovering well after grazing (C.V. Malcolm, pers. comm.).

Livestock industries

The saltbush and bluebush pastures in Australia are grazed almost exclusively by Merino sheep. The flocks are large in total (5,000–12,000), but are subdivided into up to 20 separate flocks and graze at relatively low stocking rates (see Graetz & Wilson 1984). These sheep grow medium (21–23 μm) to strong wool (i.e. not so fine = 23–25 μm) and wool constitutes about 70% of farm income.

Cattle may be grazed in conjunction with sheep, but this is usually only in favorable seasons when forage is in excess. Cattle growth is generally satisfactory (see Wilson 1980) but on average cattle produce a lower economic return than sheep on this type of land. Returns from cattle are also more variable, because no income is available in drought years, whereas some income is available from wool, albeit at a lower level than in normal years.

The sheep grazed in the chenopod pasture lands have a low productivity per hectare, but a high productivity per head. Wool production and reproduction levels are the highest of all geographical regions in Australia (Brown & Williams 1970) and production is also very stable in spite of large variations in rainfall (Roshier, pers. comm.). At least a portion of this productivity and stability comes from the presence of chenopod species that retain a higher value than grasses during the dry season and throughout the periodic droughts.

Sheep kept for wool growing are also the main animal enterprise in those parts of the agricultural regions where sown annual pastures and crop stubbles are the main source of forage. Chenopod pastures may be planted for saltland reclamation

Figure 1. Halophytic communities dominated by *Atriplex* are the basis for livestock husbandry over thousands of square kilometres of southern Australia.

on parts of these farms. In these situations the sheep have a marked nutritional gap in autumn when the pastures and crop stubbles are both sparse and of poor quality.

Nutritive value

The nutritive value of chenopod species has been difficult to determine, because of their low palatability and unusual composition. They have a high salt content and also contain secondary compounds (e.g., saponins, oxalate) that may inhibit forage intake. As a consequence livestock performance varies according to whether the plants are fed alone or in mixtures. Furthermore, value is not simply determined by quality alone. The quantity of forage produced and the timing of its use are important variables to consider in addition to protein content and digestibility. Therefore one cannot provide an absolute feeding value for these species. Value varies according to the quality of the species itself, of the alternative feed supply and the pastoral context in which the plant is used.

Analyses

A wide range of chenopod species has been analysed for crude protein and in vitro digestibility. Some typical values are given in Table 2. The most important conclusion from these analyses is that crude protein levels are high, particularly when compared with alternative forages that may be available at the same time. Initial levels may be 15% or higher in leaves, and are rarely less than 10% even during the dry season. Protein levels in fine stems, that may also be grazed, are much less at 4–6%. This may account for much of the variation between samples taken from different locations, although there is undoubtedly also variation arising from leaf age and growth conditions.

The digestibility of chenopod shrubs varies from 70% or higher, down to a low of about 50% for some species during the dry season (Wilson 1966b; Warren & Casson 1992). These values are less than for green herbaceous forages, but higher than alternative dry forages available during the dry season. A reduced amplitude of seasonal variation is one of the important features, when compared with herbaceous forages. The analyses are

Table 2. Crude protein (%), digestibility (%) and ash (%) values obtained for some species of chenopod halophytes

Species		Crude protein	Digestibility		Ash
			Dry matter	Organic matter	
A. vesicaria		11–18	52–54	44–53	18–36
A. nummularia		17–22	68–74	65–69	27–39
A. undulata	mixed	10	53	–	–
	leaf	10–18	70	–	30
A. lentiformis	mixed	12	62	–	–
	leaf	15–16	80	–	18–27
A. amnicola	mixed	99	57	–	–
	leaf	9–16	70–78	–	27
A. bunburyana	leaf	10	72		30
A. cinerea	mixed	9	60	–	–
	leaf	–	69–75	–	28
M. brevifolia		20	68	–	28
M. pyramidata		13–15	58	58	24

Data from Wilson (1966b, 1977), Malcolm et al. (1988), Warren et al. (1990), P. Morcombe (pers. comm.) and B.E. Warren (pers. comm.).

confused to an extent by expression on a dry matter basis, given that there are high levels of soluble ash in the leaf. All analyses should be placed on an ash free basis, which may reduce digestibilities by up to 5% (see Wilson 1966b), but this is rarely done. There may also be error in using *in vitro* values, that sometimes differ markedly from values determined by direct feeding (Wilson 1977). Even when these errors are overcome, there seems to be little correlation between the values obtained and forage value. For instance, *A. nummularia* has one of the highest crude protein content and digestibility values of the chenopods, yet does not perform better than any of the other species when fed.

Food intake

The results from analyses have led many observers to claim a high forage value for chenopod species in general, but this has proved to be misleading. A protein content equivalent to lucerne does not mean that forage value is equivalent to lucerne. This is mainly because food intakes are invariably lower than expected for forages of such high protein and digestibility, although the low organic matter levels also contribute to lower than expected livestock performance.

The most notable example of this is *Atriplex nummularia*, which has intakes as low as 400 g day^{-1} of organic matter (approx 550 g dry matter) in sheep at an organic matter digestibility of 65–67% (Wilson 1966b). Other saltbushes and bluebushes seem to behave the same. Thus Warren et al. (1990) recorded dry matter intakes of 400–800 g/day for 4 species (*A. undulata, A. lentiformis, A. amnicola, A. cinerea*) that had dry matter digestibilities of 53–62%. *A. cinerea* had the highest intake (780 g/day) and dry matter digestibility (60%), but this was still insufficient to maintain sheep weight and a loss of 210 g/day occurred over a 3 week period. Weight maintenance occurred in the earlier research (Wilson 1966b), but weight changes over short periods are probably unreliable because of the large fluxes of salt and water associated with feeding chenopod species. Reasons for the poor food intakes are not known definitely, but could be associated with high salt content or the presence of other compounds, such as oxalates (see below).

Le Houérou (1992) has reported that sheep become adapted to saltbush and increase their intake of forage over a 3–5 month period. Such an effect has not been looked for in Australia and is worthy of further investigation.

Value as supplements

The problems of feeding chenopods seem to be largely overcome by feeding them as a supplement or as part of forage mixtures. Thus a supplement of 284 g/day of *A. nummularia* increased the total intake of hay and saltbush by 200 g/day (Wilson 1966b). A mixture of *A. undulata* and hay had an intake of 1449 g/day, compared with intakes of only 850 g/day for hay and 615 g/day for *A. undulata* when fed alone (see Table 3) (Warren et al. 1990). This suggests that the effect of the secondary compounds is diluted and livestock performance then reflects the true nutritional value of the combined ration.

Under natural situations chenopods are eaten as a part of a herbaceous mixture. For instance, sheep consumed 10–15% saltbush in winter and 55% or more in summer when grazing on a *A. vesicaria* pasture (Wilson et al. 1969). The presence of saltbush places a minimum on forage value over the dry season of approximately 8% crude protein and 52% digestibility (Wilson &

Table 3. Food and water intakes of sheep fed on saltbushes, hay or a mixture of saltbush and hay (from Warren et al. 1990)

Diet	Hay	Hay/*A. undulata*	*A. undulata*	*A. lentiformis*	*A. amnicola*	*A. cinerea*
Feed (g/day)	850	1449	615	700	843	858
Water (L/day)	1.8	6.8	4.4	4.5	6.4	6.8
Weight (g/day)	−25	70	−225	−162	−169	−210
DMD (%)	64	64	53	62	57	60
CP (%)	3.9	7.1	10.2	12.4	9.4	9.0
Ash (%)	3.3	9.1	14.8	14.6	18.3	17.7

Graetz 1980). Weight of adult animals is maintained under these situations for many months.

Palatability

The proportion of each chenopod species eaten under natural grazing is influenced by their palatability relative to the other species present. The palatability of chenopod species varies from moderate to very low. Some ratings for palatability are given in Table 1. Moderate palatability is desirable, because this ensures that the plants are not grazed before the dry season. Low palatability, on the other hand, can be undesirable, because it may lead to low intake and hence poor performance of grazing animals. Plants such as *A. nummularia* are not grazed substantially until all alternative herbage is consumed.

Grazing value

Livestock performance in planted stands of chenopods conforms to pen feeding experience. The addition of 400 bushest ha^{-1} of *A. nummularia* has been shown to be valuable for maintaining sheep through drought periods (Leigh & Wilson 1969). Plantings of *A. lentiformis* and *A. undulata* have been useful in reducing weight loss of sheep grazing on low quality salt tolerant grasses such as mature tall wheat grass (*Agropyron elongatum*) (Warren & Casson 1992). In this example the saltbush was seeded into the tall wheat grass and provided 0.2–0.4 t ha^{-1} of higher quality forage. In another trial where sheep were grazed on individual species of saltbush, maintenance of weight in dry sheep was again evident, compared with weight loss on alternative dry grass forage. Despite differences in protein and digestibility between species, no difference in sheep weight was observed on individual species plantations of *A. cinerea*, *A. lentiformis*, *A. amnicola* or *A. undulata* (Warren 1992). Provided the bush is sufficiently palatable to be eaten, differences between species may not be important. If this is so then species can be chosen mainly on the basis of agronomic characteristics.

Where saltbush has been used for grazing lactating ewes the weight loss has been large (P. Morcombe, pers. comm.) and it would appear that these forages are unsuitable for production situations, except as a minor component of diet. It must be concluded that chenopods are only maintenance forages and should not be used as a sole ration in situations where weight gain or lactation is required. Nevertheless, maintenance can be a valuable asset during the dry season, particularly if high stocking rates can be maintained for short periods. Chenopods accumulate forage which can be used in the dry season and in plantation situations may hold up to 3000 kg ha^{-1} for use at these times.

The general conclusion is that these forages are best used as a component of a mixed diet where they can act as a protein supplement and where the salt and other secondary components that reduce feed intake present less of a problem. Laboratory analyses are of limited value in determining feeding performance, since there is little correlation with food intake or liveweight gain.

Special problems

The presence of secondary compounds that reduce forage value has been noted earlier. The ones that are known are oxalates and salt. Oxalates are present in significant quantities in some *Maireana* species (Malcolm et al. 1988). For instance, *M. brevifolia* has been reported to contain various levels between 3 and 12% total oxalates

(Malcolm 1969; P. Morcombe, pers. comm.). The variation may arise from different growing conditions and/or different analytical techniques. These may cause some kidney damage and deaths of hungry sheep (see Malcolm 1969) unless diluted by feeding with other low oxalate forages. Oxalates may also be contributors to the low palatability of these plants. However, no adverse affects have been reported from widespread feeding of natural *Atriplex* stands, so the problem should not be over-rated.

High salt content is perhaps the major negative component in *Atriplex* species. Sodium levels can be in excess of 8% of dry weight and total mineral matter above 30% (most of which is soluble) for many species, with the higher values obtained during the dry season and when grown under saline conditions (Wilson 1966a). These are precisely the situations in which these forages will be used. As a consequence intakes of sodium chloride by sheep can exceed 100 g/day. This undoubtedly has some depressing effect on food intake, although it is difficult to specify its extent. Trials with artificially added salt show that salt intakes above 80 g/day begin to reduce food intake, but the degree of effect depends on the type of food, with depression occurring earlier with low value forages than with high value forages (Wilson 1966c). B. Warren (pers. comm.) suggests an upper limit of about 100 g/day^{-1} of salt. Since this level can be easily exceeded in saltbushes, some mixing with low salt forages, such as halophytic grasses, is desirable.

The other major affect of the high salt intakes is its effect on water requirements. Water needs rise in direct proportion to the amount of salt ingested at a rate of approx 30 ml g^{-1} of sodium chloride ingested (Wilson and Hindley 1968) and leading to total water intakes up to 12 L/day for sheep and 110 L/day for cattle in summer (Wilson & Graetz 1980). Annual water requirements, including the periods of low water demand in winter, were recorded as 3300 L/year for sheep and 46,000 L/year for cattle. This places a burden on water supply systems. In addition it adds a requirement that the water be of good quality, because further salt intake cannot be tolerated. The intake of saltbush in sheep is reduced at levels as low as 0.6% NaCl and intake is halved at 0.9% NaCl (Wilson 1966a). The high salt intake also increases the frequency of drinking to at least twice a day and this limits the grazing range of livestock. It is one of the reasons why water supply is so important in the management of saltbush grazing lands (Lange et al. 1984).

The secondary consequence of increased water intakes is that the actual increase in water intake can be used to estimate the intake of saltbush by grazing livestock. If the salt content of saltbush leaf is known, the value of 27–30 ml g^{-1} NaCl in Merino sheep and 36 ml g^{-1} in British breeds of sheep can be used to determine saltbush intake. Water intakes up to 50 ml g^{-1} occur when there is free access to water in pens. A control flock fed on hay or grass is required to correct for normal water demand (Wilson & Leigh 1969).

Halophyte plantations

Plantations of chenopod species are now being increasingly used in Australia. One application is the planting of the large shrub species *Atriplex nummularia* as a dry season grazing reserve in semi-arid eastern Australia. The other is the planting of mixed halophyte species for the reclamation of saline wasteland in Western Australia.

Atriplex nummularia plantations

The plantations of *A. nummularia* are established with nursery raised 'speedlings', which are planted in 3 m rows into cultivated and ripped clay soils. These are ready for grazing after 2 growing seasons when yields of 2–3 10^3 kg ha^{-1} of edible biomass can be expected. The rainfall is 350–400 mm/year spread evenly throughout the year. These plantations are grazed during the dry autumn period when herbaceous forage is often in short supply, although they may also be lightly grazed in winter or spring. These yield about 3,000 grazing days ha^{-1} for sheep, on the shrub-pasture mix, compared with about 500 grazing days ha^{-1} for normal pasture (Condon & Sippel 1992). However, a comparison of grazing days is not a sufficient evaluation, because the plantation supplies a bulk of maintenance feed at a time of forage shortage, whereas the pasture provides a high quality feed at other times. The two types of forage are complementary.

Sheep performance in these plantations has not been well documented, but the scant information available suggests that nonlactating sheep maintain weight during at least the first month when grazing is mainly on saltbush leaf and some dry herbage is available between the bushes. Weight losses does not occur if stock numbers are limited to 2000 grazing days/ha/year. This is consistent with the pen feeding studies outlined earlier.

Economic evaluation

The critical evaluation of such plantations is economic, because they are costly to establish. Each seedling costs 15–30 cents (Australian) planted. Calculations of expected return, based on known nutritive value and costs, together with estimated yields, show a variety of returns, depending on yield and establishment cost (Wilson & MacLeod unpublished data). The assumptions for these calculations are that the saltbush has an organic matter yield of 65% and a digestibility of organic matter of 65% (Wilson 1966b), hence providing a metabolizable energy yield of 7 MJ/kg. Growth rates were assumed to increase by 0.7kg/bush/year, up to a peak yield of either 4.4 kg bush or 2,200 kg/ha.

In the first case shown in Table 4, the forage was used to replace purchased grain once every second year, on the assumption that a deficiency in herbage growth arising from low rainfall would occur at that frequency on many farms. The forage value was then set at 1 cent/MJ, equivalent to a cost of $120/tonne of oats. The returns varied from 4–24% depending on the plant density and the seedling cost. In this model, close plantings reached peak yield at an earlier stage, but are too costly, with a balance between early yield and cost occurring in the region of 1000 bushes/ha. Returns are then adequate if seedling costs can be contained at the lower end of the commercial range (i.e. 15 cents). Different returns will apply if the forage is used in other ways, but the assumption of a replacement for purchased forage every second year is a reasonable parallel with practice in the region. In this case the bush is being valued as a drought reserve even if grazed annually to maintain the shrub in a leafy state.

In the second case the extra bush forage was used to raise carrying capacity and the return was then valued at the gross margin of $12.50/sheep for each extra animal carried. In this situation returns were much lower at only 1% for dense plantings, rising to 14% at 500–1000 bushes/ha. Useful returns are therefore only achieved where the plantation fills a seasonal feed gap and the forage has a higher value.

Table 4. Economic analysis of *Atriplex nummularia* plantations established from seedlings and used as a fodder bank (Wilson & MacLeod, unpublished data). See text for assumptions

Case	Density (bushes ha^{-1})	Seedling cost (cents)	NPV[a] ($)	Payback period[a] (years)	IRR (%)
Drought feeding	3300	30	(406)	16	4
	3300	15	67	15	12
	1000	30	104	13	14
	1000	15	235	8	22
	500	30	159	10	18
	500	15	215	8	24
Higher carrying capacity	3300	30	(595)	16	1
	3300	15	(122)	16	7
	1000	30	(61)	16	8
	1000	15	69	13	14
	500	30	14	15	11
	500	15	70	12	15

[a]At discount rate of 10%.

These calculations are not definitive, because the necessary yield*age*density data is not available, but they are sufficient to show that economic value is critically dependent on planting cost and yield. Costs can be reduced without much impact on yield by planting at wider spacing than is currently practised. An alternative way of reducing cost is to sow seed directly in prepared furrows, but techniques for sowing with *A. nummularia* are unreliable.

Plantations on saline land

Plantations on saline land in Western and South Australia have been successful in obtaining valuable production from otherwise useless land. A brief summary is given here and further detail is presented in the chapter by Malcolm (1994).

These plantations cover a variety of species, such as *A. amnicola*, *A. undulata*, and *M. brevifolia* and may yield 1500–2500 sheep grazing days ha^{-1} (Malcolm & Pol 1986). Forage production from the shrub is in the order of 0.5–1.0

10^3 kg ha^{-1}, in addition to a similar amount of other herbage. However, other trials suggest that much higher yields of up to 3 10^3 kg ha^{-1} could be obtained with attention to soil preparation, spacing and nutrition (Barrett-Lennard 1992). These plantations are also used for grazing sheep during the autumn when herbaceous forage is both scarce and of poor quality. There is therefore advantage in using the forage in an integrated way with other non-saline land that supports normal pasture or crop stubble. Analyses using whole farm economic models have shown that small areas of reclaimed saltland (e.g. 50 ha on a 1280 farm) have a value of $28/ha if the yield achieved is 1 t ha^{-1} and $153/ha if the yield achieved is 2 10^3 kg ha^{-1}. Returns are also sensitive to wool prices (Bathgate et al. 1992). In keeping with the *A. nummularia* case outlined above, cost advantage can be obtained from planting the larger growing shrub species, such as *A. cinerea*, at wider spacings (Barrett-Lennard 1992). However, the more common method is to establish these plantations by direct seeding, which reduces cost to less than half that of seedlings. The methods used for direct seeding were outlined earlier.

Conclusions and research requirements

The place of chenopod species in natural plant communities of semi-arid areas is now well understood. The chenopods are grazed during times of forage shortage and assist in providing high nutritional levels on low carrying capacity lands. However, the species are readily overgrazed and they have been lost from many areas as a consequence.

Recent research on animal response shows that the highest value is obtained from these shrubs when they are fed as part of a forage mixture and when feeding is reserved until the dry season when alternative forage is scarce and of low quality. It is apparent that chenopods. whether grown naturally or sown in plantations, require association with grass forages. Establishment from seed is now established practice in South and Western Australia, with costs varying between $100 and $250/ha. Positive economic returns are obtained when the plantations are integrated into a whole farm and provide sufficient forage for maintenance of the flock during a seasonal feed gap. In saline situations the chenopods may be best planted in conjunction with halophytic grasses that are of low quality, but have a low salt content. Correct grazing management is critical for retaining yield and persistence.

The most urgent needs for research are in the development of agronomic techniques that will increase yield and reduce establishment cost. Costs of plantations may be reduced by finding the optimum return: cost for a variety of spacings, which will be sensitive to soil type, rainfall and soil salinity and by the field sowings of seed. There is also need to research the long-term outcome of plantations grown on saline ground to determine if there is a deleterious accumulation of salt in the root zone.

There is probably sufficient information available now on forage value in pen situations, but more information is still required on feeding in natural situations where integration with other forages is a feature. The possibility that there is long-term adaptation to chenopod forages, with increases in intake over 3–5 months, needs further investigation. Information on species yield in various soil and rainfall environments is still required. As research proceeds it is inevitable that specialist niches and regional situations for each species will be determined and information on ecotypes suitable for particular environments should develop. It is unlikely that one species or ecotype will be the best for all environments.

Subsequently further research will be required into the grazing management needed to enhance both yield and longevity. There may also be opportunity for selection within a species for yield and palatability.

Acknowledgements

I am grateful for critical comments made by Mr C.V. Malcolm and Dr B.E. Warren, and to UNEP for travel grant.

References

Andrew, M.H. & Lange, R.T. (1986). Development of a new

piosphere in arid chenopod shrubland grazed by sheep. Aust. J. Ecol. 11: 411–24.

Ashby, W.C. & Beadle, N.C.W. (1957). Studies in halophytes, III: Salinity factors in the growth of Australian saltbushes. Ecol. 38, 344–352.

Barker, S. & Lange, R.T. (1969). Effects of moderate sheep stocking on plant populations of a blackoak-bluebush association. Aust. J. Bot. 17: 527–537.

Barrett-Lennard, E.G. (1992). Maximizing production of *Atriplex* species. In: Davidson, N. & Galloway, R. (eds), Productive use of saline land. ACIAR Canberra, Proc. No. 42, 90–93.

Bathgate, A.D., Young. J. & Barrett-Lennard, E. (1992). Economics of revegetating saltland for grazing. National Workshop on Productive Use of Saline Land, Waite Agricultural Research Institute. South Aust. Dept. Agric., pp. 87–94.

Burbridge, N.T. (1946). Germination studies of Australian Chenopodiaceae with special reference to the conditions necessary for regeneration, II: (a) *Kochia sedifolia* FvM, (b) *K. pyramidata* Benth, (c) *K. georgei* Diels. Trans. R. Soc. South Aust. 70: 110–119.

Carnahan, J.A. (1955). Atlas of Australian Resources. Second series. Natural Vegetation. Department of National Resources, Canberra.

Chapman, E.A. & Jacobs, S.W.L. (1979). Photosynthetic responses in semi-arid environments. In: Graetz, R.D. & Howes, K.M.W. (eds.), Studies of the Australian arid zone, VI: Chenopod Shrublands. CSIRO Melbourne; pp. 41–53.

Clift, D.K., Semple, W.S. & Prior, J.C. (1987). A survey of bladder saltbush (*Atriplex vesicaria* Heward ex Benth) dieback on the Riverine Plain of south-eastern Australia from the late 1970s to 1983. Aust. Rangel. J. 9: 39–48.

Condon, R.W. & Sippel, A.L. (1992). Productivity of oldman saltbush in relation to rainfall and economic considerations. Proc. 6th Aust. Agron. Conf. Armidale. Aust Soc. Agron. pp. 150–153.

Cunningham, G.M., Mulham, W.E., Milthorpe, P.L. & Leigh J.H. (1981). Plants of Western New South Wales. NSW Govt. Printing Office, Sydney.

Eldridge, D.J., Westoby, M. & Stanley, R.J. (1990). Population dynamics of the perennial rangeland shrubs, *Atriplex vesicaria*, *Maireana astrotricha*, and *M. pyramidata* under grazing, 1980–87. J. App. Ecol. 27: 502–512.

Graetz, R.D. & Wilson. A.D. (1980). Comparison of the diets of sheep and cattle grazing a semi-arid chenopod shrubland. Aust. Rangel. J. 2 (1): 67–75.

Graetz, R.D & Wilson, A.D. (1984). Saltbush and bluebush. In: Harrington, G.N., Wilson, A.D. & Young, M.D. (eds.), Management of Australia's rangelands. CSIRO Melbourne, pp. 209–222.

Jones, R., Hodgkinson, K.C. & Rixon, A.J. (1969). Growth and productivity in rangeland species of *Atriplex*. In: R. Jones (ed.), Studies of the Australian arid zone, I: The Biology of *Atriplex*. CSIRO Division of Plant Industry, Canberra, pp. 31–42.

Lange, R.T., Nicolson, A.D. & Nicolson, D.A. (1984). Vegetation management of chenopod rangelands in South Australia. Aust. Rangel. J. 6: 46–54.

Lay, B. (1979). Shrub population dynamics under grazing: a long term study). In: Graetz, R.D. & Howes, K.M.W. (eds.), Studies of the Australian arid zone. VI. Chenopod Shrublands CSIRO, Melbourne, pp. 107–124.

Leigh, J.H. & Wilson, A.D. (1969). Utilization of *Atriplex* species by sheep. In: R. Jones (ed.), Studies of the Australian arid zone. I. The biology of *Atriplex*. CSIRO Division of Plant Industry, Canberra, pp. 97–104.

Le Houérou, H.N. (1992). The role of saltbushes (*Atriplex* spp.) in arid land rehabilitation in the Mediterranean Basin: A review. Agroforesty Syst. 18: 107–148.

Malcolm, C.V. (1969). Use of halophytes for forage production on saline wastelands. J. Aust. Inst. Agric. Sci. 35: 38–49.

Malcolm, C.V. (1994). Use of halophyte forages for rehabilitation of degraded lands (this volume, Ch. 4).

Malcolm, C.V. & Pol, J.E. (1986). Grazing and management of saltland shrubs. J. Agric. W.A. 27 (2): 59–63.

Malcolm, C.V., Clarke, A.J., D'Antuono, M.F. & Swaan, T.C. (1988). Effects of plant spacing and soil conditions on the growth of five *Atriplex* species. Agric. Eco. Environ. 21: 265–179.

Milthorpe, P.L. (1978). Some factors affecting establishment and growth of black bluebush (*Maireana pyramidata*) in western New South Wales. Proc. 1st Intern Rangel. Congr., pp. 362–365.

Mitchell, A.A. & Wilcox, D.G. (1988). Plants of the arid shrublands of Western Australia. Western Australian Department of Agriculture.

Morrisey, J.G. (1984). Arid mulga woodlands. In: Harrington, G.N., Wilson, A.D. & Young, M.D. (eds.), Management of Australia's rangelands. CSIRO Melbourne, pp. 285–298.

Noble, J.C. & Whalley, R.D.B. (1978). The biology and autecology of *Nitraria* L. in Australia, II: Seed germination, seedling establishment and response to salinity. Aust. J. Ecol. 3: 165–177.

Oxley, R.E. (1979). The perennial chenopod pasture land of Australia. In: Graetz, R.D. & Howes, K.M.W. (eds.), Studies of the Australian arid zone. CSIRO Melbourne, pp. 1–4.

Parr-Smith, G.A. (1982). Biogeography and evolution in the shrubby Australian species of Atriplex (*Chenopodiaceae*). In: Barker, W.R. and Greenslade, P.J.M. (eds.), Evolution of the flora and fauna of arid Australia. Peacock Publications, Adelaide.

Parr-Smith, G.A. & Calder, D.M. (1979). *Atriplex vesicaria* Heward ex Benth. and related species: taxonomy and variation. In: Graetz, R.D. & Howes, K.M.W. (eds.), Studies of the Australian arid zone. VI. Chenopod Shrublands. CSIRO Melbourne; pp. 17–28.

Sharma, M.L. (1976). Soil water regimes and water extraction patterns under two semi-arid shrub (*Atriplex* spp.) communities. Aust J. Ecol. 1: 249–258.

Sharma, M.L. (1982). Aspects of salinity and water relations of Australian chenopods). In: Sen, D.N. & Rajpurohit, K.S. (eds.). Tasks for vegetation science. W. Junk, The Hague, pp. 155–172.

Sharma, M.L. & Tongway, D.J. (1973). Plant induced soil

salinity patterns in two saltbush (*Atriplex* spp.) communities. J. Range Manage. 26: 121–125.

Stanley, R.J. (1983). Soils and Vegetation: An assessment of current status. In: Messer, J. & Mosley, G. (eds.), What future for Australia's arid lands? Australian Conservation Foundation, Melbourne, pp. 8–18.

Truong, P.N. (1993). Vetiver grass, its potential in the stabilisation and rehabilitation of degraded saline land (this volume, Ch. 31).

Ward, B.H.R. (1992). Shrub evaluation for rangeland. Final report to Rural Credits Development Fund. WA Department of Agriculture, 59 pp.

Warren, B.E. & Casson, T. (1992). Performance of sheep grazing salt tolerant forages on revegetated saltland. Proc. Aust. Soc. Anim. Prod. 19: 237.

Warren, B.W., Casson, T. & Ryall, D.H. (1994). Production from grazing sheep on revegetated saltland in Western Australia (this volume, Ch. 24).

Warren, B.E., Bunny, C.J. & Bryant, E.R. (1990). A preliminary examination of the nutritive value of four saltbush (*Atriplex*) species. Proc. Aust. Soc. Anim. Prod. 18: 424–427.

Wilcox. D.G. and Morrissey, J.G. (undated). Pasture plants of the Western Australian Shrublands. Bull. 4023. West. Aust. Dept. Agric., Perth.

Williams, D.G. (1979). The comparative ecology of two perennial chenopods. In: Graetz, R.D. & Howes, K.M.W. (eds.), Studies of the Australian arid zone, VI: Chenopod Shrublands. CSIRO, Melbourne, pp. 29–40.

Wilson, A.D. (1966a). The intake and excretion of sodium by sheep fed on species of *Atriplex* (saltbush) and *Kochia* (bluebush). Aust. J. Agric. Res. 17: 155–163.

Wilson, A.D. (1966b). The value of *Atriplex* (saltbush) and *Kochia* (bluebush) species as food for sheep. Aust. J. Agric. Res. 17: 147–153.

Wilson, A.D. (1966c). The tolerance of sheep to sodium chloride in food or drinking water. Aust. J. Agric. Res. 17: 503–514.

Wilson, A.D. (1977). The digestibility and voluntary intake of the leaves of trees and shrubs by sheep and goats. Aust. J. Agric. Res. 28: 501–508.

Wilson, A.D. & Graetz, R.D. (1980). Cattle and sheep production on an *Atriplex vesicaria* (saltbush) community. Aust. J. Agric. Res. 31: 369–378.

Wilson, A.D. & Hindley, N.L. (1968). Comparison of the voluntary water intakes of some Australian sheep. Fld Stn Rec. Div. Pl. Ind. CSIRO (Aust.) 7: 25–36.

Wilson, A.D. & Leigh, J.H. (1969). Use of water intakes in the estimation of saltbush intake by grazing sheep. Fld Stn Rec. Div. C Pl. Ind. CSIRO (Aust.) 8: 1–8.

Wilson, A.D., Leigh, J.H. & Mulham, W.E. (1969). A study of Merino sheep grazing a bladder saltbush (*Atriplex vesicaria*)-cotton-bush (*Kochia aphylla*) community on the Riverine plain. Aust. J. Agric. Res. 20: 1123–1136.

9. Halophytes and halophytic plant communities in Inner-Asia

ZHANG LI-YUN, XIA YANG & ZOU YUN
Xinjiang Institute of Biology, Pedology and Desert Research, Chinese Academy of Sciences, Urumqi, Xinjiang, P.R. of China

Abstract. The geographic region of Inner-Asia is located in the hinterland far away from the ocean. It is surrounded by mountains and has a typical temperate continental climate with aridity as the basic characteristic. The lowland is a salt accumulating centre. The soil is commonly highly salinized, which contributes to dominance of halophytic vegetation over wide areas. The abundant halophytic resources are of importance to local animal husbandry and have a role in soil protection. Because of increasing soil salinization the further development of animal husbandry, expansion of population and desertification of land, the plant resources have received more attention lately. In order to use the halophytic resources wisely, especially halophytic shrubs, it is urgent and necessary to conduct further research, exchange research findings and establish international cooperation.

Key words: Biogeography, Life forms, Livestock industries, Utilization

1. Location and geographic concept of Inner-Asia

The geographic concept of Inner-Asia is often confused with that of the other two terms – mid-Asia and centro-Asia but has not yet been clearly defined in the literature. In this paper the regional area of Inner-Asia refers to the inland arid zone where there are mainly desert soils. It includes most of the desert basins and part of the upper reaches of peripheral drainage systems. It lies between north latitude 35–50 degrees and east longitude 53–108 degrees. Inner-Asia borders west on the eastern coast of the Caspian sea, east on the western foot of the Helan Shan, south on the northern part of the line along the Hindu Kush mountains; – Kunlun Mountains and the Qilian mountains; and north on the hills and mountainous area in Khazakstan, mid-Altay mountains, and Gobi-Altay mountains. It extends for more than 5500 km from east to west and over 1000 km from south to north (Figure 1). Covering an area of more than 4.5 million km^2, Inner-Asia is the eastward extension of the Asia-Afro desert and includes such natural geographic units as Turan lowland, Tarim basin, Dzungarian Junggar) Basin, Alashan Plateau, Mongolian Gobi and Tianshan mountains.

Hilly areas and plains have obvious differences in elevation giving rise to pronounced vertical zonal landscapes (altitudinal zonation). The plains form many natural landscapes such as stony desert, soil desert, sandy desert and salt desert and oasis (depending on substrate). Taken together there is a recurring pattern of basin and range topography.

2. Natural environment

2.1 *Climate*

Being remote from the ocean accounts primarily for the formation of the arid desert in this area. The huge mountains and plateau in the south and middle, formed in the late orogenic movement

V.R. Squires & A.T. Ayoub (eds.), *Halophytes as a resource for livestock and for rehabilitation of degraded lands*, 115–122.

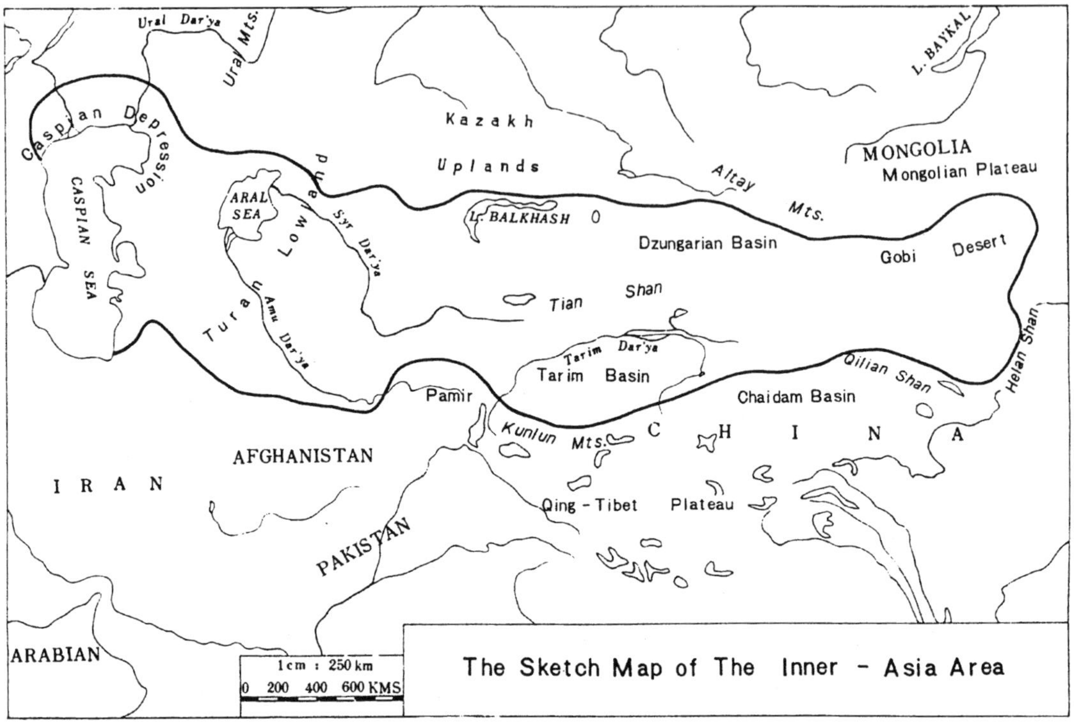

Figure 1. The plains and steppes of Inner Asia.

at the end of the Tertiary period, exert a great influence over the atmospheric circulation and reinforce the aridity resulting in sparse rainfall, cold winters, hot summers and large annual and diurnal differences in temperature. Precipitation comes mainly from the prevailing moisture-laden west wind. In the west annual precipitation tends to be non seasonal; in the east the rainfall mainly occurs in summer and autumn, often in the form of storms. There is a decrease in precipitation from west to east and year to year variability is greater in the east (Table 1).

2.2 *Soil and landscape*

Because of the sparse rainfall, strong evaporation, and high ground water tables the soil in the arid zone of Inner-Asia is commonly salinized. The lowland is a salt accumulating zone and saline soils dominate. Among the soluble salts in the soil Na^+ and SO_4^{2-} are ubiquitous. Salinity contributes much to the wide distribution of halophytic vegetation which is the most important element in natural grazing areas (rangeland) on the plains of Inner-Asia.

3. Halophytic flora, life forms and ecological groups

Generally, all plants that grow in salinized habitats including the halophilous and salt-avoiding species, are referred to as halophytes. The halophytic flora of Inner-Asia is not very rich in species but the life-forms include the five basic types of Raunkiaer's life- form system, as illustrated in Table 2.

The halophytic flora of Inner-Asia is composed of about 150 species from near 20 families. Among them, Chenopodiaceae shows the richest species diversity, totalling over 100 species in which there are not only many succulent shrubs

Table 1. Several climate parameters for the Inner Asia's plain area

	Sites	Average annual temperature (°C)	Temperature of the hottest month (°C)	Temperature of the coldest month (°C)	Yearly differences (°C)	Yearly precipitation (mm)	Yearly evaporation[a] (mm)
The West	Kizyl Arvat	16.0	32.3	−0.5	32.8	205.5	–
	Aral Sea	7.2	26.8	−11.2	33.0	101.8	–
	Kzyl Urda	9.1	26.2	−9.2	25.4	108.0	–
	Bukhara	13.7	30.2	1.7	31.9	135.0	–
	Berki	16.8	29.8	2.2	32.0	161.2	–
	Termgy	17.6	32.0	3.0	35.0	120.0	–
	Balkhash	5.0	24.0	−15.0	39.0	120.0	–
	Taklamaban	8.0	27.4	−16.0	44.1	105.3	3545.2
	Mouosen	6.1	25.8	−19.0	44.8	114.9	1943.9
The East	Hetian	12.1	25.5	−5.6	31.1	33.4	2602.2
	Kuoqiang	11.5	27.4	−8.5	35.9	17.4	2902.2
	Turfan	13.9	32.7	−9.5	42.2	16.4	2837.8
	Hami	9.8	27.2	−12.2	39.4	34.6	3064.3
	Zhuomaohu	10.0	28.2	−12.2	40.4	12.5	–
	Yemajie	7.6	26.2	−13.9	40.1	45.1	–

[a]Potential evaporation from a free-water surface. Records are not available for all locations.

Table 2. The life-form spectrum of halophytes on the southern fringe of Dzungarian Basin and in the Tarim sandy desert

Sites	Life form (in %)				
	Ph	Ch	Hk	K	T
Southern fringe of Dzungarian Basin	33.7	5.4	25.0	5.4	28.5
Tarim Sandy Desert	39.0	1.7	33.9	20.3	5.1

Ph = Phanerophyte; Ch = Chamaephyte; Hk = Hemicryptophytes; K = Cryptophytes; T = Therophytes.

and semi-shrubs, but also various perennial herbs and annuals. Chenopod-dominated communities are a characteristic of Inner-Asia's halophytic flora and vegetation. The halophytes from this family come mainly from two genera, *Salsola* and *Suaeda* (which have near 50 species just in the Dzungarian Basin and Tarim Basin in Xinjiang) and has important phyto-sociological significance.

Plants of Chenopodiaceae originated from a paleo-mediterranean flora include many which have a single genus. Most of the shrubs and semi-shrubs have a long history in Inner-Asia. There are 108 annual chenopods, most of them halophytes, in Xinjiang and there are even more in the Turan lowland in the west of Inner-Asia indicating that Inner-Asia's present ecological environment is still conducive to the evolution and variation of plants from Chenopodiaceae and that are now in a developing and prosperous phase.

Among the components of halophytic flora there are also many plants from Tamaricaceae of which *Tamarix* has nearly 20 species, mostly salt-rich shrubs. *Reaumuria soongorica* is one of the principal species, often forming widespread communities.

Though there are not many gramineal halophytes, they play a significant role in the plant community. *Phragmites communis* for example, often forms a large area of salinized meadow or becomes the dominant synusia in other community types. In addition, Leguminoseae, Compositeae, Zygophyllaceae and Salicaceae also have some significance to phytosociology and landscape in Inner Asia's halophytic vegetation as illustrated by reference to specific regions in Xinjiang.

The Dzungarian Basin is located to the northwest of the Tianshan mountains has an annual precipitation of over 100 mm and a regular seasonal distribution and snow in spring and winter. The Tarim sandy desert is in the Tarim Basin between the south of Tianshan mountains and north of the Kunlun mountains. The annual rainfall is less than 100 mm and occurs mainly in summer and autumn, often as storms. The winter

Table 3. The percentage of ecological groups of halophytes on the southern fringe of Dzungarian Basin and in Tarim sandy desert

Sites	Halo-xerophyte	Halo-mesophyte	Halo-hydrophyte
Southern fringe of Dzungarian Basin	38.4	59.6	2.0
Tarim sandy desert	23.7	72.9	3.4

temperature here is higher than in the Dzungarian Basin. The life-form spectrum (Table 3) reflects the differences in hydro-thermal conditions between the two sites.

At present, because a more accurate life-form spectrum of Inner-Asia's halophytes has not been developed, it may be useful to consider the situation in two regions in Xinjiang as indicative of the life-form spectrum of the west and east of Inner Asia. The situation on the southern fringe of Dzungarian Basin closely approximates that which applies in west Inner Asia and that in the Tarim sandy desert is typical of the situation in east Inner Asia.

It is even more difficult to classify the ecological groups of halophytes in Inner-Asia because the halophytic flora of Inner-Asia as a whole is not well documented and research on physiology and ecology of halophytes is not well developed. Therefore, the classification of ecological groups (Table 3 is based on experience and observation in the two regions in Xinjiang). Obviously, of the halophytes in the two regions, halo- mesophytes are the richest in species while halo-xerophytes take second place. Probably this pattern fits the whole of Inner-Asia.

4. The types of halophytic plant communities and their general characteristics

Though the halophytic flora of Inner Asia is depauperate the halophytic communties shows diversity, including most of the vegetation types in the steppes and plains. According to the life-form characteristic of the constructive synusiae in the communities the halophytic vegetation of Inner Asia is divided into the following seven basic types of halophytic plant communities.

1. *Succulent semi-shrub desert*
The principal species are all semi-shrubs from the family Chenopodiaceae, with diminished or specialised succulent leaves. They are distributed around the salt lakes, on the river banks and terraces and on the heavily-salinized habitats of lowlands.

The ground water table is at 1–3 m and the NaCl and Na_2SO_4 content of the top 0–20 cm of the soil can reach up to 10–30%. This makes for a high ash content (especially Na and Cl) in plants which grow there. For example, *Kalidium foliatum*, *K. gracile*, *K. cuspidatum* and *Halocnemum strobilaceum* often contain 40–50 per of ash.

The principal species in this kind of halophytic communities are *Halocnemum strobilaceum*, *Halostachys caspica*, *Kalidium foliatum*, *K. caspicum*, *K. gracile*, *K. cuspidatum*, *Suaeda physophora*, *S. dendroides* and *S. microphylla*. The communities of these plants join frequently with other communities to form a mosaic of communities with a complex distribution pattern. The height, which on rare occasion can reach more than 1 m, is 35–60 cm, foliage cover is 10–45% and floristic composition includes over 20 species. The net above-ground production available for livestock is 1000–3000 kg ha^{-1} (fresh weight).

2. *Semi-shrub and mini-shrub desert*
The plants in this type are mainly some super arid semi-shrubs and micro-semi-shrubs from the families Chenopodiaceae and Tamaricaceae. They occur on the foothills, on the floodfans and on the outwash plains in front of the mountains. The soils contain high amounts of $CaCO_3$ $CaSO_4$ and other salts. The chemical composition of elements in these dominant plants shows a moderate content of ash (20–30%) with 3–7% of Na and Cl. *Reaumuria soongorica*, *Atriplex cana*, *Camphorosma lessingii*, *Anabasis salsa*, *A. brevifolia* and *Salsola passerina* are the dominant species. The height is 10–35 cm, foliage cover is 10–30% and the floristic composition is about 25–30 species. The net fresh above-ground produc-

tion available for livestock is 500–1000 kg ha^{-1} (fresh weight).

3. *Mini-semi tree desert*
   Haloxylon ammodendron (family Chenopodiaceae) is the dominant species, its height is between 1.5–5 m. The apex of the branches is winter deciduous. With its broad ecological range this type can grow not only on sandy land but also on the lowland of the alluvial plain at the foot of mountains and on paleo-lacustrine deposits and even on the chomo-Gobi with extreme ecological conditions. The soil ranges from the non-salinized to salinized or alkalinized. The ash content is high (25–30%) with Na and Cl at 5–9% and 1.5–3% respectively. The chemical characteristics of its dominant elements is in the form $Na > N > K <> Ca > Cl$.

   The floristic composition of micro-semi-tree desert is richer but on heavily salinized soil it is about 10 species. The foliage cover is 10–25% and above-ground production available for livestock is 750–100 kg ha^{-1} (fresh weight).

4. *Annual halophytic herbosa*
   Annual halophytic plants are confined to some succulent species from the family Chenopodiaceae which can grow normally in salinized habitats and fruit once. Like all ephemerals these plants can erupt suddenly and colonize bare areas in arid lands caused by overgrazing, abandonment of cultivation and soil salinization from over-irrigation.

   Because they grow on salty soil these plants have a high content of ash and Na. For example, the content of ash and Na in *Halogeton glomeratus* are respectively 42.69 and 14.61% and in *Salicornia europaea* are 44.56% and 14.68% respectively.

   The main dominant species forming the annual halophytic herbosa are *Halogeton glomeratus*, *H. arachnoideus*, *Salicornia europaea*, *Salsola lanata*, *S. ruthenica*, *S. collina*, *S. subcrassa*, *S. nitraria*, *S. foliosa*, *S. affinis*, *Suaeda acuminata*, *Suaeda salsa*, *Petrosimonia sibirica*, *P. brachiata*, and *P. glaucescens*.

   Generally, the height of the herbosa is 20–30 cm, the foliage cover ranges between 10–80% and the botanical composition is simple with few species. The net above-ground production available for livestock is 200–500 kg ha^{-1} (fresh weight). There is a large between-year variation.

5. *Salinized scrub*
   The principal plants are some salt-tolerant shrubs from the families Tamaricaceae, Solanaceae, Zygophyllaceae and Leguminoseae. They are mainly distributed along riverbanks on desert plains, around lakes and in the lowland where the water table is high (1–5 m). The habitats are almost all salinized to various degrees, some with heavy salinization have met the criteria of salt soil. Of Inner-Asia's salinized shrubs *Tamarix* is the main genus. Plants contain about 20–30% of ash, 3.5–7.0% of Na, 2.0–2.5% of Ca and 1.7–2.0% of S. The chemical characteristic of the dominant elements is that of the type $Na > S > Ca > K > (Al)$.

   The main species are *Tamarix ramosissima*, *T. hispida*, *T. laxa*, *T. leptostachys*, *T. elongata*, *T. hohenakeri*, *T. taremensis*, *Nitraria sibirica*, *N. schoberi*, *Lycium ruthenicum* and *Halimodendron halodendron*. The communities of these plants have a foliage cover of 10–50%, species composition of 20–25 and can produce 1500–3000 kg ha^{-1} of above-ground fresh weight.

6. *Salinized meadow*
   Salinized meadows are characterised by salt-tolerant mesophilous or xero-mesophilous (mainly gramineal) forbs and grasses. Distribution is on river banks and on lowland plains where there is a shallow water table (1–3 m). Soil types include salinized meadow soils and meadow saline soil. Species are *Achnatherum splendens*, *Phragmites communis*, *Aeluropus littoralis*, *Aneurolepidium dasystachys*, *Poacynum hendersonii*, *Sophora alopecuroides*, *Alhagi sparsifolia*, *Glycyrrhiza inflata*, *G. glabra*, *Oxytropis glabra*, *Limonium gmelinii* and *Karelinia caspica*.

   In the salinized meadow the height, cover and species composition and biomass production depend on local conditions, the different plant associations and habitats and there is much variation.

7. *Salinized tugayi*

The plants of this type are mainly a few mesophytes from the families Salicaceae and Elaeagnaceae which grow as riparian communities along the banks of large rivers which the traverse the arid plains. The soil type being a salinized meadow soil or saline soil and the water table is 1.5–3.0 m deep.

The ash content in fresh branches and leaves is 10–11% while the chemical composition of dominant elements are of the type N > K > Ca(Al).

Dominant species are the riparian species *Populus euphratica*, *P. pruinosa* and *Elaeagnus oxycarpa*. Over the past 50 years the riparian communities have been ' damaged severely due to water shortage created by upstream diversion, conversion of land for cropping and cutting for lumber.

5. Evaluation of halophytes as a resource and prospects for better utilization

Halophytic communities not only dominate the vegetation but are also a valuable resource for animal husbandry in the vast expanse of the plain and desert area in Inner-Asia. For example, in the lowland districts of Xinjiang, the area of halophytic vegetation with more than 5% coverage is approximately 25 million ha. Of which 18 million ha is utilizable (equal to about 90% of the area of available rangeland of the plain in Xinjiang). In the 8 million ha of desert-zone in Mongolia, the halophytic *Anabasis brevifolia*, *Salsola passerina* and *Haloxylon ammodendron* rangelands are the dominant communities over 6 million ha (about 75% of the area of desert-zone rangeland).

In the Turan lowland and the southern area of Kazarkstan the halophytic vegetation also flourishes and become the natural forage base for animal husbandry in the region. It is evident that the halophytic vegetation is the most valued vegetation. This is especially so because of its nutritive value. The protein content and nitrogen free extract are important indices of the value of forage quality. An analysis of 100 samples from 20 halophyte species showed that the average crude protein content was 11.88% (range 4–20%). The average nitrogen free extract was 40.65% (range 27–57%). The crude protein content is similar in other desert vegetation but the nitrogen free extract of halophytes is 7% higher than non-halophyte desert vegetation (Table 4). On average, analysis of typical steppe and desert plants shows that crude protein in halophytes is 2–3% lower and nitrogen free extract is 4–14% higher than non-halophytes.

Halophytes contain large quantities of ash and are rich in mineral nutrition. The average ash content is 20% but values as high as 40% are not uncommon. *Salicornia europaea* even reaches 55% containing not only Ca and P that closely influence the production capacity of livestock but also the macroelements and micro elements essential to the growth and development of livestock. Chemical analysis of halophytes shows that minerals occur in this order Cl > NA > K > N > S(Si).

Because halophytes contain a lot of soluble salt, the livestock that live in the halophyte-dominated rangelands need no accessory salt. Because of this such rangelands are superior to mountain pasture and other non-halophyte rangeland. They are favoured by the herdsmen because most of the halophyte branches or leaves of the current season's growth will wither but not fall in the winter. The long leaves of *Phragmites communis*, *Achnatherum splendens* and *Aneurolepidium dasystachys* (family Gramineae) can remain intact in winter, even until the next year if not grazed. The withered twigs and infructescence of Chenopodiaceae also remain until the sprouting of new branches in the next year.

This characteristic of halophytes is undoubtedly beneficial to the feeding of livestock in winter. In addition, the palatability of most of the succulent halophytes such as Salsola is so poor that it is not even taken by livestock when the plants are young and tender. In winter the salt content of the dried up leaves and twigs is greatly lowered by leaching and translocation enhancing the palatability especially to camel, goats and sheep. This is the rationale for the local herdsmen using the halophytic pasture as winter-spring pasture and late autumn pasture.

Animal husbandry is traditional in Inner-Asia. the breeding of many improved breeds,such as dromedary, bactrian camel, kalkur sheep, kuche

Table 4. Nutritive composition of some halophytes from Inner Asia

Species	Phenophase	H_2O	Crude protein	Crude fat	Crude fiber	Non-nitrogenous extracted substance	Crude ash	Ca	P
Halostachys caspica	Flowering	3.38	11.44	3.70	9.86	35.69	35.93	0.52	0.18
Halocnemum strobilaceum	Fruiting	6.26	8.69	3.78	27.44	25.27	28.56	1.82	0.10
Kalidium foliatum	Fruiting	8.47	18.92	2.80	8.13	18.27	43.41	0.48	0.08
Suaeda physophora	Fruiting	8.33	18.30	1.55	20.79	35.04	16.00	0.41	0.16
Haloxylon ammodendron	Flowering	6.14	14.27	1.74	21.44	46.42	9.99	0.25	0.17
Anabasis brevifolia	Flowering	9.99	9.49	0.68	28.50	29.30	22.04	3.19	0.08
Glycyrrhiza inflata	Flowering	9.28	10.97	2.96	22.50	46.18	8.11	0.22	0.14
Phragmites communis	Flowering	7.95	12.00	3.65	34.90	31.58	9.82	0.24	0.06
Alhagi sparsifolia	Fruiting	10.27	14.35	5.33	24.80	33.59	11.21	0.90	0.12
Karelinia caspica	Flowering	5.76	11.58	5.78	20.00	45.99	10.89	0.85	0.11
Tamarix ramosissima	Flowering	9.07	10.90	1.72	13.58	53.43	11.30	1.76	0.05
Reaumuria soongorica	Fruiting	8.41	10.23	2.37	20.40	44.27	14.32	0.65	0.12

sheep, Dolan sheep, Xinjiang donkey and Rong goat is related to the local forage resources based on halophytes (see Figure 1).

Despite the long tradition of livestock raising in Inner-Asia there is scope for improving the utilization of halophytes. For a long time people in Inner-Asia have grazed the halophyte dominated rangelands. Some areas were cut for fodder but generally the palatability is poor and utilization ratio is not high. In order to improve both the authors suggest that people harvest halophytes in summer and autumn for silage or make them into various dried forages. People in northern China grind the halophytes *Alhagi sparsifolia* (Leguminosae) and *Karelinia caspica* for feeding to pigs. They use the plants in the manufacture of pellets with various forms and formulae. The idea is worth trying elsewhere.

Halophytes are adaptable to environments that contain high soil salinity, and high water tables and absorb and accumulate salt. This bio-ecological characteristic is useful in the reclamation of saline-alkaline soil. At the appropriate time each year people can transfer salt by reaping or otherwise clearing away the salt-enriched halophyte vegetation as a way to improve soil.

For nearly half a century many farmlands and irrigated oases have been established in the salinized environment in the course of the development of north west China and neighbouring countries such as Kazakstan. Recent development of the region is continuing in this difficult environment. Along with the growth of population, the continuous development of agriculture and the loss of the land occupied by industry, transport routes, civic buildings and public institutions, the reclamation and better utilization of salinized lands and saline wastelands becomes an imperative. Halophytes may have a role in reclaiming the salinized land and in providing environmental protection. The frequent winds, serious sandstorms and the strong wind and water erosion of soil do harm to agriculture and to animal husbandry. At the same time they also endanger the humans, interfere with transport and threaten water conservancy facilities. Many species of halophytes play an important ecological role in ameliorating wind damage, controlling sand and protecting the living conditions for humans. Halophyte species such as *Tamarix* and *Haloxylon* have been planted to break winds and control sand in the north of China, Kazakstan and Turkmanistan and other regions in Inner-Asia.

The firewood cut by people in the plains and steppes of Inner-Asia are chiefly halophytic shrub and tree species. As populations rise and as desertification spreads the impact of fuel wood gathering will be become greater.

Halophytes and their associated plant communities are an important kind of resource for combating desertification in Inner Asia. But for many years the resources have not been utilized fully because the management and administration of animal husbandry was diffuse and poorly targeted and the methods of land utilization was not rational. The phenomena of recurring drought, shortage of water, cutting of woody plants for fuel wood all contribute to desertification. Research on exploiting and utilization of the natural vegetation in Inner Asia would have both practical and profoundly social significance in developing the regional economy of countries of Inner-Asia. Research should focus on the following:

- Selecting the plant species that can be used as forage, food and raw materials for industry.
- Establish the foundation for artificial pasture production based on salt-tolerant plants.
- Continue work on plant introduction and evaluation for reclaiming salt-affected lands and to prevent the spread of desertification.

Inner Asia has had a long and colourful history but the overall standard of living is relatively low because the economy is not well developed. The region has a bright future because it is a vast region with a low population density, large quantities of energy and mineral resources and the benefit of lessons learned from development experience over the past 50 years. It is evident though that halophyte-dominated rangelands and the saline lands which support them will continue to form the basis of a large and important livestock industry in Inner Asia for a long time to come.

Acknowledgements

The authors thank UNEP for the opportunity to contribute to the Workshop. Our gratitude is expressed to Dr V.R. Squires for his assistance in editing this paper.

References

Nitikin, S.A. (1966). Woody plants and shrubs in deserts of the USSR. Nauka Publ., Moscow.

Petrov, M.P. (1972). Continental aspects of shrub distribution, utilisation and potentials: Asia. In: C.M. McKell, J.P. Blaisdell & J.R. Goodin (eds.), Wildlife shrubs – their biology and utilisation. USDA Forest Service General Tech Rpt. INT-1, pp, 37–50.

Squires, V.R. (1989). Farming on the desert fringe – a case study of State farms in Xinjiang, China. J. Arid Environ. 16: 1229–33.

Xia Xuncheng, Fan Zili & Chen Ruguo (1985). The oases in Xinjiang. Proceedings of the International Symposium, 7, ISEUNRAA, Urumqi, China, 7–13 August 1985, 162 pp.

Zhang Xue Zu & Ho Zhangqi (1985). The ecological characteristics of Xinjiang arid zone and its exploitation. Proceedings of the International Symposium 6 ISEUNRAA, Urumqi, China, 7–13 August 1985, 162 pp.

Additional Literature

Hou Xueyu (1982). Chinese vegetation geography and the chemical composition of dominant plant. Beijing Science Press.

Korovin, E.P. (1934). The vegetation in Middle Asia. Moscow. Tashkan, pp. 129–139.

Unatov A.A. (1950). Basic characteristics of the vegetation in people's Republic of Mongolia. 244–267. Rubl. USSR. Academy of Sciences, pp. 135–146.

Zhang Liyun (1985). The types and composition of halophytic vegetation on the edge of Dzungarian Basin. J. Arid Zone Research 1: 40–50.

Zhang Liyun & Hai Ying (1991). Some characteristics of the vegetation in the Taklamakan Desert. The Report of Scientific Exploration and Investigation in the Keliya River Valley and the Taklamakan Desert. Publ. of Chinese Science and Technology. Beijing, pp. 137–165.

Zhang Liyun, Hai Ying & Xia Yang (1991). The Therophytes and their grass-cover in Xinjiang. The collection of botanical paper in Xinjiang. Beijing Science Press, pp. 8–16.

10. Forage halophytes and salt-tolerant fodder crops in the Mediterranean Basin

H.N. LE HOUÉROU
CEFE/CNRS, Montpellier, France

Abstract. There are about 1100 species of halophytes in the Mediterranean Basin, when considered in its broadest meaning i.e. from the Aral Sea to the Atlantic Ocean. Many of those may be considered as forage species, grazed as they are by various livestock and game animals. These halophyte forage species are very diverse in terms of plant systematics, biology and ecology as well as in nutritive and grazing value. They include annual and perennial herbaceous species as well as shrubs and trees. *Chenopodiaceae* represent some 30% of them with some 15% Poaceae and 10% Fabaceae. Overall, halophytes make up in the vicinity of 10% of the flora at the country level and some 5% at the Basin level. Saline rangelands occupy in the vicinity of 50 million hectares in the Basin, most of them located to the south and the east of the Mediterranean sea. But the areas of land degraded by salinity is increasing steadily due to faulty irrigation practices. One way to rehabilitate and reclaim these is the planting of salt-tolerant forage species. Some 100,000 ha have been planted to *Atriplex* spp. in the Basin over the past 25 years or so. These include both native and exotic species, among the former are *A. halimus*, *A. leucoclada*, *A. glauca* and among the latter *A. nummularia*, *A. amnicola*, *A. lentiformis*, *A canescens*. These are capable of high productions of nitrogen-rich forage, either under rainfed or irrigation conditions including in the Mediterranean arid zone, with rainfalls as low as 150 mm yr^{-1}. Rain and Water-Use Efficiencies are high in well managed plantations and native stands, up to 10 kg DM ha^{-1} yr^{-1} mm^{-1} and 4 mg DM g^{-1} H_2O, respectively. Feeding trials have shown that livestock accustomed to this feed can make good use of this forage and retain most of the non-protein nitrogen it contains, when the ruminal flora has adjusted to this type of nutritive regime.

Key words: *Atriplex*, Classification of halophytes, Non protein nitrogen, Rain use efficiency

1. Introduction

Terrestrial halophytes are here understood as plants which, in natural conditions, are exclusively found on saline soils, or in saline environments. Saline soils are those having a soil saturation extract with an electric conductivity equal to or higher than 5 mS cm^{-1} (5 mmho cm^{-1}) in any one or in several layers(s)\horizon(s) of their profile.

The above definition does not imply that halophytes necessarily need salinity for growth and development. Under cultivation many halophytes do grow and carry high productivity without the presence of any significant salinity e.g. *Atriplex* spp., *Maireana* spp., *Tamarix* spp., *Salsola* spp., *Limonium* spp., *Puccinellia* spp., etc. these are labeled 'optional' or 'facultative' halophytes (Le Houérou 1993a). This situation, however, does not apply to all halophytes; some need saline conditions for normal growth ('obligate' or 'true' halophytes: *Halocnemum*, *Arthrocnemum*, *Salsola* spp., *Salicornia* spp., *Suaeda* spp., *Sarcocornia* spp.). Others need no salt concentration for growing, but do develop better with it (some *Atriplex*, some *Maireana*, some *Tamarix*). These are

V.R. Squires & A.T. Ayoub (eds.), *Halophytes as a resource for livestock and for rehabilitation of degraded lands*, 123–137.

called 'preferential' halophytes. Moreover, the limit between these various kinds of halophytes is not always clear. Other non halophytic species, i.e. which are found in nature under both saline and non-saline conditions, may be fairly tolerant to salinity. This is the case of a number of conventional cultivated species and their wild relatives. *Beta vulgaris* (Beet), *Gossypium* spp. (Cotton), *Hordeum* spp. (Barley) *Phoenix* spp. (Date palm), *Cynodon* spp. (Bermuda grass), *Chloris gayana* (Rhodes grass) *Festuca arundinacea* (Tall fescue), *Agropyron elongatum* (Tall wheat grass), *Medicago sativa* (Alfalfa, Lucerne), *Medicago* spp. (Medics), *Melilotus* spp. (Sweet clovers), *Lotus* spp. (Trefoils), *Trifolium resupinatum* (Persian clover), *T. yanninicum*, (Yannina subclover) etc., *T. fragiferum* (Strawberry clover), *Hedysarum carnosum* (Fleshy sainfoin) etc. many of these species, their subspecies, varieties and cultivars are able to grow normally, *under cultivation*, with soil solution ECs of 10–15 mS cm^{-1} or slightly above. They are sometimes labelled 'alternate' halophytes; but, although they are not true halophytes, they will be marginally considered in this paper. A review of salt-tolerance in alfalfa cultivars, for example, would alone require lengthy developments, not to speak of other species such as tall fescue.

2. The halophytic flora

There are perhaps as many as 6000 species of terrestrial and tidal halophytes in the world, i.e. some 2% of the flowering plants (Le Houérou 1993a). The Mediterranean climate area, from the Atlantic Ocean to the Aral Sea and the Indus Valley, includes about 1100 species i.e. about 5% of its terrestrial flora (Le Houérou 1993a).

About 1/4 of the world halophytes are *Chenopodiaceae*, 1/10 *Gramineae*, 1/20 *Leguminoseae*, 1/25 *Compositeae*, *Plumbaginaceae*, 1/33 *Aizoaceae*, *Cyperaceae*, 1/50 *Tamaricaceae*, *Zygophyllaceae*, etc. (Aronson 1989).

Halophyte families, genera and species richness are approximately as shown in Table 1.

Some 70% of these species are perennial and 30% annual or bi-annual (Le Houérou 1959, 1969, 1986, 1993a, Le Houérou et al. 1975). Among the first named a few are trees (*Tamarix* spp.), many are shrubs, particularly chamaephytes and noteworthy among the *Chenopodiaceae* (e.g. *Salicornia* spp., *Salsola* spp, *Suaeda* spp. *Atriplex* spp. etc.). The perennials also include some hemi-cryptophytes in particular grasses (e.g. *Aeluropus* spp., *Sporobolus* spp., *Puccinellia* spp., *Ammophila arenaria*, *Agropyron* spp.).

The most common annual species are the following and their relatives:

Hordeum maritimum (=*H. marinum*)	–	*Polypogon maritimum*
Lepturus cylindricus	–	*Polypogon monspeliensis*
Lepturus incurvatus	–	*Sphenopus divaricatus*
Frankenia spp.	–	*Eremopyrum orientale*
Atriplex rosea	–	*Cressa cretica*
Mesembryanthemum nodiflorum	–	*Spergularia* spp.
Aizoon canariense	–	*Melilotus* spp.

It is worth noting that 50% of the above are grasses.

3. Halophytic vegetation

Native halophytic vegetation is mainly composed of steppe-like plant communities, i.e. treeless, low shrubby vegetation with various proportions of bare ground surface. These steppes are usually dominated by dwarf fleshy shrubs of *Chenopodiaceae* mixed or not with other biological types of plants; for this reason they are often called ***'crassulescent steppes'***, opposite the other fleshy glycophytic ***'succulent steppes'*** (e.g. with cactoid *Euphorbia* spp., *Kleinia* spp., *Caralluma* spp. etc.), which, on the contrary, are not saline. Typical dominant species of crassulescent steppes are: *Halocnemum strobilaceum*, *Arthrocnemum indicum*, *Salicornia* spp. *Salsola* spp., *Suaeda* spp., *Atriplex* spp. etc. Crassulescent steppes correspond with saline soils of various types, most often Solontchaks in the Mediterranean Basin. The salinity of these soils is very variable, with various types of distribution of the salts in the profile. Water content is a capital factor in these saline soils: some may be permanently wet whilst others are quasi-permanently dry; some may have

Table 1. Approximate list of Mediterranean halophytes

Families	Genera	No. of genera having halophytes	No. of species of halophytes
Plumbaginaceae		7	301
	Limonium		280
	Limoniastrum		5
	Armeria		5
	Acantholimon		4
	Goniolimon		3
	Phylliostachys		3
	Limonopsis		1
Chenopodiaceae		45	262
	Salsola		75
	Atriplex (incl. Halimione, Cremnophyton & Blackiella)		50
	Suaeda		25
	Bassia (incl. Chenolea)		16
	Salicornia		11
	Anabasis (incl. Fredolia)		9
	Hammada		6
	Agathophora		5
	Climacoptera		5
	Corispermum		5
	Cornulaca		4
	Camphorosoma		3
	Halotis		3
	Halothamnus (incl. Aellenia)		3
	Sarcocornia		3
	Gamanthus		2
	Halocharis		2
	Halogeton		2
	Halopeplis		2
	Haloxylon		2
	Kochia		2
	Petrosimonia		2
	Polycnemum		2
	Traganum		2
	Arthrocnemum		1
	Beta		1
	Cyathobasis		1
	Cyclocoma		1
	Girgensohnia		1
	Halanthium		1
	Halimocnemis		1
	Halocnemum		1
	Halostachis		1
	Halotis		1
	Kalidium		1
	Krascheninnikovia		1
	Maireana		1
	Microcnemum		1
	Nucularia		1
	Ophaiston		1

Table 1. Continued

Families	Genera	No. of Genera having halophytes	No. of Species of halophytes
	Panderia		1
	Piptoptera		1
	Seidlitzia		1
	Sevada		1
	Traganopsis		1
Poaceae	(Gramineae)	20	50
Asteraceae	(Compositeae)	7	42
Caryophyllaceae		5	40
Zygophyllaceae		4	35
Papilionaceae	(Leguminoseae)	12	25
Cyperaceae		5	20
Frankeniaceae		2	20
Juncaceae		1	20
Tamaricaceae		2	20
Brassicaceae	(Crucifera)	5	15
Apiaceae	(Umbelliferae)	4	15
Aizoaceae	(Mesenbryanthemaceae)	5	10
Boraginaceae		3	5
Convolvulaceae		3	3
Gentianaceae		1	3
Plantaginaceae		1	3
Various families		80	326
Total		132	889

a permanent or a temporary water table or not. The classification of crassulescent steppes is therefore complex, depending as it does on 5 main edaphic factors:

a. Global ionic budget
b. Global water budget
c. Soil texture
d. Alkalinity
e. Depth

For an ecological classification of halophytes, see Le Houérou (1993a). Contrary to most other types of Mediterranean steppic vegetation, crassulescent steppes are not tied to the arid zone. They also occasionally occur in sub-humid zones ($p > 600$ mm), but commonly in the saline depressions of the semi-arid zones ($600 > p > 400$ mm) and of the hyper-arid zones ($p < 100$ mm).

4. Palatability and preference

Palatability is defined as the ratio between the amount of feed ingested by herbivores and the amount on offer, for a given period of time. Preference is the order in which forage species are selected by herbivores within a given plant community or population, or at a grazing site, at a given time (Le Houérou 1980a; Le Houérou et al. 1982). These concepts, however, are liable to a wide variability in time and space, depending, as they do, on many variables and parameters that may change with season, site, animals and other local conditions. Some of these variables are linked to the plant, others to the animal, while a third category depends on various environmental factors. For a given species palatability for a given type of animal varies with the phenological stage, the organ concerned and the season.

Moreover in almost any population, either natural or planted, of a given species of forage, there are various degrees of palatability from one plant individual to the next, ranging from highly palatable to poorly palatable. Palatability also depends on the relative abundance of the species under consideration on the rangeland; all other conditions being equal the palatability of a given taxon is inversely related to its abundance on the range, except for a few species which are specially relished in all circumstances ('ice-cream species' in american range management jargon).

Furthermore, as it has often been pointed out, there is no clear-cut relationship between palatability and feed value, only with diet selection and intake.

Besides these 'internal' plant factors here are also 'internal' animal factors which are: species and breed of livestock, age, feeding habits, physiological and health status, nutritional status (an undernourished and hungry animal is less selective than a well-fed, satiated beast). As a rule, the content of crude fibre in a forage plays an important rôle in its selection by livestock. Forages with a high fibre content are usually better accepted by cattle, then by sheep and goats; but this, in turn, depends on the proportions of the various components of fibre: cellulose, hemicelluloses, acid detergent fibre (lignocellulose), neutral detergent fibre (cellwalls) etc. But there are many other criteria for selection such as organoleptic qualities of the forage; the latter have hardly been explored in research. Mineral content may also be an important factor either limiting or favourable (usually limiting in low rainfall areas and favourable in high rainfall areas, when silica-free minerals are concerned). Finally, the overall balance of the diet plays a major role in forage selection. It has been shown (Skouri, 1975) that the amount of highly fibrous material ingested depends to a large extent on the amount of protein in the overall diet. Besides these intrinsic or internal plant and animal factors there are also extrinsic or environmental criteria for selection.

Some of those have been referred to above, such as the relative abundance of a given taxon on the range and the botanical composition of the forage available. The palatability of a given taxon of forage for a given type of animal depends, to a large extent, on the plant community or association in which this taxon is being grazed, since most of them occur in many plant associations under various ecological and bioclimatic conditions. It is a matter of competitivity or rather of 'demand and supply'. This is why a given species may be differently rated in the various plant communities.

This fact has been rightly stressed by Sarson & Salmon (1976). As a rule the palatability of a given taxon would increase with environmental aridity. There are, however, as usual, exceptions to this rule, perhaps due to unnoticed (by man) ecotypical or biochemical differences between various populations of a given taxon. An example of the first (normal) situation is *Traganum nudatum* more heavily browsed in the desert bioclimatic zone than in the arid zone. An example of the exception is *Moricandia nitens* (=*M. suffruticosa*) browsed in the arid zone of eastern Libya and Egypt and ignored in the northern Sahara of Algeria. For these reasons an assessment of palatability of browse species is always, to some degree, subjective.

It is, however, possible to reach some degree of objectivity in the rating of palatibility for a given consumption experiment with a given plant material and a given group of animals. This may be achieved by measuring the rate of use of the forage available. However the result of a given experiment can hardly be extrapolated and generalized. In Northern Africa Le Houérou (1962, 1965), Le Houérou & Ionesco (1974), Sarson & Salmon (1976) have proposed a scale of rating with 5 classes (Le Houérou, 1980a). See Table 2.

Some non shrubby perennial halophytes exhibit a fair to good palatability e.g. *Plantago crassifolia*, *Spergularia media*, *Sp. marginata*, *Hedysarum carnosum*, *Puccinellia* spp., *Spartina patens* while others are almost unpalatable e.g. *Aster tripolium*, *Heliotropium curassavicum*, *Suaeda maritima*, *Juncus* spp., *Schoenus nigricans*, *Cyperus* spp., *Scirpus* spp., *Phragmites* spp., *Typha* spp., *Arundo plinii*, *A. donax*, *Saccharum ravennae*, *Ruppia* spp., most of those are hygro-halophytes. A number of salt-tolerant forage species, that are not true halophytes are noticeable: *Festuca arundinacea*, *Papalum distichum* (=*P. vaginatum*), *Cynodon dactylon*, *Agropyrum elongatum*, *Agropyropsis lolium*, *Sporobolus ioclados*, *S. helvolus*, *Phalaris aquatica*, *Ph. truncata* among perennial grasses; *Trifolium fragiferum*, *Lotus roudairei*, *Lotus corniculatus* and *Tetragonolobus maritimus* among perennial legumes; *Melilotus alba*, *M. italica*, *Medicago ciliaris* and *M. intertexta*, *Trifolium yanninicum*, *T. resupinatum* among annual legumes (Le Houérou 1974, 1979, 1981, 1986, 1993a). *Moricandia nitens* from eastern North Africa and the Near East is a highly palatable sub-shrubby perennial with a fair tolerance to salinity, whereas its close relative *M. suffruticosa* from W. Northern Africa is usually ignored by livestock.

Most annual halophytes have a very low palatability, if any, and produce little phytomass e.g. *Hordeum maritimum*, *Palypogon*, *Sphenopus*, *Lepturus*, *Pholiurus*, *Psilurus*, *Eremopyrum*, *Frankenia*, *Aizoon*, *Mesambryanthemum*, *Cressa*, *Zygophyllum*, *Tetradiclis*, *Halopeplis*, *Halogeton*, *Schanginia*, *Suaeda*, *Salsola*, *Salicornia* etc . . . Some are toxic such as *Halogeton sativus* (Selenium toxicity).

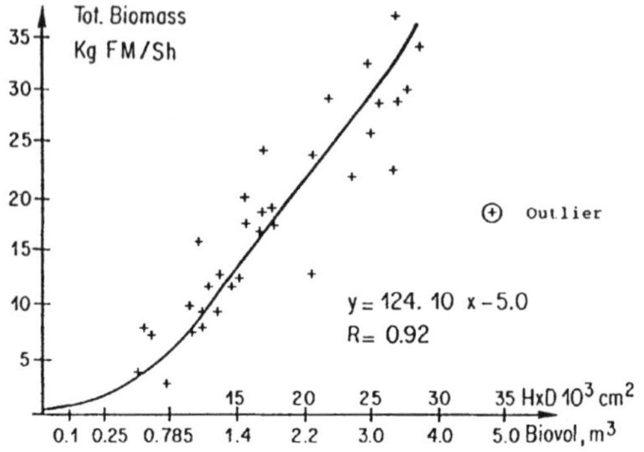

Figure 1. Allometric relations, *Atriplex nummularia* (3 years) 2000 Sh/ha.

5. Primary productivity

Native populations of halophytes may exhibit a high productivity, particularly when growing on water tables in topographic depressions and other low parts of the landscape. Dense native populations of *Atriplex halimus* may reach a total above ground phytomass of 10–15 t DM and a productivity 2000–5000 kg DM ha^{-1} yr^{-1}, of which about 50% is forage and 50% wood (Franclet & Le Houérou 1971). Crassulescent steppes of *Suaeda* spp., *Salsola* spp. *Athrocnemum*, *Salicornia* may have a biomass of 2000–5000 kg DM ha^{-1} when they are not degraded; productivity may then reach 500–2500 kg DM ha^{-1} yr^{-1}, with 40–50% forage (Le Houérou 1972, 1975; Franclet & Le Houérou, 1971, Rodin et al. 1970; Shishov, Kharin et al. 1980). But in most cases, because of permanent overstocking production does not exceed 1/5 to 1/3 of these figures. Productivity in shrubs (including halophytes) may be assessed via allometric relations between biomass an shrub size or biovolume. The relationship is shown for some shrubs between height and diamer on the one hand and biomass on the other hand. The mathematical relation is an exponential one. But the equations established in this way are only valid for a given species, in a given field, for a given season. The relations may be substantially altered depending on season and mainly on the weather conditions and grazing pressure on the site for the previous weeks and months prior to the evaluation. With this restriction in mind it is possible to assess the biomass present and the

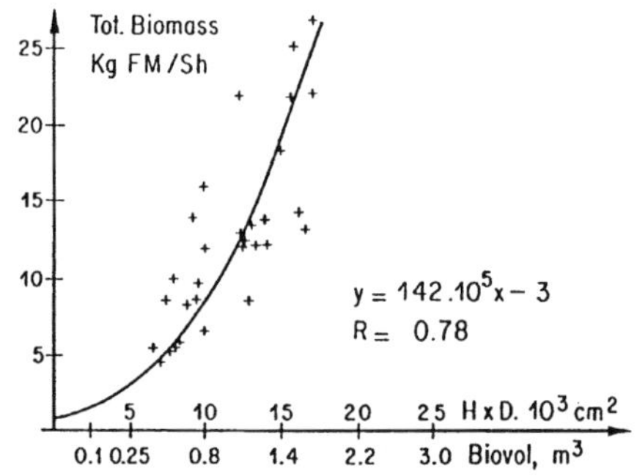

Figure 2. Allometric relations, *Atriplex amnicola* (=*A. rhagodioides*) (3 years).

forage biomass available in a quick and fairly reliable way; for details see Le Houérou et al. 1982. Figures 1–7 are shown as examples in a specific case, taken in Libya, in 1982 (Le Houérou et al. 1982).

6. Grazing and feed value

The grazing and feed value of individual halophytes and crassulescent steppes varies very widely, from virtually zero to very high. A number of species, including common and dominant ones are ignored by livestock such as *Halocnemum strobilaceum*, *Halopeplis* spp., *Salsola baryosma*, *Limonium* spp., *Limoniastrum* spp.,

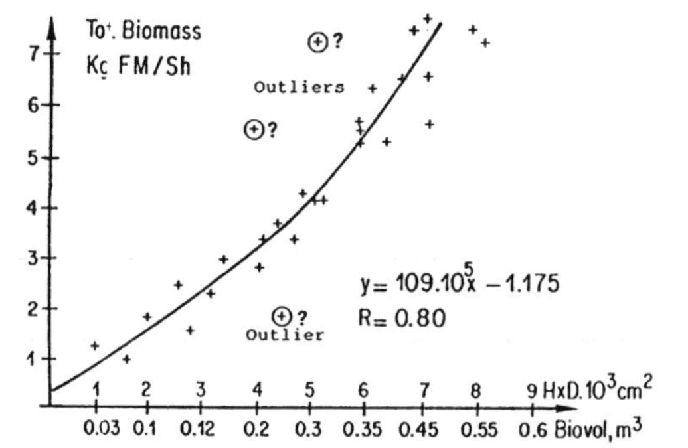

Figure 3. Allometric relations, *Atriplex mollis* (3 years) 2000 Sh/ha.

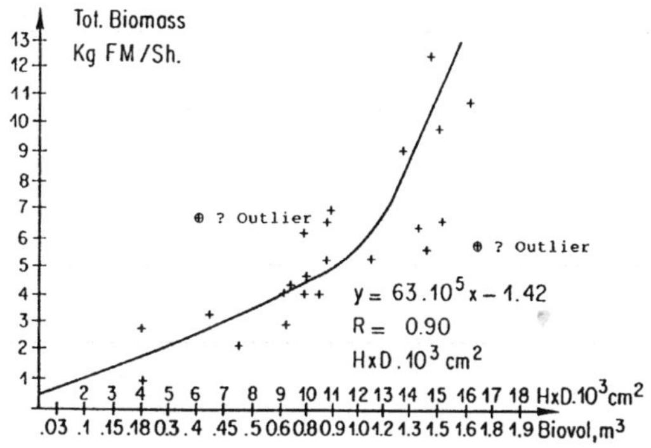

Figure 6. Allometric relations, *Atriplex canescens*, subsp. *canescens* (3 years) 2000 Sh/ha.

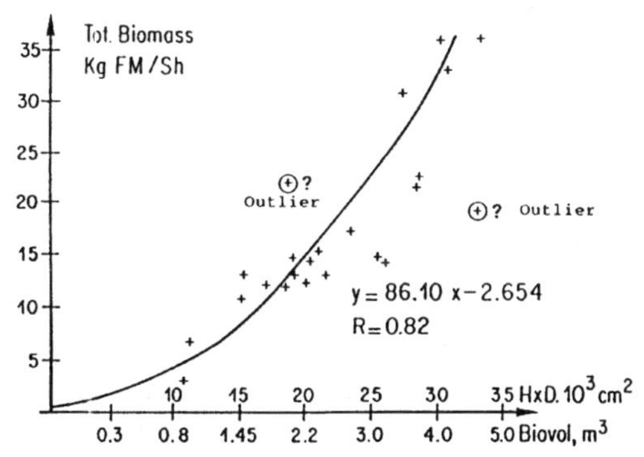

Figure 4. Allometric relations, *Atriplex halimus*, subsp. *scheinfurthii* (3 years) 2000 Sh/ha.

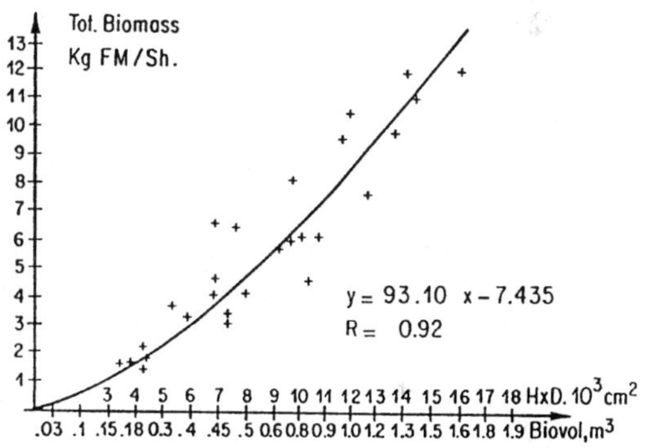

Figure 7. Allometric relations, *Atriplex undulata* (3 years) 2000 Sh/ha.

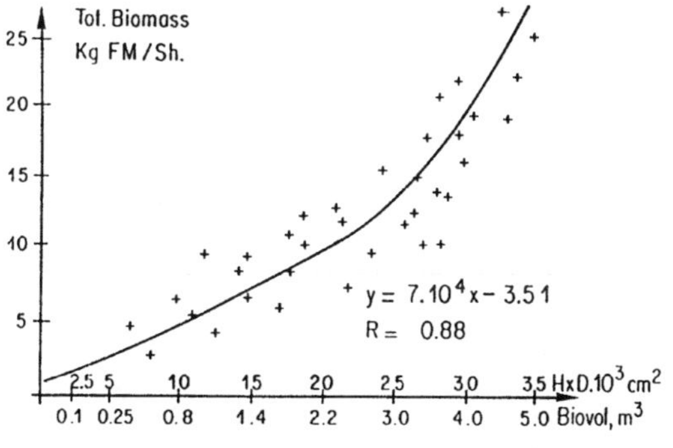

Figure 5. Allometric relations, *Atriplex canescens*, subsp. *linearis* (3 years) 2000 Sh/ha.

Mesembryanthemum spp., *Aizoon* spp., *Cressa cretica*, *Frankenia* spp. *Blackia inflata*, *Ammi visnaga*, *Centaurium spicatum etc.*

Others are more or less grazed depending on the local situation of offer and demand, on phenological stage, season etc. such are: *Arthrocnemum indicum*, *Salicornia arabica*, *Salsola tetrandra*, *S. tetragona*, *Salsola vermiculata var. villosa*, *Suaeda fruticosa* etc.

A third category should be distinguished of species which are palatable, or highly palatable, to livestock and game. Among those are *Atriplex halimus*, *A. portulaccoides*, *A. glauca*, *A. mollis*, *Salsola zygophylla*, some ecotypes of *Salsola vermiculata* (Syria), a biannual legume halophyte

Hedysarum carnosum, and a number of exotic species introduced for fodder shrub plantations:

Atriplex amnicola (Australia)
Atriplex nummularia (Australia)
Atriplex canescens (USA)
Atriplex undulata (Argentina)
Atriplex repanda (Chile)
Atriplex lentiformis (USA)
Kochia scoparia (Eurasia)
Kochia indica (Australasia)

All halophytes, however, share two common features.

1. Because of their high salt content (15–50% salts on the leaf DM) their consumption by animals is limited. But some animals, mainly sheep and camels (dromedaries), may have acquired a high tolerance to saline fodder and may then consume 100–200 g DM $kg^{-0.75}$ per day of this kind of forage without any supplement (Le Houérou 1992, 1993b; Benjamin et al. 1986; Correal et al. 1990; Otal et al. 1991). This corresponds with 1.5 to 3.0 kg DM h^{-1} d^{-1} for a 40 kg sheep and 7–14 kg DM h^{-1} d^{-1} for a 300 kg camel. As a consequence of the above, animals living on such diets need drink large quantities of water in order to eliminate the salts ingested. This water consumption may go up to 750 g H_2O $kg^{-0.75}$ in sheep under hot summer weather, that is 12 kg H_2O h^{-1} d^{-1}, for a 40 kg sheep (Wilson 1966; Le Houérou et al. 1983; Le Houérou, 1992, 1993b).

2. Because of the physiological mechanism of salt-tolerance, halophytes contain high rates of nitrogen, particularly non-protein nitrogen such as glycine betaïne and proline. Normally these compounds are not retained in the digestion processes and are eliminated in urine and faeces. About 50% of the 3–4% nitrogen contained in the halophytic green tissue DM is thus excreted.

But it would seem, from large scale experiments, that animals which are accustomed to this type of feed, and have the microorganism of their digestive tract adapted to it, can make use of this non-protein nitrogen when the diet is rich enough in energy for the microbes to break down the non-protein-protein nitrogen into their own amino-acids.

Digestibility on the dry matter may vary greatly with species, phenology and season; it may reach 70% under the best conditions and drop below 40% in unfavourable circumstances. In most cases it ranges from 50 to 60% (Le Houérou 1980, 1981, 1986, 1993b; Le Houérou et al. 1982, 1983).

Organic matter digestibility is of the order of 50% ±5 (Le Houérou 1992), while nitrogen digestibility is 65% ±5, of which only 55% is retained (Le Houérou 1980, 1992; Le Houérou et al. 1983b). Halophytes are usually fairly poor in energy and high in protein (Le Houérou 1980a, b). The available Net Energy is typically 2.5–4.0 Mj kg^{-1} DM and Metabolizable Energy 5–8 Mj kg^{-1} DM i.e. 0.36–0.57 Scandinanian Feed Units kg^{-1} DM or 0.25–0.40 kg TDN kg^{-1} DM; while conventional medium quality fodder crops have a NE of the order of 5–6 Mj kg^{-1} DM i.e. a ME 9–11 Mj kg^{-1} DM or 0.7–0.8 SFU kg^{-1} DM, or 0.5–0.6 kg TDN kg^{-1} DM. The range in halophytes is thus 50–60% of the energy value of a common fodder crop roughage (Le Houérou 1981; Le Houérou et al. 1993b).

Crude protein content is typically 10–20% on the DM, which is good to high. But, as mentioned above, no more than about 50% of these nitrogen compounds are true digestible protein (Yaron et al. 1985; Benjamin et al. 1986; Benjamin & Oren 1986; Teifert, 1989).

The nitrogen richness of halophytes may thus not be fully used by herbivores.

But in some instances it seems this non-protein nitrogen may be retained and metabolized.

There are 3 conditions under which this non-protein nitrogen may be retained and metabolized.

a. The rumen microflora has undergone a self selection over time enabling it to incorporate this nitrogen in its own substance and transform it into proteinic nitrogen.
b. There must be enough energy in the diet in order to the bacteria in the rumen to develop and multiply in order to break down these compounds and synthesize them into their own amino-acid.
c. The animals must have become adjusted to this type of diet and able to absorb substantial amounts of it (e.g. over 1 kg DM h^{-1} d^{-1} for adult sheep). Long term experiments conducted in various countries of the Mediterranean basin: Libya, Tunisia, Israel, Spain, Morocco have shown that sheep, either, penned or

browsing, can eat up to 2 kg DM h^{-1} d^{-1} of saltbush forage (*Atriplex nummularia*, *A. canescens*, *A. halimus*), without any complement. Under such conditions they can only maintain body weight, (and normal body condition) without any production. But when an *Atriplex* diet is complemented by energy-rich roughage, or grazing, or concentrate, daily body weight gains may reach 100 g h^{-1} d^{-1} in adult sheep or a carrying capacity of 3.3 sh ha^{-1} yr^{-1} in arid rangeland rehabilitated with *Atriplex* planting vs 1.0 sh ha^{-1} yr^{-1}, on conventional depleted range, or 50 kg Lamb Lwt ha^{-1} yr^{-1} vs 15 under current routine management in otherwise similar conditions (Franclet & Le Houérou 1971; Le Houérou et al. 1983, 1993; Le Houérou 1992, 1993; Benjamin et al. 1988; Correal et al. 1990; Belmonte et al. 1990; Otal et al. 1989, 1990, 1993).

Kochia spp. forage is liable to oxalate toxicity when this forage exceeds 40% of the diet for a long time, as this forage may contain 3–7% oxalates on the DM (Finley & Sherrod, 1971; Sherrod, 1971, 1973; Anaya Garduno 1993).

7. Cultivation

Few species of forage halophytes have been used under cultivation for the rehabilitation of degraded arid land, either saline or not. These are essentially exotic saltbushes in the Mediterranean Basin and also some native species (Le Houérou 1992).

Among the exotics the most widely utilized over several tens of thousand hectares each are, in decreasing order of hectareage planted:

Atriplex nummularia (Australia)
Atriplex canescens (USA)
Atriplex lentiformis (USA)
Atriplex semibaccata (Australia)
Myoporum insulare (Australia) (=*M. serratum*, *M. pictum*)

A number of other species have been successfully used on small scale trials:

Atriplex undulata (Argentina)
Atriplex repanda (Chile)
Atriplex paludosa (Australia)
Atriplex cinerea (Australia)
Kochia scoparia (Eurasia)
Kochia indica (Australia)
Atriplex amnicola (Australia)

Artificially established native species are, in decreasing order of importance,

Atriplex halimus subsp. *halimus*
Atriplex halimus subsp. *schweinfurthii*
Atriplex leucoclada
Atriplex glauca

Some other native species have been successfully established on small scale trials such as *A. mollis* (Le Houérou et al. 1982; Le Houérou & El Barghati 1982). In most cases establishment is carried out via the usual forestry technique of planting nursery-grown seedlings. This method is expensive, costing some 500–1000 US\$ ha^{-1} it can be replaced by direct sowing techniques (using, for instance, pregerminated seeds, which reduces the cost by some 60–70% (Le Houérou 1989, 1992, 1993a; Malcolm 1989; Ward 1988; Watson 1990, 1991; Watson & O'Leary 1993).

Such plantations can be successfully established on deep silty soils with an average long term rainfall above 150 mm. All native vegetation must be eliminated in order to avoid competition. Plantations are adult between the 2nd and the 4th year after establishment. With a density of 2000 shrubs ha^{-1}, in rainfed conditions, production may vary from 2000–5000 kg DM ha^{-1} yr^{-1} of which about 50% is forage and 50% wood.

Irrigated crops of saltbushes may produce 10–20 t of forage DM ha^{-1} yr^{-1} using water having a salinity of 50% sea water (Le Houérou 1992; Pasternak 1990; see Figure 8). Well managed rainfed plantations of *A. nummularia* and of *A. halimus* are known to have thrived for over 30 years. After a few years, however, rejuvenation must take place in order to keep the stands productive. This is insured by cutting-back every 3–5 years or more often.

Other sub-halophytic and salt tolerant forage shrubs and trees are used on a smaller scale than saltbushes are (Le Houérou & Pontanier 1987):

Myoporum insulare (Australia)
Prosopis cf. *juliflora* (C. America)
Prosopis chilensis (Chile)
Geoffraea decorticans (=*Gourleia chilensis*, S. America).
Haloxylon persicum (Iran, USSR)
Haloxylon aphyllum (USSR)

Table 2. Palatibility of common native and naturalized halophyte shrubs in the Mediterranean Basin

Remarks	Species	Family (1)	Ecological Wat. (2)	classification Ion. (3)	Biocl. distr. (4)	Palata-bility (5)	Parts (6)	Stock (7)
	Aellenia lancifolia (=*Halothamnus* (l)	Che	X	o/e	A	P	LT	DGS
	(=*Halothamnus suba-phylla*)	Che	X	o/e	A	P	LT	DGS
	(=*Halothamnus s.*)							
	Agatophora alopecuroides (=*Halogeton a.*)	Che	X	o	A-D	NP,To	–	–
	Alhaghi maurorum	Leg	P	o/e	D	P	LT	DGS
*	*Anabasis aphylla*	Che	X	o	A	NP-To	–	–
*	*Anabasis aretioides*	Che	X	o	A-D	pp-	T	(DG)
*	*Anabasis articulata*	Che	X	o	D	pp	T	(DG)
*	*Anabasis oropediorum*	Che	X	o	A	HP	T	ACDGS
*	*Anabasis prostrata*	Che	X	o	SA-A	NP	–	–
*	*Anabasis setifera*	Che	X	o	A-D	pp	T	DGS
	Arthrocnemum glaucum (*Salicornia macrostachya*)	Che	H	y	SA-A-D	pp	T	(DGS)
	Atriplex coriacea	Che	H	y	A-D	pp	LT	(DGS)
	Atriplex glauca	Che	X	o-e	A	P	LT	(DGS)
	Atriplex halimus	Che	M	o-e	SA-A	HP	LT	(DGS)
	Atriplex malvana	Che	H	y	A	pp	LT	(DGS)
	Atriplex mollis	Che	P	e	A	HP	LT	ACDGS
	Atriplex portulaccoides (*Halimione p.*)	Che	H	e	SA-A	HP	LT	ACDGS
	Atriplex semibaccata	Che	X	o	SA-A	P	LT	DGS
	Chenolea arabica (=*Bassia a.*)	Che	X	o	A	P	LT	DGS
*	*Cornulaca monacantha*	Che	X	o	D	HP	T	D(GS)
	Frankenia thymifolia	Fra	X	e	A-D	pp	LT	D(GS)
	Frankenia boissieri	Fra	X	e	A-D	pp	LT	D(GS)
*	*Girgensohnia oppositiflora* (=*Salsola o.*)	Che	X	o	SA-A	NP	–	–
	Halimione: cf. *Atriplex*							
	Halocnemum strobilaceum	Che	H	y	SA-A-D	NP	–	–
*	*Hammada elegans* (=*H. schmittiana* =*H. salicornica*).	Che	X	o	D	pp	T	(D)
*	*Hammada scoparia*	Che	X	o	A-D	pp	T	DGS
	Halopeplis perfoliata	Che	H	y	A-D	NP	–	–
*	*Haloxylon aphyllum*	Che	P	o	A-D	P	T	DG(S)
*	*Haloxylon persicum*	Che	P	o	A-D	P	T	DG(S)
+o	*Kochia indica*	Che	M	o	A	P	LT	CDGS
+o	*Kochia scoparia*	Che	M	o	A	P	LT	CDGS
	Kraschenninikovia ceratoides (= *Ceratoides lanata*)	Che	X	o	A	P	LT	CDGS
	Lepidium subulatum	Cru	X	o	A	P	LT	CDGS
	Limnonistrum	Plu	P	e	D	NP	–	–
	guyonianum	Plu	P	e	SA-A-D	NP	–	–
	Limoniastrum monopetalum	Plu	H/X	e	SA-A-D	NP	–	–
	Limonium spp.							
*	*Noaea mucronata*	Che	X	o	A-D	P	LT	DGS
	Nitraria retusa	Zyg	P	e	A-D	pp	LT	(DGS)
*	*Nucularia perrini*	Che	X	o	D	P	LT	D(GS)

Table 2. Continued

Remarks	Species	Family (1)	Ecological Wat. (2)	classification Ion. (3)	Biocl. distr. (4)	Palata-bility (5)	Parts (6)	Stock (7)
	Reaumuria spp.	Tam	X	o	A-D	pp	LT	DGS
	Salicornia fruticosa	Che	H	y	SA-A-D	pp	T	(DGS)
	Seidlitzia rosmarinus (=*Salsola schweinfurthii*)	Che	H	y	A-D	pp	LT	D(GS)
	Salsola baryosma (=*S. foetida*)	Che	X	e/y	D	NP	–	–
	Salsola delileana	Che	X	o	D	P	LT	DGS
	Salsola gemmascens	Che	X	o	A-D	P	LT	DGS
	Salsola gymnoschala	Che	X	o	A-D	P	LT	DGS
*	*Salsola longifolia*	Che	X	o	SA-A	NP	–	–
	Salsola tetragona	Che	H	e	D	pp	LT	D(GS)
	Salsola tetrandra	Che	H	y	SA-A-D	pp	LT	D(GS)
	Salsola sieberi.							
	var. *vesceritensis*	Che	H	e	A-D	pp	LT	DGS
	var. *zygophylla*	Che	X	o	A-D	HP	LT	ACDGS
	Salsola vermiculata							
	var. *vermiculata*	Che	X/P	o	A	P	LT	DGC
	var. *brevifolia*	Che	X	e	A-D	P	LT	DGS
	Suaeda asphaltica	Che	X	e	A-D	NP	–	–
	Suaeda monodiana	Che	H	y	D	NP	–	–
	Suaeda monoïca	Che	H	y	A-D	NP	–	–
	Suaeda fruticosa	Che	H	y	SA-A-D	P	LT	ACDGS
	Suaeda pruinosa	Che	H	y	SA-A-D	pp	LT	DGS
	Suaeda vermiculata (=*S. mollis*)	Che	M	o/e	A-D	pp	LT	DGS
	Tamarix spp.	Tam	H/p	e/y	SA-A-D	pp/NP	LT	(DGS)
	Traganum nudatum	Che	X	o	D	P	LT	DGS

Explanation
(1) *Families*: Che = Chenopodiaceae; Cru = Crucifereae; Fra = Frankeniaceae; Leg = Leguminoseae; Plu = Plumbaginaceae; Tam = Tamaricaceae; Zyg = Zygophyllaceae.
Ecological classification: (2) Wat. = water budget; X = Xerohalophytes; M = Mesohalophytes; H = Hygrohalophytes; P = Phreatohalophytes;
(3) Ion. = Ionic budget; y = Hyperhalophytes; e = Euhalophytes; o = Hypohalophytes (for definitions see Le Houérou 1993a).
(4) *Bioclimatic distribution*: SA = Semi-Arid ($P > 400$ mm); A = Arid ($400 > P > 100$ mm); D = Desert or Hyper-Arid ($P < 100$ mm).
(5) *Palatability*: HP = Highly Palatable; P = Fairly Palatable; pp = Poorly Palatable; NP = Non-Palatable; To = Toxic.
(6) *Parts consumed*: L = Leaves; T = Twigs; Fl = Flowers; Fr = Fruits.
(7) *Stock*: A = Asses; C = Cattle; D = Dromedaries (Camels); G = Goats; S = Sheep; (S) = Occasionally consumed by sheep.
Remarks: *Alternate halophyte; + Bush-like exotic annuals; o Naturalized species.

8. An outstanding halophyte fodder crop: *Hedysarum carnosum* Desf.

Hedysarum carnosum is a xero-gypso-halophyte native to the crassulescent steppes of northern Africa and particularly common in early-rain seasons in S.E. Algeria (Hodna Basin and N. watershed of Chott Melhrir, on the S. foothills and flooding zones of the Nemenchas mountains), central and southern Tunisia (Sekha Kelbia, Enfidha, Sidi el Hani, Kairouan, Soussa, Gafsa, El Hamma of the Djerid, Gabès and El Hamma of Gabès). It is rarer in W. Libya (Giosc, Jawsh, Nalut, El Wotia) and in S. Morocco. It was observed and collected by the present writer on many occasions in the course of various surveys. It is often found in cereal fields and fallows and also in crassulescent steppes in association with other halophytes such as: *Salsola tetrandra*, *Suaeda fruticosa*, *Arthrocnemum macrostachyum*,

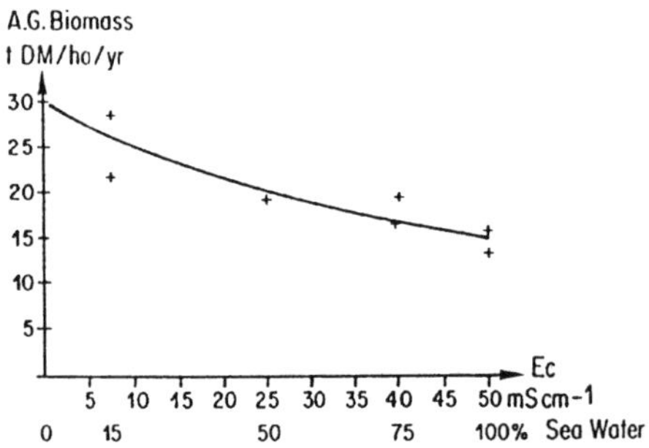

Figure 8. Productivity of *Atriplex nummularia* as a function of irrigation water salinity (Pasternak 1990).

Salsola vermiculata, *S. zygophylla*, *Atriplex halimus*, *A. glauca* etc.

Soils are always silty to loamy, gypsic and more or less saline and/or sodic, with ECe up to 40–45 mS cm^{-1} and SAR up to 80–100 m.e. L^{-1} of soil saturation extract.

Natural populations yield up to 25 tons fresh matter per ha/season in good years with early autumn rains, both in Tunisia and in Algeria (Le Houérou 1965, 1969 a, b; Le Houérou et al. 1975).

Cultivation has been carried out under both rainfed and irrigated conditions. Yields of 2–4 t DM ha^{-1} yr^{-1} have been recorded on demonstration plots of several thousand m^2 each. Under irrigation, yields of up to 50 t DM ha^{-1} yr^{-1} have been recorded on several occasions, including in real agricultural scale demonstrations of over one hectare (Le Houérou 1969a). On a particular occasion, the present author and his associates witnessed a yield of 50 tons FM ha^{-1} on a 3 ha irrigated field over a period of 6 months, with 3 complementary autumn irrigations totalling ca 100 mm, in addition to 250 mm of winter-spring rains. This took place at the state farm of El Grinn, 15 Km W of Kairouan, Tunisia, in October 1965-March 1966. The soil was a deep subsaline calcareous loam. The fodder harvested was fed fresh to a stabulated herd of friesian dairy cows over a two-weeks period, with an approximate daily consumption of 50 kg h^{-1} d^{-1} (7.5 kg DM), with virtually nil left-over and no decrease in milk production.

The proximate composition of the DM of *H. carnosum* at an early flowering stage was as follows (Delhaye 1974): Ash: 24.6%, CP 18.0; Fat: 7.7; CF: 18.5; NFE: 33.0; Ca: 3.2; P: 0.2; Mg: 0.8; K: 1.5; Na: 1.0; Fe: 0.01; Mn: 0.004. ME was estimated ca 10.5 Mj (ca 2500 Kcal.); NE: 5.25 MJ (ca 1250 Kcal.). It thus ranks as quite comparable in feed value to conventional legume fodder crops such as clovers, alfalfa, medics, albeit substantially higher in ashes and salt.

Fleshy sainfoin has been utilized in large scale range rehabilitation programmes, in conjunction with *Atriplex nummularia*, on loamy more or less saline soils in Central Tunisia (Kairouan, Enfidha, Ouaslatia etc.) Over more than 10 years and several thousand hectares.

This crop, however, is subjected to some major setbacks: having a high rate of hard seeds, it requires hulling and scarifying for good regular germination; this can also be overcome by mass selection since, as in most species in the genus *Hedysarum*, hard and soft seed are of a different colour. Such breeding programmes have been successfully carried out in Sicily on another species: *H. coronarium*. The second main drawback of *H. carnosum* is its variable ability to regrow after cutting. This requires some applied research on the relationship between season, phenological stage, height of cutting and ability to regrow.

Other sub-halophytic fodder crops, fairly to highly tolerant to salinity are the sweet clovers: *Melilotus albus*. Medic., *M. officinalis* Lamk, *M. italicus* (L.) Lamk, *M. suaveolens* Ledeb. (=pp. *M. altissimus* Thiulin).

M. italica is grown as an annual fodder crop for dairy cattle in N. Morocco (Taza and the Rif Provinces), it produces very high yields under semi-arid to sub-humid conditions and medium to fine textured soils (Villax 1963: Le Houérou 1969c). *M. officinalis* and *M. albus* are conventional biannual fodder crops, albeit little utilized in the Mediterranean Basin. *M. suaveolens* is a naturalized species in the fallows and a fodder crop in the Bale and Arussi mountains and highlands of S. Ethiopia, reaching a height of 120–160 cm at flowering stage, with a very high fodder production potential and an equally high potential for bee keeping. Fodder yields of 10–15 t DM ha^{-1} yr^{-1} have been reported under rainfed sub-humid conditions in the Arussi Highlands (Fröman, 1975). This species warrants screening trials in order to assess its fodder potential under mediterranean saline conditions.

9. Conclusions

The mediterranean flora includes about 1100 species of halophytes, some 70% of which are perennials and 30% annuals. But no more than some 50 species have a true fodder value among the perennials and almost none among the annuals. The perennial species constitute an appreciable forage resource, particularly in the most arid zones (Xerohalophytes) and in well watered depressions having a water-table (Hygrohalophytes). The latter are amenable to fairly high forage production, a fact which, in addition, is valued because these remain green and may be utilized in the worst part of the dry season when all other rangelands have long since dried out.

The use of forage halophytes is expanding as a means of reclamation of areas degraded by secondary salinity as well as in non saline depleted rangeland. Some 100,000 ha of saltbushes have been planted in the Basin to-date.

The number of species, both native and exotic utilized, so far, in large scale land rehabilitation projects is small: half a dozen species of Saltbushes. There seems to exist further possibilities particularly with species of *Kochia* and *Maireana*, *Haloxylon* and a native halophyte biannual fodder legume of high quality: *Hedysarum carnosum*. In addition to the halophytes there are a number of salt tolerant fodder species which could be utilized in land reclamation either in rainfed conditions or under irrigation using, for instance, brackish water, drainage water or treated sewage water, or seepage areas.

Perennial grasses
Festuca arundinacea
Sporobolus ioclados
Puccinellia ciliata
Chloris gayana
Paspalum distichum
Phalaris aquatica
Phalaris truncata
Elymus elongatus
Cynodon dactylon var. *hirsutissimum* & var. *villosum*

Perennial legumes
Trifolium fragiferum
Lotus coniculatus
Lotus roudairei
Hedysarum carnosum
Tetragonolobus maritimus

Annual grasses
Lolium rigidum
Hordeum sativum
Sorghum sudanense

Annual legumes
Melilotus albus
Melilotus italicus
Melilotus officinalis
Medicago ssp.
Trifolium resupinatum
Trifolium yanninicum

Large scale experiments showed that most of these are able to produce under irrigation 5 to 20 t DM ha^{-1} yr^{-1} of good quality fodder with brackish water having an EC of 5–15 mS cm^{-1}, as long as the drainage is adequate. Under rainfed conditions in semi-arid to subhumid climates (i.e. mean annual rainfall between 400 and 800 mm) these may produce 2–10 t DM ha^{-1} yr^{-1}, i.e. a Rain-Use Efficiency of 5–10 kg DM ha^{-1} yr^{-1} mm^{-1}, vs 1–3 for common rangelands.

References

Akhani, H. & Ghorbanli, M. (1993). Contribution to the halophytic vegetation and Flora of Iran. In: H. Lieth & A. Al Masoom (eds.), Towards the rational use of high salinity tolerant plants. Vol. 1, Kluwer Acad. Publ., Dordrecht, The Netherlands, pp. 34–44.

Anaya Garduno, M. (1993). Socio-economic and ecological potential of *Kochia* as a fodder crop to complement rangeland management in arid sud semi-arid regions of Mexico, In: A. Gaston, M. Kernik & H.N. Le Houérou (ed.), Proceed. IVth Intern. Rglds Congress, CIRAD, Montpellier, pp. 356–359.

Aronson, J.A. (1989). HALOPH a database on salt-tolerant plants of the world. Office of Arid Land Studies, Univ. of Arizona, Tucson, 75 pp.

Belmonte, C., Otal, J. & Correal, E. (1990). Determinacion de la capacidad de ingestion de *Atriplex nummularia* por ovejas Segurenas estabuladas en boxes. Pastos, N° extraordinario, XXX Reunion Cientif. De la SEEP, San Sebastian, Spain, pp. 363–370.

Benjamin, R.W. (1980). The use of forage shrubs in the Norte Chico region of Chile. In H.N. Le Houérou (ed.), Browse in Africa, ILCA, Addis-Ababa, pp. 261–290.

Benjamin, R.W., Barkai, D., Hefetz, Y., Lavie, Y. & Yaaron, A. (1986). The apparent digestibility of *Atriplex nummularia* and the nitrogen balance of sheep consuming it. In: A. Dovrat (ed.), Compiler: Fodder production and its utilization by small ruminants in arid regions. Inst. of Appl. Research, Ben Gurion Univ., Beer-Sheva, pp. 59–82.

Benjamin, R.W., Seligman, N.G., Forti, M. & Becker, K. (1988). The performance and feed intake, water consumption and selectivity of sheep and goats grazing saltbush – herbaceous pasture in a semi-arid area of the Negev, Israel. Inst. of Appl. Research, Ben Gurion Univ., Beer-Sheva, pp. 2–44.

Benjamin, R.W. & Oren, E. (1986). Apparent digestibility of *Atriplex barclayana* and the nitrogen balance of sheep consuming this shrub. In: R.W. Benjanim, N.G. Seligman, M. Forti & K. Becker (eds.) Analysis of animal nutrition in shrub-grassland grazing systems with special reference to semi-arid Africa, Inst. of Appl. Research, Ben Gurion Univ., Beer-Sheva, pp. 7–35.

Benjamin, R.W., Oren, E., Katz, E. & Becker, K. (1991). The apparent digestibility of *Atriplex barclayana* and its effect on nitrogen balance in sheep. *Animal Production*, (in press), 19 pp.

Breckle, S.W. (1983). Temperate deserts and semi-deserts of Afghanistan and Iran. Ecosystems of the world. Vol. 5, Elvesier Scient. Publ., Amsterdam, pp. 277–292.

Correal, E., Belmonte, C. & Otal, J. (1990). Utilization by sheep of Old Man Saltbush (*Atriplex nummularia*): Palatability, browse efficiency, voluntary intake and chemical composition. VIth meeting of the FAO-European Sub-Network on Mediterranean Pastures and Fodder Crops, Bari, Italy, 5 pp.

Finley, L.G. & Sherrod, L.B. (1971). Nutritive value of *Kochia scoparia*. J. of Dairy Sce, 54: 231–233.

Franclet, A. & Le Houérou, H.N., (1971). Les *Atriplex* en Tunisie et en Afrique du Nord. 249 p., 27 figs, 50 phot., FAO, Rome, Italie. English translation by the US Forest Service Microf. no. 17909 E, FAO Doc. Cent., Rome, Italy.

Le Houérou, H.N., (1959). Recherches ecologiques et floristiques sur la végétation de la Tunisie Méridionale. 3 vols, 510 p., 54 Tab., 4 cartes H.T., Bibl. 530, Mem. HS Inst. Rech. Sahar., Univ. d'Alger.

Le Houérou, H.N. (1962). Les pâturages naturels de la Tunisie aride et désertique. Instit. Sces Econ. Appl., Tunis, 118 p.

Le Houérou, H.N. (1965). Improvement of natural pastures and fodder resources. Report to the Government of Libya. EPTA Rep. no. 1979, AGP, FAO, Rome, 54 pp.

Le Houérou, H.N. (1969a). la végétation de la Tunisie steppique. Ann. Inst. Nat. Rech. Agron. Tun.: 42,5: 1–624, 40 figs, 40 phot, 1 carte coul. 1/500 000 (128 000 km^2).

Le Houérou, H.N. (1974). Principles, methods and techniques for range management and fodder production. The Mediterranean Basin & Tunisia, Rge Mgt & Fodder Crops Study no. 2, FAO, Rome. (3rd edition), 214 p.

Le Houérou, H.N. (1979). Resource and potential of the native flora for fodder and sown pastures production in the arid and semi-arid zones of North Africa. In: J.R. Goodin & D.K. Northington (eds.), Arid land plant resources, ICASALS, Texas Tech Univ, Lubbock, TX, pp. 384–401.

Le Houérou, H.N. (1980a). Browse in Northern Africa. In H.N. Le Houérou (ed.), Browse in Africa. ILCA, Addis-Ababa, Ethiopia, pp. 55–82.

Le Houérou, H.N. (1980b). Chemical composition and nutritive value of browse in Tropical West Africa. In: H.N Le Houérou (ed.), Browse in Africa. ILCA, Addis Ababa, Ethiopia, pp. 261–293.

Le Houérou, H.N. (1981). The feed value of *Atriplex* spp. Techn. Paper no. 13, UNTF/Lib 18, FAO & Agric. Res. Cent., Tripoli, Libya, 5 p.

Le Houérou, H.N. (1982). A list of native forage species of potential interest for pasture and fodder crops research and development Programmes. Techn. Paper no. 41, UNTF/Lib 18, FAO & Agric. Res. Cent. Tripoli, Libya, 12 p.

Le Houérou, H.N. (1986). Salt tolerant plants of economic value in the Mediterranean Basin. Reclamation & Revegetation Research, 5: 319–341.

Le Houérou, N.N. (1989). An assessment of the Economic feasibility of fodder shrub plantation. In: C.M. McKell (ed.), The biology and utilization of shrubs, Academic Press, New York, Ch. 30, pp. 603–630.

Le Houérou, H. (1992). The role of saltbushes (*Atriplex* spp.) in arid land rehabilitation in the Mediterranean Basin: A review. Agroforestry systems 18: 107–148.

Le Houérou, H.N. (1993a). Salt tolerant plants for the arid regions of the Mediterranean isoclimatic zone. In: H. Lieth & A. El Masoom (eds.), Towards the rational use of high salinity-tolerant Plants. Vol. 1. Kluwer Acad. Publ., Dordrecht, The Netherlands, pp. 403–422.

Le Houérou, H.N. (1993b). Feeding shrubs to sheep in the mediterranean Arid Zone: intake, performance and feed value. In: A. Gaston, M. Kernick & H.N. Le Houérou (eds.), Proceed. IVth Intern. Ranglds Congr. Vol. 2, CIRAD, Montpellier, France, pp. 639–644.

Le Houérou, H.N., Claudin, J., Haywood, M. & Donadieu, P. (1975). Etude phytoécologique du Hodna (Algérie). 154 p., 14 fig., 22 tabl., 1 carte coul. 1/200 000 (3 feuilles). AGS: DP/Alg/66/509, FAO, Rome.

Le Houérou, H.N., Correal, E. & Lailhacar, S. (1993). New, Man-Made Agro-sylvopastoral systems for the isoclimatic mediterranean arid Zone. In: A. Gaston, M. Kernick & H.N. Le Houérou (eds.), Proceed. IVth Internal Rglds Congr. Vol. I, CIRAD, Montpellier, pp. 383–388.

Le Houérou, H.N., Dumancic, D. & Eskileh, M. (1983). Feeding shrubs to Sheep in Libya: intake feed value and performance. 71 p., 7 figs. Techn. Paper no. 50, UNT-F/Lib 18, FAO & Agric. Res. Cent, Tripoli, Libya.

Le Houérou, H.N., Dumancic, D., Eskileh, M., Schweisguth, D. & Telahique, T. (1982). Anatomy and physiology of a browsing trial: A methodological approach to fodder shrub evaluation. 65 p., 16 figs, Techn. Paper no. 28, UNTF/Lib 18, FAO, & Agric Res. Cent., Tripoli, Libya.

Le Houérou, H.N. & El Barghati, M.S. (1982). Shrub evaluation in the Benghazi Plains. 45 p., 3 figs, Techn. Paper no. 45 UNTF/Lib 18, FAO & Agric Res. Cent, Tripoli, Libya.

Le Houérou, H.N., Ginzburger, G. & El Khodja, N. (1983). Chemical composition and nutritive value of some range plants and fodder shrubs in Libya, 14 p., 4 Tabl., Techn. Paper no. 44, UNTF/Lib 18, FAO & Agric. Res Cent., Tripoli, Libya.

Le Houérou, H.N. & Ionesco, T. (1974). Appétibilité des espèces végétales de la Tunisie steppique. AG/TUN 71/525, FAO, Rome, 68 p.

Le Houérou, H.N. & Pontanier R. (1987). Les plantations sylvo-pastorales dans la zone arid de Tunisie, Notes Techniques du MAB, no. 18, UNESCO, Paris, 81 p.

Malcolm, C.V. (1989). Forage shrub production from salt affected soils. In: C.M. McKell (ed.) The biology and utilization of shrubs, Ch. 27, Academic Press, New York, pp. 553–574.

Malcolm, C.V., Swaan, T.C. & Riding, H.I. (1980). Niche seeding for broad scale forage shrub establishment on saline soils, Internat. Symp. on salt Affected soils. Karnal, pp. 539–544.

Nour, A.M., Tageldin, A.E., Bassumi, M. & Zahran, MA. (1991). Nutritional comparative value of *Kochia indica* and some green fodders for Sheep. In: R. Choukr'Allah (ed.), Plant salinity research, New Challenges, Inst. Agron. & Vétér. Hassan II & ISESCO, Rabat-Agadir, pp. 449-445.

Otal, J., Belmonte, C., Correal, E. & Sotomayor, J.A. (1993). Evaluation of sheep production under continuous rotational grazing of a saltbush plantation (*Atriplex* spp.) in South-East Spain. in A. Gaston, M. Kernick & H.N. Le Houérou (eds.), Proceed. IVth Internat. Rglds Congr., Vol I, CIRAD, Montpellier, pp. 568–572.

Otal, J. & Correal, F. (1989). Productividad primaria y aprovechamiento ovino de *Atriplex* nummularia, arbusto forrajero introducido en el S-E espanol. Simp. Sobre Pastos, Forrajes y Prod. Anim. en Condiciones extensivas, Centr. Reg. de Inv. Agrar. de Extramadura, Badajoz, Spain, pp. 283–283.

Pasternak, D. (1990). Fodder production with saline water. Project Report 1982-Dec. 1990. Instit. of Appl. Res., Ben Gurion University, Beer-Sheva, 1973 p.

Sarson, M. & Salmon, P. (1977). Appétibilité de certaines espéces vivaces spontanées ou introduites au Maroc et en Afrique du Nord. FAO/MOR 73/016, Rabat.

Sherrod, L.B. (1971). Nutritive value of *Kochia scoparia*, yield and chemical composition. Agron J. 63: 343–344.

Sherrod, L.B. (1973). Nutritive value of *Kochia scoparia* hay, compared with alfalfa hay. J. Dairy Sce, 56: 923–926.

Shishov L.L., Kharin, N.G. & 12 others. (1980). Soil and Geobotanical studies in the pasture zone of Syrte, Libya; 3 vols, 122, 208 & 243 p. Moscow & Tripoli.

Teifert, M.A. (1989). Nitrogen compounds in *Atriplex nummularia*. In: A Dovrat, (compiler), Fodder production and its utilization by small ruminants in Arid Regions, Inst. Appl. Research, Ben Gurion Univ, Beer-Sheva, Israel, pp. 66–75.

Ward, B.H.R. (1988). Shrub evaluation for rangeland. 59 pp mimeo, WA Dept of Agric., Perth.

Watson, M.C. (1990). *Atriplex* species as irrigated forage crops. Agric. Ecosyst & Envir. 32: 107–118.

Watson, M.C. (1991). Field evaluation of *Atriplex* species using saline drainage water. In R. Choukr'Allah (ed.), Plant salinity research, New Challenges, Inst. Agron & Vétér.: Hassan II/ISESCO, Rabat-Agadir, pp. 387–395.

Watson, M.C. & O'Leary, J.W. (1993). Performance of *Atriplex* species in the San Joaquin-Valley, California, under irrigation and with mechanical harvest. Agric. Ecosyst. & Envir. 43: 255–266.

Wilson, A.D. (1966). The tolerance of sheep to Sodium chloride in food and drinking water. Austr. J. of Agric. Res. 1: 503–514.

Yaron, A., Levi, L. & Benjamin, R.W. (1985). Analysis of shrub leaves for crude protein content. In: A Dovrat (compiler), Fodder production and its utilization by small ruminants in arid regions, Inst of Appl. Research, Ben Gurion Univ., Beer-Sheva, Israel, pp. 82–92.

Zahran, M.A. (1986). Forage potentialities of *Kochia indica* in arid lands, with particular reference to Saudi Arabia, Arab Gulf J. of Scient. Res. 4 (1): 53–68.

11. Halophytes and halophytic plant communities in the Arab region

Their potential as a rangeland resource

K.H. BATANOUNY
Faculty of Science, Cairo University, Giza, Egypt

Abstract. The Arab region is unique as regards the plants and plant life and the documentation of traditional information about them. Written records date back fifteen centuries. This paper reviews the literature and focusses on the potential of halophytic communities for production of forage for livestock. The Arab countries occupy an area of about 13,652,000 km^2. Dry conditions prevail in the region. More than 60% of the area of these countries receive less than 100 mm annual rainfall. Extensive areas are natural salines. These salines vary according to their origin as well as their geographical location. More than 150 species of them dominate the plant communities there. Some of them could be considered key species in programmes to revegetate arid salines. The halophytic plant species growing in the Arab region are limited in number compared to the total number of species comprising the flora of any country in the region. The halophytes represent less than 5% of the species growing in any of the Arab countries. The geographic distribution of the various halophytes is closely related to the minimum winter temperature. Thermophilous species, which are Sudanese or tropical, are sensitive to cold winters and are confined to southern latitudes in the region. Other species tolerate low winter temperatures and they are distributed further north in the region. Different methodologies and various concepts have been adopted in the study of vegetation, including halophytes, in the Arab countries. The results of these studies are not easily compared and impose a great difficulty in use. Another problem in the study of halophytes is their confused taxonomy, especially the chenopods. Numerous synonyms are used for the same species and are published in various publications. Hence we have long lists with the same species under different names. Halophytes occur as genetically distinct growth forms or ecotypes with different degrees of palatability and salt tolerance. These differences can be exploited in revegetation programmes.

Key words: Arabic heritage, Ecotypes, Forage, Geographic distribution, Photosynthetic pathways, Salinity, Taxonomy

Introduction

The Arab countries occupy a vast belt extending from Mauritania in the West to the Gulf in the East, and from the northern borders of Iraq in the north to the extreme south of Somalia in the south. It extends between longitudes 17° E and 60° E, and between latitudes of 37°30′ N to 1°30′ S. The total area covered by the Arab countries is some 13,652,000 km^2.

The area occupied by the Arab countries exhibits wide differences as regards surface configuration, climate and vegetation. The climatic diversity in the area under consideration is reflected in the vegetation; the latter ranges from dense humid forests to dreary plantless deserts. Some countries are mainly represented by deserts, and a few have no deserts within their territories. The arid and extreme arid zones cover by far the greater part of the Arab countries. More than

V.R. Squires & A.T. Ayoub (eds.), *Halophytes as a resource for livestock and for rehabilitation of degraded lands*, 139–163.

60% of the area of these countries receive less than 100 mm annual rainfall.

Natural salines of different kinds cover considerable areas in the region with altitude ranging from −300 m below sea level to almost 200 m above sea level. These salines are inhabited by halophytes, some of which dominate well recognized plant communities of widespread occurrence. The problems of salinity are aggravated by the inadequacy of rainfall to remove the salts. The environmental conditions prevailing in the salines of the Arab region are severe and quite complex. Though the number of the halophytic species in these salines is low as compared to that of other groups of the flora of arid and semi arid zones, they contribute, in great measure, to the vegetation of vast areas in these zones.

The limits of the area occupied by the Arab countries are biogeographically rather unnatural since it consists of small sectors which belong to different floral regions. The area represents a meeting point of five main geographical regions: Saharo-Arabian, the Irano-Turanian, the Sudanian, the Mediterranean, and the Euro-Siberian regions (cf. Zohary 1973).

The meeting of the different phytogeographical regions means the presence together of different species from different regions. Responses to the prevailing temperatures, drought and saline conditions in the region will vary.

This study synthesises present knowledge of the halophytic plant communities in the Arab region. The study might be of help in assessment of the potential of the saline lands, and the halophytes they support, as rangeland resources. Knowledge of the ecology and biology of these halophytes is indispensable to make use of such a resource. These salt tolerant and/or resistant species might be the only organisms which can be propagated under the prevailing saline conditions which prevent the use of conventional crop species.

1. Land use

It has not been an easy task to obtain reliable data on the various aspects of land use in the area under consideration. Different sources have been consulted and the data collated are given in Table 1. Despite the fact that the total area of the Arab countries amounts to 13,652,000 km^2, yet only 506,000 km^2, or almost 3.4% are cultivated with field crops, vegetables and fruit trees. Irrigated agriculture is very limited in many of the countries of the region, e.g. the cultivated area in the Arab Gulf countries does not exceed 0.18% (cf. Batanouny 1990). The Arab countries with more than 30% cultivated lands with crops are Tunisia, Syria and the Gaza strip (Table 1). On the other hand, the area occupied by rangelands (or as given in some resources as meadows and permanent pastures) is almost 2,560,000 km^2, or about 18.8% of the total area of the Arab countries. The area covered by forest and woodland is about 1,370,000 km^2, i.e. about 10%.

The total land surface under use amounts to 4,100,000 or about 30%. The remainder (almost wasteland) covers 9,500,000 km^2. An extensive area is not used at all due to the environmentally harsh conditions including scarcity of water resources and salinity of considerable tracts of the region.

Besides being limited in area, the rangelands in the Arab region, are rapidly deteriorating (Batanouny, 1987, 1990). On the other hand, the population is increasing rapidly in all Arab countries (Table 1). Estimates of population in the Arab countries were 94,138,000 in 1960. This number rose to 217,807,000 in 1989 and is expected to reach 281,112,000 in the year 2000.

In view of the continuously increasing population, the increasing demands and the deteriorating resources, the gap in food and other basic needs is ever-increasing in the Arab countries. This evinces the great need for the development of further natural resources Among these resources would be the saline waste land which occupies considerable areas in the different Arab countries.

2. Plants in the Arabic heritage

The Arab region is the only area where one can trace the presence of some plant species and their uses since time immemorial. Plants and plant products are frequently mentioned in the Bible

Table 1. Area, land use and population in the Arab countries

Country	Area (100 km^2)	% of total crops	Area (1981–83) range	Population (1000s)	
				1989	2000
African Countries					
Algeria	2382	3	14	25 063	33 444
Djibouti	22	0	9	604	
Egypt	1000	2	0	54 779	63 941
Libya	1760	1	8	4 271	6 082
Mauritania	1031	–	38	1 804	2 998
Morocco	620	19	28	25 380	29 512
Somalia	638	2	46	8 552	6 671
Sudan	2506	5	24	25 008	32 926
Tunisia	146	31	20	7 930	9 429
Asian Countries					
Bahrain	669	3	6	843	693
Gaza (Palestine)	–	45	0	1 481 (1976)	–
Iraq	435	13	9	17 601	25 377
Jordan	98	4	1	3 301	6 437
Kuwait	17.8	–	8	1 967	3 007
Lebanon	10.4	29	1	2 052	3 617
Oman	212.4	–	5	1 389	1 973
Qatar	11.4	–	5	342	569
Saudi Arabia	2150	1	40	12 378	19 824
Syria	185	51	45	12 210	17 809
United Arab Emirates	777	–	2	1 455	1 939
Yemen: South	333	1	27	2 488	3 379
North	195	14	26	6 937	10 881

and the Qorán. Numerous plants have been mentioned in the Sayings (Hadith) of the Prophet Mohamed (cf. Batanouny 1986a). The classic Arabic poetry contains innumerable references to desert plants and even their habitat and phenology.

The region is the cradle of famous and very old civilizations, e.g. the Pharaonic, Assyrian and the Babylonian. Plants and plant names were depicted on the walls of the temples and on the papyri papers.

The deserts of the Arab region supported a hardy race of herdsmen. The nomadic form of pastoralism dominates. The Bedouins traverse hundreds of kilometers with their flocks to reach the verdured patches occurring at various periods of the year in different sites. For the sake of survival, the Bedouin becomes acquainted with the plant names, the season of their prosperity and their phenological aspects, and their palatability and toxicity to animals. The Bedouin since childhood is in intimate relation with the plants growing in the environment. Plants are of vital importance to the lives of Bedouins.

In the ninth century AD, the Arabs started registering their heritage of poetry, philology, religion and medicine, in addition to the immense translation from the Greek and other languages. Since then, one can read the names of a tremendous number of books and treatises dealing with botanical nomenclature, and different aspects of plant life and uses. To give a few examples, one may mention Al Asma'i (d. about 831 AD) and his 'Book of Plants' in which he included information about plants and plant life. Some useful botanical materials were included in the book of Abu Zaid Al Ansari (d. 830 AD), which was written and published by one of his disciples: Ibn Kalawih (d. 980 AD). The famous book of Abu Hanifa Ad-Dinawary (d. 845 AD) speaks about the names of plants, their habitats and life forms (trees, shrubs or forbs). He gives numerous quotations from poetry and philological interpretations of the verses.

These books, and others, classified the plants according to appropriate criteria, e.g. the presence and the absence of spines, the colour of the flower, the shape of the fruit and even the habitat features and the geographical distribution.

Though the name halophyte was given for those plants growing in salines by Pallas (cf. Schrader 1809), only early in the nineteenth century, the Arabs for more than fifteen centuries gave the name *hamdh* for such a group of plants. *Hamdh* is an Arabic word denoting plants with salts in their bodies and are growing in the *Sebakh*. The Arab lexiographers give various specifications for this group of plants, including: salty plants, plants with sour taste, succulent plants which ooze water when punctured, plants with disagreable odour, plants used for washing clothes, hands and utensils, plants which do not flower in spring and remain flourishing during summer, etc.

The above-mentioned characteristics given to the *hamdh* plants coincide with those of the halophytes as defined by the modern scientists. Arabs give examples of the *hamdh* plants as follows: *Suweid* (*Suaeda* spp.), *Shnan* (*Seidlitzia rosmarinus*. *Harm* (*Zygophyllum* spp.), *Rimth* (*Haloxylon*, and *Hammada* spp.), *Ikhreet* (*Salsola baryosma*), *Ikrish* (*Aeluropus* spp.) and many others.

The Arab differentiated between the *Khollah* (glycophytes of today) and *Hamdh* (salt plants or halophytes). They described the *Khollah* as the sweet plants, i.e. they do not contain salt in their bodies in contrast to the *hamdh* plants, which are loaded with salt. Evidently, one can easily correlate between these two terms (*Hamdh* and *Khollah*) with the halophytes and glycophytes, respectively.

The plants were valued as a natural resource by the Bedouins. They even adopted the names of plants for themselves. Men and women were named after the plants as: *Arfaj* (*Rhanterium*), *Handhalah* (*Colocynthis*), *Salama* (*Acacia ehrenbereiana*). *Samr* (*A. tortilis*). *Sudairy* from *Sidr* (*Zizyphus*), *Suwaidy* from *Suaeid* (*Suaeda*), *Morikhy* from *Markh* (*Leptadenia*) and many others.

The Arabs since time immemorial followed a system for the conservation of the rangeland (cf. Draz 1978; Batanouny 1986b). This is the *hema* system. *Hema* is an Arabic word denoting reserve. This system helped to conserve the range plants from extinction through centuries of misuse and communal grazing. No doubt, this system preserved the valuable genetic resources which are unfortunately being overexploited these days.

Arabs also gave names for particular habitats in the desert. These names have been latinized and are still used by modern scientists. The names of Arabic origin include: *hamada*, *reg*, *serir*, *sabkha*, *erg* and *wadi*. The Arabs described the *sabkha* as being saline and more or less waterlogged in some cases.

Salines are widespread in the Arab region and their area is continuously increasing due to human activities. These salines vary according to their origin and cause of formation as well as their geographical location. They could be classified arbitrarily into the following categories.

2.1 *Coastal salines*

In view of the extensive shoreline and the presence of numerous seas, oceans, lakes and gulfs in the Arab region, the coastal salines are prominent and form an important ecosystem in the region. They extend along the Atlantic Ocean, the Mediterranean Sea, the Red Sea, the Indian Ocean, the Arabian Sea, Gulf of Oman, and the Arabian Gulf. Extensive Sabkhas are present at levels slightly above or below the sea level.

The coastal salt marshes comprise different series. They may be formed of sand bars, salty flat areas, coastal swamps subject to flooding and spray by sea water, and wadi downstreams pouring their water into the sea.

The littoral belt may be under direct (inundation) or indirect (salt water table) influence of the sea water. The ground level in these coastal salt marshes is a determinative factor which seems to control the inland extent of the salt marsh and the internal organization of the vegetation units into zones. Lands subject to washing by the sea water have lower salinity in their soils than those rising above inundation.

Usually in the coastal salines, the salinity (mostly chlorides) is high at the ground surface near the shore. In the inland parts of the salt marshes, subsurface pan or pans, mostly sulphates may develop (Kassas & Zahran 1967).

The salt marsh vegetation has common features: simplicity of structure, limited number of species and notable differences in cover due to minor differences in ground level (Batanouny 1979a). Apart from the mangroves, which are absent from the Mediterranean coasts, many halophytes are able to form phytogenic mounds, which vary in size according to the plant species and the prevailing environmental conditions (Batanouny & Batanouny 1968, 1969).

The accessibility of the coastal marshes to grazing animals, especially camels, is not so frequent in the wet season or during high tide. This is evident from the browsing of the *Avicennia* plants along the Red Sea and Arabian Gulf coasts.

There is a difference between the species growing in the coastal marshes of the Mediterranean and those growing at southerly latitudes along the Red Sea coasts. In the latter marshes, thermophilous halophytes are common. These are plants which do not tolerate low winter temperatures, e.g. *Avicennia marina*, *Rhizophora mucronata*, *Bruguiera gymnorrhiza*, *Ceriops tagal*, *Halopeplis perfoliata*, *Atriplex farinosa*, *Suaeda monoica* and *Limonium axillare*. The former four species are tropical. On the other hand, halophytic species tolerating low winter temperatures occur along the Mediterranean, e.g. *Halopeplis amplexicaulis*, *Limoniastrum monopetalum* and some *Tamarix* spp.

2.2 *Great depressions and oases*

Salines and salt marshes are widely distributed in the oases of the Sahara in the African countries and East Arabia along the Arabian Gulf. Vast depressions as Qattara depression in west Egypt has been excavated to 133 m below sea level by wind erosion. The presence of the Dead Sea creates a vast area occupied with salines. In addition to salinity, this area represents a corridor with warmer climate than expected for its latitude. This leads to the occurrence of halophytes with thermophilous affinity.

The oases, being with artesian water resources and lacking drainage systems, the uncontrolled spilling of water and the flooding of extensive areas cause the salinization of these areas. In some cases, the water table is not deep or even close to the soil surface. Under the high evaporative power of the atmosphere, the salinity is continuously increasing. The common species in such habitats include *Tamarix*, *Cressa cretica* and *Juncus* spp. are common in the oases and sometimes the soil surface is covered with salt crust – hindering the growth of plants. The common species in these palm groves include: *Cressa cretica*, *Juncus* spp., *Phragmites australis* (halophytic ecotype), *Alhagi maurorum* (halophytic ecotype) and some *Tamarix* spp.

In the western and southern deserts of Iraq, there are vast areas occupied by salines. These support different halophytes including *Halocnemum strobilaceum*, *Binertia cycloptera*, *Frankenia pulverulenta*, *Aeluropus lagopoides* and *Cressa cretica*.

2.3 *Gypseous pavements*

Automorphous salines where salinity is primary and due to the salt content of the rocks are observed in the desert areas of the Arabian region. Salinity is not leached due to the scarcity of rainfall. Such type is in the form of flat plains represented by erosion pavements and plains with gypseous deposits. These pavements with surface layer of gypsum or other salts support no vegetation.

2.4 *Inland saline depression*

The saline depressions in the desert or in the wadis receive runoff water with dissolved salts. They are of limited area. Their occurrence is controlled by the topographical features. The hollows dug along the desert roads are transferred from non-saline to saline foci in the desert. This is due to the accumulation of runoff water with the dissolved salts in these man-made depressions. They support xerohalophytes.

The water table is very deep and does not exert any influence on the salinity of these salines. They are flooded periodically after rainfall and the salinity increases in summer. The xerohalophytes occurring in these depressions include *Tamarix* spp. in localities with fine alluvial deposits or *Zygophyllum album* in localities with coarse sediments.

All over the Arbian deserts, such salines occur in wadis, e.g. in Saudi Arabia (Batanouny & Baeshin 1983), Sahara (Quezel 1965) and eastern desert of Egypt (Kassas & Girgis 1964). These depressions support different halophytes as *Nitraria retusa*, *Atriplex halimus*, *Limonium infiniens* and others.

2.5 *Downstreams of wadis*

These are salines in the terminal drainage basins which occur in the deltaic plains of wadis. The salts leached by the floods are left in summer in the soil causing the rise of its salts content. The lower parts of the downstream as regards local topography may be affected by the intrusion of salt water of the sea. The plants growing there vary according to the topographical location of the wadi. Thermophilous halophytes grow in wadis at southern latitudes.

Tamarix spp. are of common occurrence in these habitats, which have fine-textured saline soils. The salt content of the uppermost layers increases during summer. The distribution of the different *Tamarix* spp. in the region is controlled by the temperature regime. The thermophilous species are confined to areas with warm winters. Some *Tamarix* species are sensitive to cold temperatures (see Table 2).

2.6 *Salines around springs and water points*

These salines are not uncommon in the region; they occur in wadis and on the rocky slopes. These springs and water points are places from which water oozes and runs down slope and forms a tricklet which soon dries only a few meters from the main source. Evaporation of this brackish or saline water leads to the salinization of the habitat. Usually hygrophilous plants grow near the source and hygrohalophilous plants away from it and further away there will be xerohalophytes. Among the common halophytes growing in such habitats are *Juncus* spp. and *Cyperus laevigatus*. These habitats were recorded by Kassas & Zahran (1962) in the Red Sea region and by Batanouny et al. (1972) in the Western Desert of Iraq as well as in Sinai (Migahid et al. 1959).

2.7 *Anthropogenic salines*

Salines formed due to human activities are of common occurrence in the region, e.g. Iraq (Buring 1960; Thalen 1979); Egypt (Schoonover et al. 1957). As a result of thousands of years of (manmade) irrigation, and lack of drainage devices, salinization took place in lowland Iraq, especially in Mesopotamia. This process is still taking place until now. Migahid et al. (1960) observed that many areas that were formerly under cultivation in the oases of the Western Desert of Egypt have become saline wastelands. Such changes were brought about by continual heavy irrigation of the cultivated land under conditions of high evaporative power of the atmosphere and the use of an improper cropping pattern. This phenomenon is of widespread occurrence in all the Arab countries. For example, irrigation from the Jordan River (approx. $500\,\mathrm{mg\,L^{-1}}$ soluble salts) at annual rate of 100 cubic meter water $1000\,\mathrm{m^{-2}}$ adds at least a net 500 mg of salts $\mathrm{m^{-2}}$ to the soil (Waisel 1972).

The widespread phenomenon of salinization due to faulty irrigation and lack of drainage resulted in the continuous increase of salinized land in the Arab region. This adds to the area of land with low economic productivity.

3. The halophytes of the Arab Region

The floras of the Arab countries comprise thousands of species, some countries have more than 2000 known species, while others have almost 300 species. However, these floras comprise some 150 halophytic species in about 55 genera and about 22 families. There is no difficulty in distinguishing between extreme glycophytes and halophytes; the problem arises at the boundaries between the two. Some species, which are not true halophytes, grow in localized sites in saline areas where the saline conditions are not too severe to prevent their appearance. After rains, which dilute the concentration of salts in the uppermost layer (seed bed), some glycophytic ephemerals germinate giving stunted plants completing their lifespan rapidly. Not all glycophytes are able to do so, but only those with capacities to tolerate slightly saline conditions at their maturity. However, these species exhibit better performance under non-saline conditions. Among the glycophytic species which grow in saline areas are: *Trigonella maritima*, *Plantago coronpus*, *P. cryspoides*,

Spergularia diandra, *Astragalus tribloides*, *Oligomeris subulata*, *Reichardia orientalis*, *Parapholis incurva*, *Senecio desfonatainei*, *Hordeum marinum*, *Launaea* spp. and many others.

Some species occur as genetically distinct growth forms or ecotypes. Among those exhibiting edaphic ecotypic differentiation are *Cyperus laevigatus* with halophytic and hydrohalophytic ecotypes, *Alhagi maurorum* with mesic and halophytic ectotypes (Batanouny et al. 1989), *Sporobolus virginicus* with dune and saline ecotypes (Blits et al. 1991) and *Liomonium pruinosum* with xerophytic and halophytic ectotypes (Batanouny 1979).

Particular varieties of the desert plant species are of halophytic nature and grow in salines, e.g. *Cyperus conglomeratus* var. *effusus*. The type species is a psammophytic one growing under desert conditions, while this variety has been recorded to be an associate of halophytic plant communities in the Red Sea littoral coast (Kassas & Zahran 1967) and Qatar (Batanouny 1981).

The majority of the halophytic species recorded in the Arabian region are perennials including a few trees, many shrubs and undershrubs and some grasses and sedges with underground perennating buds. A few annual species, which are usually succulent are recorded, e.g. *Salsola europaea*, *Binertia cycloptera*, *Suaeda aegyptiaca*, *S. salsa* and *Halopeplis amplexicaulis* (see Table 2). These annuals usually flower in late spring and summer months. On the other hand, the succulent perennial halophytes, especially the chenopodiaceous species flower in late summer and autumn. They are arido-active species and this would mean in many cases that they represent a good fodder supply during the dry season in the region. Some halophytes are browsed, e.g. *Avicennia marina* during the summer and autumn months, when the natural range plants are depleted.

There is no close relation between the flowering season and the photosynthetic pathway. Both C_3 (e.g. *Arthrocnemum* spp.) and C_4 (*Anabasis* spp.) flower in late summer and autumn. Studies on the photosynthetic pathway, and the ecological distribution of halophytes in the region (cf. Frey & Kurschner 1983; Frey et al. 1985) reveal that there is a correlation between the photosynthetic pathway, soil moisture and soil salinity. The C_4 species colonize a transition zone between C_3 hygrohalophytes and xerophytic, nonhalophytic C_3 species. However, salinity appears to be a factor of only secondary importance in this connection (Frey & Kurschner 1983). One cannot say in general that C_4 species have a selection advantage in extreme habitats. The geographical distribution of halophytes in the region is closely related to the temperature of the cold winters. Thermophilous species which are sensitive to cold winters are confined to southern latitudes in the region. Species of the same genus may show different responses to the prevailing temperature conditions, e.g. *Halopeplis perfoliata* is a thermophilous species growing along the coasts of the Red Sea, Arabian Gulf, while *Halopeplis amplexicaulis* is not sensitive to low temperature and grows at northern latitudes in the region, i.e. areas with cold winters. The same differentiation holds true with *Tamarix* spp., *Limonium* spp. and *Atriplex* spp. This phenomenon should be considered during the selection of halophytic species for propagation in the different parts of the region.

Some halophytes grow in moist saline habitats (hygrohalophytes), in the sea (marine halophytes) or under very dry conditions (xerohalophytes). This should also be considered in the revegetation schemes in the region. Halophytes have particular characteristics which enable them to evade and/or resist and tolerate salinity. Various adaptive mechanisms have been selected in halophytes during the course of their evolution. The ability of a plant to survive, grow and reproduce, or even survive without net growth, means that such a plant is endowed with particular adaptations enabling it to resist and/or tolerate salinity. Mechanisms and adaptations have been classified by different investigators (e.g. Arnold 1955; Waisel 1972; Flowers 1975; Breckle 1986; Batanouny 1992 and others). The main categories of these mechanisms and adaptations are given below:

A. Avoidance mechanisms
   1. Growth only during favourable low salinity seasons; such plants have short-time life cycles.
   2. Growth only in favourable sites with low salinity.
   3. Limitation of root growth and root absorption activity to distinct soil horizons.

Table 2. List of halophytes growing in the Arab region. Xerophytes and mesophytes with slight halophytic affinity are not included. (Chorotype: SA = Saharo-Arabian, SU = Sudanian, T = Tropical, M = Mediterranean, IT = Irano-Turanian, ES = Euro-Siberian, PL = Pluri-regional; Life Form: P = Phanerophyte, CH = Chamaephyte, H = Hemicryptophyte, T = Therophyte; Palatability: Pa = Palatable, Np = Nonpalatable). Photosynthetic types obtained from: Batanouny et al. (1988), Shomer-Ilan et al. (1981), Winter et al. (1976), Winter & Throughton (1978), Ziegler et al. (1991)

Species	Chorotype and distribution	Life form and habit	Photosynthetic type	Palatability	Remarks
CHENOPODIACEAE					
Anabasis setifera Moq.	SA, SU, IT, Arabia & N. Africa	CH, Leaf & stem succulent	C_4	Pa	Xerophytes, dominates a community
Arthrocnemum macrostachyum (Moric.) Moris & Delponte [= *A. glaucum* (Del.)] Ung.	SA, M	CH, Stem succulent	C_3	Np	Dominates a community, mainly coastal salines
A. fruticosum (L.) Moq. [= *Salicornia fruticosa* (L.)] L.	M, SA, N. Africa	CH, Stem succulent	C_3	Np	Dominates a community in coastal salines
Atriplex canescens James	Cultivated	P	–	Pa	Introduced and naturalized
A. coriacca Forssk.	SA, Arabia & N. Africa	CH	–	Pa	
A. farinosa Forssk.	SU, Arabia, Egypt, Sudan	CH	C_4	Pa	Dominates a community along the Red Sea coasts
A. halimus L.	SA, M, Arabia & N. Africa	CH, P	C_4	Pa	Variable sp., dominates a community in deserts and semi-deserts
A. leucoclada Boiss.	SA, IT	CH	C_4	Pa	Dominates a community in slight saline habitats, xerohalophyte
A. mollis Desf.	SA	CH	–	Pa	Salt bladders
A. nummularia Lindl.	Cultivated	P	C_3	Pa	Excellent naturalized fodder
A. potulacoides L. [= *Halimione portulacoides* (L.)] Aellen	M	H, prostrate Leafy	C_3	Pa	Dominates a community in coastal Mediterranean salines
A. semibaccata R. Br.	Cultivated	T	C_4	Pa	Naturalized
A. stylosa Viv.	M	CH	C_4	Pa	
A. tartarica L. [= *A. laciniata* L.]	M, IT	T	C_4	Pa	
Binertia cycloptera Bunge ex Boiss.	IT, North Arab Gulf countries	T, Leaf succulent		Np	Summer annual
Chenolea arabica Boiss.	SA	CH	C_4	Pa	Xerohalophyte
Halocnemum strobilaceum (Pall.) N. Bieb	M, IT, SA, Northern part of the region	CH, Stem succulent	C_3	Np	Dominates a community in the northern part of the region
Halogeton alopecuroides (Del.) Moq [= *Agatophora alopecuroides* (Del.)] Bunge	SA	CH, Succulent leaves with different varieties	C_4	Pa	Dominates a community in area with cold winters. Xerohalophyte
Halopeplis amplexicaulis (Vahl.) Ung. Sternb.	M, North Arabia, N. Africa	T, Succulent	C_3	Pa	Coastal hydrohalophyte, tolerates cold winters
H. perfoliata (Forssk.) Asch. & Schweinf	SU, Red Sea, Arab Gulf, S. Arabia	CH, H, Succulent	C_3	Np	Coastal hydrohalophyte, Dominates a community in coasts with warm winters
Salicornia arabica L.	PL, N. Africa	T, H, Succulent	C_3	Np	
S. europaea L. [= *S. herbacea* L.]	PL	T, Succulent	C_3	Np	

Table 2. Continued

Species	Chorotype and distribution	Life form and habit	Photosynthetic type	Palatability	Remarks
Salsola drummondii Ulbr. [= *S. schweinfurthii* auct. Botsch., *S. obpyrrifolia* Botschanzev]	PL	P, Succulent	C_4	Pa	Arab Gulf coasts, N. Arabia, Red Sea coasts, N. Africa, Dominates a community
S. inermis Forssk.	Sa	T	C_4	Pa	
S. jordanicola	Sa	T	C_4	Pa	Xerohalophyte
S. kali L.	PL	T	C_4	Np	Coastal areohalophyte
S. schweinfurthii Solms-Laub.	PL	CH	C_4	Pa	Xerohalophyte
S. sieberi Presl.	SA, N. Sahara	CH	C_4	Pa	Halogypsophilous plant with different varieties
Salsola soda L.	ES, M	T	C_4	Np	Coastal hydrohalophyte
S. tetragona Del. [= *S. pachoi* Volk. et Asch.]	SA, Sahara	CH, Succulent leaves	C_4	Pa	Xerohalophyte
S. tetrandra Forssk.	M, SA	CH, Succulent leaves	C_4	Pa	Xerohalophyte, dominates a community in coastal salines
S. vermiculata L.	SA	CH, Succulent leaves	C_4	Pa	Xerohalophyte, dominates a community
S. volkensii Asch. & Schw.	SA	T, Succulent	C_4	Pa	Summer xerohalophyte
Seidlitzia rosmarinus (Ehrenb.) Solms-Laub	East SA, N. Arabia western limit East Sinai	CH, P, Succulent	C_4	Pa	Xerohalophyte, dominates a community
Suaeda aegyptiaca (Hasselq.) Zoh. [= *Schanginia aegyptiaca* Aellen: *Suaeda hortensis* Forssk. ex Gmelin; *Schanginia baccata* (Forssk.) Moq.]	SA, widespread	T, Succulent leaves	C_4	Np	Flourishing late spring & summer. Eaten as salad with Yoghurt
Suaeda fruticosa Forssk. ex Gmelin [= *S. mesopotamica* Eig; *S. mollis* Desf.; *S. monodiana* Maire; *S. paulayana* Vierh.; *S. volkensii* C. B. Clarke; *S. pruinosa* non Lange; *S. vermiculata* auct. non Forssk. ex J. Gmelin]	SU, SA, widespread	CH, P, Succulent leaves	C_4	Pa	Dominates a widespread community. Thermophilous with tropical affinity
S. infiensis Caballero	SH, Sahara Endemic	CH	C	Pa	Oceanic and occidental Sahara, in the wadis in North African countries
S. maritima (L.) Dum	PL	T	C_3	–	
S. monoica Forssk. ex Gmelin	SU, Tropical Eastern Sector of Arab region	P, CH, Succulent	C_4	Pa	Obligate halophyte, thermophilous Dominates a community
S. pruinosa Lange	SU, SA	CH, Succulent leaves	C_4	Pa	Dominates a community
S. salsa (L.) Pallas	SA	T	–	–	–
S. vera Forssk. ex Gmelin [= *S. fruticosa* (L.) Boiss. non Forssk.]	SA	CH, Succulent	C_3	–	–

Table 2. Continued

Species	Chorotype and distribution	Life form and habit	Photosynthetic type	Palatability	Remarks
S. vermiculata Forssk. ex Gmelin [= *S. mollis* Del.]	SA, Su	CH, Succulent	C_4	Pa	Thermophilous, dominates a community
Sevada schimperi Moq. [= *Suaeda schimperi* (Moq.) Martelli; *S. vermiculata* forssk. *var. puberula* C.B. Clarke]	SA, Red Sea	CH, Succulent leaves	–	Pa	Thermophilous, Red Sea littoral salines
Traganum nudatum Del.	SA	CH, Spiny tipped succulent leaves	C_4	Pa	Halogypsophilous xerophyte, dominates a community
GRAMINEAE					
Aeluropus lagopoides (L.) Trin. ex Thwaits	SU, SA	H, Creeping Salt-secreting	C_4	Pa	Dominates a community in littoral salines, summer active
A. littoralis (Gouan) Parl.	M, IT	H, Creeping, Salt-secreting	C_4	Pa	Coastal salines
A. massauensis (Fres.) Mattei [= *A. brevifolius* Nees & Steud]	SU	H, Creeping, Salt-secreting	C_4	Pa	Coastal salines, Thermophilous
Diplachne fusca (L.) Beauv. [= *Lophocloa fusca*]	Tropical	H	C_4	Pa	Wet saline habitats, cultivated in some countries
Halopyrum mucronatum (L.) Stapf.	SU, Tropical	H	C_4	Np	Dominates a community in coastal salines in regions with warm winters
Lygeum spartum L.	M, N. Africa	H	C_3	Np	Dominates a community in N. African countries
Parapholis incurva (L.) C.E. Hubbard	SA, M	T	C_3	Pa	Borders of marshes
Paspalidium geminatum (Forssk.) Stapf.	Tropical	H, Summer active	C_4	Pa	Brackish waters
Paspalum vaginatum SW.	Tropical	H, Summer active	C_4	Pa	Littorial marshes in Djibouti
Phragmites australis Cav.	PL	H	C_3	Pa	Different ecotypes
Sphenopus divericatus (Gouan) Rchb.	M, IT	T	C_3	Pa	–
Sporobolus arabicus Boiss.	SA, SU	H	C_4	Pa	Dominates a community
S. pungens (Schrader) Kunth	SA, SU	H	C_4	Pa	Dominates a community
S. robustus	SU	H	C_4	Pa	Dominates a community in Djibouti
S. spicatus (Vahl.) Kunth	SA, SU	H	C_4	Pa	Dominates a community
S. virginicus (L.) Knuth	Tropical	H	C_4	Pa	Coastal salines
AIZOACEAE					
Aizoon canariense L.	SA, widespread	T, Succulent	C_3	Np	Eaten as salad
Mesembryanthemum nodiflorum L.	M, SA, widespared	T, Succulent	C_3, CAM	Np	–
AVICENNIACEAE					
Avicennia marina (Forssk.), Vierh.	Tropical	P, Salt secreting	C_3	Pa	Northern limit 28 N L. Dominates a community
CARYOPHYLLACEAE					
Spergularia marina (L.) Griesb.	PL	T	C_3	Np	–

Table 2. Continued

Species	Chorotype and distribution	Life form and habit	Photosynthetic type	Palatability	Remarks
COMPOSITAE					
Inula crithmoides L.	M	CH, succulent leaves	C_3	Np	Hygrohalophyte
CONVOLVULACEAE					
Cressa cretica L.	M, IT, widespread	H, Salt secreting	C_3	Np	Hygro- and xero-halophyte. Forms pure population
CYNOMORIACEAE					
Cynomorium coccineum L.	M, SA	Parasite	C_3	Np	Parasite on halophytes
CYPERACEAE					
Cyperus conglomeratus Rottb. var. *effusus* (Rottb) Kuek.	SA, SU	H	C_4	Pa	Dominates a community on maritime sands
C. laevigatus L.	SA	H	C_4	Pa	Dominates a community. Good fodder. Numerous ecotypes
Schoenus nigricans L.	M, SA	H	C_3	Np	Hygrohalophyte, dominates a community
Scirpus littoralis Schrad.	M, SA	H	C_3	Np	Hygrophilous
S. maritimus (L.) Pall.	M, SA	H	C_3	Np	Hygrophilous
FRANKENIACEAE					
Frankenia hirsuta L.	SA, M	H	C_3	Np	–
F. pulverulenta L.	ES, M, IT	T	C_3	Np	In salinized agricultural land
F. revoluta Forskk.	SA, M	H	C_3	Np	–
JUNCACEAE					
Juncus rigidus Desf. [= *J. arabicus* (Asch. & Buch) Adams; *J. maritimus* var. *arabicus* Asch. & Buch ex Boiss.	M, IT, SA	H	C_3	Np	Hygrohalophyte
J. acutus L.	M	H	C_3	Np	Hygrohalophyte
LEGUMINOSAE					
Alhagi maurorum Medik	SA	CH	C_3	Pa	Halophytic ecotype dominating a community in salines
Prosopis cineraria L. Druce	IT, M, SA; East Arabia	P	C_3	Pa	Halophytic ecotype, good fodder
P. farcta (Banks & Solms) Mac Bride	SA, IT	CH	C_3	Pa	Common weed on salinized orchards
NITRARIACEAE					
Nitraria retusa (Forssk.) Asch.	SA, SU	P	C_3	Pa	Dominates a community; forms phytogenic mounds
Nitraria schoberi L.	IT, Sahara	P	C_3	Pa	Recorded in Iraq & Sahara
PLANTAGINACEAE					
Plantago crassifolia Forssk., [*P. maritima* L.]	M	H, Succulent leaves	C_3	Pa	On maritime sands
PLUMBAGINACEAE					
Limoniastrum gayonianum	SA, Sahara	CH	C_3	Np	Dominates a community
L. infiniense (Cab.) Font Quer	SA, Sahara	CH	C_3	Np	Dominates a community
Limoniastrum monopetalum (L.) Boiss.	M	CH	C_3	Np	Dominates a community

Table 2. Continued

Species	Chorotype and distribution	Life form and habit	Photosynthetic type	Palatability	Remarks
Limonium axillare (Forssk.) Kuntze	SU	CH, Salt secreting	C_3	Np	Dominates a community in coastal salines. Thermophilous
L. carnosum (Boiss.) Kuntze	SA, coast of Arab Gulf	CH	C_3	Np	United Arab Emirates
L. cylindrifolium	SU, Djibouti	CH	C_3	Np	Coastal salines, Thermophilous
L. delicatulum (Gir.) Kuntze	M, Sahara	H	C_3	Np	On maritime saline sand
L. pruinosum (L.) Kuntze	M, SA	H	C_3	Np	Different ecotypes
L. stocksii (Boiss.) Kuntze	SU, Arab Gulf	H	C_3	Np	Coastal salines
RHIZOPHORACEAE					
Bruguiera gymnorrhiza (L.) Lam.	Tropical	P	C_3	Np	Southern latitudes, Tannins
Ceriops tagal (*Per*) C.B. Robinson	Tropical	P	C_3	Np	Djibouti, Tannins, Fuel
Rhizophora mucronata Lam.	Tropical	P	C_3	Np	Southern latitudes, Tannins
SALVADORACEAE					
Salvadora persica L.	SU	P	C_3	P	Dominates a community
TAMARICACEAE					
Tamarix amplexicaulis Ehren.	SA, SU	P	C_3	Np	Sensitive to low temperature
T. aphylla (L.) Karst.	PL	P	C_3	Np	Cultivated in many countries
T. aucherana (Decne) Baum	PL	P	C_3	Np	Extreme halophyte
T. hispida	SA	P	C_3	Np	Extreme halophyte
T. mannifera Ehren.	SA	P	C_3	Np	–
T. passerinoides Del.	SA, SU	P	C_3	Np	Dominates a community. Sensitive to low temperature
T. pyconocarpa	SA	P	C_3	Np	Extreme halophyte
T. tetragyna Ehreb.	SA, M	P	C_3	Np	Dominates a community
Reaumuria hirtella Jaub. et Spach	SA	CH	C_3	Pa	Salt secreting
TYPHACEAE					
Typha domingensis Pers	PL	H	C_3	Np	Different ecotypes
ZYGOPHYLLACEAE					
Tetradiclis salsa (Ehrenb.) Litw.	SA, IT	T	C_3	Np	On sandy salines
Zygophyllum album L.	SA (M)	CH; Succulent	C_3	Np	Dominates a community on sandy soils
Z. cornutum Coss.	SA (Sahara)	CH; Succulent	C_3	Np	Dominates a community
Z. gaetulum Emb. et Maire	SA (Sahara)	CH; Succulent	C_3	Np	Dominates a community
Z. geslini Coss.	SA (Sahara)	CH; Succulent	C_3	Np	Halogypsophilous
Z. fontanesii Webb	SA (Sahara)	CH; Suculent	C_3	Np	Dominates a community
Z. waterlotii Maire	SA (Sahara)	CH; Succulent	C_3	Np	Dominates a community

B. Evasion mechanisms
   1. Selectivity against Na^+ and Cl^-.
   2. Diversion of the salts out of the assimilation tissue.
   3. Compartmentation of salt within plant organs, tissues and cells.
   4. Retranslocation of salts and secretion by roots.
   5. Leaching of salts out of the shoot.
   6. Shedding of salt-loaded tissues and organs.
   7. Secretion of salts by salt glands and bladders.
   8. Production of adventitious roots and building mounds.

C. Tolerance mechanisms
   1. Increasing of salt resistance of tissues, cells and organelles.
   2. Production of protectants.
   3. Osmotic adjustment.
   4. Ion-specific adaptation of cytoplasm.
   5. Increase of halo-succulence, leaf and stem succulence.

Many halophytic species growing in the Arabian region dominate plant communities in the salines in different countries. These communities contribute in great measure to the vegetation of these countries.

Table 2 gives a list of the common halophytic species recorded in the Arabian region. This list does not cover all the recorded halophytes. The data presented in this table include the scientific name and its synonyms, the chorotype and the geographical distribution, the life form and the habit of the plant, the photosynthetic pathway, the palatability of the plant and some additional remarks. The data incorporated in this table are collated from numerous references as well as the personal experience and unpublished records of the author over 30 years in the region. Information about the palatability of the different species may vary in the different countries and with different animals.

4. Halophytic plant communities in the Arab Region

Not less than 70 of the aforementioned halophytic plants are either dominants of plant communities or are forming associations. The remainder are associated with these dominants. The vegetation of the Arab countries has been studied for decades; some studies concentrated on the halophytic plant communities, e.g. Tadros (1953), Tadros et al. (1958) in Mareotis, Egypt; Vesey-Fitzgerald (1953, 1957) in Saudi Arabia; Kassas (1957) in the Red Sea littoral marshes in the Sudan; Kassas & Zahran (1967) in the Red Sea littoral marshes in Egypt; Halwagy & Halwagy (1977) in Kuwait; and Batanouny & Turki (1983) in Qatar. It is not intended here to give a complete description or a census of the halophytic plant communities in the Arab region, but rather to give a brief account of the important studies and the important communities recognized. A concise description of the communities with a wide geographical distribution or with notable contribution to the vegetation will be given.

Quite different methodologies and various concepts have been adopted in the study of vegetation, including halophytes, in the Arab countries (even in the same country). The Braun-Blanquet system was adopted in some parts of the region, especially in North African countries. Other systems considering the concept of dominance and not of fidelity were applied in many parts of the region, including Egypt, the Sudan, and Arabia. The results of all the studies are not easily compared and impose a great difficulty in use. The scales given to the species in any list of the plant assemblages were different, either those of Braun-Blanquet or Domin scales. In many cases, there were cursory estimates with lists taken by hasty travellers.

Another problem in the study of halophytes in the region is their confused taxonomy, especially the chenopods. Numerous synonyms are used for the same species and are published in various publications. A glance at the list of Chenopodiaceae in Table 2 demonstrates this, e.g. in the case of *Suaeda fruticosa* and many other species. The taxonomical conflicts are not yet solved (cf. Boulos 1991; Freitag 1989, 1991). The names of species should be given with the author's names to help establish which species is being considered. This would be of help in the future when the valid names are agreed upon among taxonomists. Otherwise, we shall have long lists with the species under different names. This is an obstacle confronting making use of these species.

Figure 1. Avicennia marina trees along the Red Sea coast near Jizan, Saudi Arabia.

5. Brief description of principal halophytic plant communities

Mangal vegetation

The distribution of the mangove swamps in the Arab region is interesting. They are confined to the shores of water bodies bordering the Arabian Peninsula (cf. Batanouny 1981; Kürschner 1984; Al Hubaishi et al. 1984), the Egyptian, Sudanese, Djibouti and Somali shores (Ruwa 1994). The Mediterranean coasts are devoid of such swamps.

The recorded mangrove species in the Arab region are: *Avicennia marina* (*Shora*, *Qirm*) which extends to the most northern limit of the swamps, and other species with more southern distribution (*Rhizophora mucronata* (*Qundul*), *Bruguiera gymnorrhiza* and *Ceriops tagal*.

Avicennia marina may dominate a community, which is sometimes represented by a pure population. The dominant plant may reach a height of 8 m, especially at southern latitudes (Figure 1). Hassib (1951) writing about mangroves in Egypt on the western shore of the Gulf of Suez said, 'The plants of *Avicenna* grow on the flat muddy shores, where the water is calm and where soil is flooded with water permanently or at high tide. Generally the relief is slightly sloping towards the sea and covered by a calm water, the depth of which reaches to 120 cm. The ground is soft, deep and composed of black mud, full of organic matter . . . It extends in many places far inland along the shores and assumes the form of low, dense forest or bush. Viewed from the sea, it reveals itself as a dark, often impenetrable mass of low trees . . . The ground is covered by breathing roots which rise above the water surface as naked sticks'.

An aerial survey along the shores of Saudi Arabia along the coasts of the Red Sea from Haql in the north to Jizan in the south, showed that the occurrence of the patches of *Avicennia* swamps is mainly confined to localities where they receive fresh water and water-borne fine material from the land via the wadis draining the floods to the sea.

In Qatar, Batanouny (1981) observed that *Avicennia* community occupies localities where there is a seepage of the underground water from inland towards the Gulf. It is expected that the intermittent supply of fresh water to the habitat supporting this community would help the establishment of the seedlings.

It is to be noted that the most northern station of *Avicennia* mangrove so far known is Nabq, western shore of the Gulf of Aqaba somewhat north of 28° N (Zohary 1973). The special situation of the Gulf of Aqaba as a tropical corridor extending northwards allows the growth of tropical elements in it as well as the extension through wadi Araba. *Avicennia*, being a tropical element, is sensitive to low temperature.

The plant coverage and gregariousness are affected by grazing (cf. Kassas 1957). Where the sea water is shallow and the bottom is less muddy, camels browse the leaves and shoots of *Avicennia* and the vegetation is noticeably thinned (Ruwa 1994). Protected areas from such destructive agents, as in deep water or on the shores of reef inlets, the plant growth may be extremely dense thickets.

Rhizophora mucronata occurs at more southerly latitudes. It has been recorded in Egypt by Kassas & Zahran (1967) along the Red Sea coast with a northern limit at 22°40′ N. It has been recorded in Saudi Arabia, the Sudan and Djibouti. Kassas & Zahran (1967) noted the difference between the tidal mud of the *Avicennia marina* mangrove and that of *Rhizophora mucronata* mangrove. They state that the carbonate content, calculated as calcium carbonate, is different in both communities; being low in *Avicennia* community (4.5–19.5%) and high in the *Rhizophora* (80% or more).

Bruguiera gymnorrhiza has been recorded in the Sudan (Andrews 1950), Saudi Arabia (Migahid 1991), while *Ceriops tagal* has been recorded in Djibouti (Audru et al. 1987).

Large quantities of *Rhizophora* were formerly exported from East African countries for tannin while the local people still use mangrove tannins for preserving their fishing nets, ropes and sails. In Yemen, they use it for tanning and dyeing clothes. An Andalusian Moslem Scholar Abou'l Abbas en-Nebaty, known as Ibn ar-Roumiyah (1165–1239) wrote that mangroves were used as food, fuel, medicine for curing sore mouths and tanning leather (after Bowman 1917, given by Walsh 1977).

Halopeplis perfoliata community

This community occupies the sand beaches along the Red Sea coast, especially the southern reaches, the southern coasts of Arabia and the coasts of the Gulf of Oman as well as the southern reaches of the Arabian Gulf. Freitag (1991) states that this community occupies a higher zone behind an *Arthrocnemum* or *Avicennia* belt, south of 27–25° N replacing the more northern *Halocnemum* south of 20° N stepwise also *Arthrocnemum*. The dominant species is a Saharo-Sindian (Nubo-Sindian) extending into the Eritereo-Arabian region (Freitag 1991).

Halopeplis perfoliata is replaced in the more northern latitudes of the Arab region by another species, namely *H. amplexicaulis*. This species is a succulent annual. It has been recorded in north Egypt (Taeckholm 1974), Tunisia (Le Houérou 1969) and salines of Jordan at Al-Azraq area (Long 1957). It has been associated with such halophytes as *Arthrocnemum*, *Halocnemum*, *Zygophyllum album* and others.

For further details about the community dominated by *H. perfoliata*, see Kassas (1957) for the Sudan, Kassas & Zahran (1967) for Egypt, Batanouny (1981) and Batanouny & Turki (1983) for Qatar, Kürschner (1986) for Oman.

The rangeland forage potential of the community dominated by *Halopeplis perfoliata* is limited for two reasons: its restricted distribution as a zone within the littoral salt marshes and the unpalatability of the dominant species and most of its associates.

Arthrocnemum macrostachyum community

The community dominated by *A. macrostachyum* is widespread in the Arab region. It occurs in the littoral salt marshes (Figure 2) as well as in inland salines. It extends to more southerly latitudes than the communities dominated by *Halocnemum* and *Halopeplis perfoliata*. Also, it extends more to the north than that dominated by *H. perfoliata*. It may form an association with *Limoniastrum monopetalum* (Tadros, 1953).

Figure 2. Plant growth dominated by *Arthrocnemum macrostachyum* along the coast of the Arabian gulf, Qatar.

The associates of this community and the characteristic species of the association formed by this plant differ according to the geographical distribution. Associates and characteristic species include: *Halocnemum strobilaceum*, *Limonium axillare*, *Suaeda volkensii*, *Arthrocnemum fruticosum*, *Cressa cretica*, *Urochondra setulosa* and *Frankenia revoluta*. Annuals as *Mesembryanthemum nodiflorum* may grow in habitats supporting this community. In rainy years, the uppermost soil layer becomes less saline due to the leaching of the salts. This gives a chance for the germination of seeds of some glycophytes with halophytic tendencies or which are slightly tolerant, e.g. *Trigonella maritima* and *Parapholis incurvus*. These species are palatable and are grazed by sheep and goats.

The range capacity of this community is limited. The dominant and some associates are non-palatable. Moreover, the habitat supporting it is sometimes inaccessible due to inundation by the high tide.

For further details about this community see: Montasir (1937) for lakes north of the Nile Delta; Tadros (1953) and Tadros & Atta (1959) for the Mediterranean coastal zone in Egypt; Kassas (1957) for the Sudan; Quezel (1965) for the Sahara; Kassas & Zahran (1967) for the Red Sea littoral marshes; Le Houérou (1969) for Tunisia; Vesey-Fitzgerald (1955, 1957), Frey et al. (1984) and König (1987) for Saudi Arabia; Halwagi & Halwagi (1977) for Kuwait; Batanouny (1981) and Batanouny & Turki (1983) for Qatar; and Kürschner (1986) for Oman.

Many of the dominant species forms phytogenic hillocks formed of water-borne material (Batanouny & Batanouny 1968, 1969). The soil forming the mound body has physical and chemical attributes different from that between the mounds.

Halocnemum strobilaceum community

The plant is a stem succulent which forms phytogenic mounds (Batanouny & Batanouny 1968). This is one of the most widespread communities in the littoral and inland salines of the Arab region, mainly north of the Tropic of Cancer. The community abounds in saline flats which are usually inundated during high tides which reduces its accessibility. The distribution of the dominant species up to 55° N, where it occurs in highly continental inland marshes, indicates that it has considerable frost resistance (cf. Freitag 1991).

The associates of this community differ according to the geographical distribution. Among these associates are *Arthrocnemum macrostachyum*, *A. fruticosum*, *Halopeplis amplexicaulis*, *Salicornia europaea*, *Juncus* spp. and *Atriplex portulacoides*.

For further information about this community see: Tadros (1953) for the Mediterranean coastal zone, Egypt; Kassas & Zahran (1967) for the northern reaches of the Red Sea coast; Halwagi & Halwagi (1977) for Kuwait; Quezel (1965) for the Sahara, Le Houérou (1969) for Tunisia; Batanouny (1981) and Batanouny & Turki (1983) for Qatar; Zohary (1973) for Jordan and Palestine; Guest (1966) for Iraq and Zahran (1982) for Saudi Arabia.

The rangeland forage capacity of this community differs according to the associates. Often due to the nonpalatability of the dominant species, which forms pure stands in many cases, the range potential of this community is negligible.

Seidlitzia rosmarinus community

This community has a particular distribution in the Arab region. It occurs in North Arabia along the Red Sea and the Arabian Gulf coasts as well as in inland wadis. The eastern coast of Sinai peninsula represents its western limit. It also occurs in Palestine and Iraq.

The dominant plant is palatable and is usually overgrazed. The dried and powdered leaves are used for washing Bedouin's clothes, hair, etc. and makes a lather (like soap). The community dominated by this plants is subject to severe grazing and sometimes is replaced by a secondary vegetation dominated by *Salsola baryosma*. The plant is known in Arabic by the name *Shinan* a name sometimes given to some *Salsola* spp.

S. rosmarinus is a plant which deserves a thorough study of its reproductive ecology.

Suaeda fruticosa community

The dominant species has a lot of synonyms (Table 2) and its taxonomy is still confusing (see Boulos 1991; Freitag 1989; 1991). Boulos (1991) states that all attempts to separate *S. vermiculata* and *S. fruticosa* are so far unsatisfactory. However, in the present study I deal with *S. vermiculata* Forssk. ex J.F. Gemlin and *S. fruticosa* Forssk. ex J.F. Gemlin separately. The ideas given by Freitag (1991) were adopted. This shows that the taxonomical conflict leads to difficulties in the vegetation studies and their synthesis in one study.

S. fruticosa has a wide distribution in the Arab region and its community has been given under different names, e.g. under the name *S. volkensii* in Egypt (Kassas & Zahran 1967); *S. fruticosa* in the Sudan (Kassas 1957); *S. mesopotamica* in Iraq and Palestine. This shows that the community is widespread in almost all the Arab countries. The vast stretches occupied by this community provide good grazing ground for camels.

The rangeland forage potential of this community is considerable. The dominant, as well as the associated species, are palatable. Among the associates are: *Seidlitzia rosmarinus*, *Aeluropus lagopoides*, *Salsola cycolphylla* and *Sporobolus arabicus*.

Taxonomical and autoecological studies on this species are important. This helps make full use of this community as a forage for livestock

Suaeda vermiculata community

The community dominated by *S. vermiculata* Forssk. ex J.F. Gmelin has been recorded along the shores of the Arabian Gulf, e.g. in Qatar by Batanouny (1981) and Batanouny & Turki (1983). The dominant plant, known as *Suweid*, is grazed by camels. It forms sizeable mounds of fine textured substrate. This community could be considered in the same way as *S. fruticosa* as both species are probably the same (Boulos 1991).

Suaeda monoica community

The community dominated by *S. monoica* has a particular range of distribution. The dominant plant is thermophilous halophyte, hence the community is limited to the East Sudanian region where it forms extensive littoral marsh 'forests' especially in the vicinity of the Red Sea and the adjacent Gulfs (Zohary 1973). Its northern limits are the northern foreshores of the Dead Sea, where it is often associated with *Tamarix palaestina* and *T. nilotica*.

The floristic composition of the community varies according to the geographical distribution. In the Red Sea littoral salt marsh, the associations include: *Tamarix mannifera*, *Halopeplis perfoliata*, *Zygophyllum album* and *Heliotropium undulatum* (Kassas & Zahran 1967). In the Sudan, the associates recorded by Kassas (1957) include *Suaeda fruticosa*, *Eleusine compressa*, *Cyperus conglomeratus* var. *effusus* and *Sporobolus spicatus*.

S. monoica is a true halophyte and its growth is negatively affected by a saltfree medium (Waisel 1972). The community was found to occur on deep alluvial, fine textured, compact and salinized deposits overlain by eolian sand (Batanouny 1979b).

This community has a reasonable range forage potential, especially for camels. The dominant plant needs further autoecological studies, especially on its reproductive ecology.

Salsola tetrandra community

This community is widely distributed in North Africa along the Mediterranean coast. The dominant plant is a very efficient soil binder, and the areas occupied by the community is characterized by the mounds built by the plants.

The associates of this community are *Suaeda pruinosa*, *Suaeda fruticosa* and *Frankenia revoluta*. Annual associates appear in the wet season. Some of these annuals are glycophytes with halophytic affinity. The range potential of this community is moderate. The dominant plant as well as some associates are grazed by camels. Sheep and goats graze the annual plants during the wet season. Other *Salsola* species as *S. tetragona* and *S. drummonii* dominate communities in the different parts of the Arab region (see Quezel 1965). For further information on this community see Tadros & Atta (1958), Migahid et al. (1971) and Batanouny (1979a).

Atriplex halimus community

This community occurs Mediterranean coastal zone of the region. It occupies soils with relatively low salinity as compared to the above-mentioned communities (Tadros 1953). However, it may occur in areas exposed to salt spray (Zohary 1973). The dominant species is variable and has an interesting ecotypical makeup. It ranges in altitude from −400 to 1500 m. The community has been recorded in the desert wadis in Egypt, the lower Jordan Valley and the Dead Sea area.

Among the associates recorded in the association of *Atriplex* in the coastal Mediterranean zone are *Picris radicata*, *Bromus rubens*, *Medicago minima*, *Chenolea arabica* and *Lycium europaeum*.

The range forage potential of this community is excellent due to the palatability of the dominant species and most of the associates. The presence of annuals and perennials among the associates means possible use of this community for grazing all the year around. However, this is one of the main reasons that it has been exposed to severe overgrazing.

Atriplex halimus is a xerophyte with different varieties and ecotypes. It seems important to make use of the native varieties and ecotypes for the selection of drought and salt resistant ecotypes to be used for propagation in the salines of the arid zone in the Arab region.

There are other *Atriplex* species which dominate communities, but with limited areas, e.g. *Atriplex farinosa*, *A. leucoclada* and *Atriplex portulacoides*.

A. farinosa dominates a community in the littoral salt marshes of the Red Sea. It occupies sand bars along the shore. *A. leucoclada* dominates a community in the deserts of the region, especially in wadis. The dominant plant is subjected to severe grazing pressure. *A. portulacoides* dominates a community in the salt marshes of the Mediterranean coast. Perhaps this is the most tolerant *Atriplex* species among the ones mentioned here.

Aeluropus lagopoides community

The community occurs on saline flats with fine-textured soil along the coasts of all Arab countries along the Red Sea, Arabian Gulf, Gulf of Oman and the Arabian Sea. It is not so common in the western part of the Arab region. Usually the dominant plant forms a pure population.

Other *Aeluropus* species dominate communities in different parts of the region, e.g. *A. littoralis*, *A. messauensis* (= *A. brevifolius*). The former has a more northerly distribution, while the latter is more or less confined to areas with warm winters.

The dominant plant is a perennial salt-secreting grass with string-like creeping stem producing sterile leaf-shoots and numerous clustered short culms from the nodes. Propagation might be by cuttings.

Among the associates of this community are: *Halopeplis perfoliata*, *Arthrocnemum macrostachyum*, *Tamarix mannifera*, *Zygophyllum qatarense* and *Sporobolus spicatus*. For further readings see: Kassas (1957) for the Sudan, Kassas & Zahran (1967) for Egypt, Batanouny (1981) and Batanouny & Turki (1983) for Qatar and Kurschner (1986) for Oman.

The dominant plant and some associates are palatable. Camels usually graze this community, especially in summer.

Halopyrum mucronatum community

This community has a particular distribution in the Arab countries. As far as the author is aware, this community has not been recorded north of latitude 28° N. It occurs on the coastal sands of the Arabian Gulf (Qatar and Oman), the southern reaches of the Red Sea (south Egypt and the Sudan and Saudi Arabia). The plant forms sizeable mounds. The plant cover is relatively dense, amounting to 60% on the average. The associates include: *Sporobolus arabicus*, *Suaeda vermiculata*, *Launaea nudicaulis* and *Cressa cretica*. The range forage potential of this community is limited. However, the dominant plant is a good sand binder. For further details see: Kassas & Zahran (1967), Batanouny (1981), Batanouny & Turki (1983) and Kürschner (1986).

Sporobolus spicatus community

The community is common in the Arab region, especially in salines in the oases (cf. Migahid et al. 1960; Hassib 1951). It occurs in the littoral salt marshes of the Red Sea on both sides and also in Djibouti where the dominant plant forms phytogenic hillocks of considerable size. The range forage potential of this community is moderate. It is grazed by camels. However, the limited associates are palatable in many cases.

Sporobolus arabicus community

This community occurs in the littoral marshes east of Arabia. It has been recorded in Qatar (Batanouny 1981; Batanouny & Turki 1982).

The range forage potential of this community is moderate, especially in rainy years when the ephemeral plant growth occupies the species between the dominant plants.

Sporobolus virginicus community

It occurs in the Mediterranean zone of the region. It usually abounds in the saline habitat which is transitional between the littoral sand dunes and the salt marsh habitats (Tadros 1953). It forms almost pure populations. However, associates such as *Limonium delicatulum* grow in this community. The dominant species is palatable and could be a good fodder. This species has different ecotypes (Blits & Gallagher 1991), and this should be considered in propagation experiments.

Nitraria retusa community

This community occurs in NW Arabia including Palestine and Jordan, Egypt and some countries of North Africa. Another species, *N. schoberi*, occurs in the Sahara as well as Iraq. The dominant plant forms phytogenic mounds rising above the ground surface to a height of 2 m in many cases. In addition to its ability to form these mounds, the dominant plant is palatable. It is browsed by camels and its fruits are eaten by bustards and even human beings. The associates of this community include: *Tamarix macrocarpa*, *Aeluropus lagopoides*, *Limoniastrum infiniense*, *Cressa cretica*, *Limonium pruinosum*, *Tamarix arborea* and *T. passerinoides*. The associates vary according to the geographical distribution. For further details see: Quezel (1965), Kassas & Zahran (1967) and Zohary (1973).

Zygophyllum album community

Z. album could be considered a halo-psammophyte of the Mediterranean seashore (Zohary 1973). However, it grows also in inland salines and the oases of the Sahara. The associates vary according to the geographical location. The range forage potential of this community is limited. However, palatable associates may occur. For further information see: Montasir (1937), Tadros (1953), Migahid et al. (1960), Quezel (1965), Kassas & Girgis (1964), Kassas & Zahran (1967), Zohary (1973) and for an autecological study of the *Zygophyllum* species consult Batanouny & Ezzat (1971). All of the recorded communities dominated by this species occur north of the Tropic of Cancer.

Salvadora persica community

Records of *S. persica* indicate that its growth is mainly in moist hot places. In Arabia, it occurs mainly in southern Tihama, e.g. in Jizan, where the soil is slightly saline, the ground water is near the root zone and the weather is hot all the year around. The community occurs also on fine-textured soil in the deltas of wadis draining towards the Red Sea. In the Sudan, Kassas (1960) states that the *S. persica* belt abounds in the summer rain country of the Red Sea hills. In Egypt, it has been recorded in Sinai (Evenari & Gutterman 1973); in the Nubian Desert (Kassas & Girgis 1969–70). It occurs also in Djibouti (Audru et al. 1987). The dominant plant is a good soil binder and forms a huge mound (Figure 3). However, the continuous cutting of the plant for toothbrush making leads to its disappearance from many sites. The plant is palatable and its fruits are edible. For further information see Ma'ayergi et al. (1984).

Juncus rigidus community

This community is widespread in the Arab region. It occurs in Arabia, Iraq, Jordan, Palestine, Syria, North Africa and the Sahara. Eig (1946) recorded an association of *J. maritimus* and *Schoenus nigricans* in Palestine. Zohary (1946) recorded a community with *J. maritimus* var. *arabicus* among the halophytic plant communities in Palestine. Tadros (1953) recorded a sub-association of *J. maritimus* in the Mediterranean coastal zone of Egypt. A community dominated by *J. arabicus* was recorded in the littoral marshes of the Red Sea (Kassas & Zahran 1967) and in Qatar, Saudi Arabia and the United Arab Emirates (Batanouny, unpublished). Long (1957) recorded an association of *J. arabicus* among the halophytic plant communities in Al-Azraq area, Jordan. This association grows in habitats with high ground water table and high surface salinity. In order to clarify the taxonomical confusion, the synonyms of the dominant species are given: *Juncus rigidus* Desf. [= *J. arabicus* (Asch. et Buchen.) Adam; *J. maritimus* var. *arabicus* Asch et Buchen. ex Boiss.; *J. maritimus* f. *rigidus* (Desf.) Maire; *J. nevskii* Krecz. & Gontsch.].

Juncus acutus community

This community has narrower geographical distribution than that dominated by *J. rigidus*. It occurs in more northerly latitudes. This community inhabits sites with vary shallow water table which may be inundated. Montasir (1937) recorded this community in the lakes north of the Nile Delta. The dominant species occurs in southern Europe, Canaries, North Africa as well as parts of Arabia.

Cyperus laevigatus community

This community is widespread in the Arab region. It occurs in the coastal salines, salines bordering coastal and inland lakes and in the oases. It abounds in sites inundated by water. The species has a wide geographical amplitude in the study area. Batanouny (1979b) found that uncontrolled discharge of water and garbage from Jeddah, Saudi Arabia creates an artificial (manmade) habitat which is occupied by a dense growth of *C. laevigatus*. The plant is palatable, being grazed by sheep and goats. This species has several different ecotypes. It would be one of the promising plant species to be considered for propagation in saline land as fodder.

Figure 3. A dense growth of *Salvadora persica* in the Eastern Arabian desert, Egypt.

Tamarix spp. community

Tamarix is a characteristic species of the salines of the Arab region. It grows in almost all the countries of the region, represented by numerous species, e.g. 11 species in Lybia (Qaiser & Ghafoor 1979); 5 species in Egypt (Taeckholm 1974); 3 species in Qatar (Batanouny 1981) and 5 species in Saudi Arabia (Migahid 1988). The different species occur in different habitats comprising: salines, river banks, desert wadi beds and sands (Zohary 1973). The following species are classified by Zohary (1973) as extreme halophytes: *Tamarix amplexicaulis*, *T. passerinoides*, *T. hispida*, *T. aucheriana*, *T. tetragyna* and *T. pycnocarpa*. The majority of these are leading species in halophytic plant communities. In the large desert wadis the highly polymorphic *T. nilotica* is very common and dominant.

Due to the taxonomic difficulties in identification and typification of the different species, the recorded *Tamarix* communities are often given under the names of different synonyms. For example, the community of *T. articulata* recorded by Hassib (1951) in the coastal zone of Egypt might be named a community of *T. aphylla*. The same holds true with communities identified as those of *T. mannifera* which is synonymous with *T. nilotica* (Ehrenberg) Bunge (Qaiser & Ghafoor 1979). *T. mannifera* dominates a community in the littoral salt marshes of the Red Sea (Kassas & Zahran 1967).

Plant growth dominated by *Tamarix* species occur in the deltas of wide wadis. The dominant plant forms phytogenic mounds of considerable size. Like other woody plants in arid lands, *Tamarix* is cut for fuel and other household purposes. Zohary (1973) writes about *T. aphylla*: 'the most important among tamarisk trees, is originally a plant of deep and moist wadis in hot deserts. It is readily propagated by cuttings, grow very fast, attains an arboreal habit and a relatively great age. It grows excellently on semimobile dunes in deserts but since it requires much moisture, it rapidly impoverishes or exhausts underground water resources of its surroundings. Its huge root system may reach a depth of 20 m or more and its transpiration intensity is high.'

Summary and conclusions

The halophytic plant species growing in the Arab region are limited in number compared to the total number of species comprising the flora of any country in the region. The halophytes represent less than 5% of the species growing in any of the Arab countries. Despite this fact, some halophytes dominate plant communities which contribute in a great measure to the vegetation of these countries. More than 150 species of halophytes grow in the salines of the Arab region, almost half of them dominate the plant communities. The potential economic use of many of these species is not well known. However, some of them are reputed for their value as range plants. The nomads in the region are acquainted with these *Hamdh* plants, their life spans and palatability. Some of them could be considered key species in the propagation programmes and revegetation of the salines. It is more economical and it seems proper to make use of saline areas for grazing and fodder crops, at least for the foreseeable future. Surveys to find out useful plants started in some countries (cf. Ahmad 1987) and there is a great need for extensive and intensive studies.

Such studies are important in view of the continuous increase of the area of salines and the dwindling supplies of fresh water. The fact that wild halophytes have mechanisms enabling them to tolerate high salinity, suggest that they are useful in providing clues for combating salinity problems (cf. Le Houérou 1986).

There are some important plant communities dominated by palatable halophytes in the Arab region. These dominants could be considered as the key species for future programmes making use of halophytes. These dominants include: *Anabasis setifera*, *Atriplex halimus*, *A. leucoclada*, *Halogeton alopecuroides*, *Salsola drummondii*, *S. tetragona*, *S. tetrandra*, *S. vermiculata*, *Seidlitzia rosmarinus*, *Suaeda fruticosa*. *S. monoica*, *Aeluropus lagopoides*, *Sporobolus spicatus*, *S. arabicus*, *S. virginicus*, *Avicennia marina*, *Cyperus conglomeratus* var. *effusus*, *C. laevigatus*, *Chenolea arabica*, *Alhagi maurorum*, *Prosopis cineraria*, *Nitraria retusa*, *Plantago crassifolia* and *Salvadora persica*.

There is a great need to study the plant communities (following a standardized procedure) dominated by key palatable species. The studies should include investigations of the productivity, range capacity and range potential, ecological amplitude of the dominant species and the palatable associates, the seed bank of the natural habitats, the change in the vegetation composition as a result of grazing and trampling.

The majority of the halophytic species recorded in the Arab region are perennials including a few trees, many shrubs and undershrubs, and some grasses and sedges with underground perennating buds. The annual species among the halophytes in the region are limited in number, e.g. *Salsola europaea*, *Binertia cycloptera*, *Suaeda aegvptiaca*, *S. salsa*, *Halopeplis amplexicaulis* and *Sphenopus divericatus*. The perennial chenopods are mainly succulent and arido-active, they flower in late summer and autumn. They are a good fodder during the dry season.

A great problem confronting the efforts to make use of the halophytes, espcially the chenpods in the region is their confused taxonomy. Numerous synonyms are used for the same species in the different countries, and even in the same country in different studies. The *Salsola* and *Suaeda* species are the main problematic genera and the taxonomic problems are not yet solved (cf. Boulos 1991; Freitag 1989, 1991).

The geographical distribution of the various halophytes in the Arab region is closely related to the minimum winter temperature. Thermophilous species, which are Sudanese or tropical, are sensitive to cold winters and are confined to southern latitudes in the region e.g. *Halopeplis perfoliata*, *Salsola vermiculata*, *Suaeda schimperi*, *Aeluropus massauensis*, *Avicennia marina*, *Limonium axillare*, *L. cylindrifolium*, *Burguiera gymnorrhiza*, *Rhizophora mucronata*, *Ceriops tagal* and *Salvadora persica*. Other species tolerate low winter temperatures and thus their distribution extends to the most northern latitudes of the region, e.g. *Halocnemum strobilaceum* and *Halopeplis amplexicaulis*.

Though it seems that there is a correlation between the photosynthetic pathway, soil moisture and soil salinity, yet both C_3 and C_4 plants do well in the salines of the region. The presence of both pathways might be an ecological distribution of the resources in any habitat over the different seasons. Plants with C_4 pathways are active at a time when the C_3 plants are not under optimal conditions for their activity. However, the flowering season shows no relation with the photosynthetic pathway; succulent perennial chenopods with both pathways flower in late summer and autumn.

Some halophytic species occur as genetically distinct growth forms or ecotypes Even nonhalophytic species have some halophytic ecotypes or varieties, e.g. *Cyperus conglomeratus* var. *effusus*. Among the species which have different ecotypes are: *Cyperus laevigatus*, *Alhagi maurorum*, *Phragmites australis* and *Limonium pruinosum*. One may expect that the palatability of the different ecotypes is different. Generally, high salt tolerant ecotypes should be chosen for propagation programmes. The survey of the distribution of the halophytes in the Arab region shows that the amount of chlorides and sulphates and their ratios are factors affecting this distribution. This should be taken into consideration in the propagation programmes.

The halophytes can be classified as xerohalophytes (growing in dry saline habitats) and hygrohalophytes (inhabiting wet saline habitats). The hygrohalophytes show different responses to water logging. Some species grow only in well drained habitats, while others grow in permanently inundated habitats.

The various halophytes exhibit different responses to grazing pressure Some species are grazing resistant, while others are intolerant. This should be considered when halophytes are introduced in saline land with the intention of providing forage. The palatability of many species is not fixed. Palatability varies between sites and between seasons. Perhaps different ecotypes have different levels of palatability and the nutritive value of many halophytes is not well known and needs further studies in different seasons and locations.

The seed bank in the saline habitats has not been investigated, and the longevity of seeds in these habitats is not known. Some halophytes produce dimorphic seeds. This should be investigated to assess the germination capacity of the different types and their salt tolerance, as well as that of the emerging seedlings. The effect of trampling on seed germination under natural conditions deserves attention. Also, the effect of trampling in changing the floristic composition of the vegetation should be documented.

In view of the low number of annual halophytes, one can make use of the annual glycophytes which are recorded to grow in special sites in the rainy season in the salines. These could be used in slightly salinized localities. The autecology of the key palatable halophytes (both annual and perennial) needs further intensive investigation.

Acknowlegement

Thanks are due to UNEP for providing travel funds to attend the Nairobi workshop.

References

Abou'l Abbas en-Nabaty (1230). Introduction to '*Ibn el Beitar*' (Leclerq), V. Notices des manuscrite 23 (cited from Bowman, 1917).

Ahmad, Rafiq (1987). Saline agriculture at coastal sandy belt. Dept. of Botany, Univ. Karachi, Pakistan, 183 pp.

Al-Hubaishi, A. & Müller-Hohenstein, K. (1984). An introduction to the vegetation of Yemen. Eschborn.

Arnold, A. (1955). Die Bedeuting der Chlorionen für die Pflanzen. Bot. Stud. Vol. 2. Jena.

Aurdu, J., Cesar, J., Forgiarini, G. & Lebrum, J.P. (1987). La vegetation et les potentialites pastorales de la Republique de Djibouti. Inst. d'Elevage et des Medicines Vetrinaires des Pays Tropicaux. France.

Andrew, F.W. (1950). The flowering plants of the Sudan, Vol. I. Sudan Government.

Batanouny, K.H. (1975). Certain factors involved in the evaluation of lands in arid and semi-arid zones. Proc. Internl. Seminar Land Evaluation in Arid and Semi-arid Zones of Latin America. IILA, Rome, 4–7 February 1974, pp. 79–96.

Batanouny, K.H. (1978). Current knowledge of plant ecology in the Arab Gulf countries. Catena 14: 291–316.

Batanouny, K.H. (1979a). The desert vegetation in Egypt. African Stud. Rev. Cairo University Publ. 1:9–37.

Batanouny, K.H. (1979b). Vegetation along Jeddah – Mecca road: Pattern and process as affected by human impact. J. Arid Envi. 2: 21–30.

Batanouny, K.H. (1981). Ecology and flora of Qatar. Centre for Scientific and Applied Res. Univ. of Qatar. Qatar. pp. 245 + 124 coloured plates.

Batanouny, K.H. (1983). Human impact on desert vegetation. In: W. Holzner, M.A.J. Werger & I. Ikusa (eds.), Man's impact on vegetation. Dr. W. Junk Publishers, The Hague, pp. 139-149.

Batanouny, K.H. (1986a). Plants in the Sayings of the Prophet. Directorate of Revival of Islamic Heritage, Doha, Qatar (in Arabic).

Batanouny, K.H. (1986b). Rangelands of the Arabian Peninsula with a special reference to the history of range management, The *Hema*: An old Arabian reserve system. Proc. Second Internl. Rangeland Congress, Adelaide, pp. 234–235.

Batanouny, K.H. (1990). Rangeland ecology of the Arab Gulf countries. In: R. Halwagy, F.K. Taha & S.A. Omar (eds.), Advances in range management in arid lands. Kegan Paul Publ., London/New York, and Kuwait Foundation for the Advancement of Science, pp. 33–55.

Batanouny, K.H. (1993). Adaptation of plants to saline conditions in arid regions. In: H. Lieth & A. Al Masoom (eds.), Towards rational use of high salinity tolerant plants. Vol. I: Kluwer Acad. Publ., Dordrecht, The Netherlands, pp. 387–401.

Batanouny, K. H. & Abo Sitta, Y.M. (1979). Eco-physiological studies on the halophytes in arid and semiarid zones. I. Autecology of the salt-secreting halophyte *Limoniastrum monopetalum* (L.) Boiss. Acta Bot. Acad. Sci. Hung. 23: 13–43.

Batanouny, K. H. & Baeshin, N.A. (1983). Plant communities along Medina-Badr road across the Hejaz mountains, Saudi Arabia. Vegetatio 53: 33–43.

Batanouny, K.H. & Batanouny, M.H. (1968). Formation of phytogenic hillocks. I – Plants forming phytogenic hillocks. Acta Bot. Acad. Sci. Hung. 14: 243–252.

Batanouny, K.H. & Batanouny, M.H. (1969). Formation of phytogenic hillocks. II – Rooting habit of plants forming phytogenic hillocks Acta Bot. Acad. Sci. Hung. 15: 1–18.

Batanouny, K. H. & Ezzat, N.H. (1971). Ecophysiological studies on desert plants. I – Autecology of Egyptian *Zygophyllum* species. Oecologia (Berl.) 7: 170–183.

Batanouny, K. H. & Sheikh, M. Y. (1972). Ecological observations along Baghdad-Huseiba road, Western desert of Iraq. Feddes Repertorium 83: 245–263.

Batanouny, K.H., Hassan, A.H. & Fahmy, G.M. (1992). Eco-physiological studies on halophytes in arid and semi-arid zones II – Eco-physiology of *Limonium delicatulum* (Gir.) Ktze. Flora 186: 105–116.

Batanouny, K.H., Hassan, A.H. & Sawaf, N.A. (1991). Ecotypes of *Phragmites australis* in Egypt. Proc. Internl. Conf. on Plant Growth, Drought and Salinity in the Arab Region. The Egyptian Botanical Society, Egypt, pp. 97–114.

Batanouny, K. H., Stichler, W. & Ziegler, H. (1988). Photosynthetic pathways, distribution and ecological characteristics of grass species in Egypt. Oecologia (Berl.) 75: 539–548.

Blits, K.C. & Gallagher, J.L. (1991). Morphological and physiological responses to increased salinity in marsh and dune ecotypes of *Sporobolus virginicus* (L.) Kunth. Oecologia (Berl.) 87: 330–335.

Boulos, L. (1991). Studies on the Chenopodiaceae of Arabia. 1 – The identity, typification and distribution of *Salsola imbricata* Forssk, Kew Bull. 46 (1): 137–140.

Boulus, L. (1991). Notes on *Suaeda* Forssk. ex Scop. Kew Bull. 46(2): 291–296.

Bowman, H.H.M. (1917). Ecology and physiology of the Red Sea mangrove. Proc. Am. Philos. Soc. 56: 589–672.

Breckle, S.W. (1986). Studies on halophytes from Iran and Afghanistan. II – Ecology of halophytes and salt gradients. Proc. Royal Soc. of Edinburgh 89B: 203–215.

Buring, P. (1964). Soils and soil conditions in Iraq. Min. Agriculture, Baghdad.

Draz, O. (1956). Improvement of animal production in Yemen. Bull. Inst. Désért d'Egypte 6(2): 79–110.

Draz, O. (1978). Revival of the *Hema* system of range reserves as a basis for the Syrian range management program. Proc. first Internl. Rangeland Congress, Denver, pp. 100–103.

Eig, A. (1946). Synopsis of the phytosociological units of Palestine. Palestine J. Botany, Jerusalem series 3: 183–246.

Evenari, M.Y. Gutterman. (1973). Some notes on *Salvadora persica* L. and its uses as a toothbrush. Flora 162: 118–125.

Flowers, T.J. (1975). Halophytes. In: D.A. Baker & J.L. Hall (eds.), Ion transport in plant cells and tissues. North Holland Publ. Co., Amsterdam, pp. 309–333.

Freitag, H. (1989). Contributions to the chenopod flora of Egypt. 183: 149–173.

Freitag, H. (1991). The distribution of some prominent *Chenopodiaceae* in SW Asia and their phytogeographical significance. Flora et Vegetatio Mundi 9: 281–292.

Frey, W. & Kürschner, H. (1983). Photosyntheseweg und Zonierung von Halophyten an Salzseen in der Turkei, in Jordanien und in Iran. Flora 173: 293–310.

Frey, W., Kürschner, H. & Stichler, W. (1985). Photosynthetic pathways and ecological distribution of halophytes from the littoral salt marshes (Egypt/ Sinai, Saudi Arabia, Oman and Iran). Flora 177: 107–130.

Frey, W., Kürschner, H., El-Sheikh, A.M. & Migahid, A.M. (1984). Zonation and photosynthetic pathways of halophytes on the Red Sea coast near Tawwal (Saudi Arabia). Notes Roy. Bot. Gard. Edinburgh 42: 45–56.

Guest, E. (1966). Flora of Iraq. Vol. I: Introduction. Ministry of Agricuture, Baghdad, Iraq.

Halwagy, R. & Halwagy, M. (1977). Ecological studies on the desert of Kuwait, III: The vegetation of the coastal salt marshes. J. Univ. of Kuwait 4: 33–74.

Hassib, M. (1951). Distribution of plant communities in Egypt. Bull. Fac. Sci. Fouad 1 University, Egypt.

Jafri, S.M.H. (1977). Juncaceae. In: S.M.H. Jafri & A. El Gadi (eds.), Flora of Lybia, Vol. 29.

Kassas, M. (1957). On the ecology of the Red Sea coastal land. J. Ecol. 45: 187–203.

Kassas, M. & Batanouny, K.H. (1984). Plant ecology. In: J.L. Cloudsley Thompson (ed.), Sahara desert. Pergamon Press, Oxford, pp. 77–90.

Kassas, M.A. & Girgis, W.A. (1964). Habitat features and

plant communities in the Egyptian desert. V – The limestone Plateau. J. Ecol. 52: 107–119.

Kassas, M.A. & Girgis, W.A. (1969/70). Plant life in the Nubian desert East of the Nile, Egypt. Bull. Inst. d'Egypte 31: 47–71.

Kassas, M.A. & Zahran, M. (1962). Studies on the ecology of the Red Sea coastal land. I – The district of Gebel Ataqa and El-Galala El-Bahariya. Bull. Soc. Geog. d'Egypte 35: 129–175.

Kassas, M.A. & Zahran, M. (1967). On the ecology of the Red Sea littoral salt marsh, Egypt. Ecol. Monog. 37: 97–316.

Konig, P (1987). Vegetation und Flora in Südwestlichen Saudi-Arabien (Asir, Tihama). Diss. Bot. 101. Berlin, Stuttgart.

Kürschner, H. (1986). A study of the vegetation of the Qurm nature reserve, Muscat Area, Oman. Arab Gulf J. Sci. Res. 4: 32–52.

Le Houérou, H.N. (1969). Le vegétation de la Tunisie steppique. Ann. Inst. National Recherche Agronomique de Tunisie 42 (5): 1–622.

Le Houérou, H.N. 1986. Salt tolerant plants of economic value in the Mediterranean basin. Reclam. & Reveg. Research 5: 319–341.

Long, G.A. (1957). The bioclimatology and vegetation of eastern Jordan. FAO/57/2/1109: 97 pp. 1 map (Mimeogr.).

Ma'ayergi, H.A., Ismail, S.A., Batanouny, K.H. & Rizk, A.M. (1984). Ecological and phytochemical studies on the '*Miswak*', *Salvadora persica* L. Qatar Univ. Sci. Bull. 4: 37–44.

Migahid, A.M. (1988). Flora of Saudi Arabia. Third ed. King Saud Univ. Libraries. Riyadh.

Migahid, A.M., Shafei, M.A., Abdel Rahman, A.A. & Hammouda, M.A. (1959). Ecological observations in western and southern Sinai. Bull. Soc. Ge'og. d'Egypte 32: 165–205.

Migahid, A.M., Shafei, M.A., Abdel Rahman, A.A. & Hammouda, M.A. (1960). An ecological study of Kharga and Dakhla oases. Bull. Soc. Geog. d'Egypte 33: 277–307.

Montasir, A. H.: (1937). Ecology of Lake Manzala. Bull. Fac. Sci., The Egyptian Univ. 12: 1–50 + 5 plates.

Qaiser, M. & Ghafoor, A. (1979). *Tamaricaceae*. In: S.M.H. Jafri & A. El-Gadi (eds.), Flora of Libya. Vol. 66.

Que'zel, P. (1965). La vegetation du Sahara. G. Fischer Verlag, Stuttgart.

Ruwa, R.K. (1994). Halophytic coastal marsh vegetation in East Africa (this volume, chap. 14).

Schoonover, W.R., El Gabaly, M.M. & Hassan, M.N. (1957). A study of some Egyptian saline and alkali soils. Hilgardia 26: 565–596.

Schrader, H.A. (1809). Über Palla's Halophyta mit besonderer Rücksicht auf die Gattungen *Salsola* und *Suaeda*. Schard. Neues J. Bot. 3: 58–92.

Schomer-llan, A., A. Nissenbaum & Y. Waisel. (1981). Photosynthetic pathways and the ecological distribution of the *Chenopodiaceae* in Israel. Oecologia (Berl.) 48: 244–248.

Taeckholm, V. (1974). The student's flora of Egypt. Cairo Univ. Egypt.

Tadros, T. ill. (1953). A phytosociological study of halophilous communities from Mareotis (Egypt). Vegetatio 4: 102–124.

Tadros, T. M. & Atta, Berlanta A. (1958). Further contribution to the study of the sociology and ecology of the halophilous plant communities of Mareotis, Egypt. Vegetatio 8: 137–160.

Thalen, D.C.P. (1979). Ecology and utilization of desert shrub rangelands in Iraq. Dr. W. Junk Publishers, The Hague.

Vesey-Fitzgerald, D.F. (1955). Vegetation of the Red Sea coast south of Jedda, Saudi Arabia. J. Ecol. 43: 477–489.

Vesey-Fitzgerald, D.F. (1957). The vegetation of the Red Sea coast north of Jedda, Saudi Arabia. J. Ecol. 45: 547–565.

Waisel, Y. (1972). Biology of halophytes. Academic Press, New York, London.

Walsh, G E. (1977). Exploitation of mangal. In: V.C. Chapman (ed.), Wet coastal ecosystems. Elsevier, Amsterdam, pp. 347–362.

Zahran, M. (1982). Ecology of the halophytic vegetation of Egypt. Tasks for vegetation Science No. 2. In: D.N. Sen & K.S. Rajurhit (eds.). Dr. W. Junk Publishers, The Hague.

Ziegler, H., Batanouny, K.H., Sankhla, N., Vyas, O. & Stichler, W. (1981). The photosynthetic pathway types of some desert plants from India, Saudi Arabia, Egypt and Iraq. Oecologia (Berl.) 48: 93–99.

Zohary, M. (1947). A vegetation map of western Palestine. J. Ecol. 34: 1–19.

Zohary, M. (1973). Geobotanical foundations of the Middle East. 2 Vols. Gustaf Fischer Verlag. Stuttgart, Swets & Zeitlinger, Amsterdam.

Winter, K. & Troughton, J.H. (1978). Photosynthetic pathways in plants of coastal and desert habitats of Israel and the Sinai. Flora 167: 1–34.

Winter, K., Troughton, J.H. & Card, A. (1976). C values of grass species collected in the northern Sahara desert. Oecologia (Berl.) 25: 115–123.

12. North American halophytes: Potential use in animal husbandry

E.P. GLENN,[1] R.S. SWINGLE,[2] J.J. RILEY,[1] C.U. MOTA,[3]
M.C. WATSON[4] & V.R. SQUIRES[5]
[1]*Environmental Research Laboratory, Tucson, AZ, USA;* [2]*Animal Sciences Department, University of Arizona, Tucson, AZ, USA;* [3]*Genesis, Inc., Puerto Penasco, Sonora, Mexico;* [4]*Department of Range Science, University of Arizona, Tucson, AZ, USA;* [5]*University of Adelaide, Roseworthy Campus, Roseworthy, SA, Australia*

Abstract. North America is rich in halophyte plants with over 200 distinct species occurring in coastal and inland regions. They are not a single taxonomic group but represent many different taxa and growth forms. Emphasis has shifted from studying halophyte ecology and physiology to a serious attempt to utilize them in farming systems. Promising results have been obtained from preliminary studies on their feed value for ruminants and poultry. Satisfactory weight gains can be obtained when halophytes are fed to livestock in mixed rations as a substitute for more conventional fodders. This paper reviews the current status of halophytes in North America and summarizes the results of research on animal feeding trials.

Key words: Ash content, *Atriplex*, Cattle, Distribution, Goats, *In vitro* digestibility, Irrigation, Nutritive value, Poultry, Rangelands, *Salicornia*, Sheep, Taxonomy, Utilization

Introduction

Scientific studies of North American halophytes have shifted emphasis over the years from floristic descriptions (e.g. Reimold & Queen 1974) to ecophysiological investigations (e.g. Ungar 1992). Attempts to utilize halophytes have progressed from assessments of their value in natural rangeland settings (e.g. Goodin 1979) to agronomic experimentation (e.g. Glenn et al. 1991; Watson & O'Leary 1993). North American halophytes are now being grown on a pilot scale as crop plants in irrigation districts and even under seawater irrigation along coastal deserts. Their irrigation requirements are being determined (Miyamoto et al. 1994). Their feed value is being determined by including them as components of nutritionally-complete, formulated animal diets rather than simply as range plants (Swingle et al. 1994). At least a few North American halophytes may be undergoing the transition from wild to domesticated plants.

Number of halophyte species in North America

No comprehensive list of halophyte species has been compiled for North America. However, most portions of the continent have been surveyed for salt-tolerant vegetation at one time or another. Macdonald & Barbour (1974) and Macdonald (1977) noted 57 species of vascular plants that characterized the beach vegetation of the North American coast from Point Barrow, Alaska to the tip of Baja California, as well as 140 vascular salt marsh species. Duncan (1974) using a more inclusive definition of coastal halophyte than MacDonald and Barbour, recorded 347 species from the Atlantic and Gulf Coasts of North America. They defined a halophyte as any plant that '. . . tolerates seawater, pure or diluted'. Yensen et al. (1983) reported 24 species of intertidal salt marsh halophytes and an additional 30 species from the supralittoral zone in 15 coastal marshes in the Sonoran desert.

Ungar (1974) mentioned more than 30 inland

V.R. Squires & A.T. Ayoub (eds.), *Halophytes as a resource for livestock and for rehabilitation of degraded lands*, 165–174.

halophytes of the United States in a review that was not meant to be a comprehensive listing of species; some 200 species are mentioned in his global review of halophyte ecophysiology (Ungar 1992). Allowing for overlaps and omissions, there appear to be at least several hundred distinct coastal and inland halophyte species in North America.

New species are still being discovered (Watson & Ferren 1991). They are not a single taxonomic grouping, but represent many different taxa and growth forms of vascular plants: trees, shrubs, grasses, succulents, annuals and perennials. North American halophytes, then, represent a potentially rich source of germplasm to exploit for animal feeds (Swingle et al. 1994).

Distribution of species

Investigations of marshes and salines over broad geographical areas in Canada, USA and Mexico indicate that most halophyte genera are widely distributed across the continent and many are cosmopolitan (Ungar 1974). Many halophytic species occuring in inland saline soils have closely related species in coastal marshes, particularly in the genera *Suaeda*, *Salicornia*, *Sesuvium*, *Spartina*, *Puccinellia* and *Baccharis* (Chapman 1974). Edaphic factors, especially the relationship between saline and nonsaline soils, are key determinants of plant distribution within a locality. The development of distinct halophyte zonation within a saline habitat seems to be a response to the discontinuous distribution of salinity, depth to water table and other edaphic factors in salt-affected landscapes (Ungar 1974, 1992). Complicated classification schemes for halophytes have been devised based on these zonation patterns (e.g. Chapman 1974) and distinct physiotypes have been defined based on salt tolerance characteristics (Albert & Popp 1977) but the answer to the question 'what is a halophyte' remains elusive (O'Leary & Glenn 1994).

Mangroves (mainly *Rhizophora mangle* and *Avicennia germinans*) are the dominant vegetation in the southernmost North American coastal salt marshes. Mangrove swamps are found on the lower third of the Baja California Peninsula but extend several hundred km farther north along the warmer Sonoran coast of the Gulf of California (MacDonald 1977). *Avicennia germinans*, the most cold-tolerant mangrove, grows spottily along the Gulf of Mexico coastline of northern Mexico and USA and more prolifically on the Florida Peninsula (West 1977). South Florida has greater mangrove species diversity than elsewhere in North America. Above the mangrove line, the dominant vegetation of coastal salt marshes tends to be herbaceous, although some shrubs do occur. *Spartina alterniflora* and *S. patens* (long and short cordgrass, respectively) dominate many Atlantic coastal marshes (Duncan 1974; Reimold 1977). In the Pacific, *Spartina foliosa* is the major intertidal grass along the California coast but in British Columbia and Alaska the more cold-hardy *Puccinellia* species are primary colonists in coastal marshes (Macdonald 1977).

In the relatively dry areas of North America, especially in the west and southwest of the USA and across northern Mexico, xerohalophytes dominate the plant communities (McKell & Garcia-Moya 1989). Most are species of the Chenopodiaceae and include saltbushes and shadscales (*Atriplex*), winterfat or white sage (*Eurotia*), greasewood (*Sarcobatus*) and hopsage (*Grayia*). The Sonoran and Chihuahuan deserts lie partly in the USA and partly in Mexico and both are rich in halophytes, though the Sonoran is richer in shrub species than the Chihuahuan. In the Great Basin of USA the cold desert environment supports two halophyte-dominated associations: one dominated by *Atriplex confertifolia*, *Eurotia lanata* and *Grayia spinosa*; the other by *Atriplex corrugata*, *Atriplex nuttallii* and *Kochia vestita* (McGinnies 1972). Fourteen other *Atriplex* species occur in the region (McKell & Garcia-Moya 1989). Halophytes are also well represented in the Mohave desert where species like *Atriplex hymenelytra* occur.

The warm coastal deserts of Baja California and the Sonoran coast in the Gulf of California have their own characteristic halophyte flora, dominated in places by *Atriplex barclayana*, *Atriplex canescens*, *Atriplex polycarpa*, *Allenrolfia occidentalis*, *Atriplex lentiformis* and *Distichlis spicata*. The desert species often invade the supralittoral and high intertidal zone of coastal salt marshes and occur alongside typical marsh species from the genera *Salicornia*, *Suaeda*, *Frankenia*, *Batis* and *Mononthocloe* (Yensen et al. 1983; MacDonald 1977). The only endemic grass in the Sonoran desert is the halophyte *Distichlis palmeri*, found only in the negative estuaries of the northern Gulf of California.

Traditional use of halophytes for animal feed

Historically, some limited, direct use has been made of North American salt marshes, particularly *Spartina alterniflora* marshes (e.g. Shanholtzer 1974), for grazing cattle. *Spartina* marshes have also been cut for hay. These traditional practices have decreased with increasing protection of coastal wetlands (Hitchcock 1972). By contrast, inland halophytes, especially shrubby xerohalophytes and salt-tolerant grasses found throughout the arid and semi-arid regions of the continent at all latitudes, remain important range plants for ruminant production and for wildlife.

A considerable literature has developed on the nutritive value of North American halophytes under arid range conditions. Early investigations concluded that halophytes are generally of low acceptability and nutritive value (Cook et al. 1954) and are usually not preferred forage species even though they are grazed by livestock worldwide. High mineral content, low energy content and the presence of antinutritional factors are some of the problems associated with halophyte forages (e.g., Chatterton et al. 1971, Leigh & Mulham 1964; Goodin 1979; Gihad & El Shaer 1994).

The main value of halophytes in natural rangelands is as reserve forage plants which allow livestock to maintain weight or at least lose weight more slowly after more palatable species disappear in the dry season. However, Le Houérou (1992) has shown that *Atriplex* spp., supplemented with other feed sources, can support weight gain of sheep throughout the year in the Mediterranean Basin.

In North America, the revegetation of Lake Texcoco (Mota et al. 1980; Llerena & Tarin 1978) stands as an example of how the feed value of halophytes can be improved through innovative management. *D. spicata* is considered a poor quality pasture species, incapable of supporting ruminant weight gain (Brizuela et al. 1990). Yet by rotating cattle and sheep among paddocks, it is possible to keep the pasturage at Texcoco producing predominantly tender, new growth for grazing rather than tough, older growth, which predominates in unmanaged stands of salt grass. The salt grass paddocks support weight gain of cattle stocked at 4 head per ha during 9 months of the year and 1 head per ha during the 3 coolest months (Llerena & Tarin 1978).

Production potential of irrigated halophytes

More recent research has sought to develop halophytes as agronomic crops for brackish (e.g. Watson et al. 1987; Watson 1990; Watson & O'Leary 1993) and seawater (e.g. Glenn & O'Leary 1985; Glenn et al. 1991) irrigation in arid regions of the United States and Mexico. Considerable progress has been made in developing farming systems for these plants. Small-plot trials with a variety of halophyte species showed that biomass yields are as high as conventional forage yields even under seawater irrigation (Glenn & O'Leary 1985; Glenn et al. 1991). Annual biomass production in the range of 10–20 10^3 kg ha^{-1} was obtained from widely separated genera: *Atriplex*, *Distichlis*, *Batis* and *Salicornia*. Even discounting for high ash content, the biomass yields are promising. One species has shown particular promise as a seawater crop. *S. bigelovii*, in addition to high biomass production, yields approximately 2 10^3 kg ha^{-1} of high-quality oilseed (Glenn et al. 1991). Unlike *Atriplex*, the seeds germinate directly on seawater. Commercial production of this North American halophyte is being attempted in Saudi Arabia at a location on the northwest coast of the Arabian (Persian) Gulf (Amato 1992).

In greenhouse studies seawater has an inhibitory effect on halophyte growth rates compared to brackish water (O'Leary & Glenn 1994) but under field conditions yields appear to be more dependant upon such factors as soil type, climate and cultivation methods than upon salinity per se. For example, *Atriplex* yields have been measured under both brackish (Watson et al. 1987; Watson 1990; Watson & O'Leary 1992) as well as seawater irrigation (Glenn & O'Leary 1985). *A. len-*

tiformis and *A. nummularia* yielded 14.7 and 12.3 10^3 kg ha^{-1} under brackish (6 ppt) geothermal artesian water in Safford, Arizona (Watson et al. 1987) and 17.9 and 8.0 10^3 kg ha^{-1} under seawater irrigation in Mexico (Glenn et al. 1984). Regrowth yields of *Atriplex* furrow-irrigated with saline drainage water (approximately 12 ppt salinity) in the San Joaquin Valley were 2.2–5.3 10^3 kg ha^{-1} yr^{-1} (Watson & O'Leary 1993). The lower yields were attributed in part to the use of mechanical harvesting equipment which tended to damage the plants.

There is a difference, however, in the irrigation requirements of halophytes on brackish water and seawater. High yields are dependent upon adequate irrigation amounts. As the salinity of the water source increases, so does the irrigation requirement since a greater leaching fraction is needed to maintain a favorable salinity in the root zone. Early experiments with seawater irrigation used daily or twice daily flooding of sandy soil, essentially reproducing the natural environment of daily tidal immersion in the intertidal zone (Glenn & O'Leary 1985). By contrast, irrigation volumes using brackish water are in the range of 45–85 cm per growing season (6–12 months) (Watson et al. 1987; Watson 1990; Watson & O'Leary 1993). Up to 215 cm can be used on an annual basis. cf. conventional irrigation volumes in the same irrigation districts.

A theoretical analysis of halophyte irrigation requirements (Miyamoto et al. 1994) based in part on lysimeter data collected at Puerto Penasco, Sonora, Mexico (Miyamoto et al. in prep.), shows that brackish irrigation can be accomplished using conventional irrigation criteria whereas seawater irrigation will require high-frequency irrigation via moving booms or center pivots to minimize salt concentration in the root zone between irrigations. Nevertheless, irrigation of halophytes appears to be technically feasible over a wide range of salinities. Commercial production units in Saudi Arabia are using center pivot irrigation of *S. bigelovii* with seawater at a rate of 1.25 times the pan evaporation rate over the growing season. Total water usage per crop is approximately 3 meters of water within the range used in desert irrigation districts for high-water use crops (for example, consumptive use by alfalfa is 1.9 m yr^{-1} in Tempe, Arizona and actual application rates can exceed 3 m) (Erie et al. 1981).

In vitro digestibility of halophytes

If halophytes are to be grown as harvested crops it must be demonstrated that they can replace conventional feed ingredients (see Swingle et al. 1994). Moore et al. (1982) conducted two experiments to determine the in vitro organic matter disappearance (IVOMD) of 45 different halophyte species incubated with steer rumen fluid and acid pepsin. The test species represented a broad range of plant types collected from around the world. Some were grown under field conditions whereas others were grown in a greenhouse at three salinities: 0, 10 and 20 ppt. The results were compared to an alfalfa hay standard.

The species differed considerably in IVOMD, which ranged from 50.1–87.2%, with a mean of 70.8% (S.D. = 8.5%). By comparison, the control alfalfa IVOMD was only 64.3%. Hence, halophytes as a group compared favorably to alfalfa, and some species appeared to be far superior to alfalfa in IVOMD. The ten *Atriplex* species tested all had higher IVOMD than alfalfa.

The ash contents of the halophytes varied from 7.7–33.4%, depending upon species and salinity of the irrigation water. Both ash content and IVOMD increased with salinity; the calculated digestible organic matter content (DOMC) therefore decreased less with salinity than was expected based on ash content alone. Species that were grown under both field and greenhouse conditions (*Allenrolfia occidentalis*, *Atriplex amnicola*, *A. isatidea*, *Galenia secunda* and *Heliotropium curassavicum*) all had higher IVOMD under field than greenhouse conditions.

It was concluded that saline irrigation did not have a detrimental effect on digestibility of halophyte organic matter and that digestibility of halophytes in general was high. Several species of North American *Atriplex* as well as seed-meal, oil and straw from *S. bigelovii* have been evaluated in animal feeding trials.

Nutritional value of *Atriplex*

Worldwide more attention has been paid to *Atriplex* than to any other halophyte genus as animal feed (e.g., Le Houérou 1992). Many *Atriplex* species are already important rangeland plants and some, such as *A. nummularia*, have been introduced around the world and are cultivated commercially. The nutritive value of *Atriplex* depends upon species, growth conditions, chemical composition of the irrigation water and season of the year (Watson et al. 1987; Watson 1990; Watson & O'Leary 1993).

Many analyses show crude protein levels to be in the range of 15–22% or greater. Therefore, it has appeared logical to compare *Atriplex* to other high-protein forages such as alfalfa (O'Leary 1986). A significant portion of *Atriplex* nitrogen, however, is not in protein but in glycine-betaine and other nitrogen compounds used in osmotic regulation. The extent to which *Atriplex* nitrogen can be used for growth is controversial; it is readily assimilated, but a significant percentage may be excreted in the urine (Gihad & El Shaer 1994). *Atriplex* also differs from alfalfa in mineral content (15–30%, mainly NaCl, compared to 9–11% for alfalfa) and energy content (2.9–3.3 Mcal kg^{-1} compared to 4.5–4.8 Mcal/kg^{-1} for alfalfa).

Watson (1990) and Watson & O'Leary (1993) conducted two experiments in the San Joaquin Valley of California that shed some light on the variability of *Atriplex* yields and forage values. In the first experiment (Watson 1990) plants were hand-harvested for yield determination and analyses, whereas in the second, larger-scale experiment (Watson & O'Leary 1993), plants were mechanically harvested and baled 4 times over 27 months. They demonstrated that conventional agronomic techniques and equipment can be modified to plant and harvest *Atriplex* and that some species at least are amenable to multiple clipping (Watson & O'Leary 1993).

The best-yielding hand-harvested species, *A. barclayana* ssp. *sonorae*, produced 775.5 g dry weight per plant in two clippings of the first year, equivalent to 11 10^3 kg ha^{-1} yr^{-1} at the plant spacing used in the experiments. This and some of the other species yields were within the range of conventional forage crop yields. However, yields of mechanically-harvested plots tended to be lower (2.5–5.3 10^3 ha^{-1} yr^{-1}) but were still considered to be acceptable forage yields (Watson & O'Leary 1993).

Forage quality of most of the species, determined by proximate analyses, also differed among the two experiments. In hand-harvested plots (Watson 1990), crude protein levels ranged from 14.4–19.0% among species (mean = 16.6%); but ash was also high, 16.6–36.3% (mean = 29.7%). Based on crude protein alone, the best species appeared to be *A. cinerea*, *A. nummularia*, *A. canescens* ssp. *macropoda* and *A. vestita*. Lignin contents of the *Atriplex* species were higher than alfalfa, indicating lower fiber quality. The mechanically harvested and baled plants had lower protein levels (6–13% over all species and harvests) than hand-clipped plants in the first study (Watson 1990). Watson & O'Leary (1993) concluded that when mechanically cut and baled, *Atriplex* will not necessarily have high crude protein levels and that levels in most species depend upon harvest date.

Atriplex in blends with other feed ingredients

Instead of considering *Atriplex* as a complete replacement for alfalfa it may be better utilized in a blend with other forages and energy sources (Swingle et al. 1994). Intakes of 4 of the *Atriplex* test species grown by Watson & O'Leary (1993) were compared to alfalfa intake at a commercial cattle feedlot (Jim Elam, Agricultural Technology Inc., Solvang, California, private communication to M. C. Watson 1991). Sixty-six head of crossbred heifers were fed test diets over a period of 48 days. During this period 4 different *Atriplex* spp. were presented for 6–14 days each. Animals were fed a ration containing 20% hay (alfalfa or *Atriplex*) with the remaining ingredients made up of corn, tapioca pellets, fat and other ingredients to represent a balanced growth diet. Daily intake of the alfalfa diet was 11.3kg d^{-1} whereas consumption of *Atriplex* diets were 10.2, 9.4, 8.6 and 8.5 kg d^{-1} for *A. undulata*, *A. polycarpa*, *A. canescens* and *A. deserticola*, respectively. Hence, there were differences among species but the best species was nearly as well accepted as alfalfa in a cattle finishing diet, even though the animals were not preconditioned to eating *Atriplex*.

Wiley tested the acceptability of *Atriplex lentiformis* and *Atriplex barclayana*, produced on seawater, in Nubian goat diets (Wiley 1982). *Atriplex* dressed with 10% molasses was refused at 100%, 75% or 50% inclusion levels but a diet containing 25% *Atriplex* and 75% alfalfa was accepted. The gross energy content of *A. lentiformis* and *A. barclayana* was 3.1 and 2.9 Mcal kg^{-1} respectively, compared to 4.5 Mcal kg^{-1} for alfalfa. The energy difference was partly due to the high ash content of the *Atriplex*.

Leaching the *Atriplex* material in fresh water reduced the mineral content by one-third and improved the energy content to 3.3–3.9 Mcal kg^{-1}. However, leaching did not improve acceptability or digestibility of the material. The leaching procedure apparently removed cell-soluble nutrients as well as salts, resulting in lower digestibility per unit of organic matter remaining. *Atriplex* had greater protein digestibility than alfalfa but animals also exhibited greater nitrogen loss in the urine. Fiber digestibility was lower for *Atriplex* than alfalfa.

Wiley concluded that *A. lentiformis* was superior to *A. barclayana* in acceptability, nutrient composition and digestibility. Water consumption was higher for animals fed *Atriplex* than animals fed alfalfa, and she concluded that high ash content may ultimately limit the usefulness of *Atriplex* in areas where drinking water is a limiting factor.

De la Llata (1991) re-examined the acceptability and digestibility of *Atriplex*, using lambs fed *A. deserticola* (50:50 *mix with alfalfa*) or *A. deserticola* mixed with *A. barclayana* in complete replacement of alfalfa (all diets were 70% forage, 25% sorghum grain and 5% molasses). Energy content of the *Atriplex* material was low (3.6–3.8 Mcal kg^{-1}) compared to alfalfa (4.8 Mcal kg^{-1}) and ash content was high (19–24%). *Atriplex* crude protein levels were relatively low, 8.2–8.9%. However, acceptability of *Atriplex* was much greater in this than in the previous experiment. In fact, the intake rate of a diet containing 70% *Atriplex* was equivalent to the intake rate of the alfalfa control diet.

Water consumption increased linearly with *Atriplex* consumption, as expected due to the high salt content of the *Atriplex*, but there were no signs of distress among the animals. The author concluded that '. . . diet acceptibility and the magnitude of the digestion coefficients were sufficiently high to encourage additional research with this and other halophyte species as feed sources for ruminants'.

The above studies suggest that some *Atriplex* species can substitute for part or, in some cases, all of the forage requirement of ruminants in terms of acceptability and digestibility. Based on these results, there is reason to believe that *Atriplex* diets can support adequate weight gain, but actual animal performance was not measured in any of the experiments. Grazing studies in other parts of the world have shown that *Atriplex* can support weight gain if an additional source of energy, such as wheat straw or stubble, is available (Le Houérou 1992). Experiments are needed to confirm that North American *Atriplex* spp. blended with other forages and concentrates can support adequate long-term weight gain of animals under feedlot conditions.

Salicornia nutritional value

Although members of the same plant family, *Salicornia* and *Atriplex* have very different agronomic characteristics. *S. bigelovii* is grown as an annual crop, requiring approximately 200 days from sowing to harvest under seawater irrigation in subtropical climates (Glenn et al. 1991). As a leafless succulent, it maintains an erect posture through high turgor pressure in the stems rather than through lignification. As a result, the fiber fraction is lower in lignin and may be more digestible than other forage fibers. Chemical analyses of straw and oil are in Table 1. It is noteworthy that *Salicornia* has a lower lignin content per unit of crude fiber compared to *Atriplex* (5.6% for *Salicornia* vs 7.0–12.2% for *Atriplex* in Watson & O'Leary 1993).

The seed, representing approximately 10–15% of the harvest weight, contains approximately 30% of the total protein. Seeds are threshed from the plants mechanically. The resulting straw contains 4–6% protein. Ash content is extremely high, 30–45% of dry matter, consisting mainly of NaCl.

Table 1. Nutritional anlayses of *Salicornia bigelovii* straw and seed

Analyses (in %)	Straw		Seed	
Crude protein		5.8		31.2
Crude fiber		17.8		5.3
NDF		34.1		nd
ADF		18.3		nd
Lignin		5.6		nd
Crude fat		1.0		28.2
Fatty acid profile (% of fat):				
Palmitic	nd		8.1	
Stearic	nd		2.2	
Oleic	nd		12.5	
Linoleic	nd		74.0	
Linolenic	nd		2.6	
Ash		35.7		5.5

Values are percentage of dry weight.
Sources: Glenn et al. (1991), Glenn et al. (1992a), Moore et al. (1989).

Salicornia straw and meal in ruminant diets

Despite a higher ash content and lower crude protein level than *Atriplex*, *Salicornia* straw has proven to be a valuable forage material in formulated ruminant diets. In a 100-day weight-gain study in which Damascus kid goats were raised to slaughter weight in the United Arab Emirates, Glenn et al. (1992a) showed that untreated *Salicornia* straw could replace 50% of the Rhodes grass (*Chloris gayana*) forage in a concentrate-supplemented diet and that leached *Salicornia* straw could replace 100% of the Rhodes grass (ca. 50% of the total diet) (see also Riley et al. 1994). Leaching was accomplished by soaking straw in seawater then pressing to remove excess moisture and salt; the resultant, redried straw had a NaCl content of 11%, compared to 30% for unwashed straw.

S. bigelovii straw was substituted for alfalfa and wheat straw in another long term sheep trial conducted at a commercial feedlot in Sonora, Mexico (Dr. Arturo Madrid, Mesquital del Oro Co., Hermosillo, Mexico; unpublished results). Three sheep each were placed on 3 diets for 92 days. The control diet contained rolled wheat, molasses, cottonseed meal, mineral and vitamin premix, wheat straw and alfalfa; the two treatment diets substituted unleached *Salicornia* straw for alfalfa at 15% inclusion and for alflafa and wheat straw at 30% inclusion. *Salicornia* diets contained 10% extra cottonseed meal and 10% lower rolled wheat to balance protein levels (ca. 10%) across treatments.

Animals on *Salicornia* diets outperformed control animals, with growth rates of 95, 105 and 157 g day^{-1} for animals on control, 15% *Salicornia* and 30% *Salicornia*, respectively. Growth rates were typical of those achieved in feedlots in Sonora. Comparing treatment diets with controls, it appeared that *Salicornia* plus cottonseed meal substituted favorably for alfalfa; comparing the two *Salicornia* treatments, it appeared that *Salicornia* substitutes favorably for wheat straw as well.

De la Llata (1991) measured acceptability and digestibility by lambs of *Salicornia* forage harvested at three different cutting dates during the growth cycle. Ash content was high at all cutting dates (35–46%) but protein levels were higher than in straw (7.5–8.5%). *Salicornia* was incorporated at 30% of the diet; the control diet contained 30% wheat straw, and all diets contained 35% alfalfa hay, 33% sorghum grain and 2% molasses. All three cuttings of *Salicornia* compared favorably to the control diet in acceptability and digestibility. It was concluded that unleached *Salicornia* hay (immature plants) could replace wheat straw in lamb diets.

De la Madrid (Mesquital del Oro, Hermosillo, Mexico; unpublished results) also conducted two long-term feeding studies with *Salicornia* straw fed to Holstein male calves and steers. The first was a 96-day pilot study in which 2 calves per treatment were fed forage consisting of alfalfa hay (45%), *Salicornia* straw (40%), or *Salicornia* straw treated with 'celefos' (a commercial product to soften fibers) (40%); other feed ingredients included ground wheat, vitamin and mineral premix, molasses and cottonseed meal to balance protein levels across diets. The *Salicornia* was leached to reduce salts so ash content was relatively low (reduced from 27.3–8.9%) and salt intake was not a limiting factor for consumption. Animals had the highest weight gain and most favorable ratio of weight gain to feed intake on treated *Salicornia*; untreated straw produced equal weight gain as controls but with a higher level of feed intake. The second study was for 61 days and utilized 6 steers per treatment. Leached

Salicornia straw (ash content reduced from 27.3 to 6.7%), either celefos-treated or untreated, replaced alfalfa hay at 20% initial inclusion (reduced to 15% inclusion during the second month due to lack of material) in the two *Salicornia* test diets. Other diet components were rolled wheat (40%); molasses (10% in control, 5% in *Salicornia* diets); vitamin and mineral premix (10%); NaOH-treated wheat straw (15%); and cottonseed meal (5% in control, 10% in *Salicornia* diets to balance protein levels). Weight gain and feed consumption were equal across treatments. While preliminary, these studies reinforce the published data showing that *Salicornia* straw performs as well or better than conventional forages in diets balanced for protein. Washing the straw to remove salts increases the amount that can be added but unwashed material can be incorporated at up to 30%, a typical forage inclusion level in feedlot ruminant diets.

Protein content can be enhanced by adding *Salicornia* seeds or seed meal to the ration. In preliminary experiments, Mota (unpublished results) demonstrated that, for local Awasi-type sheep in the United Arab Emirates, protein can be supplemented by adding milled *Salicornia* seeds into the feed. He compared three test diets containing three levels of *Salicornia* straw (50, 60 and 70%) and seed (25, 20 and 15%), with a control diet containing 25% wheat bran and 50% rhodesgrass as the forage component plus corn and dates. The control diet contained 9.3% crude protein and was a typical diet for sheep in this area. The test diet containing 50% untreated *Salicornia* straw and 20% seeds had the highest protein level (12.6%) and supported weight gain of the sheep at 80 g/day^{-1} over 97 days. The other test diets, containing 10.55 and 9.94% protein, supported growth at 60 and 65 g/day^{-1}, compared to 74 g/day^{-1} for the control. The use of *Salicornia* meal as a protein source is currently being tested as well in controlled feeding trials in Tucson, Arizona (Swingle et al. 1994).

Salicornia seed meal in poultry diets

At 42% protein, *Salicornia* seed meal represents a potential primary protein source in formulated diets. Experiments with sheep, cited above, showed that the milled whole seeds were readily accepted by ruminants. However, the seeds have been found to contain 0.05% of oleanolic acid, a saponin which acts as an antifeedant compound in poultry. Glenn et al. (1991) found that NaOH treatment of the meal reduced saponin to acceptable levels for chick starter diets. A more extensive series of poultry trials was conducted in which *Salicornia* seed meal was added to diets at 0, 4, 7 and 14% in partial replacement of soybean meal (Peng, unpublished). There are several methods of removing or inactivating the saponin compound. In one treatment the meal was washed with 1 N NaOH for 30 minutes. In other treatments, cholesterol and phytosterol were incorporated into the feed to counteract the effects of the saponin on feed uptake. Untreated meal reduced chick weight gain even at the lowest level of incorporation whereas addition of cholesterol or a lower-cost phytosterol at 0.5% or 1% allowed incorporation of *Salicornia* meal at up to 14%. Washing with NaOH was only partially effective similar to previous results (Glenn et al. 1991). The most economical treatment was the addition of 0.25% phytosterol with 7% *Salicornia* meal. With phytosterol costing \$9 kg^{-1} (1986 prices), the cost of the finished diet increases by only \$20 t^{-1} at that inclusion level. Hence, *Salicornia* seed meal can be used in poultry diets in up to 30% blends with soymeal.

Conclusion

The experiments with *Atriplex* forage are not yet definitive, since long-term weight gain studies have not been conducted. Rates of intake of *Atriplex* by ruminants can approach those of conventional forage crops, but most of the experiments did show at least some reduction in intake. Energy levels are lower in *Atriplex* than conventional forages but overall *Atriplex* digestibility equals or exceeds conventional forages. As perennial shrubs, the nutritional quality of *Atriplex* is highly dependant upon the age of the plant (ratio of woody tissue to leaves) and also upon the season of the year. The ash content of *Atriplex* and other halophytes depends upon the salinity of the soil in which they are grown. Some species, such as *Atriplex canescens* ssp. *canescens*, have the ability to largely exclude sodium especially at low external salinities (Watson & O'Leary 1993; Glenn et al. 1992b).

Salicornia forage was not only well accepted by animals but it produced equal or better weight gain than conventional forages in long-term experiments conducted with goats, sheep and cattle. Although it had higher ash content than *Atriplex*, the mineral content could be reduced by leaching the material with seawater. Overall, it appeared to be a more acceptable forage than *Atriplex*, although side-by-side comparative experiments are not yet completed. *Salicornia* seed meal was an effective protein supplement for sheep but was less effective for chickens due to saponin content.

Worldwide attention has focused on *Atriplex* chiefly because the plants are grazed naturally and are important rangeland species. By contrast, succulent halophytes such as *Salicornia*, which dominate the most saline areas of the landscape, are not often well utilized by livestock. Natural stands may be avoided by free-ranging ruminants due to their high water and salt contents. Judging by *Salicornia*, however, the succulent halophytes are potential high-quality forage sources when dried to reduce water content and blended or leached to reduce salt content. Agronomically, the succulent species have the advantage that they can be germinated directly on highly saline soil and irrigated with up to full strength seawater.

The experience to date with North American halophytes is that they can be part of a complete animal nutrition program if used in blends with other feed ingredients (Swingle et al. 1994). Since livestock production in North America depends heavily upon using formulated finishing diets fed to animals in confinement, halophyte forage is consistent with normal practices of animal husbandry. Growing and harvesting halophyte biomass for animal feed is a logical extension of the increasingly intensive use of halophytes in animal husbandry.

Acknowledgements

Thanks are due to the United Nations Environment Program for a travel grant which allowed Edward Glenn and Victor Squires to attend the workshop in Nairobi.

References

Albert, R. & Popp, M. (1977). Chemical composition of halophytes from the Neusiedler Lake region in Austria. Oecologia 27: 157–170.

Amato, I. (1992). How to grow oil in the Persian Gulf. Science 258: 1574.

Brizuela, M.A., Cid, M.S., Minon, D.P. & Fernandez Grecco, R. (1990). Seasonal utilization of salt grass (*Distichlis* spp.) by cattle. Animal Feed Science Technology 30: 321–325.

Chapman, V.J. (1974). Salt marshes and salt deserts of the world. In: R.J. Reimold & W.H. Queen (eds.), Ecology of halophytes. Academic Press, New York, pp. 3–19.

Chatterton, N.J., Goodin, J.R., McKell, C.M., Parker, R.V. & Rible, J.M. (1971). Monthly variation in the chemical composition of desert saltbush. J. Range Manage. 24: 37–40.

Cook, C.W., Stoddart, L.A. & Harris, L.E. (1954). The nutritive value of winter range plants in the Great Basin. Bulletin 372. Utah State Agric. College, Logan.

De la Llata, M.M. (1991). Nutritive value of *Atriplex deserticola* and *Salicornia* forage for ruminants. PhD Dissertation, University of Arizona, Tucson, AZ.

Duncan, W.H. (1974). Vascular halophytes of the Atlantic and Gulf Coasts of North America north of Mexico. In: R.J. Reimold & W.H. Queen (eds.), Ecology of halophytes. Academic Press, New York, pp. 23–50.

Gihad, E.A. & El Shaer, H.M. (1994). Utilization of halophytes by livestock on rangelands: Problems and prospects (this volume, Ch. 6).

Glenn, E.P. (1984). Relationship between cation accumulation and water content of salt tolerant grasses and a sedge. Plant, Cell & Environment 10: 205–212.

Glenn, E.P. & O'Leary, J.W. (1984). Relationship between salt accumulation and water content of dicotyledonous halophytes. Plant, Cell and Environment 7: 253–261.

Glenn, E.P. & O'Leary, J.W. (1985). Productivity and irrigation requirements of halophytes grown with seawater in the Sonoran Desert. J. Arid Environments 9: 81–91.

Glenn, E.P., O'Leary, J.W., Watson, M.C., Thompson, T.L. & Kuehl, R.O. (1991). *Salicornia bigelovii* Torr.: An oilseed halophyte for seawater irrigation. Science 251: 1065–1067.

Glenn, E.P., Coates, W., Riley, J.J., Kuehl, R. & Swingle, R.S. (1992a). *Salicornia bigelovii* Torr.: A seawater-irrigated forage for goats. Animal Feed Science and Technology 40: 21–30.

Glenn, E.P., Watson, M.C., O'Leary, J.W. & Axel, R. (1992b). Salt tolerance and osmotic adjustment of low-sodium and high-sodium subspecies of the C4 halophyte *Atriplex canescens*. Plant, Cell & Environment 15: 711–718.

Goodin, J.R. (1979). *Atriplex* as a forage crop for arid lands. In: G.A. Ritchie (ed.), New agricultural crops. Westview Press, Boulder, CO, pp. 138–148.

Hitchcock. S.W. (1972). Fragile nurseries of the sea. National Geographic 141(6): 729–765.

Le Houérou, H.N. (1992). The role of saltbushes (*Atriplex*

spp.) in arid land rehabilitation in the Mediterranean Basin: A review. Agroforestry Systems 18: 107–148.

Leigh, J.H. & Mulham, A.D. (1964). Dietary preferences of sheep in two semi-arid pastoral eco-systems. Proc. Aust. Soc. Animal Prod. 5: 251–257.

Llerena, F.A. (1994). Massive propagation of halophytes (*Distichlis spicata* and *Tamarix* spp.) on the highly saline-alkaline soils in the ex-Lake Texcoco, Mexico (this volume, Ch. 30).

Llerena, F.V. & Tarin, M.V. (1978). Metodos de pastizacion en los suelos extremadament salino-sodico del ex-lago de Texcoco. 10 Congreso Nacional de la Ciencia del Suelo. Villahermosa, Mexico.

Macdonald, K.B. 1977. Plant and animal communities of Pacific North American salt marshes. In: V.J. Chapman (ed.), Wet coastal ecosystems. Elsevier, Amsterdam, pp. 167–191.

Macdonald, K.B. & Barbour, M.G. (1974). Beach and salt marsh vegetation of the North American Pacific Coast. In: R.J. Reimold & W.H. Queen (eds.), Ecology of halophytes. Academic Press, New York, pp. 175–233.

McGinnies, W. G. (1972). Continental aspects of shrub distribution, utilization and potentials in North America. In: C.M. McKell, J.P. Blaisdell & J.R. Goodin (eds.), Wildland shrubs – their biology and utilization. US Forest Service, Ogden, Utah, pp. 55–56.

McKell, C.M. & Garcia-Moya, E. (1989). North American shrublands. In: C.M. McKell (ed.), The biology and utilization of shrubs. Academic Press, San Diego, CA, pp. 3–23.

Miyamoto, S., Glenn, E.P. & Singh, N.T. (1994). Utilization of halophytic plants for fodder production with brackish water in subtropic deserts (this volume, Ch. 5).

Moore, J.A., Swingle, R.S., O'Leary, J.W., Glenn, E.P. & Colvin, L. B. (1982). In vitro organic matter digestibility of salt tolerant plants. Proc. Western Section, American Society of Animal Sciences 33: 319–322.

Moore, J.A., Stull, E.D. & Swingle, R.S. (1989). Evaluation of *Salicornia* forage and seed meal as potential feedstuffs for ruminants. Unpublished Report No. 89-2, Environmental Research Laboratory, Tucson, AZ.

Mota, C.U., Glenn, E.P. & Fontes, M.R. (1980). Halophytes: Potential agricultural crops for hypersaline soils and waters. In: Proceedings of the Second Inter-American Conference on Salinity and Water Management Technology, Juarez, Mexico.

National Research Council (1981). Nutrient Requirements of Goats. National Academy Press, Washington, DC.

O'Leary, J.W. (1986). A critical analysis of the use of *Atriplex* species as crop plants for irrigation with highly saline water. In: R. Ahmad & A. San Pietro (eds.), Prospects for biosaline research, Karachi Univ., Pakistan, pp. 415–432.

O'Leary, J.W. & Glenn, E.P. (1994). Global distribution and potential for halophytes (this volume, Ch. 2).

Reimold, R. J. (1977). Mangals and salt marshes of Eastern United States. In: V.J. Chapman (ed.), Wet coastal ecosystems. Elsevier, Amsterdam, pp. 157–166.

Reimold, R. J. & Queen, W. H., eds. (1974). Ecology of halophytes, Academic Press, New York.

Riley, J.J., Glenn, E.P. & Mota, C.U. (1994). Small ruminant feeding trials on the Arabian peninsula with *Salicornia bigelovii* Torr. (this volume, Ch. 26).

Shanholtzer, G.F. (1974). Relationship of vertebrates to salt marsh plants. In: R.J. Reimold & W.H. Queen (eds.), Ecology of halophytes. Academic Press, New York, pp. 463–474.

Swingle, R.S., Glenn, E.P. & Riley, J.J. (1994). Halophytes in mixed feeds for livestock (this volume, Ch. 7).

Ungar, I.A. (1974). Inland halophytes of the United States. In: R.J. Reimold & W.H. Queen (eds.), Ecology of halophytes. Academic Press, New York, pp. 235–305.

Ungar, I.A. (1992). Ecophysiology of halophytes. CRC Press, Boca Ratcon.

Watson, C., O'Leary, J. & Glenn, E.P. (1987). Evaluation of *Atriplex lentiformis* (Torr.) S. Wats. and *Atriplex nummularia* Lindl. as irrigated forage crops. J. Arid Environments 13: 292–303.

Watson, M.C. (1990). *Atriplex* species as irrigated forage crops. Agriculture Ecosystems and Environment 32: 107–118.

Watson, M.C. & Ferren, W.R., Jr. (1991). A new species of *Suaeda* (Chenopodiaceae) from coastal northwestern Sonora, Mexico. Madrono 38(1): 30–36.

Watson, M.C. & O'Leary, J.W. (1993). Performance of *Atriplex* species in the San Joaquin Valley, California, under irrigation and with mechanical harvests. Agriculture, Ecosystems and Environment 43: 255–266.

West, R.C. (1977). Tidal salt-marsh and mangal formations of Middle and South America. In: V.J. Chapman (ed.), Wet coastal ecosystems. Elsevier, Amsterdam, pp. 193–213.

Wiley, S.T. (1982). Nutritive value for goats of *Atriplex* species grown with hypersaline water. MS Thesis, University of Arizona, Tucson, AZ.

Yensen, N.P., Glenn, E.P. & Fontes, M.R. (1983). Biogeographical distribution of salt marsh halophytes on the coasts of the Sonoran Desert. Desert Plants 5(2): 76–81.

13. Halophytes as a resource for livestock husbandry in South America

R.E. BREVEDAN, O.A. FERNÁNDEZ & C.B. VILLAMIL
Centro de Recursos Naturales Renovables de la Zona Semiárida & Departments of Agronomy and Biology, Universidad Nacional del Sur, Bahía Blanca, Argentina

Abstract. Certain areas of South America, with an area of 18 million square kilometers, present suitable conditions for the growth of halophytic vegetation adapted to saline environments. This work provides a comprehensive assessment showing the distribution by country of more than 150 plant species capable of growing in salty soils and used as a forage resource for livestock. The principal type of environment where halophytes have a wide geographical distribution is represented by the semi-arid steppes which cover extensive areas of the continent. Those halophytic species considered to be a valuable resource in animal husbandry are commonly known representatives of the native flora throughout these large phytogeographical zones. Such rangelands are characterized by the utilization of natural vegetation as the sole diet of domestic animals. Many of the halophytes of these areas, such as *Atriplex* spp., *Suaeda foliosa*, *Prosopis* spp., etc have acquired a reputation as forage plants since they appear to be constantly browsed by livestock. There is a collection of salt-tolerant grasses and true halophytes which appears to have significant economic potential in animal husbandry. Of all the grasses listed, *Distichlis* spp. is the most important in terms of distribution. The aptitude of this genera as a resource of animal feed varies from one region to another and according to the type of animal using it. Natural prairies of *D. humilis* are intensively grazed by livestock in the Bolivian altiplan. Some of the species of halophytes cited in this study have been subjected to domestication programs or nutritional studies. However, in most cases there is no experimental data on their productivity, nutritive value, or management.

Key words: Biogeography, Forage yield, Life form, Salt, Utilization

Introduction

South America, with a surface of 18 million square kilometers, extends from latitude 12° N to 56° S and is 7500 km in length and 5000 km at its widest point. The altitude ranges from sea level to 7000 m in the Andes. Over such a vast territory the diversity in vertical and horizontal bio-geographical regions is of extraordinary magnitude, ranging from tropical rain forests to deserts such as that of the Puna de Atacama, temperate zones like the Pampas and the cold subantartic zones of Argentina and Chile.

In certain areas of this continent conditions allow for the presence of halophytic vegetation growing in saline environments, which according to their distinctive nature and location can be inland, coastal or marine. Estimates based on the Soil Map of the World (FAO-UNESCO 1971; Dudal & Purnell 1986) indicate that around 7.6% of the soils of South America are salt affected. Accurate data on the inland surface area covered by halophytic vegetation and its distribution at the continental level is lacking. An attempt to reveal the distribution and characterization of the saline soils of Argentina was published by Scoppa and Di Giacomo (1985). The distribution and characteristics of saline/alkaline land in Chile and its use for the production of forage and fuel has been reported by Squella (1986). The scope and

V.R. Squires & A.T. Ayoub (eds.), *Halophytes as a resource for livestock and for rehabilitation of degraded lands*, 175–199.

objective of the present chapter preclude the discussion of the vegetation of ocean salt marshes or the marine halophytic plant associations living in the intertidal zones at the interface between land and sea, such as the mangroves of Colombia, considered among the largest in the world (Cuatrecasas 1958).

In South America, as in many other parts of the world, each type of inland saline habitat contains an ample spectrum of plant forms in response to local environmental factors, and to the effect of the soil salt concentration. The present work considers two principal inland categories for halophytes: the vegetation associated with semiarid shrubby steppes, and that related to salt flats, rivers, and irrigated lands.

Semi-arid shrub steppe

The most significant type is represented by the vast geographical distribution of halophytes in association with the semiarid shrubby steppes which cover extensive territories throughout the continent. The major phytogeographical areas in South America where shrubs are present as the dominant plant form have been mapped and reviewed by Fernández et al. (1989). In these large phytogeographical areas it is known that the species of halophytes which can be considered a valuable resource for animal husbandry are common representatives of the native flora of Argentina, Chile, Bolivia and Peru. They correspond to semi-arid and arid ecosystems elsewhere in terms of their structure and function. In these areas the vegetation is commonly a combination of herbaceous and woody species in which shrubs frequently constitute the dominant plant strata. In some cases the distribution of a single species can be very wide, as is the case with *Atriplex lampa*, which extends from Bolivia in the north to southern Argentina in Patagonia (Passera & Borsetto 1989), sometimes forming almost monospecific large stands known locally as 'zampales' (Ruiz Leal 1972). One of the features characterizing these rangeland regions is based on utilization of the natural vegetation as the only source of food intake for domestic animals. Some halophytes of these areas, such as *Atriplex* spp., *Suaeda foliosa*, *Distichlis* spp., etc. have acquired a reputation as fodder plants as they appear to be constantly browsed by livestock. Abusive exploitation of the land through overgrazing has frequently led to such a degree of range degradation that over large areas the most desired species including some halophytes have become almost extinct or are found only in secluded areas, as is the case with *Atriplex repanda* in central Chile (Badilla 1975).

Halophytes of inland salines

Another category of inland environment for halophytes is represented by the vegetation growing in association with salt flats, along the courses of rivers and streams, as well as in those areas more directly disturbed by anthropogenic action such as irrigated land. Contrary to the case of the shrubby rangelands mentioned above, which extend over large territories and are rarely found in humid regions, these areas can be located in dry as well as in humid regions. They tend to be confined to limited extensions within larger phytogeographical biomes. They depict an heterogeneous group of habitats having in common the presence of salts which affect plant growth. Their size and uneven geographical distribution make it very difficult to represent them on maps on a continental scale. They are briefly discussed below.

The relatively isolated natural inland salt flats have evolved in association with the seepage of the basin-shaped topography forms of arid mediterranean zones. Originally they appeared to be identified with the drying up of salty lakes or the inflow of saline water from surrounding areas (Frenguelli 1928), the soil is frequently of alluvial origin. Some of this saline depressions land can cover thousands of hectares (Ragonese 1951). A detailed study of the saline soil basins in the Bolivian Central Altiplano and their influence on vegetation cover has been published by Salm and Gehler (1987). In humid areas they are located mostly around salty springs or on fossil salt deposits.

Halophytes in riparian habitats

Halophytic communities are also found in association with floodplain areas whenever the water of rivers and streams flows over or through the mineral substrate; the water then evaporates and the salt is left behind in the soil. Thus, isolated patches of halophytes can be found adjacent to streams and lakes in arid areas as well as in regions with humid climates and good agricultural land (Ragonese & Covas 1947). This process can be intensified due to the lack of downward percolation and when the water table is close enough to the soil surface. The depression of the Río Salado (Salty River) in the Province of Buenos Aires, Argentina, with a surface area around 4 million hectares, is a typical example of this occurrence. A detailed study of the natural vegetation of this area has been provided by Vervoorst (1967) and Leon (1975).

In many parts of South America halophytic flora is locally associated with irrigation complexes. Salinity is unquestionably a significant threat to man-made irrigation agriculture in arid and semiarid zones throughout the continent, even in areas with improved drainage procedures. Sometimes irrigated farmland has been rendered useless by salinity and waterlogging. The problem is hardly new: there are many examples of soil becoming salt-affected by anthropogenic activity, such as the farmlands of the early European settlers in the patagonian river valleys of Argentina, or of the ancient civilizations in the coastal regions of Peru. The vegetation which has colonized these areas is commonly considered to be a nuisance, interfering with the agricultural practices and irrigation systems. In general it is only of minor significance in animal husbandry. The use of selected varieties of conventional crops for greater tolerance to salinity is also becoming increasingly common in irrigated areas for the purpose of land reclamation.

Compared with other areas of the world, published information on halophytes in South America, and even more so on their utilization, is quite scarce. Much of the existing information is primarily descriptive, sometimes appearing in unfamiliar journals or in 'internal reports' of limited circulation. Relevant information can be gleaned from a number of studies on the geographical distribution and taxonomy of plants growing in saline surroundings. Several of these works are quoted in this chapter. In order to compile specific data on halophytes, a questionnaire survey was undertaken in 1991 aimed at those institutions throughout South America connected with vegetation studies and animal production (experimental stations, universities, etc.), and requesting information on halophytic vegetation types and their habitat and utilization. Of the 250 questionnaires sent out, more than 50 replies dealt specifically with halophytes. Data from the survey have been utilized extensively throughout this chapter.

Table 1 provides a comprehensive overview of the most frequently mentioned taxa, showing their geographical distribution throughout South America. Only those plants cited either in the present survey or in the bibliography as having been utilized at some stage as fodder for domesticated animals and as being adapted to a salty environment are included in the table. The rating in column 3 is to be interpreted only as a guideline to relative salt tolerance. Absolute tolerance data based on experimental work are lacking in most cases, and of course may vary for a given species in accordance with climate, soil conditions, and genetic diversity. In saline areas the growing sites of some of these species were found to overlap with those of non-halophytes, making it difficult sometimes to draw a clear-cut distinction between halophytes and non-halophytes. The ability of a plant to live in a salty environment was often cross-checked with information obtained from taxonomic reviews or phytogeographic studies. The work of Aronson (1989) was used as an additional source of reference.

Our present state of knowledge is for the most part inadequate for an exhaustive assessment of the current and potential importance of the species listed in Table 1. Their value and palatability are often a matter of conjecture based on the practical observation of browsing habits and animal preference. Except for the case of a few species, not much is known about their full nutritional value, productivity, and browsing tolerance. Further research in the field is expected to fill these gaps in our knowledge.

Table 1. Species name, life form (LF), salt tolerance (ST), geographic distribution (D) and bibliographic references of South American halophytes cited as valuable for grazing or browsing

Species name	LF[a]	ST[b]	D[c]	References
AIZOACEAE				
Sesuvium postulacastrum (L.) Linné	PH	H	ASAm	Parodi (1930)
				Barbosa*
AMARANTHACEAE				
Alternanthera ficoidea (L) Beauv.	PH	ST	VE	Pedersen (1987)
				Medina*
Amaranthus muricatus (Moq). Hieron.	PH	ST	Ar	Ragonese (1951)
				Pedersen (1987)
A. quitensis H.B.K.	AH	ST	TAm	Ragonese (1951)
				Pedersen (1987)
A. standleyanus Parodi ex Covas	AH	ST	Ar	Ragonese (1951)
				Pedersen (1987)
ANACARDIACEAE				
Schinus polygamus (Cav.) Cabrera	SH	ST	Ar	Molina Sánchez (1977)
BORAGINACEAE				
Cortesia cuneifolia Cav.	SSH	H	Ar	Boelcke (1981)
Heliotropium curassavicum L.	PH	H	Co-Ve	Ruiz Leal (1972)
			Ar	Vervoorst (1967)
				Gómez*
CELASTRACEAE				
Maytenus vitis-idaea Griseb.	T/SH	ST	Br-Pa	Boelcke (1981)
			Bo-Ar	Cantero*
				Neumann*
				Tinto*
CHENOPODIACEAE				
Allenrolfea patagonica (Moq.) O. Kuntze	SH	H	Ar	Cano (1988)
				Giusti (1967)
				Neumann*
A. vaginata (Gris.) O. Kuntze	SH	H	Ar	Cano (1988)
				Giusti (1984)
				Neumann*
Atriplex argentina Speg.	SH	H	Ar	Múlgura (1981)
				Steibel (1986)
				Neumann *
A. axillaris Phil.	SH	H	Ar-Bo	Múlgura (1981)
			Ch	Neumann*
A. chilensis Colla	AH	H	Ch	Rosas (1989)
A. chizae Rosas	SH	H	Ch	Rosas (1989)
A. coquimbana Phil.	SSH	H	Ch	Múlgura (1981)
				Rosas (1989)
A. cordobensis Gand. et Stuckert	SH	H	Ar	Sayago (1969)
				Neumann*

[a]T: tree; SH: shrub; SSH: dwarf shrub; PH: perennial herb; AH: annual herb; PG: perennial grass or grass-like; AG: annual grass or grasslike.
[b]H: halophyte; FH: facultative halophyte; ST: salt tolerant.
[c]Am: America; SAm: South America; ASAm: Andean South America; TAm: Tropical America; Ar: Argentina; Bo: Bolivia; Br: Brazil; Ch: Chile; Co: Colombia; Ec: Ecuador; Pa: Paraguay; Pe: Peru; Ur: Uruguay; Ve: Venezuela; INTR: introduced in the region.
*Data from the questionnaire survey.

Table 1. Continued

Species name	LF[a]	ST[b]	D[c]	References
A. hastata L.	AH	H	INTR	Giusti (1984) Steibel (1986)
A. lampa (Moq.) Dietr.	SH	H	Ar	Steibel (1986) Cano (1988) Neumann*
A. madariagae Phil.	SH	H	Ar-Ch	Rosas (1989) Neumann*
A. microphylla Phil.	SSH	H	Ar-Bo	Ragonese (1967) Neumann*
A. montevidensis Spreng.	PH	H	SAm	Giusti (1984) Steibel (1986) Neumann*
A. nummularia Lindl.	SH	H	INTR	Silva & Pereira (1976) Rosas (1989)
A. patagonica (Moq.) Dietr.	SH	H	Ar	Giusti (1984) Steibel (1986)
A. patula L.	AH	H	INTR	Múlgura (1982) Rosas (1989)
A. pentandra (Jacq.) Standley	SH	H	Ve	Múlgura (1984) Medina*
A. repanda Phil.	SH	H	Ch	Silva & Pereira (1976) Vera (1977)
A. rosea L.	AH	H	INTR	Giusti (1984) Steibel (1986)
A. sagittifolia Speg.	SH	H	Ar	Ruiz Leal (1972) Molina Sánchez (1977)
A. semibaccata R. Br.	PH	H	INTR	Giusti (1984) Steibel (1986) Neumann*
A. suberecta Verd.	AH	H	INTR	Rosas (1989) Steibel (1986)
A. taltalensis Johnst.	SSH	H	Ch	Rosas (1989)
A. undulata (Moq.) Dietr.	SH	H	Ar	Cano (1988)
A. valleranensis Rosas	SH	H	Ch	Rosas (1989)
Bassia hyssopifolia (Pall.) O. Kuntze	AH	H	INTR	Steibel (1986)
Beta vulgaris L.	PH	H	INTR	Giusti (1967)
Chenopodium hircinum Schrader	AH	H	SAm	Giusti (1984) Steibel (1986)
Ch. macrospermum Hooker	AH	H	SAm	Giusti (1984) Steibel (1986)
Heterostachys olivascens Speg.	SH	H	SAm	Giusti (1967)
H. ritteriana (Moq.) Ung. Stern	SH	H	Ar	Cano (1988) Giusti (1984) Neumann*
Holmbergia tweediei (Moq.) Speg.	SH	H	Ar	Giusti (1967) Neumann*
Kochia scoparia (L.) Schrader	AH	FH	INTR	Steibel (1986)
Salicornia ambigua Mich.	SSH	H	SAm	Cano (1988) Steibel (1986) Neumann*
S. pulvinata R.E. Fries	PH	H	Ar-Bo Ch	Morello (1958) Neumann*
Salsola kali L.	AH	FH	INTR	Steibel (1986)
Suaeda argentinensis Soriano	SH	H	Ar	Steibel (1986)

Table 1. Continued

Species name	LF[a]	ST[b]	D[c]	References
S. divaricata Moquin	SH	H	Ar	Giusti (1984)
				Steibel (1986)
S. patagonica Speg.	AH	H	Ar	Giusti (1984)
				Steibel (1986)
COMPOSITAE				
Aster squamatus (Spreng.) Hieron.	AH	ST	SAm	Cabrera (1963)
Baccharis acaulis (Wedd.) Cabrera	PH	H	Ar-Bo	Cabrera (1978)
B. spartioides (H. et A.) Remy	SH	H	Ar	Cano (1988)
Cyclolepis genistoides Don	SH	H	Ar	Cano (1988)
				Neumann*
Hymenoxys anthemoides (Juss.) Cass	AH	H	Ar-Br	Cabrera (1963)
			Ur	Ragonese (1951)
H. cabrerae Parker	AH	H	Ar	Cabrera (1963)
Nardophyllum obtusifolium H. et A.	SH	H	Ar-Ch	Molina Sánchez (1977)
Senecio pampae Lingelsheim	SH	–	Ar-Bo	Cabrera (1978)
S. subulatus Don ex H. et A.	SH	–	Ar	Cabrera (1971)
				Ruiz Leal (1972)
Tessaria absinthioides (H. et A.) D. C.	PH	FH	Ar-Bo	Cabrera (1963)
			Ch-Ur	Hernández (1991)
				Hunziker (1952)
T. dodonaeifolia (H. et A.) Cabrera	SH	H	Ar-Pa	Cabrera (1978)
				Ruiz Leal (1972)
				Neumann*
CYPERACEAE				
Carex phalaroides Kunth	PG	ST	Br-Ar	Barros (1960)
			Ur	
Cyperus corymbosus Rottb.	PG	ST	Ar-Br	Barros (1960)
			Pe	
Scirpus americanus Persoon	PG	ST	Am	Ungar (1974)
Sc. californicus (C.A. Mey.) Steud.	PG	ST	Am	Ruiz Leal (1972)
EPHEDRACEAE				
Ephedra ochreata Miers	SH	H	Ar	Abiusso (1962)
				Hunziker (1952)
EUPHORBIACEAE				
Colliguaja integerrima Hooker	SH	H	Ar	Molina Sánchez (1977)
FRANKENIACEAE				
Anthobryum triandrum (Remy) Surgis.	SSH	H	Ar-Bo	Ruthsatz (1978)
				Ancibor*
				Neumann*
Frankenia juniperiodes (Hieron.) Correa	SSH	H	Ar	Correa (1988)
F. microphylla Cav.	SH	H	Ar	Correa (1988)
F. patagonica Speg.	SH	H	Ar	Correa (1988)
F. pulverulenta L.	AH	H	INTR	Correa (1988)
				Gunckel (1970)
GRAMINEAE				
Agropyron elongatum (Host.) Beauv.	PG	FH	INTR	Nicora & Rúgolo (1987)
A. scabrifolium (Doell.) Parodi	PG	H	Ar-Ur	Nicora & Rúgolo (1987)
Aristida mendocina Speg.	PG	ST	Ar	Nicora & Rúgolo (1987)
Chaetotropis imberbis (Phil.) Bjorkman	PG	H	Ar-Br	Nicora & Rúgolo (1987)
Chloris berroi Arech.	PG	ST	Ar-Ur	Nicora & Rúgolo (1987)

Table 1. Continued

Species name	LF[a]	ST[b]	D[c]	References
Ch. halophila Parodi	PG	H	Ar-Bo	Nicora & Rúgolo (1987)
			Pe	
Ch. sesquiflora Burkart	PG	H	Ar	Nicora & Rúgolo (1987)
Cottea pappophoroides	PG	ST	Ar	Nicora & Rúgolo (1987)
				Ruiz Leal (1972)
				Wainstein & González (1971b)
Cynodon dactylon (L.) Pers.	PG	FH	Am	Ragonese (1967)
Diplachne uninervia (Presl.) Parodi	AG	H	TAm	Boelcke (1964)
Distichlis australis (Speg.) Villamil	PG	H	Ar	Nicora & Rúgolo (1987)
D. humilis Phil.	PG	H	Ar-Bo	Boelcke (1964)
				Bustamante & Ruiz (1988)
				Ruthsatz (1978)
D. scoparia (Kunth) Arech	PG	H	SAm	Cano (1988)
D. spicata (L.) Greene	PG	H	Am	Cano (1988)
Eleusine tristachya (Lam.) Lam.	PG	FH	Ar-Br	Cabrera (1970)
			Pa-Ur	
Eragrostis orthoclada Hackel	PG	FH	Ar-Bo	Nicora & Rúgolo (1987)
			Pa	
Festuca scirpifolia (Presl) Kunth	PG	H	Ar	Nicora & Rúgolo (1987)
				Ragonese (1967)
				Türpe (1960)
Hordeum euclaston Steud.	AG	FH	Ar-Br	Nicora & Rúgolo (1987)
			Ur	Ragonese (1967)
H. flexuosum Nees	PG	FH	Ar-Ur	Boelcke (1964)
			Br	Nicora & Rúgolo (1987)
H. halophilum Griseb.	AG	H	Pe-Ar	Nicora & Rúgolo (1987)
				Pelliza-Sbriller et al. (1984)
H. jubatum L.	PG	H	Am	Covas (1985)
				Ungar (1974)
H. stenostachys Godron	PG	H	Ar-Br	Covas (1985)
			Ur	Nicora (1978)
Lolium multiflorum Lam.	AG	ST	INTR	Boelcke (1964)
				Nicora & Rúgolo (1987)
Monanthochloë acerosa Griseb.	PG	H	Ar	Villamil (1969)
Muehlenbergia asperifolia (Nees et Meyen) Parodi	PG	H	Am	Cano (1988)
M. ligularis	PG	H	Bo	Bustamante & Ruiz (1988)
Pappophorum caespitosum Fries	PG	H	Ar	Pensiero (1986)
P. philippianum Parodi	PG	ST	Ar	Nicora & Rúgolo (1987)
Parapholis incurva (L.) Hubb.	AG	H	INTRO	Nicora & Rúgolo (1987)
Paspalum paspalodes (Mich.) Scrib.	PG	H	Ar	Cano (1988)
				Schulz (1963)
P. vaginatum Swall.	PG	H	Ar	Nicora & Rúgolo (1987)
Poa lanuginosa Poiret	PG	ST	Ar	Aronson (1989)
				Pykkö (1966)
				Somlo et al. (1985)
Puccinellia osteniana (Pilger) Herter	PG	H	Ar-Ur	Boelcke (1964)
Spartina longispica Hauman et Parodi	PG	H	Ar-Ur	Aronson (1989)
				Yensen et al. (1981)
S. schreiberi Gmelin	PG	H	SAm	Aronson (1989)
Sporobolus maximus Hauman	PG	H	Ar	Nicora & Rúgolo (1987)
Sp. pyramidatus (Lam.) Hitch.	PG	H	SAm	Wainstein & González (1971b)
Sp. rigens (Trin.) Desv.	PG	ST	Ar	Aronson (1989)
				Ruiz Leal (1972)
Sp. virginicus (L.) Kunth	PG	H	Am	Hitchcock (1950)
				Nicora & Rúgolo (1987)
Stenotaphrum secundatum (Walter) Kuntze	PG	FH	TAm	Cabrera (1970)

Table 1. Continued

Species name	LF[a]	ST[b]	D[c]	References
Trichloris crinita (Lag.) Parodi	PG	FH	Ar-Bo Pa	Cabrera (1970)
T. pluriflora Fourn.	PG	FH	TAm	Roig (1970)
LEGUMINOSAE				
Cassia crassiramea Benth.	SH	ST	Ar-Bo	Burkart (1952)
Geoffroea decorticans (H. et A.) Burkart	T/SH	FH	Ar-Ch Bo-Pa	Boelcke (1981) Delhey (1991)
Hoeffmannseggia glauca (Ort.) Eifert	PH	FH	Am	Burkart (1967) Ulibarri (1984)
H. trifoliata Cav.	PH	FH	Ar	Burkart (1967)
Lotus tenuis Waldst. et Kitt	PH	H	INTR	Montes (1988)
L. uliginosus Schkuhr.	PH	ST	INTR	Sánchez (1984)
Medicago minima (L.) Grufb.	AH	ST	INTR	Boelcke (1964) Fortunato (1984)
Melilotus albus Medik	AH	FH	INTR	Burkart (1967) Rossow (1984a)
M. indicus (L.) All.	AH	FH	INTR	Burkart (1967) Rossow (1984a)
M. officinalis (L.) Desr.	AH	FH	INTR	Burkart (1967) Rossow (1984a)
Parkinsonia aculeata L.	T/SH	ST	Am	Burkart (1952)
Prosopidastrum globosum (H. et A.) Burkart	SH	ST	Ar	Boelcke (1981)
Prosopis alpataco Phil.	SH	H	Ar	Hunziker (1952)
P. burkartii Muñoz	SH	H	Ch	Burkart (1976) Trobok (1985)
P. chilensis (Mol.) Stuntz	T	H	Ar-Ch Pe	Burkart (1952)
P. denudans Benth.	SH	–	Ar	Burkart (1952, 1976)
P. juliflora (Swartz)	T/SH	–	Am	Aronson (1989)
P. nigra (Gris.) Hieron.	T	ST	Ar	Burkart (1952)
P. pallida (Willd.) H.B.K.	T	H	SAm	Aronson (1989) Burkart (1976)
P. reptans Benth.	SSH	–	Ar	Aronson (1989) Ragonese & Piccinini (1977) Neumann*
P. ruscifolia Gris.	A	H	Ar	Burkart (1952, 1976)
P. strombulifera (Lam.) Benth	SH	FH	Ar	Cano (1988)
P. tamarugo Phil.	T/SH	H	Ch	Burkart (1976)
Trifolium fragiferum L.	PH	H	INTR	Burkart (1967) Rossow (1984b)
PLANTAGINACEAE				
Plantago lanceolata L.	PH	ST	INTR	Pontiroli (1965)
P. myosuros Lam.	AH	FH	Pe-Br Pa-Ar Bo-Ur	Pontiroli (1965)
P. patagonica Jacq.	AH	ST	Ar	Pontiroli (1965)
POLYGONACEAE				
Polygonum aviculare L.	AH	FH	INTR	Cabrera (1967) Palacios (1987) Ungar (1974)
P. brasiliense Koch	PH	H	SAm	Cabrera (1967) Palacios (1987)
P. stypticum Cham. et Schlecht.	PH	H	SAm	Cabrera (1967) Palacios (1987)

Table 1. Continued

Species name	LF[a]	ST[b]	D[c]	References
PORTULACACEAE				
Grahamia bracteata Gill. ex H. et A.	SH	H	Ar-Ch	Cullen (1984) Ruiz Leal (1972) Neumann*
Portulaca oleracea L.	AH	ST	INTR	Aronson (1989) Barbosa*
P. pilosa L.	AH	H	CAm	Aronson (1989)
P. retusa Engelm.	AH	H	SAm	Aronson (1989)
Talinum paniculatum (Jacq.) Gaert.	PH	FH	INTR	Aronson (1989) Ragonese (1951)
SOLANACEAE				
Grabowskia duplicata Arnott	SH	H	Ar-Br Ur	Ragonese (1951) Neumann*
Lycium cestroides Schlecht.	SH	ST	Ar-Br Bo-Ur	Cabrera (1983) Castillo*
L. chilense Miers	SH	H	Ar-Ch	Ruiz Leal (1972) Verettoni (1961)
L. ciliatum Schlecht.	SH	H	Ar-Bo Br-Ur	Ragonese (1951) Neumann*
L. humile Phil.	SSH	H	Ar-Ch	Cabrera (1983)
L. infaustum Miers	SH	H	Ar	Ragonese (1951) Neumann*
L. pubescens Miers	SH	H	Ar	Ragonese (1951) Neumann*
L. tenuispinosum Miers	SH	H	Ar	Cano (1988) Ragonese (1951)
TAMARICACEAE				
Tamarix gallica L.	SH	FH	INTR	Burkart (1949)
VERBENACEAE				
Acantholippia seriphioides (A. Gray) Moldenke	SH	FH	Ar	Beato & Trida (1990) Roig (1970)
ZYGOPHYLLACEAE				
Larrea nitida Cav.	SH	ST	Ar	Abiusso (1962)
Plectrocarpa tetracantha Gill. ex H. et a.	SH	H	Ar	Aronson (1989)

Salt tolerant grasses

From an examination of the grasses in Table 1 it appears that there is a collection of species that are either salt tolerant or true halophytes. Some of them, such as *Trichloris crinita*, *Hordeum stenostachys*, etc., are markedly preferred by livestock over other herbaceous species. Poor management frequently results in their replacement by less valuable species. As it is expected, these species have a large genetic variability, as indicated by their capacity to grow over a wide variety of habitats, geographical regions and climates. Knowledge about the forage potential of the grasses growing in the páramos of the Andes from Venezuela to northern Argentina, some of them known as halophytes (e.g. *Distichlis humilis*), can be gleaned from the work of Acosta-Solís (1980).

Some of these grasses could have significant economic potential as domesticated forage/fodder crops in saline environments, once their populations have been subjected to conventional plant breeding and genetic programs, or more advanced biotechnological manipulations. In most cases, little is known about their productivity and management. Their palatability is judged on the basis of the intensity of the removal by herbivores.

An interesting species with an apparent tolerance to salinity is *Trichloris crinita*. This is considered a valuable forage resource for animal husbandry in extended semiarid rangeland territories of South America (Ragonese 1951; Ruíz Leal 1972). Its geographical distribution is very wide, extending from the south of the United States and Mexico, down through Bolivia, Paraguay and Argentina (Ragonese 1951). In Argentina it is a characteristic floristic component of the phytogeographical regions of the Chaco, Monte and Southern Espinal (Fernández et al. 1989). Biotypes of this grass have been selected for productivity studies in the Province of Mendoza (Cavagnaro, personal comm.). It has also been a subject for selection at the Experimental Station of Big Spring, Texas, USA (Ragonese 1985), and its chemical composition and nutritive value have been studied by several workers (Bragadin 1959; Wainstein & Gonzalez 1971b; Díaz et al. 1972). The selection of cultivars with inherently high salt tolerance for use as a crop would be highly desirable.

Of all the grasses listed in Table 1, *Distichlis* spp. are the most important in terms of distribution, occurring in America on salty soils from Canada to southern Argentina (Cabrera & Willink 1970; Mota Urbina 1979). About 30 percent of the returned survey questionnaires and the references quoted in this work reported the presence of this genus. The potential of this genus as a resource for animal husbandry varies from one region to another and depends on the type of animal (cattle, sheep, goat, llama (*Lama glama*), alpaca (*L. pacos*) and guanaco (*L. guanicoe*)) and the particular species. Natural prairies of *D. humilis* growing in soils with a salt content higher than 2 g kg^{-1} are used by livestock in the Bolivian altiplano, being frequently found in association with *Festuca dolichophylla* (Salm & Gehler 1987). Fire is used as a management technique to induce the regrowth of *D. spicata* in the Atacama desert of Chile, favouring a higher protein content and digestibility and at the same time reducing the salt content of the forage (Lailhacar 1990). In a study of the nutritive value of forage species in the cold Argentinean semideserts of Patagonia, Somlo et al. (1985) indicate that *Distichlis* spp. (*D. scoparia*, *D. spicata*, (Boelcke, 1957)), growing in salty flats and forming rather uniform stands, constitute an important part of the cattle diet in July (22%), January (19%) and April (29%); it is of medium preference for goat and sheep. *D. scoparia* and *D. spicata* are eaten by cattle in the Río Salado depression of the Province of Buenos Aires, where humid temperate climatic conditions are prevalent, only when other more palatable species are not available (Vervoorst 1967), providing supplementary feed during adverse periods. More recently works on the cattle diet for this area have corroborated this findings (Cahuépé & Fernández Grecco 1981; Cahuépé et al. 1985). Unlike with other halophytes in South America, local information does exist on *Distichlis* spp. from the nutritional stand point (Somlo et al. 1985; Lailhacar 1990). Since many parts of South America are now experiencing similar problems of salinization to those already encountered in other parts of the world, the future role of high salt tolerant plants like *Distichlis* spp. in agriculture deserves serious research, not only as forage for animals but also as a part of the human diet. The interesting nutritional values and potential production of *Distichlis* spp. have been indicated (Mota Urbina 1979). In Mexico the government supported research has resulted in the planting of about 10,000 ha with *Distichlis spicata* pasturage in the former Texcoco lake (Mota Urbina 1979; Llerena 1994).

Chenopods shrubs and subshrubs

Although halophytes are widely distributed among the families of flowering plants, the Chenopodiaceae has by far the highest proportion (44%) (Flowers et al. 1986). With as many as 417 halophytic species all over the world (Index Kewensis), it is probably the family in which salt tolerance is most widespread and has been most thoroughly investigated. Important genera within this family are the *Atriplex*, *Suaeda*, *Kochia*, *Salsola* and *Salicornia*.

A high degree of success has been achieved in cultivating some chenopods, especially of the *Atriplex* species, in areas susceptible to summer drought or on land badly affected by salt. Many of these species not only survive but are even able to produce large quantities of biomass under saline conditions. This and the high nutritional value of these species for livestock make *Atriplex* one of the most promising genera of all economic halophytes.

Among the numerous examples of *Atriplex* native to South America, two of the most widely studied are *A. repanda* and *A. lampa*.

Atriplex repanda

This is a shrub native to Norte Chico in Chile, a region between latitudes 28°24′ and 32°13′ S. The climate is of the mediterranean type, with warm dry summers and mild wet winters; rainfall occurs mainly within a period of 4 to 6 months and ranges from 300 mm to less than 100 mm. The species was widely distributed in former times but the natural populations have been vastly reduced by overgrazing (Badilla 1975). A review of existing research on this species has been published by Olivares and Gastó (1981).

Atriplex repanda is a C_4 plant characterized by its high protein content and very good palatability. Its trunk and primary ramifications are poorly developed, but it has a highly dense foliage which stays green all the year round, even in unfavourable periods when the availability of forage of the herbaceous species is limited. This shrub is resistant to diseases and plagues.

Most of the surviving populations of *A. repanda* are to be found close to the coast, and are frost-sensitive; those populations in the foothills of the Andean Range probably have some degree of frost tolerance (Lailhacar & Padilla 1987; Le Houérou 1992). The shrub is tolerant of low levels of precipitation, and in this connection it has been suggested that the plants absorb water into the leaves and then exude it into the ground (Sudzuki 1975).

This species has been found in ruderal, saline environments, preferably coastal areas. The correlation between production and the most important soil variables was studied by Lailhacar (1976) and Gargano (1978). Badilla (1975) reported that the species can be found in a wide variety of soil types, though it is best adapted to sandy soils. It tolerates salinity levels up to 19 dS m^{-1}.

The plant exhibits growth throughout the whole year, though there are two specific periods during which the rate of growth increases significantly; the first of these is at the beginning of spring, when vegetative growth accelerates, and the second is during full flowering and the commencement of fructification. Spring growth is predominantly foliar, and the proportion of green twig growth is comparable to that of the leaves in the summer, when the fruit develops (Olivares et al. 1985).

In view of the desirable characteristics of *A. repanda* it was proposed to introduce it in the natural grasslands of the central mediterranean region of Chile, thus replacing the existing shrub stratum of low forage value with one of higher quality which in addition is preferred by the cattle (Gastó & Contreras 1972). At the moment, the herbaceous stratum in this region dries up at the end of the rainy season, producing a critical period for the sheep grazing there; it is during this period that *A. repanda* comes in useful.

In order to ensure the viability of the introduction of *A. repanda* into natural pastures, a sure and economic system must be found to carry this out. Direct sowing presents difficulties because unlike in other *Atriplex* species, the probability of seed germination in the absence of some type of pretreatment is very low (0–2%). In this connection a marked improvement has been obtained in the plant's germinative capacity by prior elimination of the pericarp that encloses the seed in a cavity and scarification of the testa (Lailhacar & Laude 1975) so that the latter no longer acts as a barrier to the passage of water (Fernández & Johnston 1978) or oxygen (Johnston & Fernández 1978).

The germination percentage in seeds having half a pericarp was found to be intermediate between that of intact seeds and that of seeds entirely without a pericarp (Cristi & Gastó 1971). Washing the fruits increased the probability of emergence (Johnston & Fernández 1979), a phenomenon which can be explained by the fact that the washing serves to eliminate the 10% saponin and 1% NaCl content (Fernández et al. 1985).

The major problem involved in introducing *A. repanda*, however, remains its difficulty in establishing itself. It is not an ecologically aggressive species, and interspecific competition can lead to the formation of weak shrubs with a slow initial development (Acuña et al. 1978). Transplanting seedlings obtained in the greenhouse gives good results; however, though there is no such guarantee of good results with direct plantation in the natural pastures, this latter does offer the advantage of being able to sow large areas within a short period of time at lower capital and labour costs (Leighton et al. 1975; Olivares & Gastó 1981). The most convenient sowing period is the second half of July, when there is still moisture in the soil and the minimum temperature begins to increase (Olivares & Gastó 1981).

The main factor regulating the productivity of *A. repanda* is plant density (Gastó & Contreras 1972). These authors obtained the highest reported yield from 18- and 30-month-old plants (7,728 and 7,777 kg ha^{-1}, respectively) together with the highest reported density (18,518 plants ha^{-1}). The ratio stems:leaves and stems:fruit increased at this density with respect to lesser densities (Gastó & Caviedes 1976). In Mediterranean herbaceous range species with *A. repanda* as a complement the optimum density would be that giving rise to the least competition between the two strata and with the lowest stem:leaf ratio (Di Marco 1973).

It is inadvisable to use the shrub throughout the whole year since it is not able to recover from more than two defoliations (Olivares & Gastó 1983). Avoiding such constant browsing, the plants can be exploited to their full advantage without suffering any damage and are able to supplement the natural pastures in periods – whenever these may occur – when the latter provide insufficient and low-quality fodder.

Leaf fall increases the total annual availability of dry material from the plants, except in those cases when it occurs during the period of vegetative growth. The highest yield of total dry material has been found to occur at the beginning of flowering (Olivares et al. 1985).

Gastó and Olivares (1979) showed that plants grazed only once a year have a higher percentage of non-productive support structures; those browsed on twice a year on the other hand have a lower percentage of support tissue, making the plants more tender and thus more preferred by the animals.

In a study carried out on natural pastures reinforced with *A. repanda*, Di Marco (1973) observed that with a density of 400 plants ha^{-1} it was possible to achieve a yield of 850 kg DM ha^{-1} in October, increasing to 2,000 kg DM ha^{-1} as of January, which is the period of maximum growth for *A. repanda*. The yield of utilizable dry material remains relatively constant from spring to the beginning of autumn at 70–75% of the total dry material provided by the shrub, decreasing thereafter to 30% during the winter months. This decrease is apparently due to the lignification of the green twigs and the pericarp of the fruit. The herbaceous stratum remains dry during the maturation period in October until after the first rains of the following year in May or June. The herbaceous stratum yielded 2,800 kg DM ha^{-1}. The animals showed a preference for this stratum (90% of their diet) for as long as consumption was not limited by availability; however, when the availability of the herbaceous stratum diminished the shrub stratum was more intensively used, and it is thus that *A. repanda* came to contribute 90% of the sheep's diet during the winter. This clearly demonstrates the importance of the species during critical nutritional periods for sheep.

The effect of using natural pastures reinforced with *A. repanda* was assessed in the various stages of the sheep's productive cycle. The number of births in herbaceous pastures was found to be 21% lower than in those supplemented with shrubs. Furthermore, the concentration of births within the first thirty days of the lambing period in the latter pastures was more than 50% higher than in the purely herbaceous pasture (Rodríguez & Gastó 1972).

Concha et al. (1977) used lambs weaned in the spring to graze on two annual mediterranean pastures, one reinforced with *A. repanda* and the other without shrubs, during the dry period (December–March). The weight of the lambs at the end of the experimental period was 41 kg in the pasture with *A. repanda* against only 34 kg in the pastures without any shrubs. This increase in liveweight was attributed to the presence of *A. repanda*.

Atriplex lampa

This is a native shrub of the arid zone in Argentina. Its area of distribution extends from 27° to 46° S, from the Andean range (2,500 m above sea level) to the Atlantic coast. It is a fairly cold-tolerant species.

In a study undertaken in a community in which *A. lampa* represented up to 40% of the vegetative cover, Passera and Borsetto (1989) recorded an aerial biomass of 1,900 kg ha^{-1}, of which 21% was adequate for forage. A close correlation was observed between aerial biomass, rainfall, and the availability of water when the plant was in its active stage of growth.

A. lampa is adapted to sandy soils, and has axonomorphous roots which develop abundant root hair at a depth of 40–60 cm. The number of ramifications decreases notably with depth, and at a depth of 3.5 m another sector of ramifications appears, thus providing the plant with a double strategy for capturing water: roots closer to the surface capture the water from light rain and the deeper ones capture that from the water table or that accumulated from more abundant rain (Passera & Borsetto 1989).

Cattle tend to consume the tender shoots and leaves, but in fields with brackish waters the consumption of this shrub is much lower.

Atriplex undulata

This is a very palatable, cold-tolerant species which develops more favourably in medium-textured deep soils, reaching its maximum productivity under these conditions. Its production potential is similar to that of *A. nummularia* and *A. halimus* (Le Houérou et al. 1982). Aronson (1985) reported that the dry matter obtained from *A. undulata* under seawater irrigation was comparable to the dry matter yields of conventional fodder plants irrigated with fresh water. This plant is susceptible to defoliation.

Suaeda spp.

Another important genus in the family of Chenopodiaceae widely distributed throughout South America is *Suaeda* spp. Pioneering research on the ecophysiology, productivity and nutritive value of the extensive halophytic steppes of *Suaeda foliosa* growing in the high plateaus (3,700 m) of the central Andes of Bolivia, have been undertaken by Bolivian researchers (Barja 1972; Rocha García 1983; Lorini et al. 1984; Alzerreca 1986; Bustamante & Ruiz 1988). The salinity of the upper soil layer is quite high, with 3.6 g kg^{-1} of soil, and is much higher at 10 cm depth. *S. foliosa* is frequently found in mixed association with grasses such as *Muhlenbergia ligularis*, *Distichlis humilis*, and others. Steppes of this sub shrub saltbush are actively exploited by assorted livestock, including goats, sheep, llamas, donkeys and cattle. This is the main activity of the area since crop cultivation is rare or non existent. Productivity of green material in *S. foliosa* can reach up to 22×10^3 kg ha^{-1} (Bustamante & Ruiz 1988). It provides a forage of good nutritive value during the dry winter season when the availability of other sources of food and water is scarce.

The genus *Prosopis*

The species of the genus *Prosopis* are trees or shrubs of varying size, rarely subshrubs, mainly xerophilous, aculeate, spiny and rarely unarmed. The genus comprises about forty-four species distributed throughout Southwest Asia, Africa, and predominantly America, from western North America right down to Patagonia, where it is found mainly in the warm dry regions of the west; the centre of polymorphism is in Argentina (Burkart 1976).

By virtue of their abundance and ecophysiological behaviour, plants of the *Prosopis* species have acquired importance in arid and semiarid regions. Some of the species produce serviceable firewood, charcoal, and fodder, and the fruit is fit for human consumption.

The *Prosopis* species flourish in places so dry that other plants find it difficult to survive there, though some species are to be found in much more humid environments. They can usually be found in areas with poor, light sandy, or rocky soil. In general they thrive on 250 mm of precipitation, though some have become adapted to zones where the annual rainfall is as low as 75 mm or less (Habit 1981). They are able to tolerate prolonged periods of drought, still managing to produce abundant fruit. Under conditions of extreme drought, the shrubby species predominate. Some of the species have a very high tolerance to salinity (Habit 1981).

Soil productivity in arid and semiarid areas is

often hindered not only by the lack of water but also by the low level of nitrogen in the soil itself. The only terrestrial species capable of both growing in saline soil and fixing N belong to the genera *Prosopis* and *Casuarina*.

Owing to the fact that the plants grow easily from seed and some are able to reproduce vegetatively, the *Prosopis* species have a high rate of survival. The following rank among the most important halophytes within the genus.

Prosopis tamarugo

Prosopis tamarugo is a native tree of the Tamarugal Pampa, which forms part of the Atacama Desert in northern Chile. The Tamarugal Pampa is a flat depression with a surface area of approximately 12,500 km^2, a gradient of 1–2% from east to west, and an altitude varying between 300–1,200 m above sea level. It runs from north to south and borders in the east on the Andean Range and in the west on the Coastal Range. This is the only area with a natural distribution of *P. tamarugo*.

The climate of the Tamarugal Pampa is desertic, and in fact it constitutes one of the driest places on earth. Precipitation in the Salar de Pintados zone of the plain is 0.3 mm yr^{-1}. The day-to-night temperature range is very wide, from 35 to −5 °C. The altitude acts as a moderating factor on the temperature, the mean annual level of which is around 17 °C. The relative humidity is low during the day, fluctuating between 10 and 30%, and is relatively high at night, particularly during the period from September to March, when it can reach up to 80–100%. The average number of clear days per year is 250. There are occasional fogs.

The Tamarugal Pampa has two clearly identifiable sectors with well-defined and differentiated soils (Habit 1981). One of these corresponds to the higher ground of the eastern section, and forms a great piedmont with coarse stratification. The layer of soil is deep, stratified, sandy, with or without organic matter, has good or even excessive drainage, alkaline saline, and is of low to very low fertility. The second sector corresponds to the western area formed by stratified soils with finer material and with depressions currently occupied by salt flats. Clayey and loamy materials are predominant, and are stratified, with a salt cover of variable thickness (a few cm to one m or more). The salts are for the most part Na, Ca, Mg and K, all deliquescent. Salt flats are located preferentially towards the south of the Tamarugal Pampa and cover an estimated area of 3,750 km^2 (Habit 1981).

The soils of the tamarugal zone are characterized essentially by sodium salts and have a surface conductivity of over 50 dS m^{-1} (Luzio & Vera 1982). In a sampling carried out in three forested sectors of the Tamarugal Pampa it was found that the electrical conductivity in the zone with the highest root density varied between 0.85 and 91.60 dS m^{-1}, the average being 29.32 dS m^{-1}. Sodium is the dominant cation (SAR = 33–687) and chloride the dominant anion. Boron concentrations are very high (8–125 ppm), particularly in surface layers (Squella 1986). Salinity decreases with depth.

The surface of the Tamarugal Pampa remains barren over vast areas. *P. tamarugo* is currently to be found in certain specific areas and is most abundant in Salar de Pintados. Former chronicles mention that the Tamarugal Pampa was at one time densely covered in tamarugos, most of which were felled at the end of the last century or the beginning of this one to satisfy the demands of the local mining industry. Woods of pure *P. tamarugo* or mixed with *P. chilensis* form a community of limited floristic diversification. Species associated with the tamarugal zone are *P. burkartii*, *Geoffroea decorticans*, *P. strombulifera*, *Tessaria absinthioides*, *Caesalpinia aphylla*, *Atriplex atacamensis*, *Euphorbia tarapacana*, *Cressa cretica* and *Distichlis spicata*. These species are found in the most xeric or most saline ecotones of the tamarugal region (Lailhacar 1990) and in greater variety in places where the groundwater table is closer to the surface.

An underground water basin underlies the entire plain at variable depths, the water being recharged from the Andean zone. The discharge of water occurs via evaporation of surface waters, salt flats, and evapotranspiration in zones with vegetation. The evapotranspiration in forested soils is lower (309 m^3 ha^{-1} yr^{-1}) than that occurring in soils without any forestation (1,590 m^3 ha^{-1} yr^{-1}) (Habit 1981).

The quality of the underground water ranges

in general from very saline to good: the salinity increases from east to west, reaching 5,000–30,000 ppm as it reaches the western edge (Peralta 1985). There is a high concentration of chloride (up to 900 ppm) and boron (2.0–4.2 ppm) in the water (Squella 1986). The use of water in the tamarugal plain has caused a decrease in the groundwater table, which in turn has led to an improvement in the quality of the soil since there is less contamination from the more superficial saline layers (Castillo 1966).

The tamarugo is a tree 10–20 m tall. Its growth period extend throughout the whole year, reaching its peak between August and November and its minimum between March and June. On the basis of the inactivity of the sap during a certain period, however, Sudzuki et al. (1973) report that the tree has a period of winter recess. The growth rate in terms of height in 8–15-year-old trees is higher in sites with a groundwater table depth of 0–4 m, and does not appear to be affected by groundwater salinity between 1 and 3%. The influence of the depth of the groundwater table diminishes with the age of trees over 15 years old.

There are four parameters considered to be relevant indicators of fodder yield: tree height, crown diameter, foliage density and the quantity of fruit within the crown projection area (Briner 1985).

Equal quantities in weight of the fruit and leaves from the tamarugo tree have been used for many years as fodder for sheep and goats. The mature fruit and dry leaves fall to the ground and are preserved in this state owing to the lack of rain and the overall dry environment. The animals are able to use this layer as feed throughout the whole year, and eventually this constitutes their exclusive diet. The production of fodder measured in terms of the weight of the fallen fruit and leaves, without taking into consideration the amount browsed, is directly proportional to the age of the tree and the size of the crown; the level of production tends to stabilize after 30 years. One important characteristic of the tamarugo is that the lower branches form a compact thorny mass, making it necessary to prune them in order to enable the forage on the ground to be harvested or browsed. The pruned branches can then be used for fuel.

Fruit production begins at 7–8 years of age, the level of production for large trees varying between 33.5 and 41.9 kg yr^{-1}. Foliage production is about 13.2 to 44.2 kg DM yr^{-1} (Squella 1986). Another study (National Academy of Science 1979) states that each mature tree yields up to 160 kg of pods, leaves, and small twigs per year. Cadahia (1970) has reported that fruit and leaf yield varies from 4,000 kg ha^{-1} in 10-year-old plantations to around 15,000 kg ha^{-1} in 35-year-old plantations.

The fruit is usually distributed evenly throughout the green canopy (Instituto Forestal 1964), and fruit yield is as follows: 15% in January, over 60% in February, 16% in March, the rest being distributed throughout the remaining months of the year (Elgueta & Calderón 1971). The greater part of leaf production – 85% of the total – occurs from July through December (Aguirre & Wrann 1985).

It is impossible to estimate the duration of the tree's productive life, though it is certainly a long-lived species, with production continuing for several hundreds of years. A successful method was developed for implanting tamarugos in the Tamarugal Pampa (Habit 1981). The carrying capacity of the forest is related to the age of the trees, and several studies have been carried out to calculate this. The capacity of an 8-year-old plantation has been reported to be 0.5 head of sheep ha^{-1}, reaching 4 head in 18-year-old plantations (Habit 1981). Other estimates take the figure up to 10 head ha^{-1} at 15 years (Instituto Forestal 1964). The sheep breeds most adaptable to this zone are the Karakul, Australian Merino, Corriedale plus Angora goats (Habit 1981).

The tamarugo has a double root system: one pivotal, made up of one or several tap roots without ramifications reaching a depth of up to 7–8 m in sandy soils, and a dense superficial root mass consisting of a ramified system of fine roots 0.40–0.80 m thick and occurring at a depth of less than one meter (Habit 1981). The extension of this root mass coincides with that of the crown of the tree, and is associated with a significant moisture content. The level of soil moisture decreases markedly outside the crown projection area, where there are no roots, and above and below the root mass (Habit 1981).

In the sites where tamarugos propagate naturally, soil salinity is considerably lower than in man-made sites. In the former case only tap roots were observed between the soil surface and a depth of over 1.40 m (Acevedo et al. 1985).

Tamarugos have very curious water requirements; they require irrigation only during the first half-year of life, during which time they are watered every 4 or 5 days in the nursery and then every 10–12 days after being transplanted to the field. Once established in the Tamarugal Pampa they grow quickly without any irrigation whatsoever.

A great deal of interest was aroused by a report of Sudzuki (1969) to the effect that the leaves of *P. tamarugo* are able to absorb tritiated water from the atmosphere and mobilize it down to the roots and to the soil in contact with these roots. This ostensibly explained why the soil immediately surrounding the dense superficial root mass has more moisture than that which is above, below, or to either side.

Despite this high moisture content in the zone of the dense root mass, the soil water potential here remains only around −2.0 MPa as a result of the high salt content. At a level of 10 cm above or below the dense root mass, the soil water potential becomes even more negative, with values of −10.0 MPa and −5.0 MPa, respectively (Acevedo et al. 1985).

On the basis of such physical parameters as dew temperature and soil and water potentials, it has been determined that the water in the dense root zone originates from the groundwater table and that the necessary physical conditions for the absorption of leaf water are not met (Acevedo & Pastenes 1980). It has been shown that the leaves practically never reach the dew point temperature, so the probability of dew formation at the periphery of the leaves is low, which makes it unlikely that the leaves absorb water in liquid form (Acevedo et al. 1985).

It is conceivable that if the leaf stomata remained open throughout the night or part of the night and there was a water vapour concentration gradient towards the inside of the leaves then it would be possible for this vapour to enter. However, the gradient has been shown to always favour the flow from the inside to the outside of the leaf (Acevedo et al. 1985). Furthermore it has been shown that the stomata remain closed during the night, making the movement of water vapour through the epidermis of the leaf virtually impossible (Acevedo et al. 1985).

On the basis of the isotopic composition of the water, the groundwater table and the atmospheric humidity, it was possible to follow the dynamics of the evaporation and transpiration processes. The results show that the water present in the zone of the dense superficial root mass and that present in the leaves, which intervenes in the transpiration process, comes from the groundwater table (Aravena & Acevedo 1985). Acevedo and Pastenes (1980) further found that the tamarugo only grows on sites where the groundwater table is between 4 and 16 m below ground.

Prosopis chilensis and *P. nigra*

These two species serve as potential sources of direct feeding through consumption of the fruit, leaves, twigs, and seedlings. The browsing of *P. nigra* by cattle is insignificant.

The legumes mature in summer and practically all of them drop to the ground – due mainly to the rain and wind – within a short lapse of time and are consumed by the animals. The availability of this fruit on the ground is relatively short-lived because it decomposes rapidly or is attacked by insects and fungi. Leaf fall of *P. chilensis* is mainly between July and November (Squella 1986).

Some rough estimates of the production of these two species in Argentina have been made, and reports indicate yields of between 5 and 100 kg fresh fruit per adult tree (Karlin & Díaz 1984). In large 50-year-old trees fruit production varies between 9.9 and 28.7 kg DM yr^{-1} (Corporación de Fomento de la Producción 1982). One of the disadvantages of fruit production in connection with these species is that it varies greatly from year to year due to climatic factors during the flowering period, such as out-of-season rains (September and October), winds, and late frosts, all of which have a detrimental effect on production (Karlin & Díaz 1984).

The availability of the fruit coincides with the main period of fodder availability, in terms of both quantity and quality. A much more efficient use of the fruit can therefore be made by storing it to be used as a food supplement in the critical periods of winter and spring. In some areas in northwestern Argentina the fruit is collected manually and kept for later use in times of forage deficit. The successful preservation of the fruit depends on reducing the moisture content and protecting them from attack by insects and fungi (Karlin & Díaz 1984).

The consumption of leaves and twigs occurs mainly towards the end of the winter, when the new grass has still not emerged in the pastures and what little there is has low quality. This source of food for animals is naturally at its optimum when the trees are of a size the animals can easily reach, i.e. when they are young and have basal growth.

The combination of *Prosopis* with Gramineae assures increased forage production. Also feasible are combinations of *Prosopis* and *Atriplex* species, for instance *A. cordobensis*, which tolerates well the shade provided by the *Prosopis* species, and which can form 2-metre-high plants (Ayerza et al. 1988).

Prosopis seedlings were screened for growth at seawater salinity concentrations. Accessions of *P. nigra/flexuosa* and *P. alba/nigra* (the taxonomy was uncertain) were included, and over 50% of the initial seedling survived in the culture solution with 3.3% NaCl. *P. chilensis* had a stable survival rate up to 3.3% NaCl and was one of the accessions exhibiting the highest growth rate (Rhodes & Felker 1988).

P. nigra thrives best in deep clay soils and tolerates alkali (Burkart 1976). It is characteristically found in degraded soils.

P. chilensis is one of the fastest-growing trees under arid conditions (Burkart 1976). Large fruits and the absence of thorns facilitate browsing.

Nutritive value as livestock feed

The two main families of forage halophytes are the Chenopodiaceae and Gramineae. There is a third group containing species of other vegetative families – for the most part shrubs or trees – which are of lesser significance as sources of forage even though they may be of exceptional importance in certain areas, as is the case for instance with the tamarugo.

Although by no means exhaustive, more information is available about the nutritional value of forage halophytes than about their productivity. The total nitrogen content of chenopods (*Atriplex* spp) ranges from about 3.31–1.80% on a dry weight basis (Table 2), whereas for most grasses it is between 1.70 and 1.10% (Table 3). Crude protein content is determined as the product of

Table 2. Nutritive value of halophytic chenopods of South America

Species	Plant part	Total N	P	Ca	Crude fiber	NDF-ADF*	Ash
					%		
Atriplex lampa[a]	Twigs	1.80	0.01	0.15	16.05		28.80
Atriplex lampa[b]	Leaves and twigs	3.15	0.21			31.7/18.7	26.6
Atriplex undulata[b]	Leaves and twigs	2.82	0.21			45.9/26.1	20.9
Atriplex repanda[c]	Leaves and twigs	3.31	0.12	1.29	8.47		22.63
Atriplex repanda[d]	Fruits	1.65	0.14	0.48	38.29		10.35
Atriplex atacamensis[c]	Leaves and twigs	2.55	0.11	0.68	9.54		29.20
Atriplex coquimbensis[c]	Leaves and twigs	2.69	0.10	0.87	11.67		18.91
Atriplex argentina[c]	Leaves and twigs	2.71	0.08	0.67	12.43		30.56
Atriplex halimus[c]	Leaves and twigs	2.10	0.11	0.87	7.93		32.65
Atriplex semibaccata[c]	Leaves and twigs	3.00	0.09	0.57	12.25		16.16
Suaeda foliosa[e]	Leaves	2.67	0.38	1.73	7.60		34.84
Suaeda foliosa[e]	Stem	2.69	0.37	1.82	14.14		30.56

*NDF, neutral detergent fiber; ADF, acid detergent fiber.
[a]Wainstein & González (1971a).
[b]Silva Colomer & Passera (1990).
[c]Squella et al. (1985).
[d]Vera (1985).
[e]Bustamante & Ruiz (1988).

Table 3. Nutritive value of grass and grass-like halophytes of South America

Species	Phenological stage	Total N	P	Ca	Crude fiber	Ash
				%		
Cottea pappophoroides[a]	Flowering	1.49	0.14	0.58	30.20	8.82
Trichloris crinita[b]	Early fructification	1.54	0.13	0.27	31.40	9.79
Trichloris crinita[c]	Past fructification	1.51	0.21	0.36		9.17
Pappophorum philippianum[a]	Past fructification	1.37	0.11	0.38	29.50	8.06
Pappophorum caespitosum[b]	Early fructification	1.70	0.14	0.28	27.70	9.71
Pappophorum caespitosum[b]	Early maturation	1.45	0.14	0.44	32.03	8.79
Aristida mendocina[b]	Early fructification	1.10	0.12	0.33	37.80	6.95
Poa lanuginosa[d]	Vegetative	1.60	0.33	0.62		
Juncus balticus[d]	Vegetative	1.52	0.28	0.35		

[a]Wainstein & González (1971b).
[b]Wainstein & González (1971a).
[c]Abiusso (1964).
[d]Somlo et al. (1985).

Table 4. Nutritive value of halophytic leguminous species of South America

Species	Plant part	Harvest time	Total N	Ash
			%	
Geoffroea decorticans[a]	Green leaves	Spring	2.88	7.9
Geoffroea decorticans[a]	Green leaves	Summer	2.98	10.6
Geoffroea decorticans[a]	Green fruits	Spring	1.04	3.2
Geoffroea decorticans[a]	Mature fruits	Summer	0.66	3.6
Prosopis chilensis[a]	Green fruits	Spring	1.76	3.8
Prosopis chilensis[a]	Dry fruits	Summer	1.41	3.8
Prosopis strombulifera[a]	Green leaves	Summer	1.74	9.5
Prosopis strombulifera[a]	Dry fruits	Fall	1.46	4.8
Prosopis nigra[b]	Shoots green		2.85	5.9
Prosopis nigra[b]	Mature fruits		1.31	4.5
Prosopis alpataco[c]	Leaves	Summer	2.91	8.3
Prosopis alpataco[c]	Fruits	Summer	1.52	3.4

[a]Lailhacar (1990).
[b]Díaz (1962)
[c]Abiusso (1964).

total N × 6.25. This value assumes that all the N calculated is digestible protein, and that the proteins have 16% N. However, the level of non-protein N (e.g. nitrates) can be very high in halophytes, giving rise to an overestimation of the crude protein level. For this reason it is preferable to use total nitrogen values (Gihad & El Shaer 1994).

The crude fiber content in *Atriplex* is lower than that in grasses (Tables 2 and 3). In general, browse species (e.g. *Atriplex*) meet the protein requirements of grazing animals but are low in energy.

The ash content of the foliage in *Atriplex* spp. is typically high, around 50% of this being made up of dissolved salts, mainly NaCl. Among atriplexes, *A. repanda* is one of the species attracting most interest because of its nutritive value and the fact that is preferred by animals over other species; this latter fact is probably related to its relatively low salt content (Le Houérou 1992).

The higher the salt content in their diet, the greater is the water intake required by animals. The acceptable amount of salt in the diet of different types of animal varies according to the concentration of salt in the water and the environmental conditions. This high salt content of the foliage is a serious limitation to the use of the

Table 5. Nutritive value of halophytic shrubby species of South America

Species	Phenological stage	Plant part	Total N	P	Ca	Crude fiber	Ash
					%		
Ephedra ochreata[a]	Vegetative	Twigs	0.99	0.20	1.94		9.93
Larrea nitida[a]	Vegetative	Twigs	3.23	0.28	2.83		12.90
Lycium tenuispinosum[b]	Budding		1.76	0.13	4.41	30.44	20.24
Lycium chilense[a]	Fructification	Shoots	2.21	0.13	1.37		9.32
Lycium chilense var. *chilense*[c]	Past fructification		2.08	0.12	1.43	25.33	13.03
Lycium chilense var. *minutifolium*[c]	Flowering		2.25	0.08	0.96	30.25	10.87

[a]Abiusso (1962).
[b]Wainstein et al. (1979).
[c]Wainstein & González (1971b).

Table 6. Seasonal changes in nutritive value of leaves and twigs of *Atriplex undulata*

Harvest time	Total N	P	NDF	ADF	Ash
			%		
Spring	2.82	0.21	45.9	26.1	20.9
Summer	2.38	0.14	38.1	18.8	20.8
Fall	2.64	0.11	42.3	19.5	27.8
Winter	2.46	0.19	29.8	18.4	24.0

Adapted from Silva Colomer & Passera (1990).

Table 7. Chemical composition (in %) of fruits and leaves of *Prosopis tamarugo*

Components	Fruit	Leaves
Dry matter	91.6	90.3
Total N	1.68	1.74
Crude fiber	29.7	15.2
Acid detergent fiber	46.8	24.8
Cells walls	54.4	31.2
P	0.05	0.03
Ca	0.18	1.39
Ash	3.80	9.80

Adapted from Corporación de Fomento de la Producción (1985).

Table 8. Dry matter digestibility and feed intake of *Prosopis tamarugo*

Species	Fruit	Leaves
Dry matter digestibility (%)		
Sheep	42.4	41.4
Goats	68.0	49.9
Feed intake (g day^{-1})		
Sheep	1339	542
Goats	1525	821

Adapted from Corporación de Fomento de la Producción (1985).

Atriplex spp. for consumption by animals (Wilson 1994).

The total nitrogen level in the leaves and twigs of *A. repanda* is two times higher than in fruits (Table 2).

The concentration of total nitrogen in the leaves of shrub or arboreal legumes is very high (Table 4) and approaches that of high quality herbaceous forage legumes such as alfalfa. The fruits of these species have lower total nitrogen values but in general are adequate for those categories of animals with high nutritional requirements such as young growing animals and are even eaten in preference to the leaves. The higher the level of tannins in the fruit (in the case of *P. nigra*), the lower the tolerance and digestibility of the proteins; however, high tannin levels also mean greater protection against attack by insects and birds, so the fruit can be conserved longer.

Halophytic shrubs has a fairly high concentration of total nitrogen ranges from 3.23–0.99% (Table 5).

In general all the species contain sufficient Ca, though in grasses the level is lower than in other plant groups (Table 3).

In the majority of halophytes studied the levels of P are below the recommended values (Tables 2, 3, 5) (National Academy of Sciences 1984). The ratio Ca:P can normally be taken as a good

Table 9. Daily consumption of dry matter, digestible energy and digestible protein of *Prosopis tamarugo* by goats at various physiological stages

Class and weight	Dry matter consumption	Digestible energy	Deficit or excess	Digestible protein	Deficit or excess
kg		kcal		g	
Maintenance and dry goats					
27.2	1.3	2744	−56	78.8	+1.8
45.4	1.7	3600	−100	103.4	+17.4
Lactation (16 weeks)					
27.2	1.8	3823	−577	109.5	−4.5
45.5	2.0	4226	−1074	125.6	−10.4
Kid growth and production					
18.0	1.1	2329	−771	73.0	−8.7
36.3	1.5	3177	−623	91.1	+4.8

Adapted from Corpación de Fomento de la Producción (1985).

indicator of the level of availability of P, and is generally found to be adequate. Since grasses are usually deficient in P but high in energy, a combination of shrubs and grasses should provide a reasonably balanced diet.

There is no major seasonal variation in the concentration of nitrogen and phosphorus in the atriplexes (Table 6); in grasses, however, the concentration drops as the season advances.

The majority of species considered here have relatively high levels of total nitrogen and intermediate levels of crude fiber helping to provide and adequate protein and energy content in animal diet.

Prosopis tamarugo

Chemical analysis was performed on a representative sample of the annual output of tamarugo plants (Table 7) (Corporación de Fomento de la Producción 1985). On the basis of this analysis the tamarugo fodder can be classified as roughage with a high fiber content, particularly in the case of the fruit, and low in nitrogen. The Ca:P ratio is inadequate. The data thus indicate that in order to provide a balanced diet for herds of sheep and goats it is necessary to supplement this forage with another.

Data on digestibility and feed intake is presented in Table 8 (Corporación de Fomento de la Producción 1985). Digestion coefficients are very low except for the case of fruit in the goat trial. The figures show that goats digest the tamarugo fruit and leaves to a greater extent than sheep do. Differences in digestibility affect voluntary intake, and goats therefore consume more than sheep.

One striking feature noted both with sheep and goats is that their intake of leaves is significantly lower than that of fruit, even though the former has a lower crude fiber content. In a trial set up to determine the selectivity of tamarugo fodder by goats it was established that the greatest consumption corresponds to fruit (68.7% of intake on a dry matter basis).

Table 9 gives a theoretical evaluation of the extent to which tamarugo fodder meets the nutritional requirements of goats during the various physiological stages of their development (Corporación de Fomento de la Producción 1985). The energy content in tamarugo fodder is very low, and of course this deficiency becomes more critical during the most demanding stages of physiological growth such as late pregnancy and lactation. The protein deficit is less marked.

For utilization as fodder for sheep the digestibility of the fruit and leaves of the tamarugo must be considered generally low. Energy, and to a certain extent protein supplementation, should be utilized in the nourishment of goats during late pregnancy and lactation. The mineral deficit (calcium, phosphorus etc.) should also be corrected by the ad libitum administration of mineralized salt and bone meal.

Acknowledgement

Thanks are due to Consejo Nacional de Investigaciones Científicas y Técnicas of Argentina and UNEP for travel grants for REB and OAF, respectively, to participate in the Nairobi workshop. We greatly appreciate the many collaborators of South America that answered our questionnaire survey.

References

Abiusso N.G. (1962). Composición química y valor alimenticio de algunas plantas indígenas y cultivadas en la República Argentina. Rev. Investigaciones Agrícolas 16: 93–247.

Abiusso, N.G. (1964). Composición química y valor alimenticio de algunas plantas indígenas y cultivadas en la República Argentina, II. Rev. Investigaciones Agropecuarias Serie 2, Biol. Prod. Veg. 1: 311–338.

Acevedo, E. & Pastenes, J. (1980). Distribución de *Prosopis tamarugo* Phil. en la Pampa del Tamarugal (desierto de Atacama). III Congreso Internacional de Zonas Aridas. La Serena, Chile.

Acevedo, E., Sotomayor, D. & Zenteno, V. (1985). Foliar tissue water parameters of *Prosopis tamarugo* Phil. In: Habit, M.A. (ed.), The current state of knowledge on *Prosopis tamarugo*. FAO, Regional Office for Latin America, Chile, pp. 257–262.

Acosta-Solís, M. (1980). Plantas indígenas para forrajicultura tropandina. Rev. Academia Colombiana Ciencias Exactas Físicas Naturales 15: 57–87.

Acuña, P., Avendaño, J., Lailhacar, S. & Neumann, E. (1978). Efecto de la competencia de las terófitas (*Bromus mollis* y *Medicago polymorpha*) en el crecimiento inicial de *Atriplex repanda* Phil. Avances Producción Animal 1: 31–40.

Aguirre, J.J. & Wrann, J. (1985). The genus *Prosopis* and its management at the Tamarugal Pampa. In: Habit, M.A. (ed.), The current state of knowledge on *Prosopis tamarugo*. FAO, Regional Office for Latin America, Chile, pp. 3–32.

Alzerreca, H. (1986). Evaluación preliminar de praderas nativas y forrajes introducidos, zonas Turco, Cosapa, Toma-Toma y el Chorro. Consultant Report, Oruro, Bolivia.

Aravena, R. & Acevedo, E. (1985). The use of environmental isotopes oxygen-18 and deuterium in the study of water relations of *Prosopis tamarugo* Phil. In: Habit, M.A. (ed.), The current state of knowledge on *Prosopis tamarugo*. FAO, Regional Office for Latin America, Chile, pp. 251–256.

Aronson, J. (1985). Economic halophytes: A global review. In: Wickens, G.E., Goodin, J.R. & Field, D.V. (eds.), Plants for Arid Lands. George Allen & Unwin, London, pp. 177–188.

Aronson, J. (1989). Halophytes. A data base of salt tolerant plants of the world. Office of Arid Lands Studies. University of Arizona, Tucson, AZ.

Ayerza, R., Díaz, R. & Karlin, U. (1988). Manejo de *Prosopis* arbóreos en sistemas de producción ganaderos en el Chaco seco de Argentina. Gaceta Agronómica 8: 140–156.

Badilla, S.I. (1975). Características ecológicas y fitosociológicas de *Atriplex repanda* Phil. Thesis Lic. Agr. Universidad de Chile, Santiago.

Barja, G. (1972). Unpublished Report. Ministerio de Asuntos Campesinos, La Paz, Bolivia.

Barros, M. (1960). Las ciperáceas del Estado de Santa Catalina. Sellowia 12: 182–450.

Beato, R.O. & Trida, C.A. (1990). Digestibilidad in vitro de algunas especies arbustivas de la provincia de La Pampa. Rev. Fac. Agronomía (Univ. Nac. La Pampa) 5: 37–63.

Boelcke, O. (1957). Comunidades herbáceas del norte de la Patagonia y sus relaciones con la ganadería. Rev. Investigaciones Agrícolas 11: 5–97.

Boelcke, O. (1964). Forrajeras. In: Parodi, L.R. (dir.), Enciclopedia Argentina de Agricultura y Jardinería, Vol. 2, Part 2. Acme Agency, Buenos Aires, pp. 707–787.

Boelcke, O. (1981). Plantas vasculares de la Argentina nativas y exóticas. Hemisferio Sur, Buenos Aires.

Bragadín, E.A. (1959). Las pasturas en la Región de los Llanos. Rev. Agronómica Noroeste Argentino 3: 289–334.

Briner, C. (1985). Phenotypic characterization of the tamarugo biotypes at the Tamarugal Pampa. In: Habit, M.A. (ed.), The current state of knowledge on *Prosopis tamarugo*. FAO, Regional Office for Latin America, Chile, pp. 221–225.

Burkart, A. (1949). Notas sobre algunas plantas exóticas, adventicias o naturalizadas en la República Argentina. Lilloa 15: 19–26.

Burkart, A. (1952). Las leguminosas argentinas. Acme Agency, Buenos Aires.

Burkart, A. (1967). Leguminosae. In: Cabrera, A.L. (dir.), Flora de la Provincia de Buenos Aires, Part 3. Colección Científica INTA, Buenos Aires, pp. 394–647.

Burkart, A. (1969). Gramíneas. La familia botánica de los pastos. In: Burkart, A. (dir.), Flora ilustrada de Entre Ríos (Argentina), Part 2. Colección Científica INTA, Buenos Aires, pp. 1–551.

Burkart, A. (1976). A monograph of the genus *Prosopis* (Leguminosae subfam. Mimosoideae). J. Arnold Arboretum 57: 219–249 & 450–525.

Bustamante, Z. & Ruiz, M.C. (1988). Nutrientes del cauchi (*Suaeda foliosa*), forrajera del altiplano central de Bolivia. Ecología Bolivia 12: 29–37.

Cabrera, A.L. (1963). Compositae. In: Cabrera, A.L. (dir.), Flora de la Provincia de Buenos Aires, Part 6. Colección Cientifica INTA, Buenos Aires, pp. 1–443.

Cabrera, A.L. (1967). Polygonaceae. In: Cabrera, A.L. (dir.), Flora de la Provincia de Buenos Aires, Part 3. Colección Científica INTA, Buenos Aires, pp. 57–81.

Cabrera, A.L. (1970). Gramíneas. In: Cabrera, A.L. (dir.), Flora de la Provincia de Buenos Aires, Part 2. Colección Científica INTA, Buenos Aires, pp. 1–624.

Cabrera, A.L. (1971). Compositae. In: Correa, M.N. (dir.), Flora Patagónica, Part 7. Colección Científica INTA, Buenos Aires, pp. 1–451.

Cabrera, A.L. (1978). Compositae. In: Cabrera, A.L. (dir.), Flora de la Provincia de Jujuy, Part 10. Colección Científica INTA, Buenos Aires, pp. 1–726.

Cabrera, A.L. (1983). Solanaceae. In: Cabrera, A.L. (dir.), Flora de la Provincia de Jujuy, Part 8. Colección Científica INTA, Buenos Aires, pp. 292–493.

Cabrera, A.L. & Willink, A. (1980). Biogeografía de América Latina. 2nd ed. Ser. Biol. Monogr. 13. Organización Estados Americanos, Washington.

Cadahia, D. (1970). Informe sobre el Plan Forestal Ganadero Pampa del Tamarugal. Study of FAO for IDB. CORFO, Chile.

Cano, E. (1988). Pastizales naturales de La Pampa, I: Descripción de las especies más importantes. Convenio AA-CREA-Provincia de La Pampa.

Castillo, U. (1966). Profundidad, sentido de escurrimiento y calidad química de la superficie freática del agua subterránea del Salar de Pintados. Report Corporación Fomento de la Producción, Departamento de Recursos Hidráulicos, Santiago. Chile.

Cauhépé, M.A. & Fernández Grecco, G.R. (1981). Dieta de vacunos en pastoreo sobre un pastizal natural de la depresión del Salado. Producción Animal 8: 85–95.

Cauhépé, M.A., Hidalgo, L.G. & Galataire, A. (1985). Aplicación de un índice de valorización zootécnica en pastizales de la Depresión del Salado. Rev. Argentina Producción Animal 5: 681–690.

Concha, R., Silva, M., Bonilla, S. & Cabrera, R. (1977). Uso del *Atriplex repanda* como refuerzo de una pradera natural mediterránea semiárida pastoreada con ovinos en períodos secos. Avances Producción Animal 2: 11–22.

Corporación de Fomento de la Producción (1982). Características básicas del tamarugo y el algarrobo. Soc. Agrícola Corporación de Fomento de la Producción, Santiago, Chile.

Corporación de Fomento de la Producción (1985). Nutritional assessment of tamarugo and algarrobo. Sheep and goat metabolic profiles at Tamarugal Pampa. In: Habit, M.A. (ed.), The current state of knowledge on *Prosopis tamarugo*. FAO, Regional Office for Latin America, Chile, pp. 71–115.

Correa, M.N. (1988). Frankeniaceae. In: Correa, M.N. (dir.), Flora Patagónica, Part 5. Colección Científica INTA, Buenos Aires, pp. 161–166.

Covas, G. (1985). Las especies pampeanas del género *Hordeum*. Apuntes para la flora de La Pampa 96: 381–388.

Cristi, A. & Gastó J. (1971). Alteraciones ambientales y del fruto en la germinación de *Atriplex repanda* Phil. Universidad de Chile, Facultad de Agronomía, Boletín Técnico 34: 25–40.

Cuatrecasas, J. (1958). Aspectos de la vegetación de Colombia. Rev. Academia Colombiana Ciencias Exactas Físicas Naturales 10: 221–264.

Cullen, D.A.S. de (1984). Portulacaceae. In: Correa, M.N. (dir.), Flora Patagónica, Part 4a. Colección Científica INTA, Buenos Aires, pp. 168–195.

Delhey, R. (1991). El chañar (*Geoffroea decorticans*, Leguminosae): Etnobotánica y utilización. Parodiana 6: 337–362.

Díaz, H.B. (1962). Especies arbóreas más comunes de la zonas ganaderas de la provincia de Tucumán que sirven de alimento al ganado. Turrialba 12: 195–199.

Díaz, H.B., Lagomarsino, E.D. & Prete, I.R. (1972). Determinación de la digestibilidad de especies forrajeras naturales comunes y de algunas cultivadas en la zona semiárida del noroeste argentino. Rev. Agronómica Noroeste Argentino 9: 55–68.

Di Marco, A.O. (1973). Consumo y preferencia ovina estacional al aumentar la intensidad del pastoreo en un bioma biestratificado can *Atriplex repanda*. Thesis Magister, Universidad de Chile, Santiago.

Dudal, R. & Purnell, M.F. (1986). Land resources: Salt affected soils. Reclam. Reveg. Res. 5: 1–9.

Elgueta, S.H. & Calderon, S. (1971). Estudio del tamarugo como productor de alimento del ganado lanar en la Pampa del Tamarugal. Instituto Forestal, Informe Técnico 38, Chile.

FAO-UNESCO (1971). Mapa Mundial de Suelos, Vol. 4: América del Sur. UNESCO, Paris.

Fernández, G. & Johnston, M. (1978). Efecto de la testa en la germinación de *Atriplex repanda* Phil., I: Entrada de agua a la semilla. Phyton 36: 97–102.

Fernández, G., Johnston, M. & Olivares, A. (1985). Rol del pericarpio de *Atriplex repanda* en la germinación, III: Estudio histológico y químico del pericarpio. Phyton 45: 165–171.

Fernández, O.A., Bóo, R.M. & Sánchez, L.F. (1989). South American shrublands. In: McKell, C.M. (ed.), The biology and utilization of shrubs. Academic Press, New York, pp. 25–59.

Flowers, T.J., Hajibagheri, M.A. & Clipson, N.J.W. (1986). Halophytes. Quart. Rev. Biol. 61: 313–337.

Fortunato, R. (1984). *Medicago*. In: Correa, M.N. (dir.), Flora Patagonica, Part 4b. Colección Científica INTA, Buenos Aires, pp. 244–251.

Frenguelli, O.G. (1928). Acerca del origen de los salares de la región de los desiertos de la Puna y de Atacama. Anales Soc. Arg. de Estudios Geográficos GAEA 3: 167–186.

Gargano, A.O. (1978). Influencia de algunas variables de suelos del Norte Chico en el crecimiento inicial de *Atriplex repanda* Phil. Thesis Magister, Universidad de Chile, Santiago.

Gastó, J. & Caviedes, E. (1976). Interferencia intraespecífica de *Atriplex repanda* en el secano mediterráneo de Chile. Universidad de Chile, Facultad de Agronomía, Boletín Técnico 41: 3–18.

Gastó, J. & Contreras, D. (1972). Bioma pratense de la región mediterránea de pluviometría limitada. Universidad de Chile, Facultad de Agronomía, Boletín Técnico 35: 3–29.

Gastó, J. & Olivares, A. (1979). Analisis cuantitativo de la arquitectura de *Atriplex repanda* Phil. Ciencia Investigación Agraria 6: 105–113.

Gihad, E.A. & El Shaer, H.H. (1994). Utilization of halophytes by livestock on rangelands: Problems and prospects (this volume, Ch. 6).

Giusti, L. (1967). Chenopodiaceae. In: Cabrera, A.L. (dir.),

Flora de la Provincia de Buenos Aires, Part 3. Colección Científica INTA, Buenos Aires, pp. 81–127.
Giusti, L. (1984). Chenopodiaceae. In: Correa, M.N. (dir.), Flora Patagónica, Part 4a. Colección Científica INTA, Buenos Aires, pp. 99–137.
Gunckel, L. (1970). Revisión sistemática de las especies chilenas del género *Frankenia* L. An. Museo Historia Nat. Valparaíso, Chile 3: 7–52.
Habit, M.A. (1981). *Prosopis tamarugo*: arbusto forrajero para zonas áridas. FAO, Plant Production and Protection Publication, No. 25, Rome.
Hernández, P. (1991). Estrategias de crecimiento y reproducción de *Tessaria absinthioides*. Maleza de distritos de riego. Thesis Magister, Universidad Nacional del Sur, Bahía Blanca, Argentina.
Hitchcock, A.S. (1950). Manual of the grasses of the United States, 2nd. ed., revised by A. Chasse, USDA, Washington, DC.
Hunziker, J.H. (1952). Las comunidades vegetales de la cordillera de La Rioja. Rev. Investigaciones Agrícolas 6: 167-196.
Instituto Forestal (1964). Posibilidades de reforestación en la Pampa del Tamarugal. Report to Corporación de Fomento de la Producción, Chile.
Johnston, M. & Fernández, G. (1978). Efecto de la testa en la germinación de *Atriplex repanda* Phil., II: Determinación de la intensidad respiratoria. Phyton 36: 103–109.
Johnston, M. & Fernández, G. (1979). Rol del pericarpio de *Atriplex repanda* en la germinación, I: Efecto del lavado de los frutos en agua. Phyton 37: 145–151.
Karlin, U. & Díaz, R. (1984). Potencialidad y manejo de algarrobos en el árido subtropical argentino. Programa Nacional de Recursos Naturales Renovables, Sec. Ciencia y Técnica, Ministerio de Educación y Justicia, Argentina.
Lailhacar, S. (1976). Effect of soil parameters on the components of biomass production in *Atriplex polycarpa* (Torr.) Wats. and *Atriplex repanda* Phil. Ph.D. Thesis. University of California, Davis.
Lailhacar, S. (1990). Evaluación nutritiva de los recursos forrajeros nativos y naturalizados de la I Región. Avances Producción Animal 15: 61–80.
Lailhacar, S. & Laude, H.M. (1975). Improvement of seed germination in *Atriplex repanda* Phil. J. Range Manage. 28: 491–494.
Lailhacar, S. & Padilla, F. (1987). Selection de provenances d'*Atriplex repanda* Phil., une ressource fourragere pour la prairie de la zone a climat méditerranéen aride du Chile. Proc. 5th. Meeting FAO-European Sub-network on Mediterranean Pastures and Fodder Crops, pp. 68–72.
Le Houérou, H.N. (1992). The role of saltbushes (*Atriplex* spp) in arid land rehabilitation in the Mediterranean Basin: A review. Agroforestry Systems 18: 107–148.
Le Houérou, H.N. & El Barghati, M.S. (1982). Shrub evaluation in the Benghazi Plains. Tech. paper No. 45, UNTF Lib. 18, FAO, Agr. Res.Center, Tripoli, Libya.
Leighton, G., Gastó, J. & Cristi, A. (1975). Establecimiento de arbustos forrajeros nativos y exóticos de Chile: Incidencia de la época, profundidad y dosis de siembra en secano mediterráneo semiárido. Ciencia Investigación Agraria 2: 41–52.
León, R.J.C. (1975). Las comunidades herbáceas de la región Castelli-Pila. Comisión Inv. Científicas (Argentina), Monografía 5: 73–109.
Llerena, F.A. (1994). Massive propagation of halophytes (*Distichlis spicata* and *Tamarix* spp.) on the highly saline-alkaline soils in the ex-Lake Texcoco, Mexico (this volume, Ch. 30).
Lorini, J., Geyger, E. & Liberman, M. (1984). Ecofisiología de algunas halófitas en un ambiente especial del altiplano central de Bolivia. Ecología Bolivia 5: 1–28.
Luzio, W. & Vera, W. (1982). Caracterización de los suelos en los ecosistemas de la I Región. In: Análisis de los ecosistemas de la I Región. CORFO-SACOR, Universidad de Chile. Santiago, Chile, pp. 45–119.
Molina Sánchez, D. (1977). Especies autóctonas de la flora patagónica. Boletín divulgación, INTA-EERA Bariloche.
Montes, L. (1988). Revisión bibliográfica. *Lotus tenuis*. Rev. Argentina Producción Animal 8: 367–376.
Morello, J. (1958). La Provincia fitogeográfica del Monte. Opera Lilloana 2: 1–155.
Mota Urbina, J.C. (1979). Determinación del rango de tolerancia al ensalitramiento por el pasto salado *Distichlis spicata* (L.) Greene, en suelos del ex Lago de Texcoco. Ciencia Forestal 4: 21–44.
Múlgura de Romero, M.E. (1981). Contribución al estudio del género *Atriplex* (*Chenopodiaceae*) en la Argentina, I. Darwiniana 23: 119–150.
Múlgura de Romero, M.E. (1982). Contribución al estudio del género *Atriplex* (*Chenopodiaceae*) en la Argentina, II: Las especies adventicias. Darwiniana 24: 49–68.
Múlgura de Romero, M.E. (1984). Contribuciones al estudio del género *Atriplex* (*Chenopodiaceae*) en la Argentina, III. Darwiniana 25: 235–253.
National Academy of Sciences (1979).Tropical legumes: Resources for the future. Washington, DC.
National Academy of Sciences (1984). Nutrient requirements of beef cattle, 6th. ed. Nat. Acad. Press, Washington, DC.
Nicora, E.G. (1978). Gramineae. In: Correa, M.N. (dir.), Flora Patagónica, Part 3. Colección Científica INTA, Buenos Aires, pp. 1–563.
Nicora, E.G. & Rúgolo de Agrasar, Z.E. (1987). Los géneros de gramíneas de América Austral. Hemisferio Sur, Buenos Aires.
Olivares, A. & Gastó, J. (1981). *Atriplex repanda*. Organización y manejo de ecosistemas con arbustos forrajeros. Ciencias Agricolas No. 7, Universidad de Chile, Santiago, Chile.
Olivares, A. & Gastó, J. (1983). Epoca y frecuencia de utilización en *Atriplex repanda* (Phil.). Avances Producción Animal 8: 29–35.
Olivares, A., Johnston, M., Fernández, G. & Pagliaricci, H. (1985). Características del crecimiento en *Atriplex repanda* Phil. Phyton 45: 173–181.
Palacios, R. (1987). *Polygonum*. In: Burkart, A.R. (dir.), Flora ilustrada de Entre Ríos, Part 3. Colección Científica INTA, Buenos Aires, pp. 69–89.
Parodi, L.R. (1930). Ensayo fitogeográfico sobre el partido de Pergamino. Rev. Fac. Agronomía Veterinaria (Univ. Bs. As.) 7: 65–271.
Passera, C.B. & Borsetto, O. (1989). Aspectos ecológicos de

Atriplex lampa. Investigación Agraria. Producción y Protección Vegetales 4: 179–198.

Pedersen, T.M. (1987). Amaranthaceae. In: Burkart, A.R. (dir.), Flora ilustrada de Entre Ríos, Part 3. Colección Científica INTA, Buenos Aires, pp. 160–203.

Pelliza-Sbriller, A., Bonino, N.A., Bonvissuto, G.L. & Amaya, J.N. (1984). Composición botánica de la dieta de herbívoros silvestres y domésticos en el área de Pilcaniyeu (Río Negro). IDIA 429–32: 63–73.

Pensiero, J.F. (1986). Revisión de las especies argentinas del género *Pappophorum* (Gramineae – Eragrostoideae – Pappophoreae). Darwiniana 27: 65–87.

Peralta, F. (1985). Water resources at the Tamarugal Pampa. In: Habit, M.A. (ed.), The current state of knowledge on *Prosopis tamarugo*. FAO, Regional Office for Latin America, Chile, pp. 53–64.

Pontiroli, A. (1965). Plantaginaceae. In: Cabrera, A.L. (dir.), Flora de la Provincia de Buenos Aires, Part 5. Colección Científica INTA, Buenos Aires, pp. 331–342.

Pyykkö, M. (1966). The leaf anatomy of East Patagonian xeromorphic plants. Ann. Bot. Fennici 3: 453–622.

Ragonese, A.E. (1951). La vegetación de la República Argentina, II: Estudio fitosociológico de las Salinas Grandes. Rev. Investigaciones Agrícolas 5: 1–233.

Ragonese, A.E. (1967). Vegetación y ganadería en la República Argentina. Colección Científica INTA 5, Buenos Aires.

Ragonese, A.E. (1985). Forrajeras nativas. In: El desarrollo de las forrajeras en la región pampeana. Technical Meeting INTA-EEA Pergamino, pp. 9–26.

Ragonese, A.E. & Covas, G. (1947). La flora halófila del sur de la Provincia de Santa Fé. Darwiniana 7: 401–496.

Ragonese, A.E. & Piccinini, B.G. (1977). Consideraciones sobre las Salinas de Mascasín (La Rioja, San Juan, República Argentina). Darwiniana 21: 49–60.

Rhodes, D. & Felker, P. (1988). Mass screening of *Prosopis* (mesquite) seedlings for growth at seawater salinity concentrations. Forest Ecol. Manage. 24: 169–176.

Rocha Garciá, E. (1983). Estudio agronómico nutritivo y análísis de semilla de cauchi (*Suaeda foliosa*) en dos épocas de siembra. VII Reunión Nacional de Pastos y Forrajes and V Reunión Nacional de Ganadería, Potosí, Bolivia, pp. 205–209.

Rodríguez, E. & Gastó, C.J. (1972). Estudio del efecto del consumo de pradera natural bistratificada con *Atriplex repanda* en el período de encaste de ovinos. Report Departamento Producción Animal, Facultad de Agronomía, Universidad de Chile.

Roig, F. (1970). La Reserva Forestal de Ñacuñán. Deserta 1: 5–239.

Rosas, M. (1989). El género *Atriplex* (Chenopodiaceae) en Chile. Gayana 46: 3–82.

Rossow, R. (1984a). *Melilotus*. In: Correa, M.N. (dir.), Flora Patagónica, Part 4b. Colección Científica INTA, Buenos Aires, pp. 252–256.

Rossow, R. (1984b). *Trifolium*. In: Correa, M.N. (dir.), Flora Patagónica, Part 4b. Colección Científica INTA, Buenos Aires, pp. 268–282.

Ruiz Leal, A.R. (1972). Flora Popular Mendocina. Deserta 3: 9–296.

Ruthsatz, B. (1978). Las plantas en cojín de los semidesiertos andinos del Noroeste Argentino. Darwiniana 21: 491–539.

Salm, H. & Gehler, E. (1987). La salinización del suelo en el altiplano central de Bolivia y su influencia sobre la cobertura vegetal. Ecología Bolivia 10: 37–48.

Sánchez, M.I. (1984). *Lotus*. In: Correa, M.N. (dir.), Flora Patagónica, Part 4b. Colección Cientifica INTA, Buenos Aires, pp. 237–238.

Sayago, M. (1969). Estudio fitogeográfico del Norte de Córdoba. Bol. Acad. Nac. Ciencias Córdoba 46: 123–427.

Schulz, A.G. (1963). Plantas y frutos comestibles de la región chaqueña. Rev. Agronómica Noroeste Argentino 4: 57–83.

Scoppa, C.O. & Di Giacomo, R.M.S. (1985). Distribución y características de los suelos salinos y/o alcalinos en Argentina. IDIA 437–440: 49–77.

Silva, F. & Pereira, C. (1976). Aislación y composición de las proteínas de hojas de *Atriplex nummularia* y *A. repanda*. Ciencia Investigación Agraria 3: 169–174.

Silva Colomer, J. & Passera, C.B. (1990). The nutritional value of *Atriplex* spp. as fodder for arid regions. J. Arid Environm. 19: 289–295.

Somlo, R., Durañona, C. & Ortiz, R. (1985). Valor nutritivo de especies forrajeras patagónicas. Rev. Argentina Producción Animal 5: 589–605.

Squella, F. (1986). Forage and fuel production from salt affected wasteland in Chile. Reclam. Reveg. Res. 5: 41–48.

Squella, F., Meneses, R. & Gutiérrez, T. (1985). Evaluación de especies forrajeras arbustivas bajo condiciones de clima mediterráneo árido. Agricultura Técnica 45: 303–314.

Steibel, P.E. (1986). Las Quenopodiáceas de la Provincia de La Pampa. Rev. Fac. Agronomía (Univ. Nac. La Pampa) 2: 13–38.

Sudzuki, F. (1969). Absorción foliar de humedad atmosférica en tamarugo, *Prosopis tamarugo* Phil., III: Absorción foliar de agua atmosférica en tamarugo. Universidad de Chile, Facultad de Agronomía, Boletín Técnico 30: 10–23.

Sudzuki, F. (1975). Captación y economía del agua en plantas que viven en ambientes de desierto. Universidad de Chile, Facultad de Agronomía, Boletín Técnico 40: 38–47.

Sudzuki, F., Botti, C. & Acevedo, E. (1973). Relaciones hídricas del tamarugo (*Prosopis tamarugo* Phil.) en la localidad de Canchones. Universidad de Chile, Facultad de Agronomía, Boletín Técnico 37: 3–23.

Trobok, S. (1985). Fruit and seed morphology of Chilean *Prosopis* (Fabaceae-Mimosoideae). In: Habit, M.A. (ed.), The current state of knowledge on *Prosopis tamarugo*. FAO, Regional Office for Latin America, Chile, pp. 227–241.

Türpe, A.M. (1960). Las especies argentinas de *Festuca* (excluídas las patagónicas). Darwiniana 15: 189–283.

Ulibarri, E.A. (1984). *Hoffmannseggia*. In: Correa, M.N. (dir.), Flora Patagónica, Part 4b. Colección Científica INTA, Buenos Aires, pp. 220–223.

Ungar, I.A. (1974). Halophyte Communities of Park County, Colorado. Bull. Torrey Bot.Club 101: 145–152.

Vera, P. (1977). Composición química y mineral de *Atriplex repanda y Atriplex lentiformis*. Ciencia Investigación Agraria 4: 141–145.

Verettoni, H. (1961). Las comunidades halófilas de la región de Bahía Blanca. Author's edition.

Vervoorst, F.B. (1967). La vegetación de la República Argentina: Las comunidades vegetales de la Depresión del Salado (Pcia. Buenos Aires). Serie Fitogeográfica INTA 7, Buenos Aires.

Villamil, C.B. (1969). El género *Monanthochloë* (Gramineae). Kurtziana 5: 369–391.

Wainstein, P. & González, S. (1971a). Valor nutritivo de plantas forrajeras del este de la provincia de Mendoza (Reserva Ecológica de Ñancuñán), I. Deserta 2: 67–75.

Wainstein, P. & González, S. (1971b). Valor nutritivo de plantas forrajeras del este de la provincia de Mendoza (Reserva Ecológica de Ñancuñán), II. Deserta 2: 77–85.

Wainstein, P., González, S. & Rey, E. (1979). Valor nutritivo de plantas forrajeras de la provincia de Mendoza, III. Cuaderno Técnico (IADIZA) 1: 97–108.

Wilson, A.D. (1994). Halophytic shrubs in semi-arid regions of Australia: Value for grazing and land stabilization (this volume, Ch. 8).

Yensen, N.P., Fontes, M.R., Glenn, E.P. & Felger, R.S. (1981). New salt tolerant crops for the Sonoran Desert. Desert Plants 3: 111–118.

14. Halophytic coastal marsh vegetation in East Africa

R.K. RUWA
Kenya Marine & Fisheries Research Institute, Mombasa, Kenya

Abstract. The Indian Ocean coasts of the East African mainland and the Madagascar subcontinent have estuarine and brackish water coastal marshes that support significant vegetation which is predominantly mangrove. The anthropogenic impacts on mangrove vegetation in all these countries are similar and are mostly due to urbanization, sewerage disposal, solid waste and toxic waste disposal, threats due to oil pollution, dredging operations, damming of rivers, salt and prawn farming, and over cutting of wood for building materials and for fuel. This paper reviews the distribution and utilization of coastal marsh vegetation of the East African region with special reference to mangroves.

Key words: Agroforestry, Aquaculture, Biodiversity, Fuelwood, Irrigation, Mangroves, Review, Salinity, Salt marsh

Introduction

The vegetation along the coasts of the East African mainland is predominantly a forest savanna mosaic in some instances alternating with dry savanna, except in Somalia where it is classified as sub-desert steppe (Figure 1).

A feature of the coastline is the extensive area of estuarine and brackish water coastal marshes that support significant vegetation which is predominantly mangrove (Figure 2).

Because of the aridity, nomadic grazing is a common land use, notably in Kenya and Somalia on the East African mainland and on parts of Madagascar, especially the north and south of the island where the vegetation is typically wooded steppe (Figure 3).

The rapid rise in human population in the East African region requires increased food production, shelter and fuel wood. Climatic conditions do not favour rainfed agriculture in most of the region and therefore food production can only be increased through irrigation and aquaculture.

The region is not well endowed with fresh water supplies and a significant proportion of the underground water is saline. In some areas the meagre supply of freshwater is reserved for domestic use. Water potentially available for irrigation is saline. Successes have already been reported in some parts of the world where cultivation, of halophytes on saline soils or using saline water has been undertaken (Glenn & Watson 1993). Most success has been from using indigenous plant species. The halophytic coastal marsh vegetation of East Africa represents a major resource.

This paper reviews the distribution and utilization of coastal marsh vegetation in the East African region with special reference to the mangrove communities.

Coastal marsh vegetation

Coastal saltmarsh plants have been categorized into various associations (Schimper 1891) whose structural distinctiveness however may vary regionally depending on climatic factors (Macnae

V.R. Squires & A.T. Ayoub (eds.), *Halophytes as a resource for livestock and for rehabilitation of degraded lands*, 201–210.

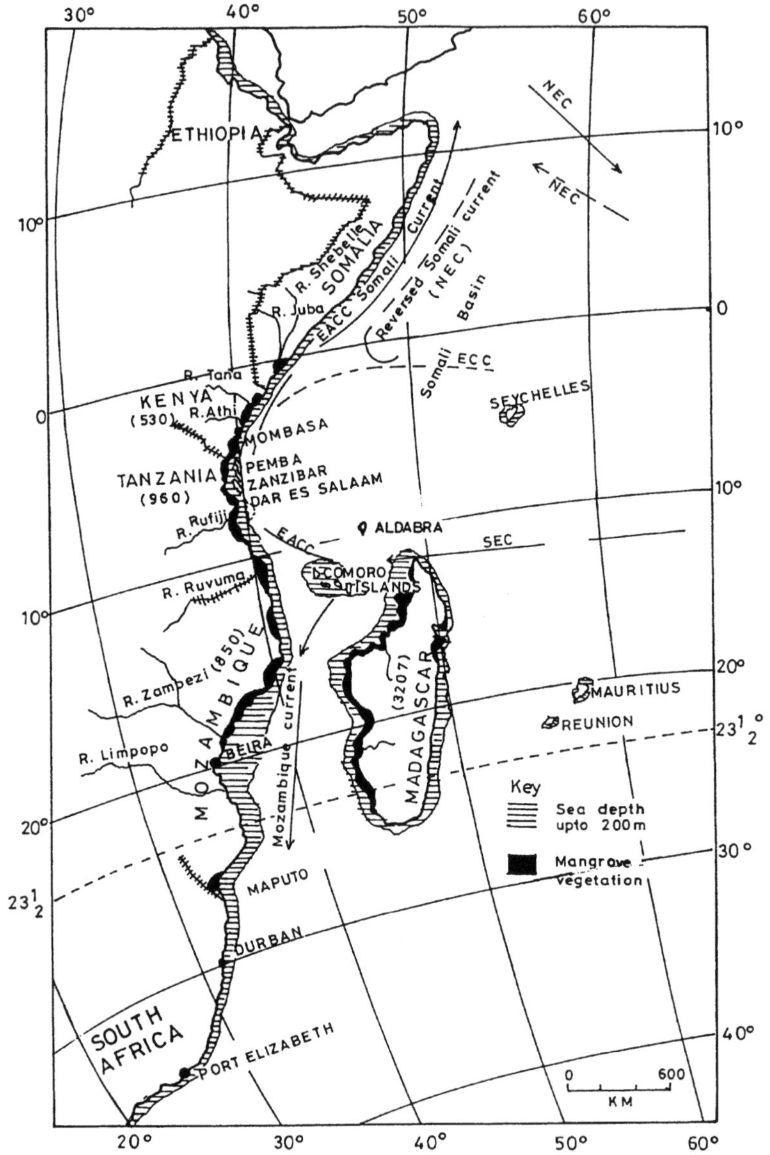

Figure 1. Distribution of mangrove vegetation along the coasts of the African mainland facing the Indian Ocean and its islands. The total vegetation cover (km^2) in each country is shown by the number in parenthesis.

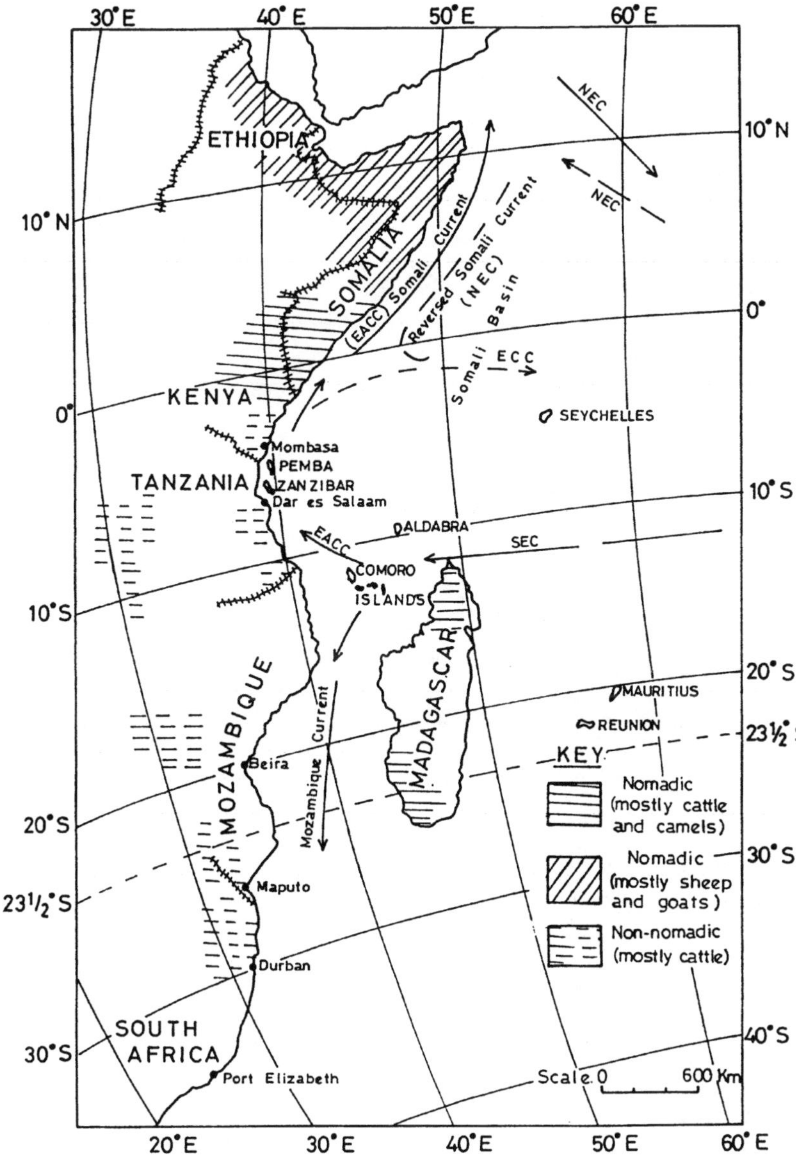

Figure 2. Patterns of livestock keeping in the East African region.

1968). According to Schimper (1891) there are four categories namely: (a) the mangal i.e. mangrove tree species and non-woody herbaceous plants; (b) the Nypa Association which occurs to the landward and upstream of a mangal and is dominated by the rhizomatous *Nypa* with isolated trees e.g. *Heritiera*; (c) the Barringtonia Association which may be dominated by either *Barringtonia* when it occurs behind a Pes-caprae Association, or by *Barringtonia*, *Heritiera* spp if the association occurs behind a mangal; (d) the Pes-caprae Association which is dominated by *Ipomea pes-caprae* Roth and may be mixed with *Casuarina* and other woody halophytes. Macnae (1968)

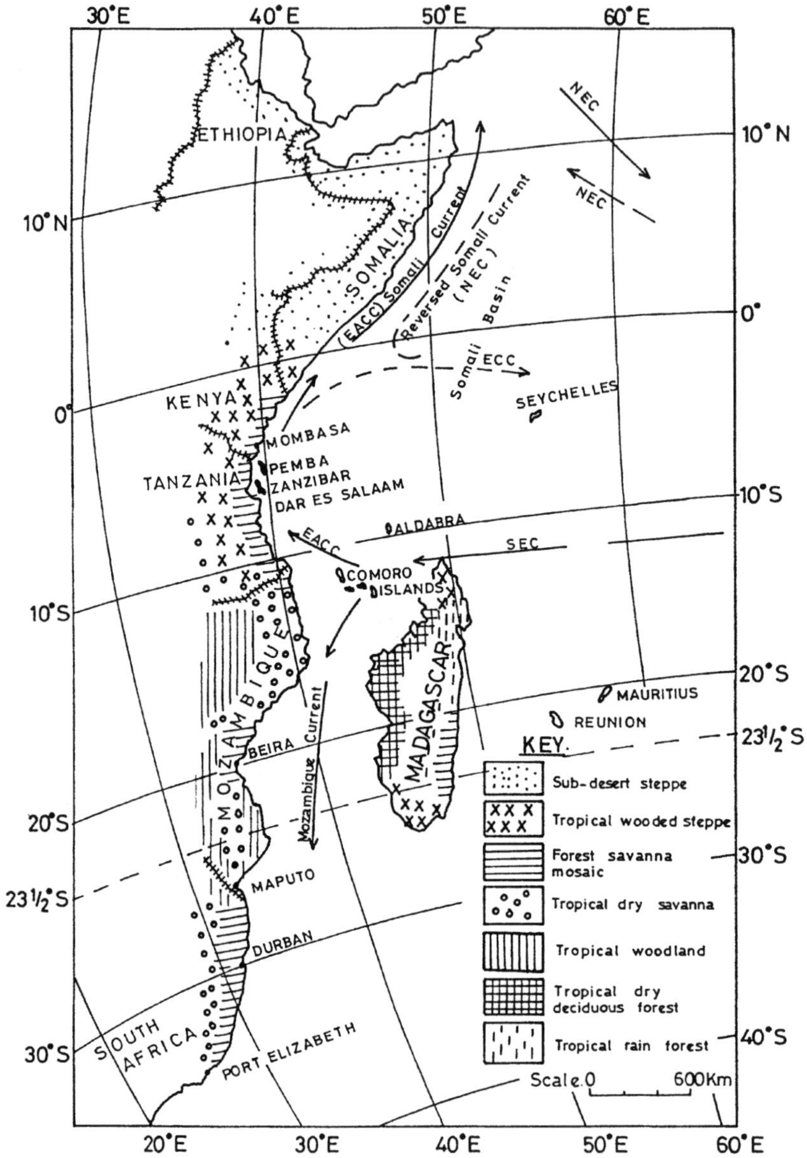

Figure 3. Types of terrestrial vegetation in the East African region.

introduced a further category, the Saltwort Association which is typical in situations where evaporation comes to exceed availability of freshwater. The Saltwort association is dominated by bushes or low shrubs of species of perennial *Arthrocnemum* and the annual *Salicornia*.

Unlike the mangal which is always littoral the Pes-caprae and barringtonia associates are always supralittoral according to the shore terminology of Lewis (1964). The Nypa association tends to occur from the ordinary spring tides and above extreme high water mark and is therefore more supralittoral in its occurrence.

Macnae (1968) recognized two associations for

the Indo-west Pacific which includes the Eastern Africa region. The two distinct associations are:

1. The Ipomea pes-caprea Association: It extends from semi-desert shores to ever wet areas where rainfall may be abundant at all seasons. In humid tropics this may be associated with a Barringtonia associations;
2. The Mangal as the principal vegetation with the Nypa association, Barringtonia association and Saltwort association categories as mangal associates: The Nypa association and Barringtonia association are typical mangal associations in areas under freshwater domination where evaporation and transpiration are more than balanced by a high rainfall and/or supply of freshwater from large rivers. In arid areas however the Nypa and Barringtonia associations are not distinct in their structural composition. The saltwort association occurs where evaporation and transpiration are in excess of the supply of fresh water.

The mangal

The estuarine and brackish water coastal marshes in East Africa predominantly support mangroves. Mangroves are evolutionarily terrestrial plants that adapted to life in estuarine brackishwater and saline biotopes (Chapman 1976). They are still not fully aquatic plants and survive in semi-terrestrial or semi-aquatic conditions that are well provided for in the intertidal zone, making them true intertidal flora with respect to their rooting whereas their canopy is always above the high water tide and therefore exhibiting a terrestrial bias. *Avicennia* and *Rhizophora* genera are argued to be the first mangrove genera to evolve (Chapman 1976).

The adaptations of mangroves to estuarine brackishwater and saline biotopes have been variously well documented (e.g. Chapman 1977a) and from this knowledge of their certain terrestrial and aquatic qualities, clues into the practical possibilities of how to undertake mangrove agro forestry have been derived. From our knowledge of the natural history of mangroves it is necessary that consideration of mangroves as a resource should be done from two points of view. First as a land based resource and secondly as a water based resource. The foregoing review was done.

Geographical distribution

There are eight species of mangroves in the Eastern African region. These are *Sonneratia alba* J. Sm., *Rhizophora mucronata* Lam., *Bruguiera gymnorhiza* (L.) Lam K., *Ceriops tagal* (Perr.) C.B. Rob., *Xylocarpus granatum* Koening, *Avicennia marina* (Forst.) Vierh, *Heritiera littoralis* (Dryand.) Ait. and *Lumnitzera racemosa* Presl.

The distribution of these species and the total mangrove cover in each country in the region is summarized in Table 1. The ocean current systems shown in Figure 1 are responsible for the distributions of the mangrove species and other halophytes with floating seeds in the region.

Except for *Heritiera littoralis* the small islands and atolls off the coasts of East Africa mainland support the same species of mangroves encountered on the coasts of the East African mainland and the Madagascar coastline. Seychelles, being a country composed of several islands and an atoll supports a higher number of mangrove species than other country or islands. The Aldabra Atoll has the highest species richness of mangrove species and supports all the species in Table 1 except *Heritiera littoralis* (Macnae 1971).

Along the coasts of the East African mainland, disparities in the distribution of species of mangroves is evident just as similarly shown in the mangrove vegetation cover from country to country (Figure 1). Mozambique and Tanzania support the most significant mangrove vegetation cover on the mainland but the largest mangrove cover in the region is found in Madagascar. The Durban area (South Africa) which represents the southern most extent of the mangrove distribution supports only 3 species (Day & Morgans 1956). The southernmost extent of the mangrove distribution is inhibited by the change of the climatic conditions to sub-tropical type (Chapman 1977a). The arid areas in Somalia supports a very poor mangrove vegetation structure which is confined around the estuary of R. Juba/Shebele (Figure 1). The lack of significant mangrove vegetation may be explained by the lack of estuarine or brackishwater conditions to provide the lower salinities required by the seeds to root. Like with other halophytes, mangrove seeds require salinities lower than seawater for them to produce roots and grow (Ungar 1982) but thereafter the tolerance to higher salinities increases with age

Table 1. Distribution of mangrove species and cover in the East African region: Seychelles (Sey), Mauritius (Mau), Reunion (Reu), South Africa (S. Af. Durban), Madagascar (Mad.), Mozambique (Moz), Tanzania (Tan), Kenya (Ken) and Somalia (Som) including crab and prawn fisheries production. MSY = maximum sustainable yield

	Sey	Mau.	Reu.	S. Afr.	Mad.	Moz.	Tan.	Ken.	Som.
1. *Sonneratia alba*	+	–	–	–	+	+	+	+	–
2. *Rhizophora mucronata*	+	+	–	–	+	+	+	+	+
3. *Bruguiera gymnorhiza*	+	+	–	+	+	+	+	+	–
4. *Ceriops tagal*	+	–	–	+	+	+	+	+	–
5. *Xylocarpus granatum*	+	–	–	–	+	+	+	+	–
6. *Avicennia marina*	+	–	–	+	+	+	+	+	+
7. *Heritiera littoralis*	–	–	–	–	+	+	+	+	–
8. *Lumnitzera racemosa*	+	–	–	–	+	+	+	+	–
Total number of species	7	2	–	3	8	8	8	8	2
Total Mangrove area (km^2)	–	–	–	–	3207	850	960	530	
Annual MSY (tonnes) for crabs	–	–	–	–	5760	1530	1728	1057	–
Annual landings of prawns (tonnes)	–	–	–	–	1700	1000	979	350	–

Source: Mangrove species: Seychelles (Macnae 1968, 1971); Mauritius (Fagoonee 1990); South Africa (Day & Morgans 1956); Madagascar (Macnae 1968); Mozambique (Macnae & Kalk 1962); Tanzania (Chapman 1977b), Kenya (Isaac & Isaac 1968), Somalia (Chapman 1977b); Mangrove area: Madagascar, Mozambique and Tanzania (SW10P 1989a); Kenya (Doute et al. 1981); Crab and prawn fishery yields: Madagascar, Mozambique, Tanzania and Kenya (SW10P 1989a,b; 1990).

of the plant. The ecological significance of this inhibition is that root damage during the dispersal period is avoided (Ungar 1982). In the semi-arid Kenyan coast it has been observed that most of the major mangrove vegetation cover occurs far away from estuaries but such areas receive submarine groundwater discharges that create the required lower salinities for rooting of the seeds and development (Ruwa and Polk 1986). The coastal and maritime halophytes and the saltmarsh saltwort associations have not yet received much attention in this region except for records in Mozambique by Macnae and Kalk (1962). The supralittoral fringe supports *Juncus* spp, clumps of *Acrostichum* spp. *Chenolea* spp. and *Sporobolus* spp. Where a pes-caprae Association occurs, it is characterized by *Ipomea pes-caprae* L. (Sweet), *Vigna* spp, *Casuarina* spp and *Pandanus* spp. The saltwort associations in this region are mainly composed of the following: *Sesuvium* spp, *Salicornia* spp and *Arthrocnemum* spp.

Socio-economics

Structural labour divisions found in communities living around mangrove forests allow, to varying degrees, members of such communities to generate incomes from consumptive and non-consumptive use of mangrove resources. For example, not everyone is a woodcutter, a charcoal burner, a honey gatherer, a boat builder, fisherman or a livestock farmer. In such communities there are different experts for these different occupations and who ever demands their services pays, thus minimizing competition if everybody was a Jack of all trades. At a such humble organization centered on the mangroves it is almost axiomatic to postulate that economic derivatives from man-mangrove interaction have been nurtured and cherished since the time such communities started living around such biotopes. A further initimate man-mangrove interaction is the demarcation of certain portions of mangrove forests for use as tribal shrines for worshipping, giving offerings and where traditional medicinemen sometimes treat their clients for special ailments.

Impacts

The mangrove biotopes in Eastern Africa are threatened by: (a) pollution mostly from industrial wastes, solid wastes and sewage disposals from urban areas, oil pollution from ships and storage tanks in mangrove creeks; (b) physical destruction of the habitat: through overcutting of mangroves which consequently increase shore erosion; clearing of mangroves or creation of salt farms and prawn ponds; reclamation of salt marshes for housing estates, harbour development, etc.; (c) diversion of rivers to create dams for production of electricity has caused salinity stresses in estuaries. The activities that presently threaten mangrove vegetation are not the historical types of activities practiced by village communities that have interacted and coexisted with mangroves (Ruwa 1985) but new developments unrelated to the lifestyles of the village communities e g. industries, hotels, harbours that have brought an influx of immigrants to provide labour etc. Such activities have transformed the villages to urban areas causing increased exploitation of mangroves for wood and degradation of the environment. The degradation of the environment further reduces the chances of successful rehabilitation of destroyed areas and reforestation of new areas.

In situ and exsitu management of mangroves

The mangrove forests in the East African region are not adequately protected by any form of law and there are no management guidelines being practiced except for a few pockets of mangroves that benefit from protection through tribal religious practices that prohibit exploitation of trees from shrines. The mangrove cutters come from villages surrounding the mangrove forests although the middlemen who buy the mangrove wood include residents from elsewhere. Honey gatherers freely keep their beehives in the forests.

There is need that management strategies are geared to:

a) discourage environmental degradation of marshes e.g. through pollution, overexploitation of mangroves, unplanned reclamations etc.;
b) providing legal provisions that will allow prosecution of abusers of marsh biotopes and make them responsible for rehabilitation of such places they have destroyed even if that has happened by accident;
c) maximize economic benefits for surrounding villagers, e.g. using profits from park charges and allowing non-consumptive type of uses like honey gathering, aiding non-destructive type of aquaculture projects like oyster farming, cage culture, algae culture etc. so that they can appreciate the extensive values of mangroves;
d) encourage replanting programmes by villagers by providing them with seedlings and providing tokens of appreciation to those who will look after their trees carefully;
e) initiate agroforestry programmes aimed at commercial exploitation for fodder and other forest products. Besides agroforestry, initiate plantations of herbaceous halophytes, e.g. *Sesuvium*, *Salicornia*, *Juncus* etc. The most preferred mangrove species for fodder by livestock is *Avicennia marina* which has further advantages due to having wider salinity tolerance and temperature ranges;
f) introduce and encourage use of mangrove species and other halophytes for landscaping, thereby also allow their ex-situ preservation while at the same time enhancing panoramic views. Tree heights of mangroves can even be controlled by using irrigating water of appropriate salinities. Being evergreen and halophytic, mangroves are pleasant plants for landscaping in semiarid and arid area where vegetation is scanty and soil salinities are prone to changes when irrigation is practiced. Such strategies will ensure substainable use of mangrove vegetation and other halophytes while allow marsh biotopes to maintain their functions, encourage eco-tourism with economic benefits to surrounding populations arising from agroforestry.

Towards agroforestry

There are no agroforestry plantations using mangrove species in this region. Overcutting indicates that the output can not meet the demand. The high demand for mangrove wood for various uses is a clear manifestation that mangrove wood is popular and therefore marsh biotopes that support them should not be considered as wastelands. Both terrestrial and intertidal biotopes can be used for mangrove plantations but terrestrial plantations require irrigation. In most coastal areas, the underground water is saline and can be used for irrigation of halophytic plants. Saline underground water aquifers are therefore important resources in the farming of halophytic plants. Both inland and semi arid areas with sa-

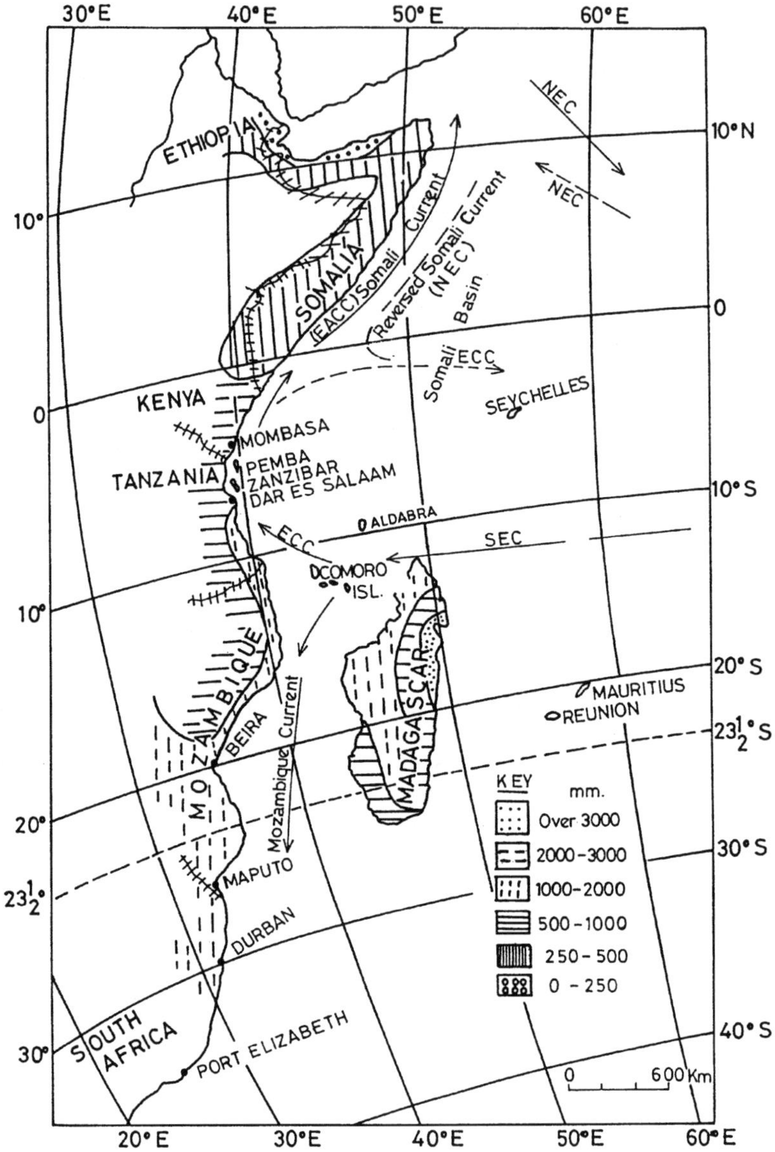

Figure 4. Rainfall zones in the East African region based on mean annual rainfall.

line or brackish water can be utilized for farming of halophytes.

Nomadic grazing is common especially in Kenya and Somalia on the African mainland and in some parts of Madagascar.

On Madagascar Island the west coast supports most of the mangrove vegetation which is characterized by tropical dry deciduous forest. The eastern side supports a tropical rainforest and a tropical woodland savanna.

Potential areas for fodder cultivation. The coastal

semi-arid and arid areas from northern coast of Kenya to Somalia where the vegetation is too sparse for adequate grazing offer high potential areas for fodder cultivation and grazing by livestock. This will reduce nomadic pastoralism and preserve the delicate vegetation structures in these drylands. The herbaceous halophytes and the woody halophyte *Avicennia marina* are suitable fodder plants for these areas because of their wide temperature and salinity tolerance ranges. The semi-arid western coast of Madagascar has also a high potential for fodder cultivation and grazing thereby also reduce dangers of habitat destruction through nomadic pastoralism on the island and to maintain the uniqueness of its biodiversity.

Potential areas for timber plantations. The halophytic mangroves grow luxuriantly to large and tall sizes in places with high rainfall (e.g. Macnae 1968). In arid and semi-arid areas they are dwarfed and remain mostly as shrubs and do not provide high quality wood due to high soil salinity but are, however, useful for fodder and landscaping. For timber production, irrigation of plants with saline water is unsuitable if there is not enough freshwater to flush out any build up of salts in the soil. For mangroves, successful plantations for production of timber can be located in the intertidal zones where the tide helps to flushout salts daily. In the Eastern Africa region the tides which are described as macrotidal because the tidal range is more than 3 m (Stoddart 1971) low gradient shores provide large area for planting of mangroves.

The potential areas for planting mangroves for timber which have adequate rainfall in this region are coasts of Southern Tanzania, Mozambique and eastern Madagascar (Figure 4). The preferred species for wood products in this region are: *Rhizophora mucronata*, *Bruguiera gymnorhiza* and *Ceriops tagal*. In saline soils in the supralittoral zone and further inland *Casuarina* spp and *Eucalyptus* can be grown. *Casuarina* spp and *Eucalyptus* spp thrive well in this region and can also be grown in the semi-arid areas such as those in Kenya where they equally provide wood of good quality.

Besides wood production, plantations of mangroves in the intertidal zone provide further advantages e.g. (a) the trees increase the species richness or biodiversity in the intertidal zone (Ruwa 1990); (b) the increase in mangrove vegetation cover increase crab and prawn catches (Martosubroto and Naamin 1977) as also indicated in Table 1; (c) reduce shoreline erosion.

Conclusions

The need to foster the cultivation of coastal halophytic plants to provide fodder and wood requirements to meet demands placed by burgeoning human populations.

Because halophyte farming is new to the region, some pilot projects should be set up. These will form the nuclei for demonstration farms in the region. There is a clear need for international cooperation at both global and regional level to ensure the transfer of appropriate technology. Rewards for such efforts will not only be the increase in food, fodder, fuelwood, timber and honey production but also enhanced fisheries production.

To generate more public awareness about halophytes evergreen herbaceous or grass halophytes can be used for landscaping purposes in streets and public parks. These can be irrigated with saline water. Other ways of enhancing public awareness is through organized ecotourism activities in mangrove forests by conducting guided educational walks on boardwalks and building bird watching platforms for birdwatchers. It is also important that legal provisions are made to protect and preserve mangrove forests from overexploitation and environmental degradation so that people or institutions that abuse such biotopes are made to rehabilitate them at their own cost.

References

Chapman, V.J. (1976). Mangrove biogeography. Proc. Symp. Mangrove, Honolulu, 1974, vol. 1, Univ. Florida, pp. 3–22.

Chapman, V.J. (1977a). Introduction. In: V.J. Chapman (ed.), Wet coastal ecosystems. Amsterdam, Elsevier, pp. 1–29.

Chapman, V.J. (1977b). Africa B. The remainder of Africa.

In: V.J. Chapman (ed.), Wet Coastal Ecosystems. Amsterdam, Elsevier, pp. 233–240.

Day, J.H. & Morgans, J.F.C. (1956), The ecology of South African estuaries, Part 7: The biology of Durban Bay. *Annals of the Natal Museum* 13(3): 259–312.

Doute, R.N., Ochanda, N. & Epp, H. (1981). A forest inventory of Kenya using remote sensing techniques. KREMU, Technical Report series No. 30. Nairobi, Kenya Rangeland Ecological monitoring Unit.

Fagoonee, I. (1990). Coastal marine ecosystems of Mauritius. Hydrobiologia 208(1–2): 55–62.

Glenn, E.P. & Watson, M.C. (1993). Halophyte crops for direct salt water irrigation. In: H. Lieth & A. Al Massoom (eds.), Towards the rational use of high salinity tolerant plants, Vol. 1, pp. 379–385. Kluwer Academic Publishers, Dordrecht.

Isaac, W.E. & Isaac, F.M. (1968). Marine botany of Kenya coast, 3: General account of environment, flora and vegetation. Journal of East Africa Natural History Society 27(1): 2–28.

Lewis, J.R. (1964). The ecology of rocky shores. London, Hodder & Stoughton, 2nd ed., 323 pp.

Macnae, W. (1968). A general account of the fauna and flora of mangrove swamps and forests in the Indo West Pacific region. Advances in Marine Biology 6: 73–270.

Macnae, W. (1971). Mangroves on Aldabra. Phil. Trans. R. Soc. Lond. B 260: 237–247.

Macnae, W. and Kalk, M. (1962). The ecology of the mangrove swamps at Inhaca Island, Mozambique. J. Anim. Ecol. 50: 19–34.

Martosubroto, P. & Naamin, N. (1977). Relationship between tidal forests (mangroves) and commercial shrimp production in Indonesia. Mar. Res. Indonesia 18: 81–86.

Ruwa, R.K. (1985). Human habitations along the Kenyan shoreline and their impacts on marine environment. Post (Kenya) 10(2–3): 23–25.

Ruwa, R.K. (1990). The effects of habitat complexities created by mangroves on macrofaunal composition in brackishwater intertidal zones at the Kenyan Coast. Discovery and Innovation 2(1): 49–55.

Ruwa, R.K. and Polk, P. (1986). Additional information on mangrove distribution in Kenya: Some observations and remarks. Kenya Journal of Science and Technology (B) 7(2): 41–45.

Schimper, A.F.W. (1891). Die Indo-Malayische Strandflora. Bot. mitt. Trop. Jena. Vol. 3, 204 pp.

South Western Indian Ocean Programme (1989a). Major mangrove areas in the South Western Indian Ocean (SWIO) region. SWIOP Fisheries bulletin 30: 1.

South Western Indian Ocean Programme (1989b). Proceedings of the workshop on the management of the shallow water shrimp fishery of Northwest Madagascar. FAO/UNDP RAF/87/008/D R/50/89/E: 95 pp.

South Western Indian Ocean Programme (1990). Proceedings of the Workshop the management of the shallow water shrimp fishery of Tanzania. FAO/UNDP/RAF/87/008/OR/58/90/E: 94 pp.

Stoddart, D.R. (1971). Environment and history in Indian Ocean reef morphology. In: B. R. Rosen (ed.), Symposium of Zoological Society. London 28: 3–38.

Ungar, I.A. (1982). Germination ecology of halophytes. In: D.N. Sen and K.S. Rajpurohit (eds.), Tasks for vegetation science, Vol. 2. Dr. W. Junk Publishers, The Hague, pp. 143–155.

15. Halophytes of the Indian desert-communities of the various 'ranns'

S.K. SAXENA
Central Arid Zone Research Institute, Jodphur, Rajasthan, India

Abstract. The vegetation of the saline depressions in Rajasthan, locally called 'ranns', is described and the relationship with salinity and utilization pressure explored. The results of a vegetation survey are presented.

Key words: Land use, Salinity, Soils, Vegetation survey

Introduction

India has a significantly large arid zone (Figure 1). Western Rajasthan and Gujurat are the principal arid zones constituting 61 percent and 20 percent respectively. The districts in Rajasthan wholly in the arid zone are Ganganagar, Bikaner, Jaisalmer, Barmer, Jodphur and Churu (Subramaniam 1988). The deserts of western India and the neighbouring region of Pakistan share many of the same conditions (Dakshini 1972; Bodla et al. 1994).

Indian desert is well dotted with saline depressions or playas which are locally called 'Ranns'. They exist in different shapes and sizes and are widely distributed in the arid tract of western Rajasthan. They cover only a small fraction of the total area.

The surfaces of all these depressions are nearly level with less than 1% slope. They collect salts and evaporites (Ghose & Singh 1968). A few of them have some economic use e.g. small or large scale salt mining.

A few halophytic species, which can tolerate the salinity are capable of colonizing these special habitats. The overall plant productivity is very much restricted due to low water availability and high salinity.

Natural salinity

Ranns are the areas where the salinity has been built up over a long period under natural processes. The source of salts is from weathering of rocks and from accumulation of salt in the soil profile under low rainfall conditions and high evaporation. Lack of water does not permit these lands to be reclaimed.

According to Ghose (1964) and Ghose & Singh (1968) the salt basins are the relics of a well integrated drainage system. Along with the sediments of different composition, mineral salts (detached in soluble form from the parent material) were carried downstream along the channels and precipitated according to their solubility. Sodium and calcium salts are deposited in the lower reaches and at confluences of rivulets i.e. Pachpadra and Thob.

Soils

The soils of these depressions are quite variable. They are mostly deep to very deep and have a dominant colour of brown to dark greyish brown, often with a thin salt encrustation at the surface. The soils are loam to clay loam, silty loam and

V.R. Squires & A.T. Ayoub (eds.), *Halophytes as a resource for livestock and for rehabilitation of degraded lands*, 211–215.

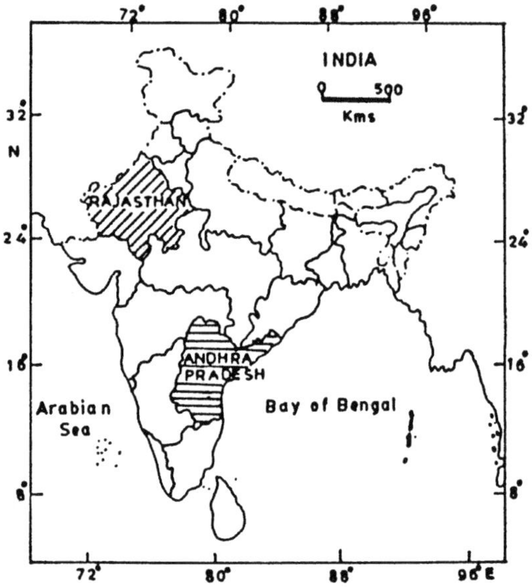

Figure 1. India has a large arid zone. Rajasthan is the biggest section.

few patches with sandy loam. In many cases, the substrate is largely variable with respect to size and proportion of coarse fragments. The soils are highly calcareous throughout the profile. In some cases, thick layers of lime concretions are recorded (Shankarnarayan & Kolarkar 1987). The pH of rann soils generally ranges from 8.2–8.7 and in few cases it reaches 9.0–9.2, while the SAR ranges from 35–95. The soils of the ranns are quite poor in nitrogen status, low to medium in phosphate, medium to high in potash where as Fe, Mn, Zn and Cu are well supplied (Joshi et al. 1988a,b).

The salt content is highly variable from place to place and it is variable within the the rann (18 to above 100 mmhos EC). During the monsoon season, the salinity of the surface soil decreases. Many weeds and grass seed germinate at this time and colonize the area. The salinity becomes highest during the summer months as the salts are brought to surface by capillary action.

The surface soil (0–15 cm) contains the lowest salinity whereas it reaches maximum concentration below 90 cm (i.e. Malhar). Sodium chloride and sulphate are by far the dominant salts but bicarbonates are prominant in some sites.

Present land use

Mining of sodium salt continues on many ranns. In many ranns there are widely scattered small to large mounds of soils, due to old salt mining. The majority of ranns are barren in the centre but there is colonization of the periphery where blown sand is deposited from the neighbouring fields or sand dunes. Many pioneer succulent and halophytic species colonize this modified habitat. Many are found vegetated with trees or shrubs (or both). The common species are *Salvadora oleoides*, *S. persica*, *Capparis decidua*, *Prosopis juliflora*, and *Tamarix* spp. These are invariably grazed or browsed.

Vegetation survey of ranns of western Rajasthan

Almost every district of the Indian desert has saline depressions or alkaline/saline soils. Out of the eleven districts forming the arid area of western Rajasthan, significant ranns may be seen in the district of Barmer, Bikaner, Jaisalmer, Jodhpur and Nagaur (Figure 2). The plant communities recorded on them are described below. Basal cover at ground level was estimated with standard techniques.

Plant communities by district

A. **Barmer.** This district covers an area of 87.0 km^2 under various ranns.
  1. *Sporobolus marginatus – Desmostachya bipinnata* 3% cover (Rann of Kutch).
  2. *Prosopis juliflora – Capparis decidua* 60–120 plants ha^{-1} (Utterlai).
  3. *Desmostachya bipinnata – Gnephalium pulvinatum* 2.0% cover (Utterlai).
  4. *Sporobolus marginatus – Chloris virgata* 2.0–3.5% cover (Thob).
  5. *Suaeda fruticosa – Sporobolus marginatus* 5–8% cover (Pachpadra Salt Basin).
  6. *Prosopis juliflora – Acacia jacquemontii* 40–75 plants ha^{-1} (Pachpadra Salt Basin).

B. **Bikaner.** The saline depressions of this district cover an area of 9.7 sq. km within various ranns.
  7. *Peganum harmala – Aeluropus lagopoides* 2.0% cover (Lunkaransar).

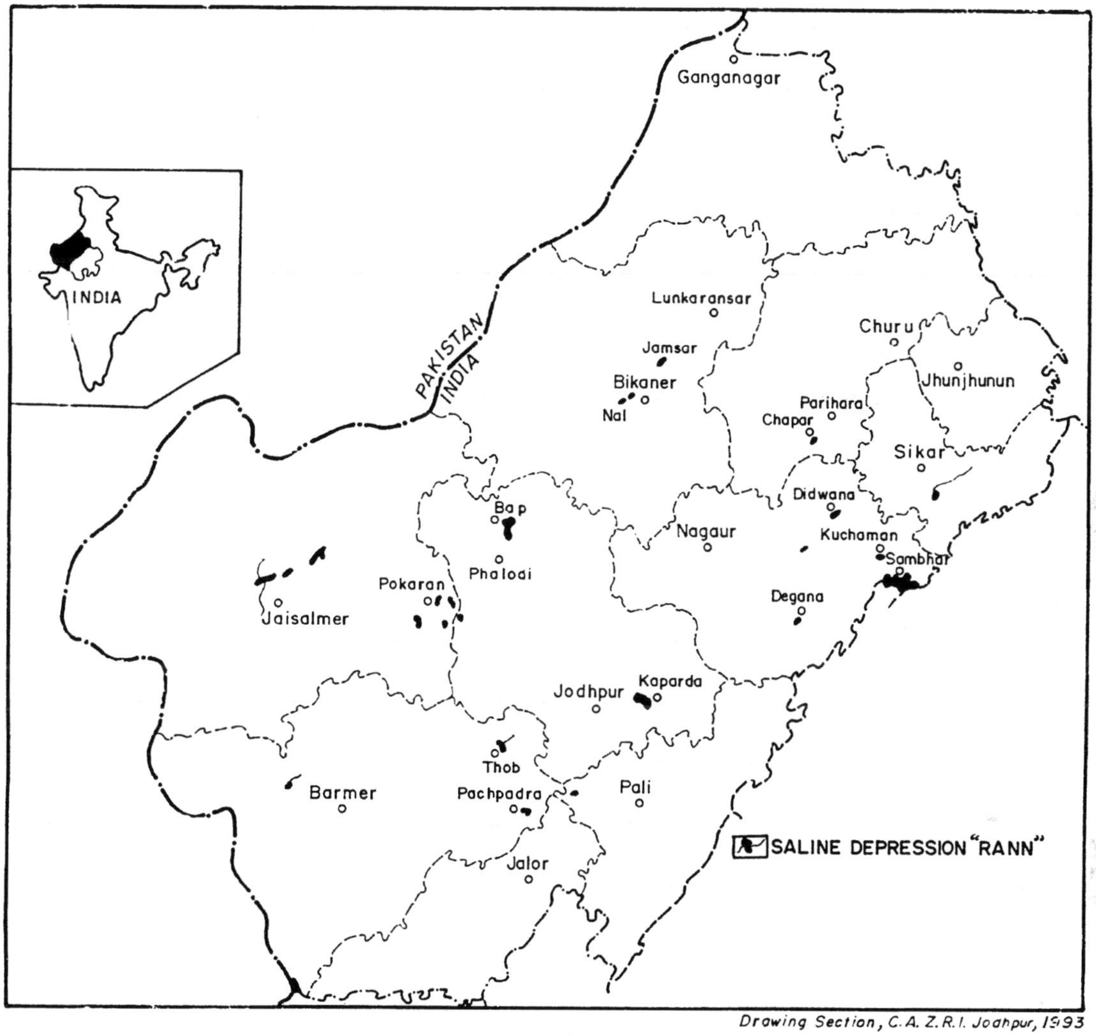

Figure 2. Distribution of saline depressions or 'ranns' in western Rajasthan, India.

8. *Sporobolus marginatus – Eleusine compressa* 3–5% cover (Jamsar & Nal).
9. *Suaeda fruticosa – Sporobolus marginatus* 9.8% cover (N.W. of Bikaner).

C. **Jaisalmer.** The ranns of Jaisalmer district are scattered among the rocky outcrops. All of them are fed by a minor channel. They cover an area of 125.4 km^2 (Anon 1992).

10. *Sporobolus marginatus – Aeluropus lagopoides* 1.8% cover (Pokaran).
11. *Sporobolus marginatus – Eleusine compressa* 2.5–4.0% (Kharia rann).
12. *Aeluropus lagopoides – S. marginatus* 2.0–3.8% cover (That rann).
13. *Sporobolus marginatus – Lasiurus sindicus* 3.5–6.0% cover (Metha rann).
14. *Sporobolus marginatus – E. compressa – Cressa cretica* 4.5–5.5% cover (Lawan).

D. **Jodhpur.** The major ranns of this district cover an area of about 1.56 km^2 (Anon.

1982). There are several other small patches distributed in the district.

15. *Suaeda fruticosa* – *Salsola baryosma* 7.8% cover.
16. *Haloxylon salicornicum* – *S. marginatus* 5.2% cover.
17. *Fimbristylis* – *Cyperus* sp. on water logged area (all three at Malhar & Kaparda rann).
18. *Sporobolus marginatus* – *Chloris virgata* 3–6% (Dedia & Bilara).

E. **Nagaur.** Saline depressions and salt lakes occupy an area of 165.7 km^2 (Anon. 1989). Sambhar lake, where sodium chloride is extracted, is deeper while others are shallow and remain dry for the major part of the year.

19. *Sporobolus marginatus* – *Desmostachya bipinnata* 2.0–5.0% cover (Sambhar Lake).
20. *Cressa cretica* – *A. lagopoides* 2.0–3.5% cover (Kuchaman).
21. *Eleusine compressa* – *S. helvolus* – *D. bipinnata* 2.0–7.0% cover (Didwana).
22. *Sporobolus marginatus* – *Suaeda fruticosa* 3.0–7.5% cover (Jaswantgarh).
23. *Desmostachya bipinnata* – *S. fruticosa* 2.5–4.5% cover (Nawa).

Colonization of the ranns

Grasses are the first to establish. Succulent plants and other pioneer plants soon follow. The halophytic grasses are *Sporobolus marginatus*, *S. helvolus*, *Eragrostis diarrhena*, *Chloris virgata*, *Desmostachya bipinnata*, *Cenchrus setigerus*. *Eleusine compressa*, *Aeluropus lagopoides*, *Trianthema pentandra*, and *T. triquetra*. Other common colonizers are *Portulaca oleracea*, *Zygophyllum simplex*, *Cressa cretica*, *Gnaphalium pulvinatum*, *Solanum surattense*, *Cyperus rotundus*, *Cyperus* sp. *Sesuvium sesuvioides* and *Boerhavia elegans*.

Halophytic chenopod shrubs species demonstrate considerable tolerance to salinity and species like *Suaeda fruticosa*, *Salsola baryosma*, *Haloxylon recurvum*, *H. salicornicum* grow on the white incrustation of salt. Shrubs and trees such as *Salvadora oleoides*, *S. persica*, *Prosopis juliflora*, *Tamarix aphylla*, *Acacia nilotica* and *Capparis decidua* are invariably found on the outer fringes of the ranns.

Management

The Ranns exhibit low productivity due to existing salinity conditions and high grazing pressure from bovines. The forage production ranges from 300–500 kg ha^{-1}. These areas, when simply protected from grazing can produce 2–3 t ha^{-1} in a two to three years period.

Production can be increased further by adopting some improved techniques of soil conservation which change the micro-relief (raised bunds and other mechanical interventions). This will create area of low salinity on the slope and top of the bunds/beds. Here, highly productive perennial grasses and other forage species, like *Dichanthium annulatum*, *Cenchrus setigerus*, could be raised, via rooted slips, during the monsoon. Halophytic browse (top feed) species like *Haloxylon salicornicum*, *H. recurvum*, *Suaeda fruticosa* etc. could be introduced in the area between the raised bunds.

These grasslands may be grazed up to 75% utilization for sustained production. Although Ranns represent a fraction of the total area of Western Rajasthan they are the regular grazing grounds for livestock and hence any improvement effort will be rewarded by higher forage production.

Acknowledgements

The author is grateful to Director and Head of Resource Survey and Monitoring Division, CAZRI, Jodhpur, for providing facilities and to the Australian Development Assistance Bureau, Canberra for a grant as support to attend the Workshop. Thanks to Dr V.R. Squires for help in editing the manuscript.

References

Anon. (1970). Resource Survey Report. Chohtan (Barmer) CAZRI, Jodhpur.
Anon. (1974). Basic Resources Bikaner. CAZRI Jodhpur.
Anon. (1982). Basic Resources Jodhpur. CAZRI Jodhpur.
Anon. (1984). Annual Report. CAZRI Jodhpur.
Anon. (1989). Natural Resources Nagaur. CAZRI Jodhpur.
Anon. (1992). Integrated Natural Resources Jaisalmer. CAZRI Jodhpur.

Bodla, M.A., Shabbir Baig, M. & Shamsi, S.R.A. (1994). Forage production from salt-affected and water-logged areas in Punjab, Pakistan: A case study (this volume, Ch. 18).

Dakshini, K.M.M. (1972). Continental aspects of shrub distribution, utilization and potentials: Indian subcontinent. In: C. McKell, J.P. Blaisdell & J.R. Goodin (eds.), Wildland shrubs – their biology and utilization. USDA Forest Service, General Tech. Rpt. INT-1.

Ghose, B. (1964). Proc. Symp. Probl. Indian Arid Zone, Jodhpur, pp. 79–83.

Ghose, B. & Singh, S. (1968). Symp. Arid Zone, Jodhpur, pp. 54–57.

Joshi, D.C., Singh, N. and Kolarkar, A.S. (1988a). Tropical Agric. 64: 41–48.

Joshi, D.C., Singh, N. & Kolarkar, A.S. (1988b). J. Indian Soc. Soil Sci. 36: 833.

Shankarnarayan, K. & Kolarkar, A.S. (1987). Proc. Intl. Symp. Salt-affected soils. CSSRI Karnal, India, pp. 77–92.

Subramaniam, A.R. (1988). Agroclimate of Indian arid zones. In: P.W. Unger, T.V. Sneed, W.R. Jordan & R. Jensen (eds.), Challenges in dryland agriculture: A global perspective. Proc. Intl. Conf on dryland farming, Amarillo/-Bushland, Texas Texas Agric. Exp. Station.

16. Banni grassland and halophytes
A case study from India

S.K. SAXENA
Central Arid Zone Research Institute, Jodphur, Rajasthan, India

Abstract. The biophysical and biological attributes of the Banni grassland, within the rann of Kutch in south west India, are described. The results of a vegetation survey are presented. The effects of grazing management and impacts of recurring drought on the ecology of the region are assessed. Measures to mitigate the effects are outlined.

Key words: Arid, Buffalo, Cattle, Grazing management, Impeded drainage, India, *Prosopis*, Salinity

Introduction

The Banni area is a slightly raised land surface in the Great Rann of Kachchh (Kutch). It is located in the north of the Kachchh mainland (23°19′ N to 23°52′ N and 68° 56′ E to 70°32′ E). It was formed in a shallow sea from stream-borne sediments and covers about 2500 km^2. The entire Banni area is almost flat (>1% slope) and averages 3–5 m above mean sea level (m.s.l.). High salinity and impeded surface and subsurface drainage characterize 96% of the entire Banni area. Animal husbandry is the dominant land use.

Geography of Banni

There are 46 small villages scattered over the area with a population (1981 census) of 14,842. The inhabitants are locally called the Maldharis. There are two broad groupings; the Bari Banni and Choti Banni. The former has 33 villages and the latter 16 villages. Bherindiyara, the largest village, is centrally located and is only 4.5 m above mean sea level. Six villages in Choti Banni and one in Bari Banni are abandoned because of:

- salinity ingress in the northern part,
- lack of communication and transport facilities, and
- lack of potable water.

Physical environment

Climate. The Banni area is arid (mean annual rainfall < 200 mm). Rainfall is erratic, undependable and poorly distributed. Sometimes heavy downpours bring about desalinization of surface layers and promotes higher seed germination of grasses and forbs. High temperatures (up to 45 °C) with large diurnal variation (15–18 °C) are common in March to May. Monsoon rains prevail from June to September. Cooler weather occurs from December to February when the region experiences continental winds. January and February experiences fog and heavy dew fall.

Soils. Banni soils consist of recent alluvium mixed, in places, with aeolian sandy deposits and over all present a younger formation. The marshy land and mud flat cover about 7850 ha area. Soils are deep to very deep, dark greyish brown (10 YR 3/2) to yellowish brown (10 YR 5/6), indicating variation from clay to coarse textured soils in discontinuous patches. They are invariably calcareous. Silty loams cover 61.0% of the area fol-

V.R. Squires & A.T. Ayoub (eds.), *Halophytes as a resource for livestock and for rehabilitation of degraded lands*, 217–222.

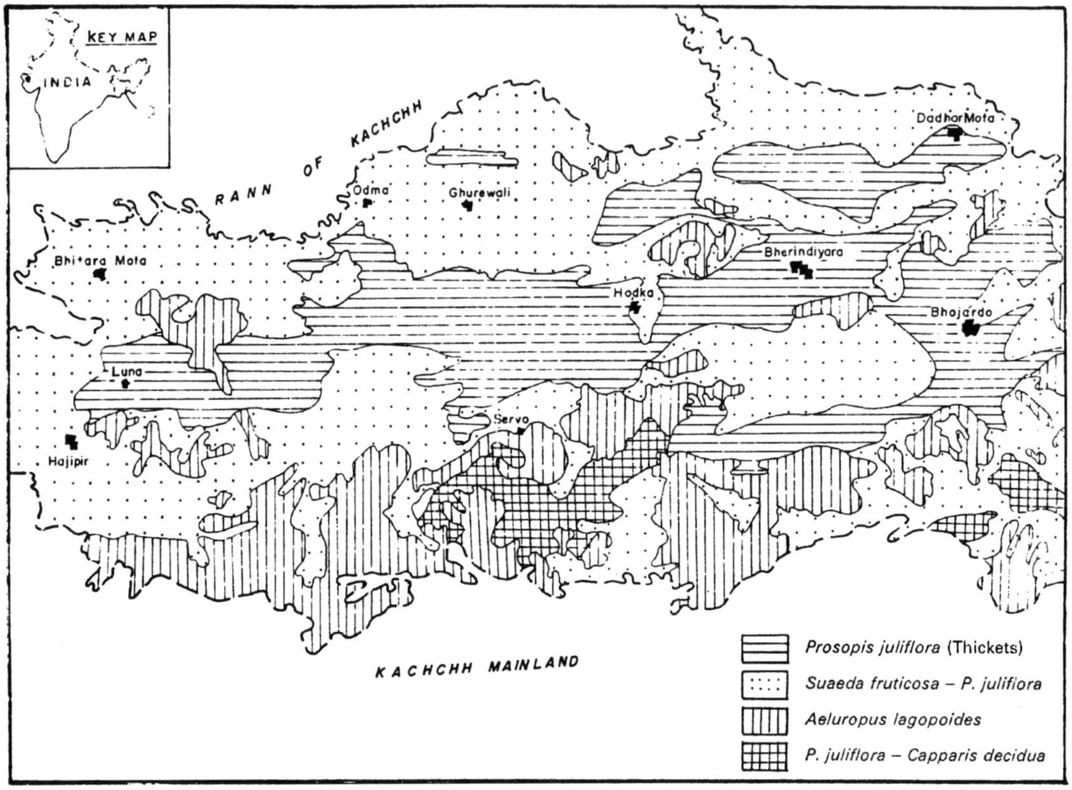

Figure 1. Halophytic communities of Banni, Western India.

lowed by silty clay loam (28.0%) and silty soils (8%). The rest is loamy sand. Soils are saline to highly saline (10.0–15.0 mmhos cm^{-1} of EC). More than 50% of the area contains about 15.0 mmoh cm^{-1}, while about 40% contains 3–15 mmoh cm^{-1} and 10% contains 1–3 mmohs cm^{-1}. The pH varies between 7.1 to 8.5; with a few spots as high as 9.0. These soils have 22.9 to 67.1% water holding capacity. Based on this variation, the vegetation colonization differs and it could be categorised into five zones (see below).

Water resources. The subsurface water is highly saline and hence no wells are dug in the area. The rain water is collected in large open tanks and used for human and animal consumption. During summer months shallow wells are dug in dry tank bed to get percolated water with moderate salinity (EC of water varies from 10,000 to 15,000 mmohs or more).

Biotic environment

During the 1960s Banni had more than 70,000 animals, but continuous droughts of sixties and seventies reduced the population to 50,000. The present composition of the animal population is 45% buffalo (22,174), 20% cattle (9625), 33% sheep and goats (16,502) and the remainder are camels, horses and asses (1983 census). Kankrej and Thani are famous breeds of cattle. On average, each well-to-do Maldhari family maintains 35–50 cattle and earns Rs. 6000 to 10,000 ($US190–320) per month (Bharara 1987).

In good rainfall years there is large scale immigration of sheep, goat and cattle from adjoining

Figure 2. Camels grazing banni grassland dominated by *Suaeda fruticosa*.

Talukas and districts. Revenue from grazing rights provides additional income to Maldharis, but causes excessive grazing pressure as there are no grazing norms.

Many of the more nutritive perennial grass species have succumbed to this pressure and survive only within the *Prosopis juliflora* thickets. Poor availability of good grasses and recent dairy development has virtually changed the livestock composition. The presence of coarse grasses and halophytic shrubs suit the buffalo more as they can easily shift over to the less palatable material than can cattle. During summer buffaloes can be maintained on the new growth of *Suaeda fruticosa* and pods of *P. juliflora*.

Impact of consecutive droughts

Recent consecutive drought years (1986 to 1988) had adverse effect on the perennial grass clumps. Excessive grazing, coupled with continuous droughts, have virtually eliminated the established forage producing grass species. Lack of forage and large scale animal migration caused many casualties. Animals that remained had high mortality rates because of a heavy intake of non-palatable material, containing alkaloids etc. (eg. leaf of *P. juliflora*).

Sustenance of livestock

Though vast areas of Banni had earlier provided an ideal region for animal husbandry, successive drought years and excessive animal pressure have virtually left *P. juliflora* and *Suaeda fruticosa* as the only perennials in the Banni grassland. *S. fruticosa* became the main top feed for buffalo and camel during March to June. In the Banni area *Suaeda* grows continuously and provides younger tender growth as browse which is much preferred by buffaloes, camels and goats. Ripe pods of *P. juliflora* are collected on a large scale, boiled and fed to milch animals.

Figure 3. Suaeda fruticosa – Sporobolus marginatus community (Pachpadra).

Ecological changes

During 1950s the Banni was predominantly a good grassland, where distantly scattered salt tolerant low trees and shrub species (10–15/ha) like *Salvadora oleoides*, *Salvadora persica*, *Acacia nilotica* and *Capparis decidua* were found. Trees and shrubs with large crowns were the ideal places for animal resting during summer months. They were also the source of top feed.

Drought-hardy and salt tolerant exotic species of *Prosopis juliflora* was introduced by the State Forest Department in the 1960s, especially along the southern fringes of the Banni. Seeds of *P. juliflora* were provided to Maldharis for propagation. In a span of ten years this potentially useful species was established over extensive areas.

Large herds of sheep and goat browsing on pods distributed *P. juliflora* through their faeces. Now dense thickets (2070 plants ha^{-1}) have developed, replacing the valuable grass species, especially around the watering points and settlements. In many cases even animals cannot enter these dense thickets.

The dense canopy, profuse low level branching of *P. juliflora* and presence of an inhibitor in its leaves (allelopathy) do not allow germination and establishment of high forage producing species. Besides, the indigenous species, referred to earlier, are suppressed or even eliminated.

Growth of *Suaeda fruticosa* when growing in association with *P. juliflora* is also restricted. Patches of *S. fruticosa* without competition from *Prosopis* have considerably larger crown diameters.

Vegetation of Banni

The distribution of halophytic vegetation types is shown in Figure 1. During resource surveys of 1989–90 the following plant/grass communities were recorded in various micro-habitats of the Banni.

1. Deep to very deep sandy loam to clay soils

Figure 4. Banni grassland dominated by *Aeluropus lagepoides*.

with low salinity are dominated by *P. juliflora* thickets providing a density of 2316, 1016, 3875, 4308 and 5858 ha^{-1} at Sargu, Hodka, Dumado, Dhorda and Bhirandiyara villages, respectively. Here they grow 5–6 m in height with multiboles of 20–35 cm girth.

2. Low-lying flat lands which remain dry for the larger part of the year where elevation is <3 m above mean sea level with moderate to high salinity. The dominant community is *P. juliflora – S. fruticosa*. There are 30 and 2070 plants ha^{-1} respectively at Lalwara with a crown cover of 5.4%. The bushy form of *P. juliflora* predominates with an average height of 70 cm. *Suaeda* clumps ranged in size from 0.06 m^2 to 0.44 m^2.

3a. There are pure stands of *Suaeda fruticosa* (1814 to 4120 plants/ha) on flats with high to very high salinity between Hudco and Sagra. Here the deep soils are heavy and clayey and remain moist for the greater part of the year. An average *Suaeda* was 31 cm high and covered 0.34 m^2 (18.5% ground cover). Other species included the coarse perennial grass *Aeluropus lagopoides*. During lean periods (March–June) they provide green browse material to cattle. Figures 2 and 3.

3b. The low lying flat muddy areas with high salinity and deep clayey soils are almost bare, Fig. 4. The built-up of sand ridges due to water/wind action, allows the establishment of an *Aeluropus lagopoides* community. Sites at Loriya, Odma, Sawa and between Bhirandiyara to Loriya were surveyed. There were 51–68 plants m^2, providing 125 to 625 kg ha^{-1} of above-ground biomass and 1% ground cover. Other species were *Aristida funiculata*, *Sporobolus marginatus* and *Suaeda fruticosa*.

4. Raised flat areas (3–5 m above sea level) with very deep fine sandy clay loam soils with moderate salinity support *P. juliflora – Capparis decidua* community (Sabu village) with 100 plant ha^{-1} plant density.
   Here *P. juliflora* generally attained a height of 266 cm while *Capparis* was >250 cm and

provided 10.5% crown cover. The same community at Dhorado, due to more salinity, had poor growth conditions, (*Capparis* 70 cm and *P. juliflora* 150 cm) and provided 1.3% crown cover. Ground flora provided 0.85% cover and 150 kg ha^{-1} of dry forage.

Some of the lowlying and waterlogged areas in mud flats where water stagnates for the greater part of the year are well colonized by hydrophytic species like *Cyperus*, *Eleocharis*, *Scirpus* and *Fimbristylis*. When green, they are also grazed by buffaloes under stress. On the high salinity zones coarse perennial grasses of low palatability like *Eurochondra setulosa*, *Sporobolus pallidus*, *S. quancilotus* are recorded as widely scattered clumps.

Forage production

In 1983, a good rainfall year, the production potential of excellent (low salinity) Banni grassland at a site 3 km south of Bherandiyara village was assessed. The dominant species were *Cenchrus ciliaris*, *Desmostachya bipinnata* and *Sporobolus helvolus*. Clumps of these grasses were well established (60–80 cm diameter) with an average height of 90–120 cm.

There were 80 *P. juliflora* plants/ha in the grassland. Associated grasses and weeds were *Dichanthium annulatum*, *Digitaria adscendens*, *Dactyloctenium sindicum*, *Cynodon dactylon*, *Eleusine compressa*, *Cressa cretica*, *Bothriochloa pertusa*, *Eragrostis ciliaris*, *Chloris virgata*, *Indigofera cordifolia* and *Zornia diphylla*.

Above ground biomass was estimated by harvesting 10 quadrats, each of 3 × 1 m. A line intercept method was used to assess basal cover by laying 10 lines each of 10 m. Ground cover was 8% and oven-dry biomass (forage production) was 3.5 t ha^{-1} but in 1988, a drought year, biomass was reduced to 0.5 t ha^{-1}.

Measures for improvement

- Control of immigration of animals.
- Development of fodder banks in good rainfall years.
- Deepening of water storage tank to increase watering facilities.
- Construction of Pucca bund to check ingression of sea water.
- Reseeding of grasslands with nutritive and high production grasses.
- Introduction of palatable halophytic species to increase top feed (browse) production.
- Systematic felling of thickets of *Prosopis juliflora* for improvement of forage.
- Pod collection of *P. juliflora* pods on an extensive scale coupled with feeding to milch animals (after boiling) to avoid mass scale regeneration through faeces.

Acknowledgements

The author is grateful to the Director and Head of Resource Survey and Monitoring Division, CAZRI, Jodphur for providing research facilities; to AIDAB for a travel grant to attend the Workshop in Nairobi and to Dr V.R. Squires for editing the manuscript.

References

Anon. (1988). Sub-committee for salinity ingress in Banni area. Government of Gujarat.

Bharara, L.P. (1987). Socio-economic survey of Banni area Kutch (Gujarat). CAZRI, Jodhpur.

Kar, A. (1992). In: Sharma, P.R. (ed.), Applied geom. in tropics. Banaras Hindu Univ. Varanasi.

17. Potentials for cultivation of halophytic crops on saline wastelands and sandy deserts in Pakistan to overcome feed gap for grazing animals

R. AHMAD,[1] S. ISMAIL,[1] M.A. BODLA[2] & M.R. CHAUDHRY[2]
[1] *Biosaline Research Laboratories, Department of Botany, University of Karachi, Karachi, Pakistan;*
[2] *International Water and Salinity Research Institute, Lahore, Pakistan*

Abstract. An overview of the situation in Pakistan is given. There is potential to grow forage crops on saline lands and in sandy deserts. This would help to alleviate the current feed gap. Results from co-operative research shows that high levels of biomass production can be achieved with halophytes.

Key words: *Atriplex*, Buffalo, Cattle, Goats, Grasses, Pasture legumes, *Prosopis*, Salt tolerant grasses, Sheep

Introduction

Pakistan lies between 24–37° N latitude and 61–76° E longitude. Its geographical location is shown in Figure 1. The total land surface of 79.61 m ha^{-1} is populated by 112.05 million inhabitants. The climate varies from arid to semi-arid. In the southern part of the country (coastal belt, lower Indus plains and adjoining deserts), the rainfall, being unpredictable, ranges between 90–200 mm $year^{-1}$. Average annual temperature varies between 22–28 °C, with maxima often reaching 45–50 °C, and minima from 4–8 °C. Rainfall in northern region (central Punjab and foot-hills of Himalayan mountains) is comparatively high (350–600 mm per annum) where average temperature varies between 23–26 °C. The maximum temperature may reach to 38 °C and minimum below 4 °C. Ombrothermic diagrams of various ecological regions are given in Figure 2.

The status of Pakistan's cultivable land, and extent of salinity/alkalinity and waterlogging is given in Table 1. Locations of saline areas and sandy deserts are also depicted on the map (Figure 1). Estimates of the area of saline and waterlogged land under rainfed area are not available, although soil salinity and waterlogging exists there as well.

In addition, there is a coastal sandy belt (450 km long) and three inland barren sandy deserts where, sufficient amounts of underground saline water are present to allow cultivation of useful salt tolerant plants. Modern methods have made it possible to use saline water in irrigation for obtaining economically feasible amount of food, forage and fuel, from halophytic plants in these barren sandy deserts (Ahmad et al. 1986).

Being basically an agricultural country, there exists a good livestock industry in Pakistan but the present population of livestock is insufficient to cater for the needs of the people. Hence a moratorium of two days a week has been enforced in slaughtering the grazing animals throughout the country.

Trifolium alexandrianum (berseem) and *Medicago sativa* (alfalfa) are the two main leguminous crops which supply the bulk of fodder in the country. Among the monocots, oat, barley, maize, sudax and mott are also cultivated for fodder on a limited scale. Because of the importance of cash crops like rice, cotton, sugarcane, etc., cultivation of fodder crops becomes of secondary importance and is shifted towards less fertile and marginal lands where productivity is low. There appears to be two periods when supply of fodder becomes insufficient during the year (Figure 3). Lack of

V.R. Squires & A.T. Ayoub (eds.), *Halophytes as a resource for livestock and for rehabilitation of degraded lands*, 223–230.

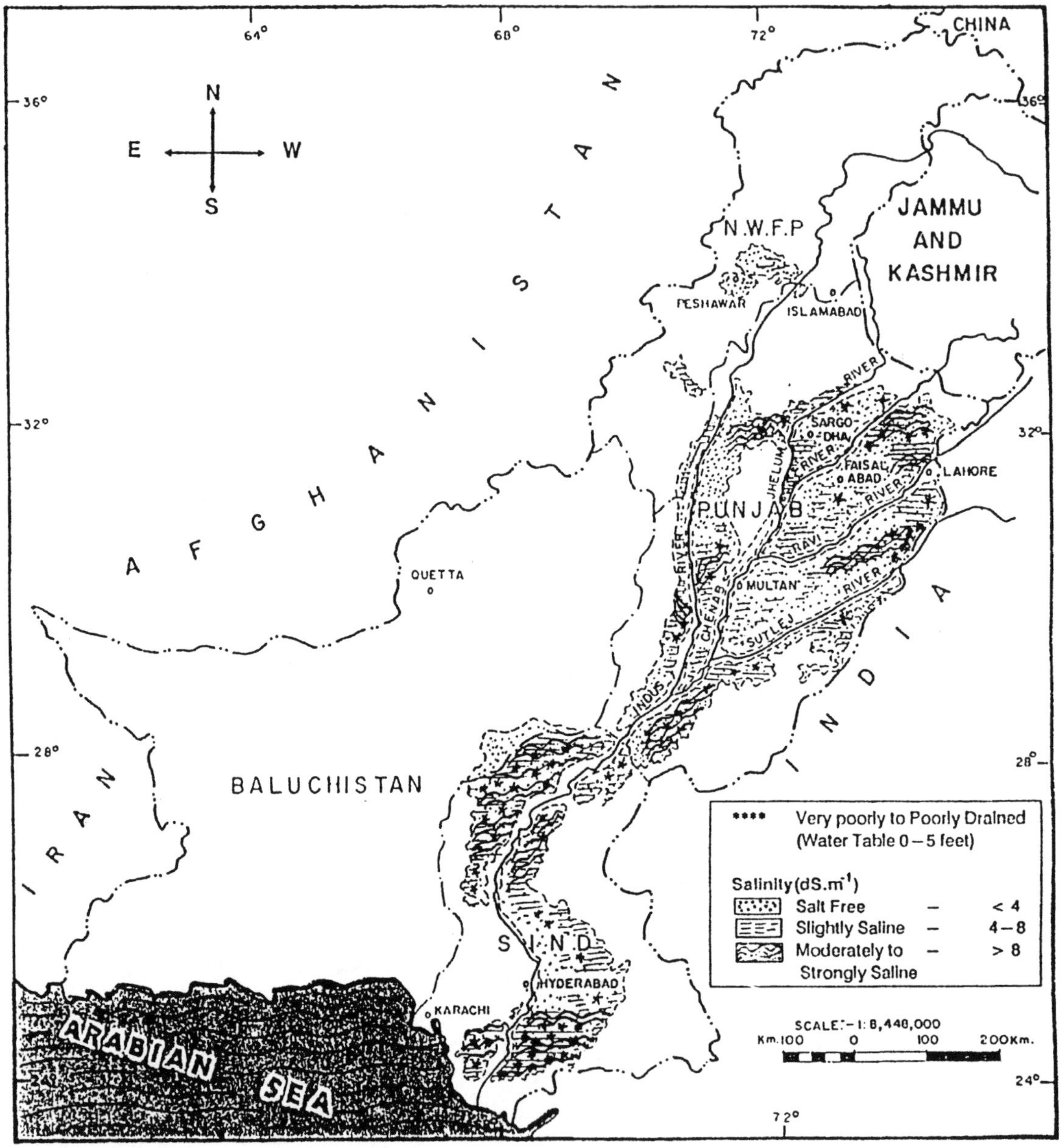

Figure 1. Salinity and water table map (irrigated areas of Indus Plain).

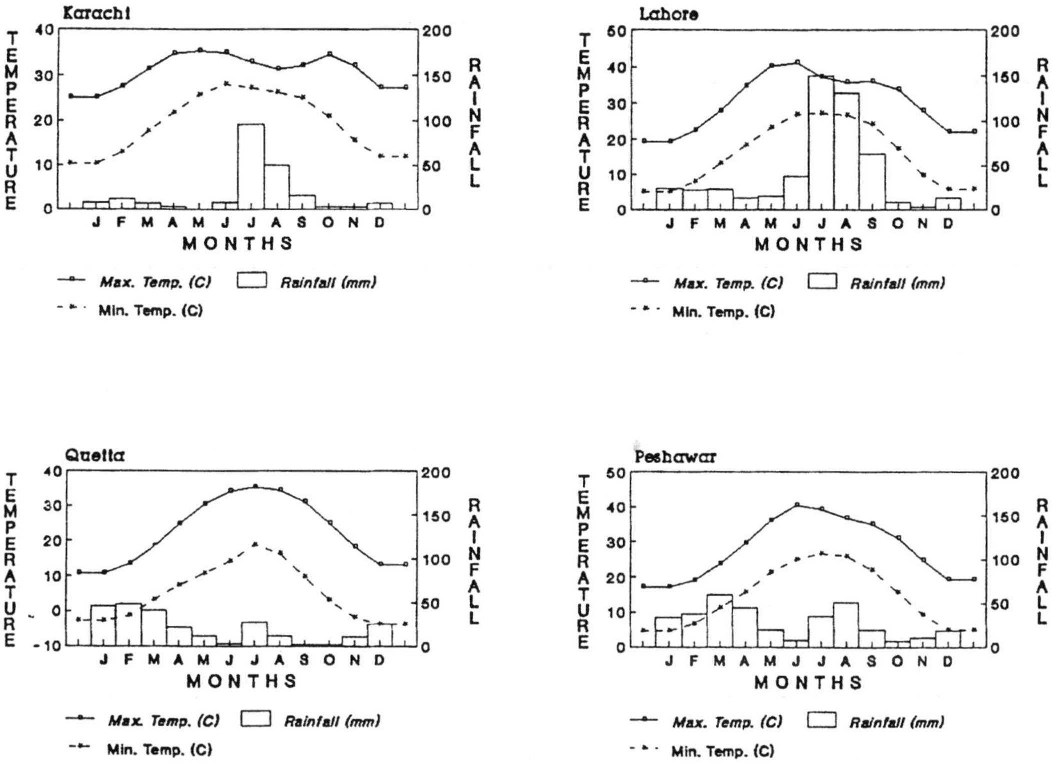

Figure 2. Ombrothermic diagram of different locations in Pakistan.

Table 1. Extent of salinity and waterlogging in Pakistan

Geographical area:	79.61 10^6 ha
Forests	3.46 10^6 ha
Cultivated area:	20.66 10^6 ha
Canal Commanded Area: (irrigated by canals)	13.96 10^6 ha
Salt-affected area (64% of canal commanded area)	
-moderate: (EC: 8–15 $dS\ m^{-1}$)	3.72 10^6 ha
-severe (EC > 15 $dS\ m^{-1}$)	3.08 10^6 ha
-saline/alkaline:	2.10 10^6 ha
Waterlogged area, within salt-affected area (50% of canal commanded area)	
Water table:	
0–150 cm	2.48 10^6 ha
150–300 cm	4.51 10^6 ha

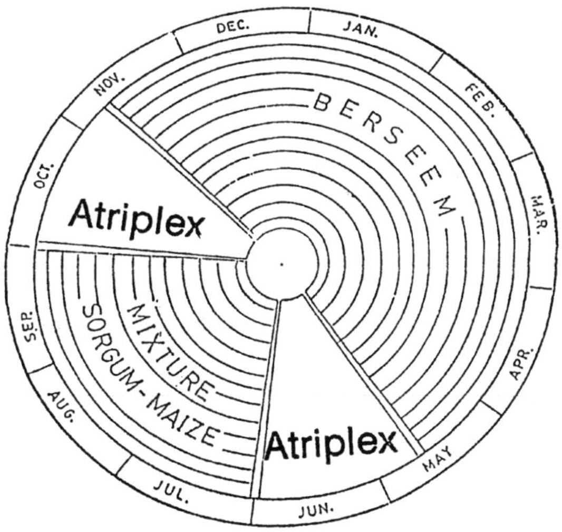

Figure 3. *Atriplex* species, grown on scleric scale, have potential to fill seasonal feed gap. □ deficit period; ▤ supply period.

Table 2. Nutrient requirement and availability for livestock in Pakistan

Animal category	Number (in millions)	Requirement (million tons)	
		Total digestible nitrogen	Digestible protein
Cattles	17.05	22.83	1.97
Buffaloes	13.07	21.04	1.97
Sheep & Goats	54.77	13.03	1.08
Others	4.21	6.32	0.52
Total	*89.10*	*63.20*	*5.53*
Nutrient availability (million tons)		37.55	3.95
Nutrient deficiency (million tons)		25.65	1.58
Nutrient deficiency (%)		40.58%	28.55%

Table 3. Feed resources in Pakistan

Sources	Contribution (%)
Crop residues	50.70
Fodder, range grazing	37.85
Cereal by-products	6.10
Oil cake, meals and animal proteins	2.35
Other (post harvest grazing, etc.)	3.00

feed is the principal reason that more livestock cannot be raised. Information about the livestock population and their feed resources in the country is given in Tables 2 and 3.

Utilization of grazable salt bushes and salt tolerant grasses grown at saline wastelands and barren sandy deserts could fill this feed gap and solve the problem. Halophytic fodders, in the case of stall feeding, are generally supplemented with leguminous and gramineous feed which comprise <40% of the ration. The effect of sodium is diluted if molasses or cotton seed cake is mixed while feeding such rations to the cattle. Research work undertaken at different Institutes of the country has been quite fruitful in this respect and progressive farmers are using their saline land for growing halophytic fodder. A brief account about cultivation of salt tolerant plants for feed under saline environment is given below.

Salt tolerant grasses

Some grasses have shown great potential under saline conditions and produce economically feasible biomass for grazing sheep, goats and cattle. *Sporobolus marginatus*, *Desmostachya bipinnata*, *Cynodon dactylon* are the most common grasses of saline/alkaline wastelands and barren sandy areas and could easily be raised for fodder. The biomass produced by *Panicum turgidum*, *Sporobolus arabicus* and *Leptochloa fusca* in the experimental fields which were irrigated with underground water of high salinity (EC: 18–20 dS m^{-1}) on the coastal sandy belt of Pakistan is given in Figure 4 (Ahmad et al. 1986).

Kallar grass (*Leptochloa fusca*) is considered to be a highly salt-tolerant palatable grass (Malik et al. 1986). It can withstand both salinity/sodicity and waterlogging. It can withstand salinity as high as that of EC: 10–15 dS m^{-1} without showing any significant loss in biomass. It is palatable to buffaloes, goats and sheep equally. Progressive farmers are cultivating this grass on their partially waterlogged saline/sodic land in Punjab to feed their buffaloes. It is also used for stall feeding by obtaining four cuttings per year. A weekly irrigated field produces about 42.1 t ha^{-1} green matter (19.4 t ha^{-1} dry matter). Its cultivation has helped in reclamation of soil as well. Depending on the prevalent salinity regime, farmers can select *Leptochloa* or other suitable grasses for raising fodder from their saline wasteland.

Salt tolerant grazable legumes

Different species of *Sesbania*, *Leucaena* and *Prosopis* have shown promising growth under various

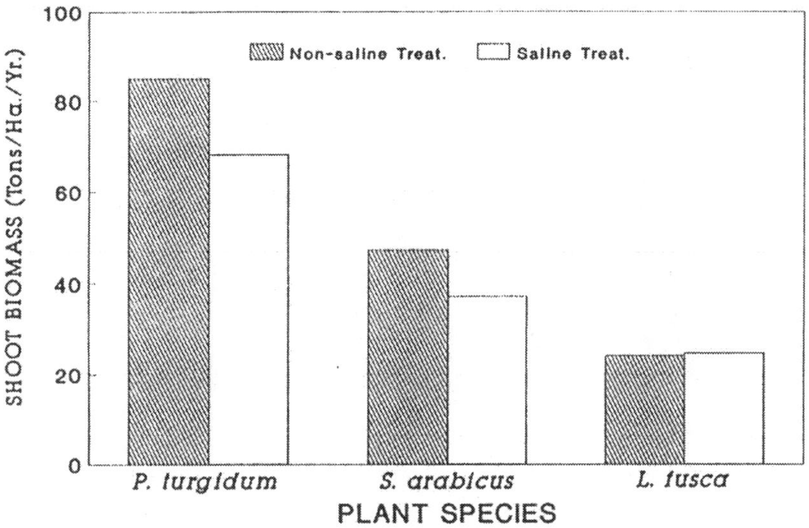

Figure 4. Biomass productivity of some grass species irrigated with underground saline water (EC: 18–20 dS m^{-1}) on sandy strata.

salinity regimes. Foliage from plants of the first two genera is used as animal feed, whereas, only fruits of *Prosopis* spp. are taken (except that of *P. cineraria* in which leaves are also fed to the cattle).

There is a common practice of growing *Sesbania aegyptica* and *S. bispinosa* on saline fields. Ploughing them down when their stems have reached a height of about 1 m provides green manure and improves the condition of saline soil. Foliage of these species (which has a high protein content) is sold as fodder in the market. *S. formosa* is another fast growing salt tolerant plant. Cattle relish its feathering leaves, soft stem and long pods and is also an excellent green manure.

In our field trials, it reached to a height of about 350 cm in eight months under irrigation with saline water (EC: 17 dS m^{-1}) at coastal sandy belt. Reduction of biomass in comparison to its control (irrigated with good quality water) was very little. Protein content ranged between 25–45 mg g^{-1} on dry weight basis. Cultivation of different *Sesbania* species/lines/accessions for providing fodder from salt affected wasteland is in practice on a limited scale in the Punjab and Sind provinces. Biomass production of *S. sesban*, *L. leucocephala* and *S. aegyptica* grown on sandy soils and irrigated with water of different salinity is given in Figure 5.

Saltbushes

Certain species/lines of grazable saltbushes (*Atriplex* sp.) have recently been introduced to Pakistan, via PARC–ACIAR Coordinated Project No. 8619 for testing their growth on salt-affected wastelands and with saline water irrigation on barren sandy areas. Their adaptation trials were conducted by three universities (Agriculture University, Peshawar; Agriculture University, Faisalabad and University of Karachi) and one Research Institute (Pakistan Council of Research in Water Resources, Bahawalpur) located under different ecological and climatic regions. Out of the sixteen species/lines of saltbushes (*Atriplex*) and bluebushes (*Maireana*) tested in these trials, four saltbushes namely *A. lentiformis*, *A. amnicola*, *A. undulata* and *A. cinerea*, proved most promising under local halo-xeric conditions (Qureshi 1992).

Biomass production of these saltbushes is given in Figure 6. It may be pointed out that these species/accessions of *Atriplex* are being grown in

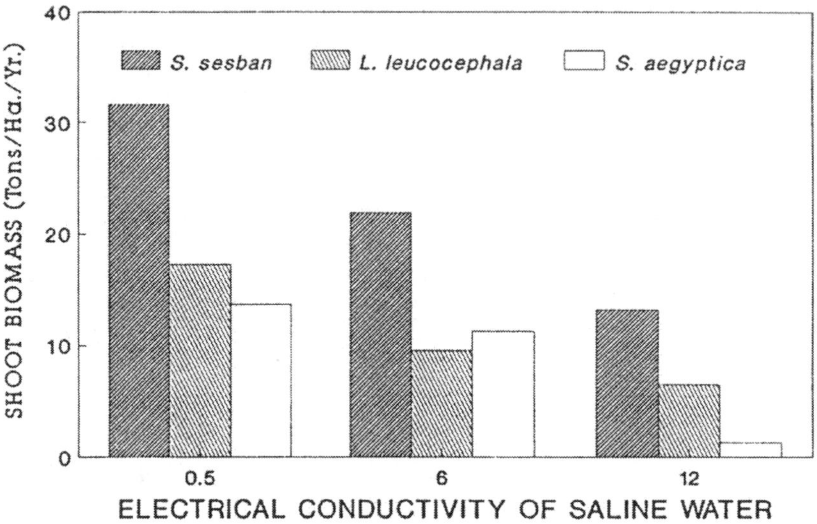

Figure 5. Biomass productivity of some legume species irrigated with water of different electrical conductivity (dS m^{-1}).

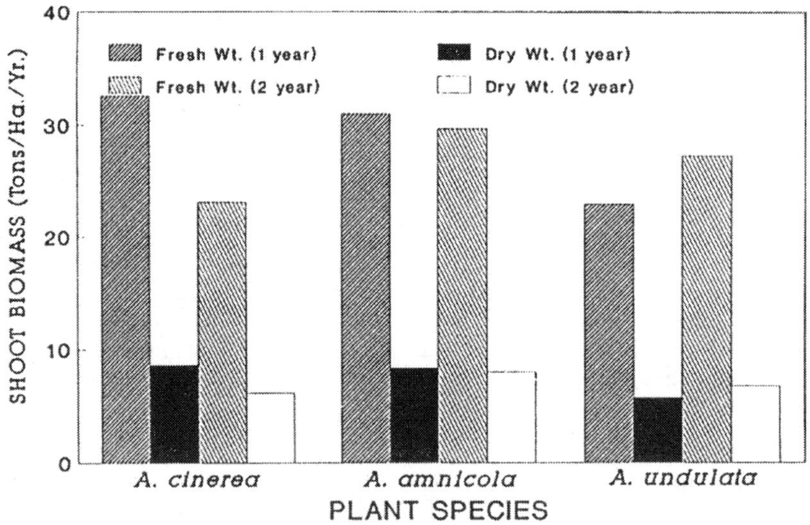

Figure 6. Biomass productivity of some salt-bushes at saline soils (EC_e: 70–125 dS m^{-1}), sodic soils (SAR: 60–100), and partially waterlogged (1–2 m) soils of Sujawal (Sind).

Table 4. Fresh biomass in fodder beet cultivars irrigated with normal and saline water (EC: 15–18 dS m^{-1}) at coastal sandy belt (harvested after 75 days)

Cultivar	Fresh weight/plant		Dry weight/plant	
	Control	Saline	Control	Saline
SHOOT				
Productiva	118.40 ± 10.89	88.12 ± 3.57	23.07 ± 1.80	22.24 ± 1.35
Polygroningia	87.73 ± 4.39	129.69 ± 7.25	19.42 ± 0.86	24.35 ± 1.10
Monoval	95.16 ± 4.94	106.31 ± 6.50	21.73 ± 1.44	22.81 ± 1.16
ROOT				
Productiva	139.62 ± 10.60	112.68 ± 10.09	30.35 ± 2.11	28.38 ± 2.33
Polygroningia	132.93 ± 9.66	136.24 ± 6.38	34.32 ± 3.45	30.89 ± 1.00
Monoval	137.17 ± 6.97	131.49 ± 5.87	56.69 ± 10.84	28.74 ± 1.95

Australia, USA, and other countries on saline land for feeding sheep, goats and cattle during the gap period when supply of conventional fodder is significantly reduced. Australian experience in growing these grazable saltbushes (*Atriplex* spp.) at saline soil and feeding to sheep for at least 4–6 months in a year (when good quality fodder is not available) provide guidance for the others (Malcolm 1986; Warren et al. 1994). Some species of *Atriplex* have produced 8–17 t dry weight ha^{-1} when irrigated even with 40 ppt sea water (O'Leary 1988).

The above mentioned plants are capable of providing fodder from areas of comparatively higher salinity regimes (EC: 15–30 dS m^{-1}) where many other salt tolerant plants fail to produce sufficient grazable biomass. The agricultural extension service has started to persuade the farmers to raise these forage species on their saline wastelands or on barren sandy areas using saline water irrigation for producing feed during the gap period.

Fodder beet

Fodder beet (*Beta vulgaris* L.) is another very promising crop which has good potential for growing as a stock feed under saline environments. It is grown as supplementary cattle feed in the Netherlands, and many other countries. The top can be used for grazing and roots could be stored for stall feeding. Performance of fodder beet under saline conditions is increased by adding potassium in a fertilizer mix. It is successfully grown using saline water in irrigation on sandy soils.

Our trials on growing different spp./lines of fodder beet on the coastal sandy belt of Pakistan using saline water (EC: 17 dS m^{-1}) for irrigation have given promising results. Since the seeds are susceptible to salinity at germination, the seedlings are transplanted at three leaf stage. The crop is capable of producing good amount of forage and tuberous roots within a period of four months (Table 4). A system is being developed for sowing beet through a strip of mulch soaked with good quality water on sandy soil which would avoid the process of transplantation. Seedling after emergence could be directly irrigated with saline water after they have reached the three leaf stage.

In addition to the above mentioned plants, there are many other salt tolerant plants cultivated for the supply of food, fodder and fuel under saline conditions. *Distichlis spicata* is cultivated under hypersaline conditions in Mexico as cattle feed (Llerena 1994). A good number of salt tolerant plants could be selected from the data base record prepared by Aronson (1989). The bibliography of forage halophytes and trees for salt affected lands, prepared by Ismail et al. (1990) is a very valuable publication for this purpose. Information about utilization of saline areas of Western Rajastan through suitable plant introduction is available in the literature (Sen et al. 1985). Irrigation with highly saline underground water for cultivation of cereals, vegetables and fodder on different types of sandy soil has given encouraging results (Gupta & Gupta 1987).

Halophytic fodder plants are generally more tolerant to salt than the most salt-tolerant cultivars of conventional crops. Hence, through proper cultivation methods, they are capable of providing a good amount of fodder from saline wastelands or through saline water irrigation at barren sandy deserts.

Acknowledgement

The authors are indebted to Dr. Sadaqat Hanjra, Professor, University of Agriculture, Faisalabad, for providing information regarding feed requirements and available resources in the country. UNEP provided travel funds to enable Rafiq Ahmad, and M.A. Bodla to attend the Nairobi Workshop.

References

Ahmad, R. Ismail, S. & Khan, D. (1986). Use of highly saline water for irrigation at sandy soils. In: R. Ahmad & A. San Pietro (eds.), Prospects for Biosaline Research Proc. US-Pakistan Biosaline Research workshop, Karachi, Pakistan, pp. 387–411.

Aronson, J.A. (1989). HALOPH – A database of salt-tolerant plants of the world. Office of Arid Land Studies, University of Arizona, Tucson, 77 pp.

Gupta, S.K. & Gupta, I.C. (1987). Management of saline soils and waters. Oxford IHB Publishers, New Delhi.

Ismail, S., Malcolm, C.V. & Ahmad, R. (1990). A bibliography of forage halophytes and trees for salt-affected land: Their use, culture and physiology. Department of Botany, University of Karachi, Karachi, Pakistan, 258 pp.

Llerana, F.A. (1994). Massive propagation of halophytes (*Distichlis spicata* and *Tamarix* spp.) on the high saline-alkaline soils in the ex-Lake Texcoco, Mexico (this volume, Ch. 30).

Malcolm, C.V. (1986). Rainfed halophyte forage production on salt-affected soils. In: R. Ahmad & A. San Pietro (eds.), Prospects for Biosaline Research Proc. US-Pakistan Biosaline Research Workshop, Karachi, Pakistan, pp. 541–551.

Malik, K.A., Aslam, Z. & Naqvi, M. (1986). Kallar grass: A plant for saline land. NIAB Publication, Faisalabad, 93 pp.

O'Leary, J.W. (1988). Saline environments and halophytic crops. In: E.E. Whitehead, C.F. Hutchinson, B.N. Timmermann & R.G. Varady (eds.), Arid Lands Today and Tomorrow. Proc. Int. Res. Dev. Conf., Tucson, AZ, pp. 773–790.

Qureshi, R.H. (1992). Minutes of PARC-ACIAR Annual Review Meeting, Faisalabad.

Sen, D.N., Jhanb, R.B. & Bhandari, D.C. (1985). Utilization of saline areas of Western Rajasthan through suitable plant introduction. Proc. National Symp. on Evaluation of Environment (Supplement in Geobios), pp. 348–360.

Warren, B.W. Casson, T. & Ryall, D.H. (1994). Production from grazing sheep on revegetated saltland in Western Australia (this volume, Ch. 24).

18. Forage production from salt-affected and water-logged areas in Punjab, Pakistan
A case study

M.A. BODLA,[1] M. SHABBIR BAIG[2] & S.R.A. SHAMSI[3]
[1]*International Waterlogging and Salinity Research Institute (IWASRI), Lahore, Pakistan*
[2]*Ecology Unit, Soil Survey of Pakistan, Lahore*
[3]*Botany Department, University of the Punjab, Lahore*

Abstract. Persistent high watertable and consequent salinity and sodicity are among the major ecological factors rendering more and more soils unfit for agriculture. Such areas however, could be utilized on a sustainable basis for a variety of land-uses including saline pasturing and saline forestry. The study deals with the delineation, qualification and quantification of halophytic vegetation covering 1.14 m.ha saline-sodic and waterlogged areas of the Punjab province. An assessment of the forage and livestock production potential of these areas both under unmanaged and managed environments is presented. In all, twenty three 100 m^2 freshly laid out enclosures, each representing a homogenous landscape ecological unit, were studied in three bioclimatic regions: arid, semi-arid and sub-humid variants of sub-tropical continental monsoonal climate within Punjab. In arid and semi-arid environments, herbage and browse production inside the protected areas was about double that in the grazed area. In sub-humid region it was thirteen times that of the unprotected counterparts. In drier zones, area requirement per animal unit decreases from 3–4 ha under unmanaged conditions to 1.5–1.6 ha under managed conditions, whereas in wetter zones it is from 8.3–0.6 ha. The number of animal units which can be fed under controlled conditions when halophytes are incorporated into the system was about double in arid and semi-arid sites and thirteen times as many in sub-humid sites.

Key words: Animal units, Climatic zones, Feeding, Grazing capacity, Halophytes

Introduction

The whole of western Pakistan i.e. Baluchistan, Sind, Thar and Punjab is classified as in the arid zone (Figure 1). Punjab is one of the four Provinces of Pakistan covering 20.8 million ha in arid, semi-arid and sub-humid subtropical continental monsoon rains regions (Figure 2).

The alluvial and generally fertile soils are irrigated by a 37,000 km network of canal system. About 50% of water is lost as seepage and evaporation. This has raised the watertable and caused consequent salinity and sodicity. In spite of intensive and expensive anti-waterlogging and salinity measures an area of 1.14 million hectare is affected by waterlogging and salinity and rendered unsuitable for conventional agriculture.

These areas however, can be economically utilized for a variety of land-uses including among others livestock production through growing of ecologically adapted forage plants (grasses, shrubs and trees).

Most of the work conducted so far on natural vegetation of salt-affected and waterlogged areas in Pakistan is localized and qualitative in character with few exceptions (Hussain 1968; Khan 1971; Baig 1977, 1980, 1981, 1985; Alferaz 1982; Malik et al. 1983; Mahmood and Bhatti 1983; Sheikh 1986; Sandhu et al. 1986; Ahmad 1987, 1989; Mahmood 1987; Muhammad 1987, 1988; Hussain et al. 1989; Bodla 1988, 1990, 1991).

Although the development potential of these lands now in pastoral use is low, their vast extent and the large number of livestock they support makes them a highly valuable resource.

V.R. Squires & A.T. Ayoub (eds.), *Halophytes as a resource for livestock and for rehabilitation of degraded lands*, 231–234.

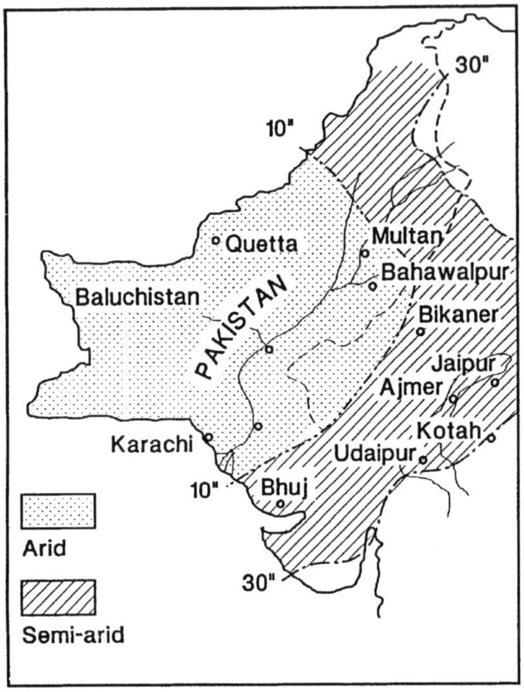

Figure 1. Map of Pakistan and part of northern India showing the delineation of the arid zone.

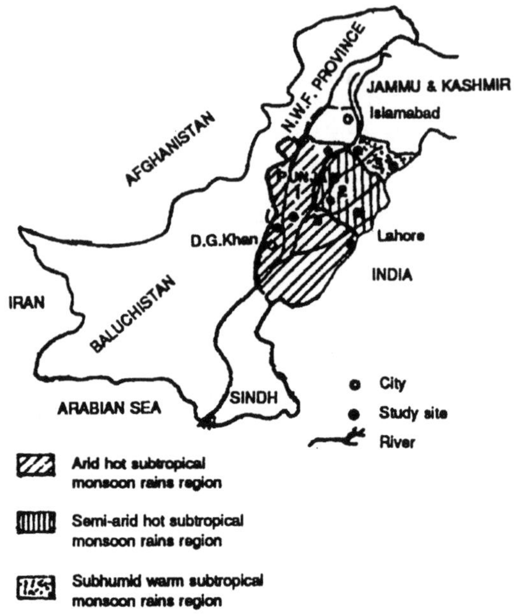

Figure 2. Map showing location of study sites.

Materials and methods

The study was conducted in three different climatic zones (Baig 1985) of Punjab, Pakistan viz: Arid Hot Subtropical Monsoon Rains Region, Semi-arid Hot Subtropical Monsoon Rains Region and Sub-humid Warm Sub-tropical Monsoon Rains Region. Each region was further divided into various landscape ecological units based on physiography, soils, hydrology and vegetation. Initially, a detailed vegetation survey of each landscape unit (involving nine localities in three climatic zones) was carried out. The vegetation was classified into associations and sub-associations by using Braun-Blanquet (1964) approach of phytosociology and correlated with site factors.

Permanent enclosures of 100 m^2 each were installed at 23 sites to note the ecological recovery in time and space. Each enclosure was further divided into 1 m^2 quadrats which were harvested in post-winter and post-monsoon seasons.

The productivity of unprotected area was also assessed concomitantly. Fresh, as well as dry matter, yields (gm m^{-2}) were computed and possible take off per ha as well as per zone was estimated to calculate grazing capacity in terms of hectares per animal unit using the formula of Thalen given in Baig (1977).

The factor not so far taken into account is the relative palatability of forage in different agroecological zones. The animals eventually eat almost all the plant material, if nothing better is offered. Therefore all the available forage has been considered palatable. The proper use factor, already applied allows the animals to select the best 60–70% of the available forage.

Results

In all, nineteen plant associations (named after dominant species), from variously salt-affected and waterlogged areas in three agro-ecological zones were identified including *Aeluropus lagopoides*, *Aeluropus lagopoides – Suaeda fruticosa*, *A. lagopoides – Phragmites karka*, *S. fruticosa –*

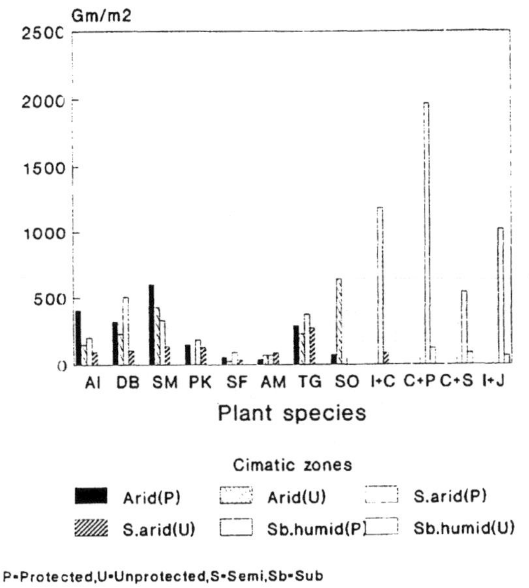

Figure 3. Summarized forage production (DM) from saline and waterlogged areas.

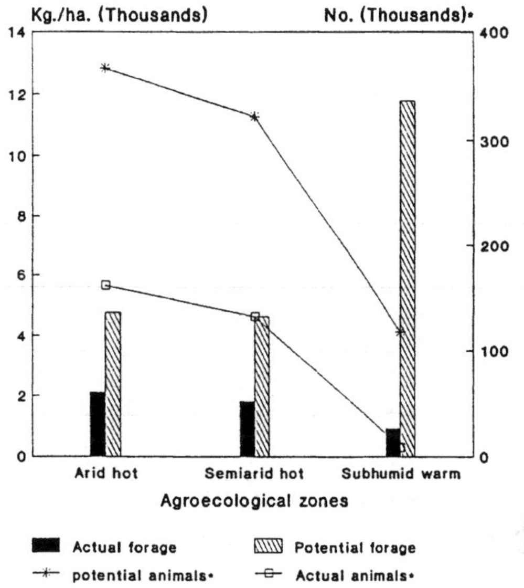

Figure 4. Actual and potential forage production and possible number of animals.

Desmostachya bipinnata, *Cynodon dactylon*, *Sporobolus arabicus*, *Eleusine flagellifera*, *Peganum harmala*, *S. fruticosa*, *Tamarix gallica*, *T. gallica – Prosopis glandulosa – S. fruticosa*, *Prosopis cineraria*, *Salvodora oleoides*, *Scirpus mucronatus*, *Scirpus maritimus*, *Paspalum paspalodes*, *Cyperus rotundus*, *Phragmites communis*. These associations are further classified into 47 subassociations having various numbers of species.

The dominant species protected to assess their forage production include *A. lagopoides* (*AL*), *D. bipinnata* (*DB*), *S. mucronatus* (*SM*), *P. karka* (*PK*), *Imperata cylindrica* (*IC*) *C. dactylon* (*CD*), *P. paspaloides* (*PP*), *Sporobolus arabicus* (*SA*). *Juncus Sp* (*J*). *S. fruticosa* (*SF*), *Alhagi maurorum* (*AM*), *T. gallica* (*TG*) *and S. oleoides* (*SO*)

The annual forage production (total of post-winter and post-monsoon harvests) of the species is represented in Figure 3.

The forage production of *A. lagopoides*, *S. fruticosa* and *S. maritimus* is higher in arid climate as compared to semi-arid environment, whereas *D. bipinnata*, *A. maurorum* and *T. gallica* yield more production in semi-arid zones.

Analysis reveals that the herbage and browse production in arid and semi-arid regions is almost equal. In subhumid region, the browse production is nil, the absence of browse species is correlated to marshy habitat, not favourable for the growth of shrubs and trees. However, the margins of marshlands, allow shrubs/trees to grow but they are heavily browsed by the animals in the early stages of their growth. The elimination of browse species from the saline patches is due to the excessive cutting by the local people for fuel purposes.

The quantitative assessment of actual gross forage production shows that all the areas in different climatic zones are under heavy grazing pressure. In arid and semi-arid zones, 3–4 ha are required to fulfill the feed requirement of an animal unit (one mature cow). The situation is worst in sub-humid zone where 8.3 ha are required for one animal unit under an unprotected scenario.

Inside the enclosure the annual forage production in arid and semi-arid is almost double that outside the enclosure while in the sub-humid zone it is thirteen times higher (Figure 4).

In arid and semi-arid zones, per animal unit area requirement from protected grazing lands decreases from 3–4 ha to 1.5 to 1.6 ha while in sub-humid zone, the area requirement per animal unit decreases from 8.3 ha to 0.6 ha.

The potential number of animal units which can feed from unprotected conditions is 161,778 in arid zone, 132,077 in semiarid zone and 8,507 in sub-humid zone. Under protected conditions these numbers may be doubled in arid and semi-arid conditions and increased thirteen times in sub-humid zones (Figure 4).

It is concluded that proper management of the area may increase its gross productivity 2–13 times, depending upon its location in different climatic zones. The wetlands in sub-humid climate are much more productive than the saline areas in arid and semi-arid climate.

Recommendations

1. Salt tolerant and economic plant species like, *Acacia nilotica*, *Tamarix aphylla*, *T. gallica*, *Prosopis cineraria*, *P. juliflora*, *Salvadora oleoides*, *Phoenix dactylifera* and *Eucalyptus* sp. may be planted in salt affected areas.
2. On wetlands, grass species like *Cynodon dactylon*, *Aeluropus lagopoides*, *Sporobolus arabicus*, *Scirpus mucronatus*, and *Cyperus rotundus* may be generated to develop wetlands as pastures.
3. The revegetated area may initially be properly protected and used through rotational grazing.
4. Permanent marshlands should not be reclaimed but developed for inland fisheries and recreational purposes such as for shooting game birds.

References

Ahmad, R. (1987). Saline agriculture at coastal sandy belt. Final research report, Bot. Dept. Uni. Karachi Pakistan.

Ahmad, R. (1989). Research on forage shrubs in Pakistan. Paper presented at Inaugural meeting of ACIAR project 8614, Islamabad Pakistan.

Alferaz, A.C. (1982). Fodder crops and pastures. Livestock Dev. Center, Bahadur Nagar, Pakistan, Assignment Report FAO, 32 p.

Bodla, M.A. (1988). Ecological studies on natural vegetation of salt-affected and waterlogged areas in the Punjab. Sample site Pind Dadan Khan. Int. Waterlogging and Salinity Res. Inst. Pub.2.

Bodla, M.A. (1989). Ecological studies on natural vegetation of salt-affected and waterlogged areas in the Punjab – sample site Dera Ghazi Khan. Int. Waterlogging & Salinity Res. Inst. Pub. No. 6.

Bodla, M.A. (1991). Ecological studies on natural vegetation of salt-affected and waterlogged areas in the Punjab – sample site Kot Adu/Diara Din Pinnah (Thal). Int. Waterlogging and Salinity Res. Inst. Pub. No. 23.

Braun Blanquet, J. (1964). In: Robert H. Whittaker (ed.) Classification of plant communities. Dr. W. Junk Publ., The Hague, Boston, 1978, pp. 18.

Hussain, F., Qureshi, M. & Shaukat, S. (1989). Vegetation of waterlogged and saline areas of Hazro, District Attack (Pakistan) Pak. Jr. Agri. Res. 10 (2): 158–162.

Malik, M.N. & M.I. Sheikh (1983). Planting of trees in the saline and waterlogged areas. Pak. Jr. For. 33 (1): 1–17.

Muhammed, N. (1989). Range land management in Pakistan. Centre for integrated mountain development Kathmandu, Nepal, 193, pp.

Sandhu, G. R. & Qureshi, R.H. (1986). Salt affected soils of Pakistan and their utilization. Rec. Reveg. Res. 5: 105–113.

Shabbir Baig, M. (1977). An integrated approach for evaluation of rangelands in arid zones. Pak. Soil Bull. 12, Soil Survey of Pakistan.

Shabbir Baig, M., Akram, M. & Hussain, M.A. (1980). Possibilities for range development in Cholistan Desert as reflected by its physiography and soils. Pak. Jr. Forest 30 (2): 61–71.

Shabbir Baig, M. (1981). Vegetation classification for evaluation of rangelands in arid zones. Pak. Soil Bul. 15. Soil Survey of Pakistan, 57 p.

Shabbir Baig, M. & Qamar Ali (1985). Vegetation ecological zones of Pakistan. Proc. XII Int. Forum on Soil Taxonomy and Agrotechnology Transfer, Lahore, Pakistan.

19. Prospects for saltbush (*Atriplex* spp.) as animal feed in irrigated and marginal lands in Sudan

MOHOMMED AHMED MOHOMMED KHAIR
Agricultural Research Corporation, P.O. Box 126, Wad Medani, Sudan

Abstract. Previous studies on saltbush and the degree of salinity and sodicity in Sudan soils were reviewed. Due to its high dry matter yield, high crude protein percentage and high leafiness, saltbush could be a good forage crop in the Sudan. The prospects of growing the plant within the crop rotations of the irrigated areas and/or to rehabilitate the marginal saline/sodic soils are discussed.

Key words: Crude protein, Forage crop, Nile river, Salinity, Sodicity

Saline and sodic soils in the Sudan

Sudan is a large country extending from latitude 4° N to a point very close to the Tropic of Cancer. As a consequence, different climates prevail in the different parts of the country. This includes the very humid tropics in the south to a completely arid zone in the north. In between, regions with different rainfalls prevail. The degree of the soil salinity and/or sodicity in the country seems to follow an opposite pattern to that of the rainfall. The saline/sodic soils (% total area) increases from the south to the north. Table 1 shows the saline/sodic soils (% of total area) in some surveyed soils in the different parts of the country (Nachtergaele 1976). The saline/sodic soils constituted only 13% in upper Nile state in the south of the Sudan. The corresponding figures in the central and northern Sudan are over 80, 76 and 68% in the Central, Khartoum and northern states respectively.

In addition to the Sudan Gezira Scheme, the Central state includes the areas along the banks of the white and the blue Niles. Most of the soil, in the white Nile areas are salt affected and consequently are characterised with low crop production. Even in the Gezira Scheme, sizeable areas are either sodic or saline. Most areas around Khartoum are undercropped due to their salinity and sodicity. The most common crop in these areas is a forage sorghum variety which is a moderately salt tolerant crop (Khair & Jarrell 1987). The same picture exists in the northern state. The net result of the salinity/sodicity problems is that large areas are left bare.

In addition to their limited returns, those bare soils favour desertification especially in the White Nile where vast areas of clay soils along the White Nile banks are being covered with sands.

Saltbush in the Sudan

Saltbush, *Atriplex* spp. were reported to be found naturally in the Red Sea area of the eastern state (Andrews 1950). Queensland saltbush, *Atriplex muelleri*, gave 10 tons DM/ha at the Gezira Research Farm. Two experiments on saltbush were conducted in the Gezira Research Station during 1930s (Greene & Snow 1939). In the first experiment, four different species viz *Atriplex hortensis* (Massey), *Atriplex halimoides*, *Atriplex semibaccabta* and *Atriplex muelleri* were compared. Table 2 shows the CP% in the different parts of the different species. Interspecific variations exist in both CP% and in the leafiness (leaves as % of

V.R. Squires & A.T. Ayoub (eds.), *Halophytes as a resource for livestock and for rehabilitation of degraded lands*, 235–237.

Table 1. Soils with some degree of salinity and/or sodicity in some surveyed areas in the Sudan

Area	Total area Feddan*	None saline and none sodic soil (%)	Soils with some degree of salinity and/or sodicity (%)
Upper Nile	131,000	85	15
Central State			
South Gezira	37,000	19	81
W. Nile Schemes	92,000	17	83
Central gezira	–	4	96
Geneid	43,000	16	84
North Gezira	–	2	98
Khartoum State	103,000	24	76
Northern State	487,000	32	68

Adapted from Nachtergaele (1976, Table 3).
*One Feddan = 0.42 ha.

DM). The CP% of all species in the leaves was invariably surpassed by that of *Atriplex hortensis*. Overall, *Atriplex semibaccata* had higher CP% than the other species. Regarding the leafiness, *Atriplex semibaccata* had the highest leafiness followed by *Atriplex halimoides*.

In the second experiment, the role of Queensland saltbush (*Atriplex muelleri*) in crop rotation was studied. The treatments comprised continuous saltbush, i.e., saltbush after saltbush in comparison with saltbush grown after cotton. The growth of the continuous saltbush was markedly poorer. The CP% of the continuous saltbush was almost half of that grown after cotton (Table 3). Moreover, leafiness of the saltbush after cotton was higher than that of the continuous saltbush.

Discussion

Despite the interspecific variations in the CP% in the different parts (Table 2), the leaves of the species with the lowest CP% compares closely with earlier growths of tropical forage legumes in the Sudan (Khair 1991) and alfalfa (Khair 1992a) and of temperate forage legumes (Abd-Elmoneim et al. 1990). The CP% in the stem of all species also compares closely to the leaves of forage sorghum (Khair 1992b). Regarding the leafiness, that of *A. halimoides* and *A. semibaccata* compares closely with that of tropical forage legumes (Khair 1991).

The purpose of introduction of this plant to the Sudan Gezira Scheme was to eliminate the sodium from the soil. Sodium is added continuously by the irrigation water. Saltbush takes up one fifth of its dry weight as sodium (Greene & Snow 1939), it was calculated that due to the extended crop rotation which was adopted in the 1930s, a single crop of the saltbush every 13 years would be enough to eliminate the sodium which was added to the soil over such a period by the irrigation water. Currently, as the cropping system is more intensified, more water is being applied and therefore, the need for the saltbush is even higher. In addition to its reclamation role, the higher dry matter yields, CP% and leafiness makes the saltbush a good forage crop. This seems to be in harmony with the recent idea of integration of animals with field crops in the irrigated schemes.

As discussed earlier, vast areas in the Sudan,

Table 2. Crude protein (CP%), P_2O_5 and dry matter (% of the whole) of different parts of four species of saltbushes

		Species			
		A. hortensis	*A. halimoides*	*A. semibaccata*	*A. muelleri*
Leaves	CP%	25.75	20.75	21.63	19.81
	P_2O_5	0.73	0.47	0.57	0.46
	D.M%	0.8	41.0	55.0	24.0
Stem	CP%	8.94	7.56	21.19	6.63
	P_2O_5	0.36	0.26	0.36	0.25
	D.M%	52	40.0	45.0	28
Seed	CP%	28.26	15.81	–	9.38
	P_2O_5	1.04	0.50	–	0.34
	D.M%	20.0	19.0	–	48

Adapted from Greene & Snow (1939, Table 25).

Table 3. Crude protein (CP%) in the different parts of Queensland saltbush *Atriplex muelleri* grown after cotton (C–S) or after saltbush (S–S)

Plant parts		CP%	DM (% total)
Leaves	C–S	22.13	39.0
	S–S	11.00	30.0
Stem	C–S	8.75	39.0
	S–S	4.44	28.0
Seeds	C–S	12.75	32.0
	S–S	6.31	42.0

Adapted from Greene & Snow (1939, Table 24).

are considered marginal due to their salinity and/or sodicity. In such areas, saltbush could be a suitable plant. Saltbush is a C_4 plant, high in CP% and drought tolerant. Henceforth it can play an important role in the rehabilitation of those lands to stop the desertification and to ensure adequate returns in terms of animal production.

In conclusion, saltbush is a plant with marvellous prospects in the irrigated areas as feed for livestock. It has a role for the removal of the sodium added by the irrigation water. In the marginal lands it can serve as animal feed and to rehabilitate the marginal lands and henceforth minimise or stop desertification.

References

Abd-Elmoneim, A.M., Khair, M.A. & Rihawi, S. (1990). Effect of genotypes and plant maturity on forage quality of certain forage legume species under rainfed condition. J. Agronomy and Crop Science 164: 85–92.

Andrews, F.W. (1950). The flowering plants in the Anglo-Egyptian Sudan, Vol. 1. Buncle and Co., Arbroath, Scotland.

Greene, H. & Snow, O.W. (1939). Soil improvement in the Sudan Gezira. Journal of Agricultural Science 29: 1–34.

Khair, M.A.M. & Jarre, W.M. (1987). Responses of *Sorghum bicolor* (L.) Moench cv. (Abu Sabeen) to saline media. Sudan Agricultural Journal 12: 118–137.

Khair, M.A.M. (1991). Seasonal growth and quality of tropical forage legumes under irrigation. Annual Report, Gezira Research Station. Was Medani, Sudan.

Khair, M.A.M. (1992a). Effects of variety, and seed rate on the yield and quality of alfalfa. Annual Report, Gezira Research Station, Wad Medani, Sudan.

Khair, M.A.M. (1992b). Forage yielding ability of certain grain and forage genotypes of sorghum. Annual Report, Gezira Research Station, Wad Medani, Sudan.

Nachtergaele, F.O.F. (1976). Studies on saline and sodic soils in Sudan. Technical Bulletin No. 24. FAO/UNDP project for strengthening the soil survey administration; SUD/71/553.

20. Halophytes and desertification control in Iraq

F. AL-FARRAJII[1] & M.R. AL-HILLI[2]
[1]*Department of Desertification Control, Irrigation Ministry, Baghdad, Iraq;*
[2]*Department of Biology, Science Faculty, University of Baghdad, Iraq*

Abstract. Halophytic communities are mostly confined to large depressions and plains around the marshes in southern Iraq. The arid regions are characterized by mobile pseudo-sand dunes which threaten roads, irrigation canals, and drains and other infra-structures. Secondary soil salinity is also a problem in the irrigation areas. The work of the Department of Desertification Control is described. The principal task is to control shifting 'sands'. Mechanical methods have proved successful. These are described.

Key words: *Atriplex* plantations, Irrigation, Pseudo-sand, Range management, Sand dune fixation, Soil survey

Introduction

Iraq is located between longitudes 38°42′ E and 48°32 E and between 29°27′ N and 37°23′ N (see location map). It covers an area of about 444,500 km. More than half of this area comes strictly under an arid zone.

The sand dunes in the country are divided into two categories: Active sand dune (Pseudo-sand, Hummocky, high salinity, waste land, bad land, severe wind erosion) and sand land (sand and mainly quartz, severe wind erosion, free to slightly saline, waste land and bad land). This is some what ironical because it was man himself who created the desert by poor and mis-management of nature (Al-Farrajii 1983).

About 780,000 ha of alluvial and more or less saline soils were located between Shatt Al Diwania and Mabkharat Al-Nasar (location map No. 1). On the East of Al-Masab Alam (drain canal) have been exposed to severe degradation. Sand dune formation and accelerated process of desertification. This is due to mis-use of the range land (severe grazing, uprooting and cereal crop cultivation) and irrigation with saline water and inadequate drainage system.

The system of grazing management and consequently the deteriorated condition of the range land is pro affected by present social organization. Man exploits the range lands with a very heavy stock (mainly camels) using two main ways: transhumant system which is more common and sedentary range land usage. This process consequently will lead to active mobile massive dune and to the formation sands hummocks (barchans), which endanger both the old and the new drain canals under the effect of being dumped with sand particularly on an area of about 120 km long at Al-Fajir city.

Activities of reclamation project were concentrated so far on fixation of sand dune by mechanical ways through covering them with clay loamy soil excavated from the interdune areas by heavy soil moving equipment then planting and reseeding the protected area with *Tamarix*, *Prosopis Atriplex* and with native shrubs such as *Seidlitzia* sp. and *Cornulaca* sp.

Climate

The climate of Iraq in general is an arid or semi-arid type having very hot summer and mild winter.

V.R. Squires & A.T. Ayoub (eds.), *Halophytes as a resource for livestock and for rehabilitation of degraded lands*, 239–248.

LOCATION MAP

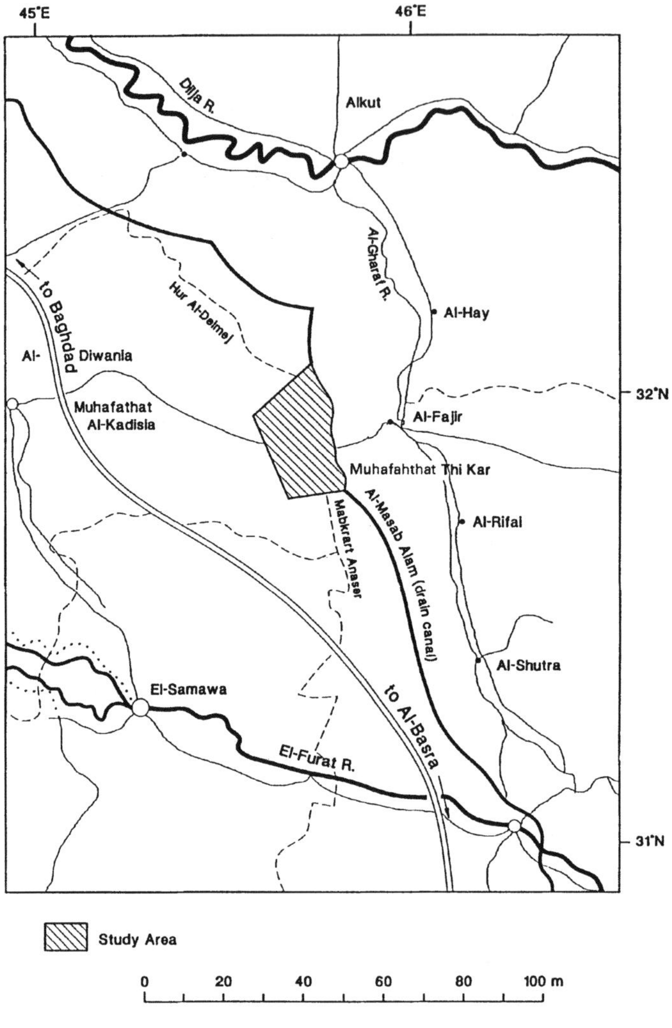

The precipitation is erratic and fall during cold season. The average rainfall is around 125 mm. The differences in rainfall distribution from year to year or within the rainy season has a direct influence on the natural vegetation as well as on the cropping systems.

The mean minimum temperature does not fall below 5 °C. However the absolute minimum can go down below 0 °C in December and January. During winter frost may occur on an average of 15 nights per annum. The mean maximum temperature is very high (more than 40 °C) from June to September. Maximum daily temperature exceed 35 °C on an average of 150 days annually.

The daily evapotranspiration (ETP) rates vary from 1.5 mm in the winter months reaching its peak during summer months with a value of 9 mm. The total yearly E + T is around 2000 mm (according to Baghdad Airport station) (Gaddas 1990).

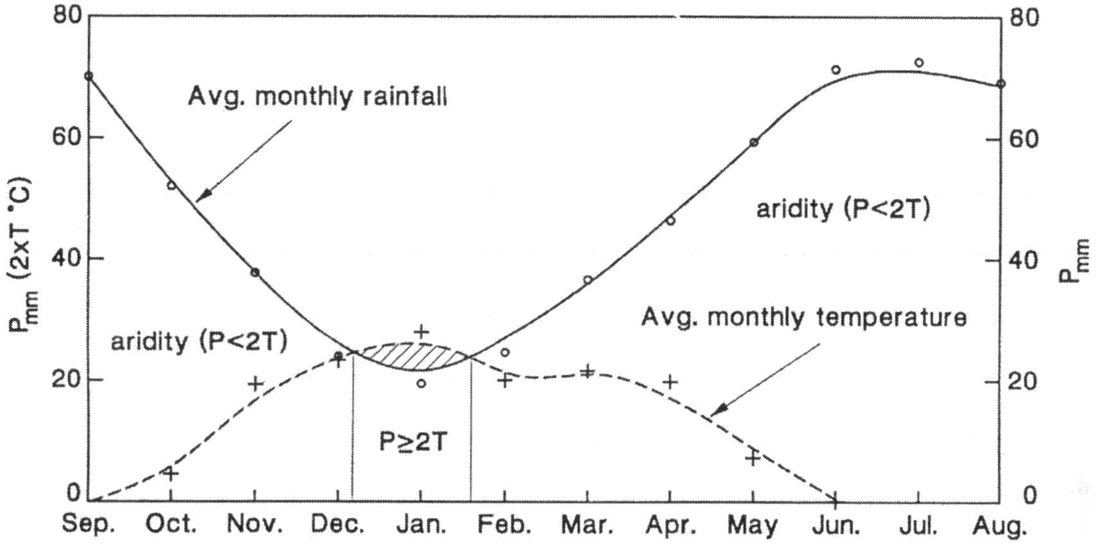

Figure 1. Aridity according to Bagnouls and Gaussen's Formula – Al Hay Station.

The dry season is very long. It lasts more than ten months according to Gaussens formula (Figure 1).

Wind speeds reach their maximum during July, these winds are generally from N.W. (see Figure 2). They do not produce the severe dust storms. Dust storms occur more frequently during Autumn with an average frequency of 20 days per annum. Thunder storms accompanied by high winds and dust occur with an average frequency of 12 days per annum and tend to be more frequent during the spring months causing significant damage to crops and soil by wind erosion. The result is the formation of hummocks (barchans) and mobile sand dunes making a new harsh life conditions for the people.

Soil

The soil of Iraq has been studied from time to time by many workers in the field, Buring (1960) had studied these soil in detail and also had taken into account all available findings.

According to new report, prepared for the

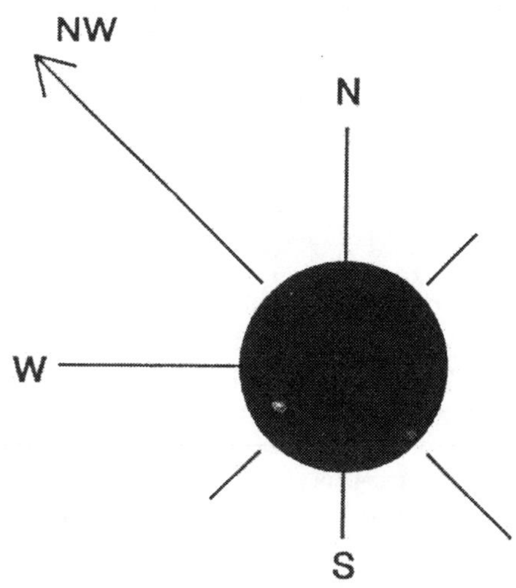

Figure 2. Compass card showing the wind directions and the number of windy days – Al Hay Station.

Flora of Iraq, there are in reality four main zonal physiographic divisions of the soils in the country (Guest 1966) namely:

1. The Mesopotamian plain formed by sedimentation of rivers, carrying particles of the soil from the higher regions and depositing them in the lower basin of the Euphrates and Tigris. The soil is characterized by salinity and partly by alkalinity.
2. The desert soil characterized by poor organic matter and divided into three sub-zones, (i) the Jezira region consisting mostly of gypsiferous soil, (ii) the north-western desert consisting mainly of lime-stone, (iii) the southern desert consisting of lime-stone and gravelly sand soil.
3. The upland and hilly region mainly occupy the high plain and foot hills on the north and north-east. The soil of this region consists of many kinds of elements which varies from red soil to brown soil. Part of the upland and foot hills consists of shallow soil overlying gypsum, mud, sand-stone and gravel and some parts are deeply eroded.
4. The mountain soil of the north, consists mainly of lime-stone except the peaks. The mountain region in general is very rough, broken and stony and comprises of lithosol, gravels, broken stones and shallow soil. In between the mountain chains the soil is brown or deeply brown and rich in organic humus matter.

The region comprises a nearly flat plain of fine textured soil with a chequered history of range land use, flooding and irrigation. Most of the soils are salt affected to some degree. Grazing, fuel collection in the past and present agricultural use has been responsible for the removal of most of the perennial vegetation from a large portion of the region. It was not caused by agricultural use of land but by the mis-use of these land. As a result an extensive area of the plain has been degraded to pseudo-sand which has been piled up by the wind forming mobile dunes. The soil in the dunes is virtually the same texture as the underlying hard soil (silty loam or silty clay loam). The dunes are advancing on towards roads, canals, agricultural lands and other strategic points, over an extensive area.

Halophytes (saline) plants

Halophytes are plants which are able to grow on salt soils. High concentration of salt in the soil solution has been regarded in the past as making it difficult for halophytes to absorb water. Glyciophytes which grow on saline soils also accumulate salts in the cell sap where salt concentration is almost as high as that of the soil solution. True halophytes shows enhancement of growth in salt solution.

Xero-halophytes are found most often on lands where the top sandy soil is slightly saline and not continuously wet, local precipitation is the main source of fresh water. These habitats, usually are located in the semi-desert areas within the marshes and bordering the Ahwar on its southern reaches.

Salsola jordanicola and *Salsola incanescens* are the main dominants, while *Bienertia cyclopra*, *Seidlitzia rosmarinus*, *Aizoon hispanicism*, *Heliotropium sapinum*, and *Frankenia pulverulenta* occupy large flat depressions where the top-soil is soft and powdery or sandy with minute salt crystals, the sub-soil is silty and relatively hard. The greater part of Iraqi halophytes consists of monoregional plants.

Halophytic communities

This habitat type includes large depressions within the region (most common on Euphrates and Shatt Al-Gharraf lower reaches) and the plains around the Marshes (Al-Ahwar). These are complexes in which shallow water-table and high salinity prevail. The source of surface water to depressions is mainly run-off collecting from neighbouring high ground; and while the source of water is either by occasional flood or spring rain showers. The depressions in general showed a high surface salinity of solonchak type characterized by almost complete absence of soil structure, high amount of sodium and calcium chlorides dark in colour, and was often wet throughout the year. The plains on the other hand, had white salt crust in the dry season, these solonchak soils were less saline, and might have resulted from leached solonchak, but the proportion of exchangeable sodium was high.

Table 1 The standing crop biomass ($g\,m^{-2}$) of the above-ground parts of the *Suaeda vermiculata* Community. Suq Ash-Shuyukh, 1973 (Al-Hilli 1977)

Stand No.	*Suaeda vermiculata*	Cover value (%)	Average height of plant (cm)
1	1138.6	95	85
2	896.2	100	110
3	905.9	100	90
4	689.6	100	90
5	430.2	60	60
6	616.0	70	85
7	927.2	100	120
8	785.6	90	100
9	618.1	80	90
10	517.8	100	100

1. *Suaeda vermiculata community. Suaeda vermiculata* L., is a succulent under-shrub 60–120 cm dark green, drying black, branching from the base, branches whitish, leaves short-petioled succulent, lower leaves obovate – oblong the upper ovoid to spherical, locally called 'Swaida'. This saharo – Arabian species was confined to the middle and southern part of the country, but more common in the south. The *Suaeda* shrubs in its proper habitat were dense and crowded and formed paches of more than 6–8 m – wide and 60–120 cm high, in which no other plant species beside the dominant was observed, its vegetation cover could range from 60–100%.

The soil was dark, loose, structureless to a depth of 25 cm below which the soil was more moist, moderately compact and impregnated with white salt veins, thin crusts (2–3 mm) of fine material and salt crystals cover the loose surface layer in the small catchment areas within the site (Al-Hilli 1977).

Soil through all depths was very highly saline (saline-alkaline group) and its salinity was of chloride type. The texture was mainly silt loam and silt clay loam in which silt particles were the dominant fraction. The soil was brownish in its dry state, and in its wet state the first two layers were dark yellowish brown while the third layer was dark brown, (Al-Hilli 1977).

The biomass of the standing crop of the above ground (new green twigs) parts of *Suaeda vermiculata* community was examined in late September 1973. Ten stands of 20 m intervals were clipped, data illustrated in Table 1. The dry weight value of the different stands ranged from 430.2 $g\,m^{-2}$ to 1138.6 $g\,m^{-2}$ with a mean value of 752.5 $g\,m^{-2}$, (Al-Hilli 1977).

Suaeda vermiculata in this zone was grazed by large camel herds and was also used as fuel by the natives.

2. *Tamarix gallica community. Tamarix gallica* L., is a perennial undershrub locally called Tarfha. It is 40–110 cm high, glaucescent, glabrous, leaves half clasping at base, ovate-deltoid, petals 5, disk 10-lobed, stamens 5, inserted between the lobes. The analysed site was in the Elbn-Adum locality, about 6 km N. Suq Ash-Shuyukh.

Tamarix grew in clumps of 4–6 m wide and 45–110 cm high, under each clump a small mound around (20–30 cm high) was formed of dead leaves, litter and sand materials. In certain localities of Suq Ash-Shuyukh *Suaeda vermiculata* was in association with *Tamarix gallica*.

The soil inhabited by this community was covered with salt crust of about 0.5 cm. Under which the soil was moist moderately compact and more or less homogeneous in colour as well as texture to a depth of 49 cm, salt veins appeared below the 25 cm depth (Al-Hilli 1977).

The substrata inhabited by *Tamarix* at this site was very highly saline (saline alkaline), its salinization was of chloride type and its texture was loam on the surface and silt loam beneath. The colour was pale brown in dry state (Al-Hilli 1977).

It may be noted that *Tamarix* is cut for firewood on a large scale throughout the country, a prosperous markets for home-made 'Tarfha' charcoal is still active in several places up to present day. Table 2 shows that the yield of the different stands ranged from 217.9 $g\,m^{-2}$ to 967.3 $g\,m^2$ with a mean value of 589.9 $g\,m^2$.

3. *Aeluropus lagopoides community. Aeluropus lagopoides* L. Trin. is a perennial grass with densely tufted culms, sending up spaced erect shoots from long, prostrate, widely spreading stems; stolons and/or rhizomes present, the latter covered with hairy overlapping scales: culms up to 15 cm tall. In the eastern sector of the Ahwar region the plant called 'Flue Wee' while in other areas of the region it is called 'Ichrish' or 'Shir-

Table 2. The standing crop biomass of the above-ground (new green twigs) parts of *Tamarix gallica* Community, Suq Ash-Shuyukh, Summer 1973 (Al-Hilli 1977)

Stand No.	*Tamarix gallica*	*Cressa cretica* (dry weight g m^2)	*Alhagi mannifera*	Cover value (%)	*Tamarix* height
1	848.8	50.7	9.2	70	90
2	645.2	–	–	40	40
3	629.4	–	–	50	85
4	583.3	–	–	60	80
5	771.0	1.2	–	60	110
6	540.6	–	–	50	85
7	217.9	–	–	30	72
8	967.3	11.6	–	50	85
9	357.0	–	–	40	75
10	340.8	–	–	45	45

Table 3. The standing crop biomass of the above-ground parts of the *Aeluropus lagopoides* Community, Suq Ash-Shuyukh, Summer 1973 (Al-Hilli 1977)

Stand No.	*Aeluropus lagopoides*	*Cressa cretica* (dry weight g m^{-2})	Cover value	*Cressa* individually
1	84.9	–	30	–
2	71.0	–	35	–
3	96.2	–	35	–
4	51.9	–	25	–
5	44.9	–	15	–
6	83.8	1.2	40	3
7	73.6	1.7	15	6
8	37.1	13.0	20	14
9	50.0	3.7	20	5
10	42.3	2.7	25	12

raib', in Surayfa the word 'Shirraib' is applied to *Aeluropus littoralis*. *A. lagopoides* is a widely distributed grass in the middle and southern parts of the country up to 350 m, alt. (Al-Rawi 1964).

The dry, hard soil of this site was covered with salt crust. In low-lying land the grass was covered with salt dust. Animal usually avoided *Aeluropus* and *Cressa* at this season (summer) due to salt content, but active grazing starts a short time after the first rain shower (Al-Hilli 1977).

The site characterised by a highly saline soil of chloride salinization type through all depths, but the texture was variable, it was silt loam, loam, clay loam and silt clay. The saline alkaline soil of this site showed a relatively high level of CEC and ESP.

The dry weight value of the different stands ranged from 37.1–96.2 g m^{-2} with a mean value of 63.6 g m^{-2} as is shown in Table 3.

Aeluropus is a valuable fodder plant especially in semi-desert areas or the plains around and within the Ahwar region. It is grazed mainly by sheep and goats after the plant growth had been washed with early shower. Natives sometimes wash the harvested grass to feed their starving animals. Guest (1933) listed this species among the valuable grazing plants of Iraq.

Halophytic grasses thriving where little else will grow and providing useful pasturing. When desert grazing is scarce but water accessible, Shepherds sometimes gather this grass and wash it to remove the salt after which it is fed to the starving sheep.

4. *Cressa cretica community*. *Cressa cretica* L. is a perennial, dicotyledon, deciduous herb locally called 'Shir Raib' widely distributed species in Iraq, it is reported to occur in several districts from north to south Iraq (Al-Rawi 1964). *Cressa* in the Ahwar region grew on solonchak soil. In

Table 4. The standing crop biomass of the above-ground parts of the *Cressa cretica* Community, Suq Ash-Shuyukh, Summer, 1973 (Al-Hilli 1977)

Stand No.	*Cressa cretica*	*Alhagi mannifera*	*Aeluropus lagopoides*	Cover value
	(dry weight g m^{-2})			
1	13.5	29.9	4.4	32
2	70.4	–	–	16
3	35.5	81.1	–	20
4	55.8	–	–	23
5	47.1	–	–	14
6	5.4	77.8	9.3	19
7	11.2	–	0.8	9
8	31.7	–	–	27
9	10.0	36.8	16.3	10
10	13.5	22.8	2.1	7

general its dense, homogeneous growth abounded on saline soil with shallow water table. *Cressa* individuals were noticed to grow with *Suaeda vermiculata* zone, with *Tamarix* spp., with *Prosopis farcta* under date palm zone. *Cressa* and *Aeluropus* jointly formed the ground vegetation cover under *Alhagi mannifera* growth, (Al-Hilli 1977).

Al-Hilli (1977) concluded, the saline dry land inhabited by *Cressa* had a salt crust cover under which a thin layer of loose soil abounded, the soil became hard and more compact at greater depth.

Cressa grow vigorously in saline soil. It spreads over all the middle and lower zones in Iraq. Thriving in the arid situation or in moist places by the wadies in the desert or in irrigated lands. Table 4 shows that the productivity of various stands ranged from 5.4–70.4 g m^{-2} with a mean value of 29.4 g m^{-2}.

5. *Seidilitzia rosmarinus*. Perennial shrubs belong to chenopediaceae height 90–80 cm. Stem straight, branches succulent white, blooming in winter season, growing in saline soil, and in hot desert grazed by camel. Bedouin use it as a detergent for washing their clothes and bodies instead of soap.

6. *Prosopis juliflora*. Shrub or small trees, up to 13 m, or more in favourable conditions, but generally 3–8 m, occasionally cultivated in gardens in the desert region of Iraq as an ornamental or hedge. It is deep rooted and it can develop tap roots to 20 m. It is drought and saline resistant and could be established where the water table is within reach. It can maintain itself without irrigation. Though its branches are likely to be cut back by frost it makes a quick recovery after winter. It grow quickly reaching a height of about 8 m in 8–10 years, and is abundantly cultivated in gardens and avenues especially, just across the south-east frontier. However, it is liable to wind damage when mature and its leaflets and other litter are a nuisance in gardens; nor will flowers and other plants flourish below it.

The leaves and young shoots can be grazed by cattle and the pods are of excellent stock feed, the spongy wall of the pod contains a highly nutritious syrup. The branches yield a gum similar to gum Arabic and the bark can be used as a tanning agent. The timber is hard and heavy and makes good charcoal.

7. *Atriplex* spp. These small shrubs rarely reach more than 2 m come mainly from Australia. The genus comprises about 200 species. The plants are drought resistant and can thrive well even under severe conditions. The Australian Salt bushes belong to this genus. *Atriplex* were imported to the Middle East, because of its drought resistance and also for high nutritional value as a sheep fodder. It is widely distributed in the arid and semi-arid zones.

The main species widely cultivated in Iraq for above purposes are *Atriplex nummularia*, *Atriplex lentiformis*, and *Atriplex halimus*. Recently other species brought from Australia like *Atriplex amnicola* and *Atriplex undulata* to be cultivated in Iraq. Al-Farrajii (1983), concluded that *Atriplex nummularia* could survive in salty soils at very high levels of water table.

Problems of desertification in Iraq

Desertification as a spread of desert-like conditions in arid or semi-arid areas up to 600 mm rainfall is due to man's influences or to climatic change.

There are three distinctive types of desertification problem in middle and south of Iraq:

1. Sand dunes cover extensive area of the land either immediately in, or just up wind (N.W.) of the intended path of the new Main Drainage Canal and Highway Roads presently under construction. These dunes threaten and invade these projects.
2. Wind eroded soils essentially devoid of topsoil and vegetation exist adjacent to the sand dunes and covers even more extensive areas of the Al-Masab Alam landscape these lands are still being actively eroded and appear to be a major source of the sand being deposited in the dune area.
3. Fallow or poorly vegetated salt affected and poorly drained agricultural soils cover even more the landscape in the Thi-Qar Governorate than either of the two previously described areas. They are primarily cropped using a winter grain/fallow system, much of these soils are visually salt affected.

About 25% of the agricultural land in Iraq is affected by active sand dune movements (Annual Abstract of statistics 1971). The total area of active sand dune in Iraq is more than 750,000 ha (Buring 1960). Recently this area increased.

The sand dune in Iraq manifest themselves in different forms ranging from low hummocks and sand ridges to small barchans. The dunes reach heights from 3–6 m. The general movement of the dunes in Iraq is in a south-easterly direction because of prevailing north and north-westerly wind.

There are three main locations and areas of sand dunes in Iraq:

1. Around Baiji and Samarra, north middle of Iraq.
2. Area between the two main rivers Tigris and Euphrates (Kut, Qadisia, Muthana and Thi Qar Governorates).
3. A sand-belt more or less parallel with Euphrates on the western side.

The blowing away of fertile soil and sand dune movement has a number of consequences such as; the engulfing of agriculture fields and grazing lands, destruction of irrigation and drainage systems. Direct affect on roads and railway-line, villages, etc. This has led to lower productivity of livestock industries and high maintenance costs of irrigation, drainage and communication systems.

It is therefore, seen that shifting sand particles have resulted in an imbalance situation between the human and animal population on the one hand and the plants, water and land resources on the other.

Due to this growing problem. The Iraqi government has pooled its resources to stop further advancement of dunes through the Desertification Control Department.

The aims of this Department are:

1. To stabilize the wind eroded land causing the dune formation.
2. To protect roads, drains, canals, agricultural land and other facilities from windblown materials.
3. To produce forage and fuel.
4. To improve the environment for human habitation and for wild life.

Sand dune fixation in Iraq

According to the sand dune movements many strategic projects (Main Drainage Canal, Highway roads, Railway-lines and others) are directly affected by this movement. Dunes must be fixed in order to protect these projects from sand movement.

1. *Mechanical method*

Sand dunes in middle and south of Iraq are really pseudo-sand dunes. The fine textured material is not sand and it is misleading to refer to it as sand. So a clay soil covering has been shown highly successful. The source of clay soil (from interdune areas) is moved onto dunes by heavy equipment. Areas treated up to 8 years ago are still completely stable and natural vegetation has stabilized the area. In north of Iraq, where the sand is real, this method cannot be used. Protection of the dune area from grazing by fencing and planting shrubs and trees, especially using cutting of *Tamarix articulata*, which gave successful results in Baiji.

2. *Petroleum and chemical method*

To reduce the erodibility, spraying of the soil with bituminous emulsion, commonly called sand dune stabilization oil, or with a synthetic rubber or latex or other chemical products can be done. The methods are used just after the first rain shower, following sowing of area with drought and salinity resistant species. The spray forms a thin layer over the dune's surface. It has high permeability for water and the germinating seedlings can emerge through it. This thin layer also reduces the evaporation from the soil, so at the same time, it keeps a high soil moisture content. This been used in Iraq in a limited way. There are difficulties cost, and the side affect (which causes pollution).

3. *Biological method*

This method improves the soil texture and the water content by addition of organic matter. It appears that the biological method is only successful when it is combined with mechanical methods for sand dune fixation.

Revegetated dunes, which have been covered with clay soil, and the flats between have developed good covers of annuals, especially after protection from grazing. The main species observed growing in the clay flats are *Cornulacca*, *Seidlitizia* sp. and *Anabasis*. These provide surface cover which leads to reduced wind erosion beside trapping windblown pseudo-sand.

Reafforestation of the treated areas by mechanical methods, especially after natural vegetation establishment by using drought and saline resistant species, gave a good result to reduce windblown sand to the minimum. *Tamarix*, *Atriplex*, *Prosopis* and other species are commonly used for this purpose on a wide scale. In the middle and south of Iraq 40,000 ha of active sand dunes area have already been treated with both mechanical and biological methods and it becomes completely covered with natural vegetation and shelter belts. Millions of shrubs and trees (such as *Prosopis* spp. *Atriplex* spp. and *Tamarix articulata*) are cultivated in this area. All these plants having a good fodder value for livestock.

The future plan for the Desertification Control Department is to stabilize an area of active sand dunes (about 80,000 ha) to protect the Third River. Both mechanical and biological methods will be used to plant these areas with drought and saline resistant plants (*Prosopis* spp., *Tamarix* sp. and *Atriplex* spp.), which have a good fodder value.

Two million forage seedlings are established each year to plant degraded lands. Three big nurseries are already established in middle and south Iraq and in the western desert. Beside that the Department is looking after the development of natural vegetation in order to provide protection for the environment and food for livestock.

Conclusion

1. Continue the activities of sand dunes stabilization by improving the present techniques and testing other methods which cost less and are quicker.
2. Improve the grazing management and monitor both vegetation and flock management.
3. Include forage cultivation in rotation with barley and use windbreaks in the cultivated areas.
4. A global survey is recommended as soon as possible for whole regions affected by sand dunes invasion. This survey will include essentially vegetation forage survey, the land use/-land capability; socio-economical aspect and range management.

References

Al-Farrajii, A.H. (1983). Drought resistance evaluation of six desert species commonly used for sand dune fixation in Iraq. PhD Thesis, Ghent University, Ghent, Belgium.

Annual Abstract of Statistics (1971). By the Central Statistics Organization of the Ministry of Planning, Iraq.

Al-Hilli, M.R. (1977). Studies on the plant Ecology of Ahwar region in Southern Iraq. PhD Thesis, University of Cairo, Cairo, Egypt.

Al-Rawi, A. (1966). Wild plants in Iraq with their distibution. Division of Botany and Range Management, Agricultural Ministry, Baghdad, Iraq.

Buring, (1960). Soil condition of Iraq. Ministry of Agriculture, Baghdad, Iraq.

Gaddas, N. (1990). Range improvement of range management. Consultancy Report, UNDP/FAO Project, IRQ/87/006, FAO, Rome.

Guest, I. (1966). Flora of Iraq, Agricultural Ministry, Baghdad, Iraq.

Malcolm, C.V. (1990). Agronomy of salt affected lands. Consultancy Report UNDP/FAO Project, IRQ/87/006, FAO, Rome.

21. Feed value of some halophytic range plants of arid regions of Iran

A. KOOCHEKI & M.N. MOHALATI
College of Agriculture, Ferdowsi University of Mashhad, Iran

Abstract. The nutritive value of 12 halophytic range plants from arid regions of Iran was evaluated in terms of digestibility, crude protein, crude fibre and ash content. The crude protein and ash content of the chenopod species was rather higher than other species but crude fibre content of these species were lower than others. Higher ash content of chenopods was probably related to their salt content. Dry matter digestibility (DMD) of the species ranged from 41.5–78.5%. The higher organic matter digestibility (OMD) and digestible organic matter in dry matter (D-value) of chenopod species could be related to their lower crude fibre content. There was a high correlation between DMD, D-value and OMD. However, no correlation was found between digestibility and crude protein or crude fibre.

Key words: Crude fibre; Crude protein, Digestibility; Nutritive value; Salt tolerant plants.

Introduction

The total salt affected area of Iran is estimated to be 15% of the land surface or about 24500 km^2 (Figure 1), most of it is located in arid and semi-arid regions of the country (Pazira 1991). In this area nomadic pastoralism, based on extensive grazing of camels, sheep and goats is normally practised. Herds are mixed flocks of sheep and goats but sheep production is the main enterprise. Sheep are raised for meat, milk, wool or hide production and ewe milking is widely practiced Milk is traditionally consumed in the form of products such as yogurt and cheese. The predominant food production systems are based on small grain crops and ruminant grazing for meat and often those are combined into various agropastoral (crop and grazing) systems.

Halophytes constitute a significant part of the local flora in many places and they have been grazed or browsed by animals for a long time. In many cases the halophytes represent supplementary or emergency feed during unfavourable conditions, since the animals graze them only when other, more desirable forages, are not available. However, the nutritional value of some halophytes is relatively good, and has been recognised for many years. Some of these plants (e.g. *Atriplex* shrubs) are used for rangeland renovation on saline soils (Malcolm 1994; Kernick 1986). The high salt content of halophytes offers some constraint to their utilisation. There are ways to overcome that problem, such as blending the halophytes with other components in prepared feed mixes (Swingle et al. 1994). It has also been reported (Moore et al. 1982) that salinity has no negative effects on digestibility.

The purpose of this study was to compare nutritive value of some halophytic rangeland species found in the arid areas of Iran.

Materials and methods

Twelve of the dominant halophytic species from an area in the southern part of Khorasan province (Figure 2) of rangelands namely: *Kochia prostrata*, *Salsola rigida*, *Haloxylon persicum* (shrub), *Seidletzia rosmarinus*, *Aellenia subaphylla*, *Aeluropus repens*, *Calligonum polygonoides* (shrub),

V.R. Squires & A.T. Ayoub (eds.), *Halophytes as a resource for livestock and for rehabilitation of degraded lands*, 249–253.

Figure 1. The database of desert salt affected land in Iran. About 15% of the land surface (24,500 km^2) is salt affected.

Suaeda sp (shrub), *Aristida pennata*, *Astragalus kahiricus*, *Londosia eryanta* and *Noaea mucronata* were collected here at flowering stage. Each species was sampled at three sites. There were 3 replications. For herbaceous plants, the whole of the above ground parts were collected but for shrub species only leaves and twigs were harvested. Samples were dried separately at 70 °C for 48 hours and milled to pass mesh 40. Dried material was used for chemical analysis and nutritional evaluation. Crude protein and crude fibre was determined by conventional methods (Meleon & Pomoyanz 1973).

Digestibility of dry matter and organic matter was determined by the two-stage *in vitro* method (Jones & Hayward 1973) using cellulose and pepsin enzymes. In order to control the precision of this method some standard samples of known *in vivo* digestibility were included in the determination.

Results and discussion

Table 1 shows crude protein (CP), crude fibre (CF) and ash contents of the species. The crude protein content of chenopods was highest and

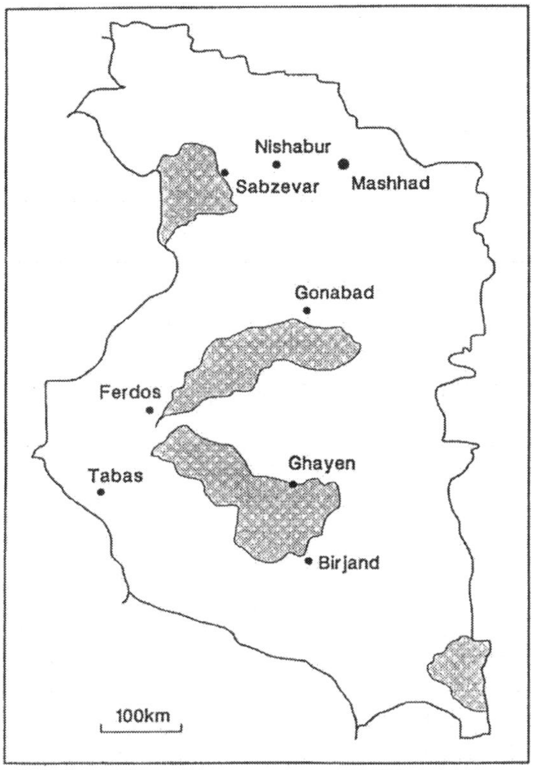

Figure 2. Location map showing salt affected land in Khorasan province of Iran.

Table 1. Percentage crude protein (CP), crude fibre (CF) and ash content of Iranian halophytic rangeland species*

Species	CP	CF	Ash
Aellenia subaphylla	19.16 a**	21.00 d	14.37 a
Suaeda sp.	17.76 b	17.00 f	9.14 e
Haloxylon persicum	16.30 c	16.06 h	9.77 d
Kochia prostrata	16.06 cd	17.63 e	6.10 i
Londosia eryanta	15.63 d	22.03 b	11.13 b
Noaea mucronata	14.76 e	22.80 bc	7.69 f
Salsola rigida	14.66 e	16.30 gh	11.39 b
Astragalus kahiricus	12.66 f	20.90 d	10.33 c
Aeluropus repense	12.26 fg	22.30 c	14.17 a
Seidlitzia rosmarinus	11.80 g	17.33 ef	7.16 g
Calligonum polygonoides	11.20 g	16.96 fg	7.35 fg
Aristida pennata	8.20 h	24.13 a	6.85 h
LSD 0.01	0.56	0.62	0.29

*Mean of three samples.
**The same letter in each column indicates that the numbers are not significantly different at the 1% level.

grasses lowest. *Aellenia subaphylla* a prostrate chenopod bush, was 19.2 percent and that of *Aristida pennata* normally an upright grass used for sand dune fixation was 8.20%. Values for other species were between 11.2–17.76, although not significantly different in some cases (Table 1). The protein content of *Kochia scoparia* was reported to be 23.1% and that of an *Atriplex* species 23.8% (O'Leary, 1985). However, in other reports mean crude protein content for some chenopods (including genus *Salsola*, *Atriplex* and *Anabasis*) was recorded as 13% (Le Houérou 1980). Jones & Hodgkinson (1970) reported that some halophytes compare well with the conventional irrigated forage such as alfalfa. High protein content of some halophytes has been reported in the literature (O'Leary 1985).

Crude fibre content was 24% for *Aristida pennata* and 16% for *Haloxylon persicum*. Other species have crude fibre content of 16 to 23% (Table 1). This seems to be in accord with other findings (O'Leary 1985) in which fibre content of most *Atriplex* species ranged from 15–24%. However, in some rare cases fibre content of up to 40% was reported. In general, fibre content of *Chenopodiaceae* was relatively lower than other species from other families. This is also reported in the literature (O'Leary 1985).

Ash content was from 6% (*Kochia prostrata*) up to 14% (*Aellenia subaphylla*). Ash content of up to 34% has been recorded for *Salsola tetrandra* (Le Houérou 1991). It has been pointed out that some saltbushes take up nearly one-fifth of their dry weight as ash ingredients, largely common salt. True yield of halophytes should be expressed as ash-free weight since ash content can be significant (Wilson 1994).

Dry matter and organic matter digestibility and D-value (digestible organic matter in dry matter) has been shown in Table 2. Dry matter digestibility of the species ranged from 78% (*Salsola rigida*) to 41% (*Aristida pennata*). D.M.D. of two grass species *Aristida* (41%) and *Aeluropus* (56%) was rather low and is expected to be higher compared with those of other grasses reported elsewhere (Spedding & Dickmans 1972). However, low digestibility of these two species seems to be related to their higher fibre content and other possible 'antiquality' compounds produced in their tissue in response to harsh conditions of

Table 2. Percentage dry matter digestibility (DMD) and organic matter digestibility (OMD) and D-value of Iranian halophytic rangeland species

Species	DMD	OMD	D-value
Salsola rigida	78.53 a**	90.90 a	69.06 a
Seidlitzia rosmarinus	68.80 b	80.90 b	64.03 b
Londosia eryanta	68.13 b	80.90 b	60.76 d
Kochia prostrata	66.60 c	78.93 c	62.90 c
Suaeda sp.	65.26 d	76.40 d	58.90 e
Haloxylon persicum	64.50 d	73.13 e	56.50 f
Aellenia subaphylla	63.00 e	69.00 f	51.06 g
Aeluropus repens	56.16 f	67.00 g	48.90 h
Noaea mucronata	46.06 g	51.36 h	41.26 i
Calligonum polygonoides	45.80 g	51.56 h	41.10 i
Astragalus kahiricus	45.60 g	48.13 g	37.10 k
Aristida pennata	41.53 h	49.03 i	38.76 j
LSD 0.01	0.78	0.81	0.58

*Mean of three samples.
**The same letter in each column indicates that the numbers are not significantly different at the 1% level.

their habitat. Salinity *per se* has been reported not to affect digestibility and plant irrigated with saline water were slightly higher in *In vitro* digestibility (Moore et al. 1983; Traver et al. 1983). However, palatability may be reduced by salinity (Wilson, 1977). Dry matter digestibility of two shrub species, *Haloxylon persicum* and *Calligonum polygonoides*, were 65.5 and 45.8% respectively. It should be noted that a higher proportion of samples collected from these species were leaf and twigs and therefore, a higher digestibility would have been expected. A rather lower digestibility for this type of shrub has also been reported elsewhere (Le Houérou 1980; 1991). Except in a few cases, organic matter digestibility showed similar trends with those of dry matter digestibility. This was also true for D-value. Higher organic matter digestibility and D-value of chenopod species in this study could also be attributed to their lower crude fibre content. In other words lower cell wall and thus higher cell contents of these species is responsible for their higher organic matter digestibility.

Although D-value and OMD showed a similar trend, the former was lower than the latter. For instance organic matter digestibility of *Salsola rigida* was 90.9% while its D-value was 69%, a difference of almost 22 percentage units. This was respectively 48 and 37% for *Astragalus kahiricus*, a difference of 11 percentage units.

Higher correlation co-efficients found between D-value and DMD ($r = 0.974$) and between D-value and OMD ($r = 0.988$) and between DMD and OMD ($r = 0.980$) are in line with data reported in the literature (Spedding & Dickmans 1972). There was no significant correlation between CP and D-value ($r = 0.470$). There is evidence (Minson et al, 1960; Tilly & Terry 1963) which indicate no relation between digestibility and N-content, because they are associative rather than causative. No significant correlation between crude fibre and herbage digestibility found in this study ($r = 0.36$) seems to be not unusual because there are reports (Pigden 1969) which indicate a wide range of correlations between crude fibre and digestibility ($r = 0.50$ to $r = -0.94$). Van Soest (1975) also found a negative correlation between these two factors.

In general, a rather higher protein content and high digestibility of halophytes, together with a reasonable dry matter productivity under the harsh condition of arid regions, seems to justify more research on the potential use of these plants as animal feed in salt affected areas of Iran.

References

Jones, D.I.H. & Hayward Margaret, V. (1973). A cellulose digestion technique for predicting the dry matter digestibility of grasses. J. Sci. Fd. Agric. 24: 1419–1424.

Jones, R. & Hodgkinson, K.C. (1970). Root growth of rangeland chenopods. Morphology and productivity of *Atriplex nummularia* and *Atriplex vesicaria*. In: Jones, R. (ed.), The biology of *Atriplex*. CSIRO, Canberra.

Kernick, M.D. (1986). Forage plants for salt affected areas in developing countries. Reclamation and Revegetation Research 5: 451–459.

Le Houérou, H.N. (1980). Browse in Africa. International Livestock Center for Africa. Addis Ababa, pp. 55–82.

Le Houérou, H.N. (1991). Feeding shrubs to sheep: intake, performance and feed value. Proceeding of 4th International Rangeland Congress, Montpellier, France.

Malcolm, C.V. (1972). Establishing shrubs in saline environment. Technical Bulletin 14. Development of Agriculture in Perth, Western Australia.

Malcolm, C.V. (1994). Use of halophyte forages for rehabilitation of degraded lands (this volume, Ch. 4).

Meleon, C.E. & Pomoyanz, Y. (1973). Food analysis laboratory experiments. The Avi Publishing Company Inc., pp. 94–96.

Minson, D.J., Rayward, W.F. & Harris, C.E. (1960). Studies

on the digestibility of forage. VIII the digestibility of S37 cocksfoot, S23 ryegrass and S24 ryegrass. J. Br. Grassld. Soc. 15: 174–180.

Moore, J.A., Swingle, R.S., O'Leary, J.W., Glenn, E.P. & Colvin, L.B. (1982). In vitro organic matter digestibility of salt tolerant plants. Proc. Western Sec. Am. Soc. Anim. Sci. 33: 319–322.

O'Leary, J.W. (1985). Saline environment and halophyte crops. In: Whitehead, E.E., Hutchinson, C.F., Timmerman, B.N. & Varady, R.G. (eds.), Arid lands, today and tommorrow. Proceeding of an International Research and Development Conference. Tucson, AZ.

Pazira, A. (1991). A brief review of soil salinity and alkalinity problems in Iran and recommended methods of improvement and reclamation. Zeiton 100: 12–14.

Pigden, W.J. (1969). In: Campbell, J.B. (ed.), Experimental methods for evaluating herbage. Canadian Department of Agriculture, Publ. 1315, pp. 52–72.

Spedding, C.R.W. & Dickmans, E.C. (eds.), (1972). Grasses and legumes in British Agriculture. Bulletin 49. Commonwealth Agricultural Bureaux.

Traver, L.D., Preston, R.I. & Goodin, J.R. (1983). Digestibility of *Atriplex canescence* as affected by saline conditions during plant growth. Proc. Western Sec. Am. Soc. Anim. Sci. 34: 201–202.

Tilly, J.M.A. & Terry, R.A. (1963). A two-stage technique for the *in vitro* digestibility of forage crops. J. Br. Grassld. Soc. 18: 104–111.

Van Soest, P.J. (1975). Physiological aspects of fibre quality. Symposium of Society of Chemical Industry, London.

Wilson, A.D. (1977). The digestibility and voluntary intake of leaves of trees and shrubs by sheep and goats. Aust. J. Agric. Res. 28: 501–508.

22. Use of edible shrubs in pasture improvement under Mediterranean environment in northern Syria

A.E. OSMAN[1] & M.A. SHALLA[2]
[1]*ICARDA Alleppo, Syria;* [2]*Steppe Directorate Alleppo, Syria*

Abstract. Rangelands in Syria originally supported woodland and shrub vegetation. They now support, mainly short grasses and forbs (occasionally small populations of annual legumes). This degradation is also accompanied by soil erosion. Planting of native shrubs *Atriplex halimus* and *Salsola vermiculata* may help not only in reversing land degradation but also in providing reserve feed in drought years. On-going research is to determine productivity of rangelands improved by seeding with edible shrubs, monitoring the effects of three stocking rates on biomass production and on sustainability of livestock production, and measuring the economic return.

Key words: *Atriplex*, Economic analysis, Forage utilization, Rehabilitation, Stocking rates

Introduction

Out of 18.5 million ha, the total area of Syria, 10.2 million is considered as rangeland. These rangelands receive no more than 200 mm of rainfall and are not suitable for cultivation but rather support native pasture which is utilized by an increasing number of livestock. The 1963 estimate of sheep population in Syria was 4.3 million but by 1988 this had risen to 13.3 million head. It is estimated by the Syrian Ministry of Agriculture that 75% of the total sheep population use the rangelands during at least part of the year. Syria, is like many other of west Asia and north Africa in that rangelands have suffered considerable deterioration over the years. The principal contributing factors are:
- uncontrolled and prolonged grazing;
- increased livestock populations;
- uprooting the shrubs for fuel;
- encroachment of cereal cultivation onto rangelands.

In addition, a more important factor lately is the use of mechanized transportation (a tractor, trailer and water tanker) to carry fodder and water to the rangeland (Treacher 1991). This 'kit' allows the livestock owners to transport sheep, feed and water to remote parts of the rangelands and allows the flock to stay longer – thus accelerating rangeland destruction. Rangelands in Syria originally supported woodland and shrub vegetation (Emberger 1956). There has been severe land degradation which is accompanied by soil erosion. They now support short grasses, and forbs (occasionally some legumes). Annual biomass production averages 400 kg ha^{-1} (PFLP 1986) in contrast to the expected 1,200 kg ha^{-1} of Mediterranean rangelands in general (Le Houérou and Hoste 1977). Under a similar rainfall regime in Libya, yields in excess of 2000 kg ha^{-1} have been reported (Ginztburger 1986).

Range rehabilitation

Following research on edible shrubs by the Syrian Ministry of Agriculture in collaboration with regional and international organizations during the period 1974–85, the Syrian government started an extensive program of range rehabilitation (Figure 1). At the beginning, they were growing seed-

V.R. Squires & A.T. Ayoub (eds.), *Halophytes as a resource for livestock and for rehabilitation of degraded lands*, 255–258.

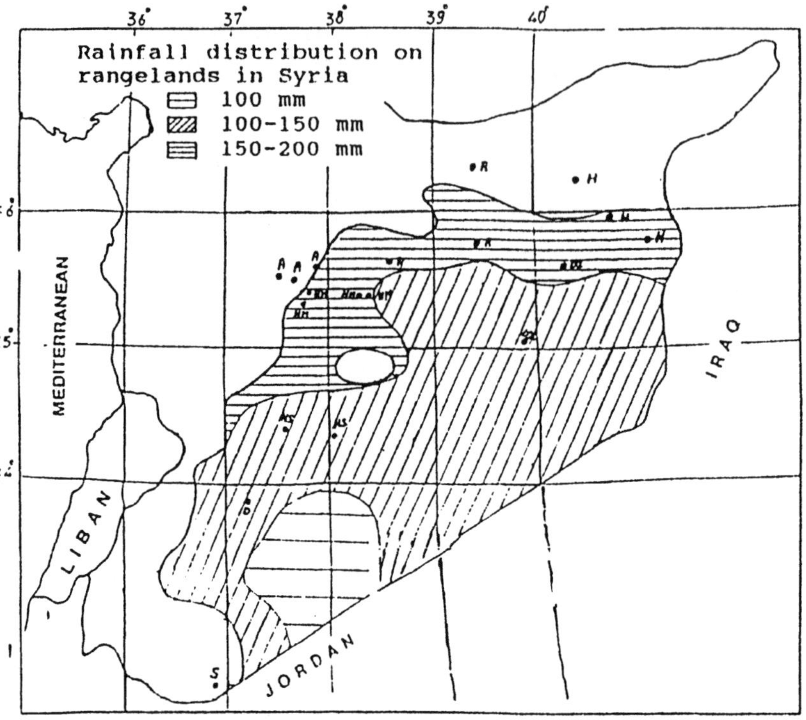

Figure 1. Shrub plantations in the Syrian provinces: Alleppo (A), Damascus (D), Hassaka (H), Hama (HM), Hems (HS), Dir El Zor (DZ), Raga (R) and Sueda (S).

lings of exotic species *Atriplex nummularia* and *A. canescens*. Draz (1978) reported that the Australian species, *A. nummularia* is adapted to marginal areas of the steppe in Syria. Later work has shown that the native shrubs *A. halimus* and *Salsola vermiculata* perform even better than *A. nummularia*. They can be established by direct seeding and, more importantly, they are capable of self-reseeding. The total area to date of shrub plantations in government reserves is 47,747 ha. In addition, there are 14,125 ha established by direct seeding. The latter method is expected to become more important and supersede the seedling establishment method because of cost savings that it offers. In a recent study by ICARDA in two Provinces (Hama and Aleppo) the cost of seedling on one hectare ranged between 2,225 and 2,300 Syrian pounds (1 $US = 42 S.L.) while the cost of direct seeding on the same area was 200–345 Syrian pounds (Table 1).

Management

Although shrub plantations have not been put to systematic use by livestock (except in experimental plots) the main use has been for seed production. However, in two drought years 1989 and 1990 one of the reserves (Maragha, Alleppo province) was opened for grazing in the months June–November (160 days in the first year and 100 days in the second).

During these periods the shrub reserve (6,500 ha) was used by a government flock of 11,000 sheep. Assuming one feed unit per sheep a day, the total feed units provided by the shrub reserve is estimated at approximately 3 million (Table 2). In 1989 the shrub reserve consisted of 5,800 ha established from seedlings and 700 ha by direct seeding. Using the 1985 and 1986 figures for establishing the reserve (1,100 S.L by seedlings and 150 S.L ha^{-1} for direct seeding the

Table . Cost of establishing one hectare of edible shrubs (in Syrian pounds)[a] using seedlings or direct seeding methods in two provinces

Items	Seedlings		Items	Direct seeding	
	Alleppo	Hama		Alleppo	Hama
Land preparation	500	180	Cultivation	95	170
Transport	25	60	Seed collection[b]	50	120
Planting	200	450	Seed transport	5	5
Irrigation	800	800	Sowing	50	50
Seedlings (400 ha^{-1})	700	810			
Total	2225	2300		200	345

[a]$US 1 = 42 Syrian.
[b]15 kg of seed collected.

Table 2. Estimated feed units utilized by sheep out of the shrub reserve at Maragha, Alleppo Province during 1989 and 1990[a]

	1989	1990
No. of sheep	11,000	11,000
Grazing days	165	100
Feed units[a]	1,815,000	1,000,000

[a]Assuming one feed unit/sheep per day.

overall cost of this reserve would be 6.5 million Syrian pounds. On current prices the total cost would be 13.0 million Syrian pounds.

To obtain detailed information on shrubs a joint field study (ICARDA, Syrian Ministry of Agriculture and ACSAD) was started at Maragha in 1989/90 on 100 ha of land improved by direct seeding of *Salsola vermiculata* and *A. halimus*.

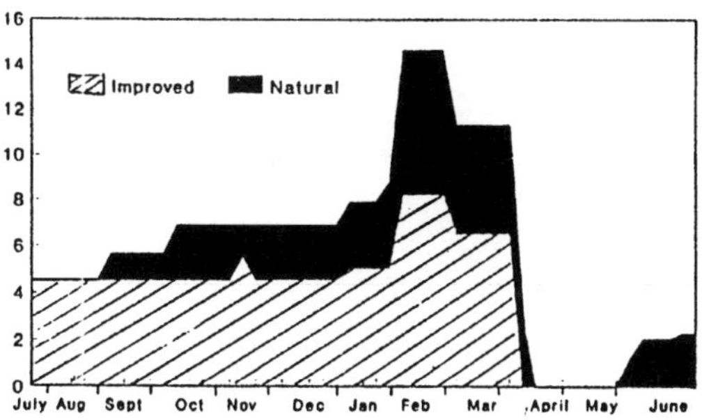

Figure 2. Total feed consumption in Mj/animal per day by sheep grazing on two types of pasture at Maragha, northern Syria in 1990/91 season.

A comparison was made with 100 ha of native unimproved pasture. The objectives were:
- to determine the productivity of rangelands improved by seeding with edible shrubs;
- monitor the effects of stocking pressure at three rates on production of biomass and on sustainability of livestock production;
- measure the economic return. Three stocking rates, one sheep per 2.25 ha, one sheep 1.5 ha^{-1} and 1 sheep 0.75 ha^{-1}, on both improved and native areas were achieved by grazing 8 ewes/plot with plot sizes of 18, 12 and 6 ha. Each treatment was replicated three times. All plots were securely fenced.

Animals were introduced into the experimental plots on February 1 1990. Thereafter, the ewes grazed the plots from early morning to sunset, being sheltered at night. Animals on both pastures were supplemented in late pregnancy and early lactation with barley, cottonseed cake and hay when liveweights would otherwise not be maintained by grazing alone.

The seeded pastures maintained the sheep all year long but the native unimproved pasture provided only 81 days in 1990 and 112 days of grazing in 1991. The only period when animals grazing the shrubs required more supplementation than those on native pasture was in April and May of the first season when the accumulation of salts on the shrubs during wind storms caused the pasture to be closed temporarily. During the rest of the first season and throughout the second season, requirements for supplementary feed were greater on native pasture, especially in winter (Figure 2). The experiment is planned to continue for 5 years (to 1994) but there is already evidence suggesting more favorable impact of shrubs on soil stabilization and erosion control especially wind erosion (PFLP 1990/1991).

References

Draz, O. (1978). Revival of the hema system of range reserves as a basis for the Syrian range development program. In: Proc First Int, Rangeland Congr, Denver, pp.100–103.

Emberger, L. (1956). Study of vegetation and land use. In: FAO Report of the 1st meeting of the FAO working party on the development of grazing and fodder resources of the Near East, Cairo, FAO, pp. 1–27.

Gintzburger, G. (1986). Seasonal variation in above ground phytomass of an arid rangeland in Libya. J.Range Manage 39: 348–353.

Le Houérou, H.N. & Hoste, C.H. (1977). Rangeland production and annual rainfall relations in the Mediterranean basin and in the African Sahelo-Sudanian zone. J. Range Manage. 30: 181–189.

PFLP, Pasture, Forage and Livestock Program (1986). Annual Report, ICARDA III. En.

PFLP, Pasture Forage and Livestock Program (1990/1991). Annual Report ICARDA -039, July 1992.

Treacher, T.T. (1991). Linkages between steppe and cultivated areas through livestock systems. Implications for research and development to establish more productive and stable systems. Colloque International Deveir des Steppes d'Arabie et du Bilad Ach. Cham. Paris, 10–13 April 1991.

23. The potential economic use of halophytes for agricultural development of southern Morocco

R. CHOUKR-ALLAH,[1] G.W. JONES[2] & L. KENNY[1]
[1] *Institut Agronomique et Veterinaire, Hassan II, Agadir, Morocco;*
[2] *Centre for Arid Zone Studies, University College of North Wales, Bangor, Gwynedd, UK*

Abstract. Salinity has substantial adverse social and economic effects in Morocco causing deteriorating agricultural productivity in many irrigated perimeters. A bio saline approach has been considered to establish forage production in arid rangelands in the southern part of Morocco using saline irrigation water. A collaborative project involving the United Kingdom and Morocco, aims to select halophytic species adapted to the local conditions, and to determine the optimal methods for their establishment and their suitability as forage. This paper reviews the work conducted on the introduction, development and use of halophytes as fodder, and as trees for revegetation. Trials are also underway on the introduction of ornamental plants to be irrigated with saline water and used for landscaping and gardening.

Key words: Biomass production, Direct seeding, Root growth, Salinity, Screening, Utilization

Introduction

It is estimated that 7% of the world's land surface is salt-affected (Dudal & Purnell 1986). In the southern part of Morocco, where rainfall is less than 100 mm and most of the ground water has a salt level exceeding 5 g L^{-1}, the agricultural system is based on animal grazing. In an attempt to improve pasture, explore the use of a range of halophytic species and increase the economic use of saline water, a collaborative program was established between Institut Agronomique et Veterinaire at Hassan II University, Agadir, Morocco and Centre for Arid Zone Studies, University of North Wales, Bangor, United Kingdom.

In these studies, growth and performance of several species of *Atriplex* were evaluated under irrigation application with saline water and multiple harvests. The aim of this study is to establish the basic agronomic parameters for propagation culture and the suitability of these plants as forage for livestock.

Plant establishment

Direct-seeding to establish *Atriplex* is more practical and economical in large areas than transplanting. For most *Atriplex* species, however, establishment using the direct-seeding method is currently limited by the availability of a source of seed.

Many so-called halophytes are known to be intolerant to high salinity during germination and early seedling growth. Also a great variation in growth has been observed in stand establishment from direct seeding. Due to the fact that the only available water in the area is brackish water and in order to establish a more uniform field stand from the best clone, we undertook several studies on the development of macro propagation techniques involving rooting of stem cutting under saline conditions (see also Stringi et al. 1994).

From a technical standpoint stem cuttings are relatively easy and can be even cheaper than nursery grown seedlings, but success depends upon the genotype, the growth conditions and the tech-

V.R. Squires & A.T. Ayoub (eds.), *Halophytes as a resource for livestock and for rehabilitation of degraded lands*, 259–261.

Table 1. Effect of salinity on root growth on stem cuttings of *A. halimus* (*A.h.*) and *A. nummularia* (*A.n.*) harvested on adult shrubs and grown under plastic house conditions

	NaCl (g L^{-1})	*A.h.*	*A.n.*
Root numbers	0	27.1	6.0
	10	14.4	1.6
Root length	0	11.4	1.0
(cm)	10	4.7	0.6

Table 2. Effect of watering regimes on biomass production, Na+, K+, crude protein content and digestibility

	L.	B.	G.	H.	D.
FW I.	425.6	432.7	150.6	196.6	367.3
(*g/plt*)					
NI.	225.0	226.0	103.3	88.3	225.0
DW I.	95.6	89.7	40.7	40.6	47.0
(*g/plt*)					
NI.	47.3	64.0	35.3	20.3	0.26
Na[a]I.	0.12	0.19	0.14	0.22	0.26
(*g/g DW*)					
NI.	0.13	0.18	0.12	0.16	0.23
K[a]I.	0.037	0.047	0.020	0.037	0.036
(*g/g/DW*)					
NI.	0.036	0.047	0.024	0.041	0.032
CR I.	16.0	17.0	17.5	14.0	15.1
(%)					
NI.	15.5	15.7	11.8	11.8	12.5
IVDMD I.	69	66	63	63	65
(%)					
NI.	71	63	59	66	66

[a] Water extraction, DW: Dry matter, FW: Fresh weight, CR: Crude protein, IVDMD: *in vitro* digestibility, I. Irrigated, NI: Non irrigated. B: Barclayana. H: Halimus, G: Glauca, D: Deserticola, L: Lentiformis.

nique used. Not all the species are easy to propagate especially under saline conditions. A distinction should be made between difficult-to-root species such as *Atriplex nummularia* and easy-to-root ones such as *Atriplex halimus* (Table 1). For a given species, even among the difficult ones, a tremendous improvement in rooting stem cuttings can be accomplished by clonal selection. The difference between clones in terms of the average root number found with *Atriplex halimus* and *Atriplex nummularia* was about 80 and 85% respectively. The second more important factor in rooting stem cuttings of *Atriplex* is the growth conditions particularly relative humidity. Among propagators, high humidity (90–100%) is generally a prerequisite for the rooting of woody plant species. It is not so with *Atriplex* species. It is even harmful under saline conditions. The highest percentage of rooting with the best quality of roots are obtained at 40–70% RH. In addition cuttings tolerate the adverse effect of salinity better under low humidity regimes.

As to the quality of the cuttings, two main selection criteria have to be considered when harvesting the cuttings: 1) the size of the cutting, and 2) its position along the shrub. Single node stem cuttings generally give satisfactory results. There is no need for a bigger size (4–6 nodes per cuttings). Basal cuttings are however more amenable to rooting than the terminal ones. Hormonal treatments using IBA at 3000 ppm significantly improve the rooting especially when saline water is used for irrigation.

When saline water containing 10 g L^{-1} of NaCl was used for irrigation cuttings performed better than seedlings. The latter produced an average of 14 roots per seedling while the average with cuttings was about 25 roots. But the seedlings produced a longer root system.

Biomass and productivity

Field trials were conducted from March 89 through March 90 in a site located at about 2 km east of Tantan city.

Soil profile to 50 cm was a sandy loam, and irrigation water applied monthly contained about 10 g L^{-1} of salt. Forage yields, distribution of dry matter in leaves and stems and forage chemical composition was evaluated for *A. barclayana*, *A. lentiformis*, *A. glauca*, *A. deserticola* and *A. halimus* planted at a density of 10,000 plants ha^{-1}. Total dry and fresh matter production varied from one species to another and from the first clipped harvest to the fourth harvest. Dry matter production varied between 6 to 13 t ha^{-1} yr^{-1}. *A. barclayana* and *A. lentiformis* gave the highest yields. We also undertook a study to compare irrigated and non irrigated production of *Atriplex* species.

The results of our experiment show clearly that the productivity values, in general, are substantially higher for irrigated *Atriplex* even though the response to irrigation varied from one species to another (Table 2). There was 100% increase in the dry matter production of *A. lentiformis* and *A. halimus* under the irrigated regime. However, *A. glauca* showed an increase of only 15% in dry matter production. In general, irrigation had more effect on fresh weight biomass than on the dry weight production.

The crude protein content of the 5 species of *Atriplex* fall in the range of 12–17% (Table 2). This compares well with a conventional forage such as alfalfa, with a range of 12–22%.

There were substantial differences among species in terms of their response to the irrigation with saline water. *A. glauca*, *A. halimus* and *A. deserticola* had a substantial increase in their protein content when irrigated. However, *A. lentiformis* and *A. barclayana* showed a slight increase under irrigated regime.

The digestibility of the *Atriplex* species produced under the irrigated regime did not show any significant difference with the rainfed production. The in vitro organic matter disappearance of all species tested ranged from 59–71%. Standard alfalfa hay has a digestibility of around 64.3%. The salt content and particularly the $Na+$ level varied substantially between species.

A. deserticola and *A. halimus* accumulated more $Na+$ in their shoot than the others. It should be pointed out that *A. barclayana* and *A. lentiformis* presented the highest K/Na level and were the most productive species under both irrigated and rainfed conditions.

Conclusions

This study shows that forage yield and quality vary with species and period of harvest. The irrigated *Atriplex* had much higher yield than under rainfed conditions. The most promising *Atriplex* species for the Tantan area is *A. barclayana* based on its productivity and protein content and its high ratio of leaves to stem.

In spite of this there are some problems associated with the use of *Atriplex* as an animal feed, notably the high ash content of the tissue (Gihad & El Shaer 1994).

References

Dudal, R. & Purnell, M.F. (1986). Land resources: Salt affected soils. Reclamation and Revegetation Research 5: 1–9.

Gihad, E.A. & El Shaer, H.M. (1994). Utilization of halophytes by livestock on rangelands: Problems and prospects (this volume, Ch. 6).

Stringi, L., Giambalvo, D., Accardo, A. & Amato, G. (1994). Characterization of progeny clones belonging to natural Sicilian *Atriplex halimus* populations (this volume, Ch. 33).

24. Production from grazing sheep on revegetated saltland in Western Australia

B.E. WARREN, T. CASSON & D.H. RYALL
Department of Agriculture, Sheep Industries Branch, Katanning, WA, Australia

Abstract. Results are presented from a field experiment on severely salt affected land in which four species of *Atriplex* were planted alone or in combination and then grazed by sheep. Performance of the sheep was assessed in terms of liveweight change. Preliminary examination of the data suggest that, on the mixed saltland pastures utilised in this experiment (1078 to 3213 kg total DM/ha and 0 to 1200 kg saltbush DM/ha), about 300 kg DM, including 90 kg saltbush DM, per sheep was necessary at the beginning of grazing for liveweight maintenance over the 9 week grazing period.

Key words: *Atriplex*, Forage, Grazing sheep, Liveweight change, Saltland

Introduction

Dryland salinity is a major problem in Australia (Truong 1994) with about 2 million ha of agricultural land now salt affected (Barrett-Lennard et al. 1986). About 500,000 ha of this land is in Western Australia (Figure 1). Recent emphasis has been on revegetating these lands with salt tolerant plants for grazing or browsing. The area being revegetated with saltbushes (*Atriplex* spp.) and salt tolerant grasses (*Agropyron*, *Puccinellia*) is increasing.

However, there is little production data from animals grazing salt tolerant forages. Warren et al. (1990) found that provision of a low salt forage with saltbush markedly improved sheep performance. More recently Warren & Casson (1992) showed sheep grazed on a low quality tall wheat grass (*Agropyron elongatum*) based pasture lost twice as much weight (80 g/day) as sheep grazed on the same low quality grass plus a small amount of saltbush, or on dry pasture supplemented with 250 g oats per day. Both experiments showed a complimentary effect between saltbush and low salt, low quality forage.

Because of the increasing importance of saltbush in WA grazing systems, this work has been extended to a new field experiment with grazing sheep.

Grazing plots (22 × 0.9 ha) of four saltbush species (*Atriplex undulata*, *A. lentiformis*, *A. amnicola*, *A. cinerea*) planted alone or as a combination of the four, plus an untreated control (T6 × R[3 or 4]) have been established on moderately saline land. Adult wether sheep, grazed in common on dry pasture and supplemented with 200 g oats/head/day from early March, were allocated to all plots (8/plot) and grazed for a period of 7 or 9 weeks in late autumn, 1992. Dry matter (DM) available from the bushes and from the understory was measured at the commencement of grazing (Table 1). Sheep liveweights were monitored while grazing in common and then 6 times while grazing saltland (Figure 2).

Although the DM available from saltland pasture components differed ($p < 0.05$), sheep performance was similar on all saltbush species ($p > 0.05$) and liveweights after 9 weeks were the same as at commencement of grazing ($p > 0.05$). However, the sheep grazing dry pasture alone lost weight at about 80 g/day until they were removed from the plots after 7 weeks grazing.

V.R. Squires & A.T. Ayoub (eds.), *Halophytes as a resource for livestock and for rehabilitation of degraded lands*, 263–265.

Table 1. Mean dry matter (DM) available from saltbush* (*Atriplex* spp.), samphire (*Halosarcia* sp.) and understory grasses to sheep grazing revegetated saltland near Katanning, Western Australia

Treatments	Saltbush kg DM/ha	Understory kg DM/ha	Samphire kg DM/ha	Total kg DM/ha
Control	0**	1438	22	1460^{a}
A. amnicola	662^{cd}	1413	254	2329^{c}
A. cinerea	408^{bc}	1163	198	1769^{abc}
A. lentiformis	563^{c}	1061	33	1658^{ab}
A. undulata	926^{d}	1305	3	2233^{bc}
Mixed species	255^{ab}	1214	108	1577^{ab}

* Edible material (leaf and small stem only), does not include stems above approximately 2 mm dia.
** Different superscripts (a-d) within columns indicate significant differences ($p < 0.05$).

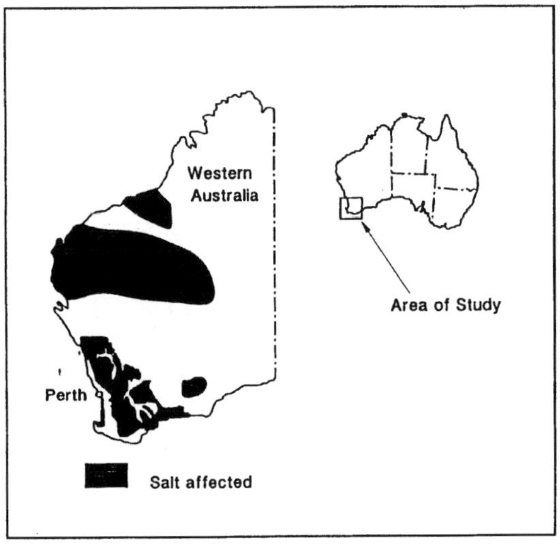

Figure 1. Distribution of salt-affected land in Western Australia and the location of the study area.

At the end of 9 weeks all sheep were removed from saltbush plots as leaf material had been consumed and many animals were losing considerable weight. When the liveweight change of the sheep over the saltland grazing period was examined over plots, there were differences which related to total DM on offer and the saltbush component.

Preliminary examination of the data suggest that, on the mixed saltland pastures utilised in this experiment (1078 to 3213 kg total DM/ha and 0 to 1200 kg saltbush DM/ha), about 300 kg DM, including 90 kg saltbush DM, per sheep was necessary at the beginning of grazing for liveweight maintenance over the 9 week grazing period (Equations (1) and (2)).

$$\text{Liveweight change (g/day)} = -247 + (1.68 \times \text{kg total DM}) - (0.00036 \times [\text{kg total DM}]^2) \quad (r = 0.54) \quad (1)$$

$$\text{Liveweight change (g/day)} = -84 + (1.83 \times \text{kg saltbush DM}) - (0.01 \times [\text{kg saltbush DM}]^2) \quad (r = 0.82) \quad (2)$$

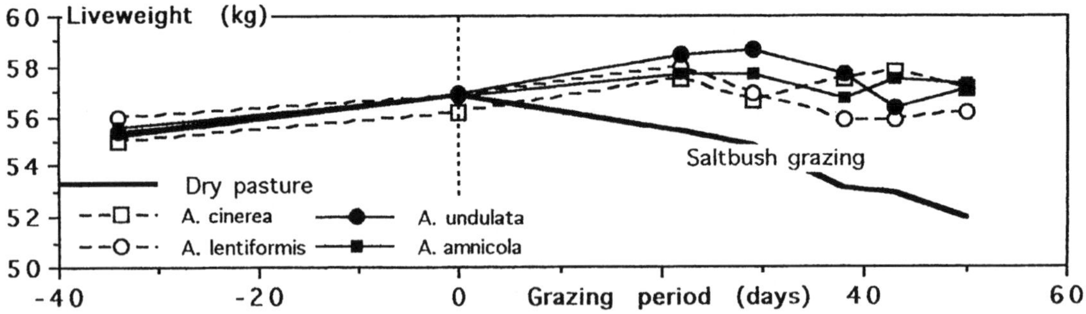

Figure 2. Liveweight (kg) of sheep grazed on dry pasture and on revegetated saltland over the summer/autumn period in 1992 at Katanning, WA.

Further analyses are being undertaken to determine the effects of grazing saltland pasture on wool growth and quality, DM intake, diet selection (using n-alkanes), water consumption and tissue water retention and chemical composition of the forage. These will be reported in a later paper.

Acknowledgements

One of us (BEW) received a travel grant from the Australian Development Assistance Bureau (AIDAB) to attend the Nairobi Workshop.

References

Barrett-Lennard, E.G., Malcolm, C.V., Stern, W.R. & Wilkins, S.M. (eds). (1986). Forage and fuel production from salt affected wasteland. Elsevier, New York.

Truong, P.N.V. (1994). Vetiver grass, its potential in the stabilisation and rehabilitation of degraded saline land (this volume, Ch. 31).

Warren, B.E., Bunny, C.J. & Bryant, E.R. (1990). A preliminary examination of the nutritive value of four saltbush (*Atriplex*) species. Proc. Aust. Soc. Anim. Prod. 18: 424–427.

Warren, B.E. & Casson, T. (1992). Performance of sheep grazing salt tolerant forages on revegetated saltland. Proc. Aust. Soc. Anim. Prod. 19: 237.

25. Impact of halophytes on animal health and nutrition

M.T. BAYOUMI & H.M. EL SHAER
Desert Research Center, Mataria, Cairo, Egypt

Abstract. The phytochemistry of Egyptian desert plants, including those of the coastal marshlands are being studied. Some plants are toxic. Toxicity may result from a number of secondary metabolites. In other cases there are anti-nutritional substances which affect palatability and intake. Treatments, such as ensiling of halophytes, may improve their utilization.

Key words: Antinutritional substances, Camels, Egypt, Goats, Intake, Marshes, Palatability, Phytochemistry, Rabbits, Sheep, Sinai, Toxicity

Introduction

There are many records dealing with the distribution of halophytic natural vegetation in Egypt, especially on the littoral salt marsh on the north western Mediterranean coast (Tadros 1956; Girgis 1973) along the Deltaic Mediterranean Coast (Anon 1992) on the Red Sea from Suez to Sudan-Egyptian border (Kassas & Zharan 1967) and in Sinai (Anon 1981). The common halophytic plants are two mangrove species *Avicennia marina* and *Rhizophora mucronata* in addition to, *Aeluropus* sp., *Alhagi maurorum*, *Arundo donax*, *Arthrocnemon glaucum*, *Atriplex halimus*, *Cressa cretica*, *Cynodon dactylon*, *Cyperus rotundus*, *Desmostachya hipinnata*, *Frankenia laevis*, *F. pulverulenta*, *F. revoluta*, *Halocnemon strobilaceum*, *Halimione protulacoides*, *Halopeplis perfoliata*, *Halopyrum mucronatum*, *Imperata cylindrica*, *Inula crithmoides*, *Juncus acutus*, *J. rigidus*, *Kochia indica*, *Limoniastrum monopetalum*, *Limonium axillare*, *L. pruinosum*, *Lygeum spartum*, *Mesembryanthemum nodiflorum*, *Nitraria retusa*, *Parapholis marginata*, *Phragmites australis*, *Salicornia fruticosa*, *S. herbacea*, *Salsola tetrandra*, *S. vermiculata*, *Scirpus litoralis*, *Spergularia marina*, *Sporobolus virginicus*, *S. spicatus*, *Suaeda vera*, *S. monoica*, *S. pruinosa*, *S. salsa*, *Typha australis* and *Zygophyllum album*.

Halophytic plants vary considerably in their palatability to different animal species, and in their nutritional value. Several species could be used as animal feed although the quantity of active principles which would be of medicinal or poisonous importance varies.

This paper examines the importance of phytochemistry in relation to the utilization of halophytic plants in the Egyptian desert.

Phytochemical and medicinal importance

Information about the chemical composition and the characteristics of the active principles for many of these plants is still incomplete. We need to know this to make best utilization of such halophytic plants for livestock (Gihad & El Shaer 1994).

It is desirable to know whether the substance that produce the remedy (or signs of poisoning in animals) is an alkaloid, glycoside or steroid or if it contains nitrate or belongs to some other category. Knowledge of this sort leads to a more rational treatment of the affected animals than

V.R. Squires & A.T. Ayoub (eds.), *Halophytes as a resource for livestock and for rehabilitation of degraded lands*, 267–272.

can be developed by mere observation alone. It becomes possible to detect the poison in the tissues of the animals thus leading to a diagnosis in obscure or doubtful cases.

Poisonous plants can greatly affect animal production in several ways. It can affect grazing strategies for livestock operation and forage utilization. Poisonous (or potentially poisonous) forages have always been a component of rangelands. There are many plants capable of producing toxic metabolites, including some palatable plants (Sperry 1964).

Therefore grazing management is recognised as the manipulation of livestock to obtain a desired performance. This definition can play an important role in reducing livestock deaths from poisonous plants. The presence of some plants that can be poisonous to animals on the rangeland does not necessarily mean that poisoning is imminent, but it means that proper management and continuous awareness must be considered if poisoning is to be prevented.

Toxicity of an ingested plant may not be determined solely by the amount of toxic substances ingested by a particular animal. Rather toxicity is affected by numerous factors such as rate of ingestion, types and rates of microbial transformations in the rumen, rates of gastro-intestinal absorption, rates and pathways of biotransformation in gut tissues, liver and kidney and effects of enzyme induction or inhibition (Hathcock 1986).

The compounds of poisonous plants that can cause adversity in ruminants are subjected to chemical transformations not only by rumen microbes but also by enzymatic changes in gut mucosa, liver and other tissues (Smith 1992). Some chemical changes increase toxicity of plant compounds and some cause detoxification. Nevertheless, animal management or treatment protocols may alter rumen microbial toxification and detoxification (Hathcock 1986).

It has been suggested that rumen-microbial detoxification of poisonous plants might be as important as rumen microbial degradation of cellulose for the evolutionary level and ecological expansion of ruminants (Van Soest 1982).

Microbial populations change gradually with prolonged, increasing exposure to toxic substances, and such gradual adaptations allow ruminants to increase their tolerance of some poisonous plants. It has been established that microbial adaptation provide detoxification of plant toxicants such as nitrate, nitrite, oxalate, sulfate and sulfide, some alkaloids (e.g. mimosine) and cyanogenic glycosides (Daniel et al. 1989).

Over the past two decades collaboration between ecologists and chemists has begun to indicate that many secondary metabolites and other chemical nutrients influence livestock by a combination of deterrence and toxicity (Meyer & Kazarov 1991).

Phytochemistry of some Egyptian desert plants

Nitraria retusa, as one of the most important palatable halophytic shrubs, has received most attention in relation to its nutritional and medicinal importance. Alkaloids were well represented (0.033% of crude alkaloids). Toxic alkaloids such as hermol alkaloid (26 mg%) were isolated (Shalaby et al. 1977; Anon 1981). These have a distinct hypotension action on human peripheral blood. Tannins, saponins, alkaloids and sterol were also found. Flavonoids were found in the plants which were collected from an elevated plain but not from those from a saline depression. A resin was present as a secondary product (Anon 1981).

Despite the presence of these secondary metabolites *Nitraria retusa* as well as *Suaeda fruticosa* and *Atriplex nummularia* are always over grazed by sheep, goats and camel (El Shaer 1981). On the other hand, there are several halophytes such as *Atriplex halimus*, *Tamarix mannifera*, *Salsola tetrandra* and *Tamarix aphylla* which are considered less palatable. Where as, there are numerous species mainly *Arthrocnemon glaucum*, *Halocnemon strobilaceum*, *Salicornia fruticosa*, and *Zygophyllum album* which are totally unpalatable.

The following halophytes which are widely distributed in Sinai (Anon 1981) were screened for phytochemical components.

1. *Arthrocnemon glaucum*
2. *Halocnemon strobilaceum*
3. *Juncus acutus*
4. *J. subulatus*
5. *Limonium pruinosum*

6. *Nitraria retusa*
7. *Phragmites communis*
8. *Cressa cretica*
9. *Atriplex halimus*
10. *A. leucoclada*
11. *Halopeplis perfoliata*
12. *Salicornia fruticosa*
13. *Salsola tetrandra*
14. *Suaeda pruinosa*
15. *Suaeda monoica*
16. *Tamarix aphylla*
17. *Tamarix mannifera*
18. *Zygophyllum album*

Tannins were present in all plants while flavonoids were in plants No. 5, 9, 13, 14, 17 and 18. Many of the plants tested contained saponins except No. 1, 2, 3, 4 and 7. Only few plants (3, 8, 9 & 12) contained alkaloids. Resins were found in plant No. 6 as well as 8, 9, 11, 15 and 18.

Phytochemical screening of some halophytic forage plants growing naturally in Deltaic Mediterranean Coastal Belt of Egypt (Anon 1992) revealed the presence of triterpenoids and sterols, tannins and alkaloids in *Aetheorthiza bulbosa*, *Asparagus stipularis*, *Pancratium maritimum*, *Juncus subulatus*, *Cyperus capitatus* and *Schoenus nigricans*. Saponins were found in *Asparagus* and *Pancratium* and flavonoids in several plant species.

The medicinal uses of halophytes for veterinary medicine by bedouins were recorded in Sinai (Bailey & Danin 1981). *Atriplex halimus*, *A. leucoclada*, *Frankenia revoluta*, *Nitraria retusa*, *Salsola tetrandra*, *Salsola vermiculata* and *Suaeda aegyptiaca* are used for treatment of internal parasites. *Zygophyllum dumosum* is used for treatment of mange and *Tamarix nilotica* for sore eyes and for treatment of scratch or abrasions in camels. *Mesembryanthemum forsskalei* is a plant from which bread is made.

Toxicological importance

Nitrate is one of the secondary products in plants which were intensively investigated as a first record for Sinai plants in Egypt (Bayoumi & Ahmed 1983). The natural vegetation in Sinai was classified, according to its nitrate content, into four groups (Table 1).

i) nitrate-free plants;
ii) plants sometimes free of nitrate but which occasionally have different levels, depending on locality;
iii) plants which consistently contain nitrate;
iv) plants containing high nitrate level of toxic nature.

Palatable plants of negative and of low nitrate content which are not harmful to animals are suitable to improve and develop ranges in Sinai. In addition, toxic plants containing high levels of potassium nitrate like *Salsola tetrandra* and *Zygophyllum album* (1.955 and 1.881 gm% dry matter respectively), could be added to the list of poisonous plants of Egypt. However, most of these plants are of low palatability (El Shaer 1986) so losses from poisoning rarely occurs.

Extensive studies have conducted in Desert Research Center in Egypt on the survival of animals feeding on toxic plant *Frankenia pulverulenta* which is distributed in Siwa Oasis in Western Desert (Bayoumi 1978) and *Halocnemon strobilaceum* in Sinai (Bayoumi et al. 1990). *Frankenia pulverulenta* contained 2 alkaloids, 5 glycosides and 2 phytosterols. The palatability of mixed diet with the plant offered to mice decreased as the concentration was increased. A single dose of 1.75 gm per 100 gm BW represented the acute toxicity for mice. This may be due to salt accumulation in the plant.

Frankenia pulverulenta was added to the list of poisonous plants in Egypt that produce clinical signs of puffiness and swelling of the face in mice. Intake was correlated to the kidney dysfunction (due to 25.9% ash), decrease of body weight and body fat percent and increased mortality as intakes rose.

Utilization of *Halocnemon strobilaceum* in Sinai was investigated (Bayoumi et al. 1990) in an attempt to determine the survival of sheep and goats and their ability to maintain themselves when fed this salt marsh plant as sole constituent of their diet. The performance of animals was very poor as digestible matter intake (DMI) was low (39.1 and 34.0 gm per $kg^{0.75}$ BW) for sheep and goats respectively. The animals tended to lose considerable weight due to lower intake but no serious health problem had been observed.

Table 1. Nitrate content of halophyte vegetation in Sinai.

Plant	KNO_3 gm%	Family	Locality
I Plant free of nitrate			
Arthrocnemon glaucum	Zero	Chenopodiaceae	El Bradawill
II Plants sometimes free and occasionally contain nitrate			
Tamarix mannifera	Zero	Tamaricaceae	El Arish-Lehfan
Tamarix mannifera	0.058	Tamaricaceae	Abu Suweira
Tamarix aphylla	Zero	Tamaricaceae	El Arish-Lehfan
Tamarix aphylla	1.219	Tamaricaceae	Wadi Gharandal
Zygophyllum album	Zero	Zygophyllaceae	El Arish-Lehfan
Zygophyllum album	0.180	Zygophyllaceae	El Bradawill
Zygophyllum album	0.317	Zygophyllaceae	El Bradawill
Zygophyllum album	1.881	Zygophyllaceae	El Bruk
III Plant containing nitrate			
Alhagi maurorum	0.245	Leguminosae	Wadi El Tur
Alhagi maurorum	0.361	Leguminosae	Ayon Musa
Alhagi maurorum	0.426	Leguminosae	Wadi Gharandal
Atriplex nummularia	0.022	Chenopodiaceae	Wadi Bruk
A. leucoclada	0.361	Chenopodiaceae	Wadi Bruk
Halocnemon strobilaceum	0.137	Chenopodiaceae	El Bardawill
Salsola tetrandra	0.058	Chenopodiaceae	Wadi Sudr
Salsola tetrandra	0.202	Chenopodiaceae	Wadi Sudr
Salsola tetrandra	0.216	Chenopodiaceae	Abu Suweira
Salsola tetrandra	0.909	Chenopodiaceae	Abu Suweira
Salsola tetrandra	0.923	Chenopodiaceae	Wadi Gharandal
Salsola tetrandra	0.678	Chenopodiaceae	Wadi El tayiba
Salsola tetrandra	1.955	Chenopodiaceae	Wadi Bruk
Suaeda fruticosa	0.844	Chenopodiaceae	Abu Suweira
Juncus acutus	0.382	Juncaceae	El Bardawill
Juncus acutus	0.534	Juncaceae	Wadi Gharandal
IV Plants contain high nitrate of a toxic nature			
Zygophyllum album	1.881	Zygophyllaceae	El Bruk
Salsola tetrandra	1.955	Chenopodiaceae	Wadi Bruk

Table extracted from Bayoumi & Ahmed (1983).

Nutritional importance

Trials had been done on feeding of *Halocnemon strobilaceum* to sheep and goats in its fresh state, air dried as well as ensiled with broiler litter (El Shaer et al. 1990). Sheep and goats were not able to maintain themselves on fresh and dried plants (Table 2) due to the low intake. This may have resulted from high ash and lignin content (El Shaer 1981). Ensiling halophytes with broiler litter and molasses improved animal acceptability and thus intake and nutrient utilization was improved.

Livestock benefitted from ensiling a number of the less unpalatable halophytes, except *Zygophyllum album*. Sheep and goats were able to cover their full maintenance requirements or a significant part of it from (in descending order) *H. strobilaceum* and *A. halimus*. *T. aphylla* and *Z. album*. Total digestible crude protein intake from *A. halimus*. *H. strobilaceum* and *T. aphylla* silage, exceeded maintenance requirement Table 2.

The nutritive value of two salt tolerant halophytes, *Juncus subulatus* and *Diplachne fusca* which are growing naturally in non-cultivated salt-affected lands in the Deltaic Mediterranean Coastal belt of Egypt has been evaluated (Anon 1992). Digestibility of the two plants compared favourably to the commercial ration despite their lower palatability to the rabbits. It is possible to include up to 40% of *Juncus* and *Diplachne* in rabbit feed mixture as a substitute for clover hay.

Table 2. Feeding values of some halohytic plants (gm/day/kg$^{0.75}$)

Plant	Animal species	DMI at 10-day				Silage intake[a]	
		Fresh	Dry	Silage	DM	DCP	TDN
Tamarix aphylla	Goat	8.782	nil	18.4	49.2	2.93	17.3
	Sheep	6.69	nil	16.3	43.3	3.20	18.7
Atriplex halimus	Goat	15.8	21.0	19.8	58.5	5.62	32.5
	Sheep	19.4	13.2	34.1	65.0	6.37	32.2
Zygophyllum album	Goat	2.94	nil	10.2	28.2	1.62	8.81
	Sheep	2.12	nil	19.7	32.7	1.95	9.64
Halocnemon strobilaceum	Goat	14.2	8.23	25.8	59.0	4.56	29.8
	Sheep	18.2	9.45	30.9	59.5	5.57	28.7

Cited from El Shaer et al. (1990)
[a]Silage intakes at 30 day period.

Conclusion

Much work has been done on the distribution of various halophytic plants in Egypt. However, there is a lack of biological information on halophytes in relation to phytochemical, pharmacological, toxicological and medicinal importance. There is a need to conduct a systematic study for all plants used in native medicine or folk medicine on which there is as yet no chemical research in order to make best use of them. In addition attention could be paid to all species from any genus containing even one species of known usefulness.

The results clearly demonstrate the potential of ensiling the aerial parts of halophytic plants, especially those of low palatability. Ensiling can increase the available quantities of animal feed resources and reduce feed costs. Increasing output of animal products will improve food security and contribute to human welfare in the region.

References

Anon. (1981). Agricultural and water investigations of Sinai. In: A.M. Ahmed (ed.), Part IV. Plant Ecology. Report by Desert Institute for Dames & Moore.

Anon. (1992). Revegetation of salt affected lands of the deltaic Mediterranean Coastal, Egypt. Project No. 9009, Phase II, Final report, Foreign relation co-ordination unit Supreme Council of Universities, Egypt.

Bailey, C. & Danin, A. (1981). Bedouin plant utilization in Sinai and the Negev. Economic Botany 35(2): 145–162.

Bayoumi, M.T. (1978). Phytochemical and toxicological studies of *Frankenia pulverulenta* L. Desert Inst. Bull. ARE 28(1): 303–314.

Bayoumi, M.T. & Ahmed, A.M. (1983). Nitrate content of natural vegetation in relation to prospect of animal production in Sinai peninsula. World Rev. Animal Production 19(1): 25–22.

Bayoumi, M.T., El-Shaer, H.M. & Assad Fawzia (1990). Survival of sheep and goats fed salt marsh plant. Journal of Arid Environments 18: 75–78.

Daniel, S.L., Cook, H.M., Hartman, P.A. & Allison, M.J. (1989). Enumeration of anaerobic oxalate-degrading bacteria in the ruminal contents of sheep. FEMS Microbiol. Ecol. 62: 329.

El Shaer, H.M. (1981). A comparative nutrition study on sheep and goats grazing southern Sinai desert range with supplements. PhD Thesis, Fac. Agric. Ain Shams Univ., Egypt.

El-Shaer, H.M. (1986). Seasonal variation of the mineral composition of the natural pasture of Southern Sinai grazed by sheep. Proc. 2nd Egyptian British Conf. on Animal & Poultry Production, August 26–28, Bangor, Gwynedd, UK.

El-Shaer, H.M., Salem, O.A., Khamis, H.S., Shalaby, A.S. & Farid, M.F.A. (1990). Nutritional comparison studies of goats and sheep fed broiler litter ensiled with desert shrubs in Sinai. Proc. Inter. Goat Production Symp. Oct. 22–26, Tallahassee, FL., USA, pp. 70–74.

Gihad, E.A. & El Shaer, H.M. (1994). Utilization of halophytes by livestock on rangelands: Problems and prospects (this volume, Ch. 6).

Girgis, W.A. (1973). Phytochemical studies on the vegetation of the Ras El Hikama-Mersa Matruh coastal plain, Egypt. J. Bot. 16(103): 393–410.

Hathcock, J.N. (ed.) (1986). Nutritional toxicology. Vol II. Academic Press, New York.

Kassas, M. & Zahran, M.A. (1967). On the ecology of the Red Sea littoral salt Marsh, Egypt. Ecological Monographs 37(4): 297–315.

Meyer, M.W. & Karazov, W.H. (1991). Deserts. In: T. Palo & C. Rabbins (eds.), Plant defenses against mammalian herbivory. CRC Press, Boca Raton.

Shalaby, A.F., Etman, M.A., Yossef, M., Habibi, A.M. & Amer, Kh.F. (1977). A chemical investigation of *Nitraria retusa* Forssk. Desert Inst. Bull. ARE 27(2): 199–204.

Smith, G.S. (1986). Gastrointestinal toxification and detoxification of plant compounds by ruminants. J. Range. Manage. 42: 25.

Sperry, O.E., Dollohite, J.W., Hoffman, G.O. & Camp, B.J. (1964). Texas plants poisonous to livestock. Texas Agric. St. Bull. 1028.

Tadros, T.M. (1956), An ecological survey of the semi-arid coastal strip of the western desert of Egypt. Bull. Inst. Desert d'Egypt. 6: 28–56.

Van Soest, P.J. (1982). Nutritional ecology of the ruminant. O & B Books, Corvallis.

26. Small ruminant feeding trials on the Arabian peninsula with *Salicornia bigelovii* Torr.

JAMES J. RILEY,[1] EDWARD P. GLENN[1] & CARLOS U. MOTA[2]
[1]*Environmental Research Laboratory, University of Arizona, Tucson, AZ, USA;* [2]*Genetica y Sistemas de Ingenieria Solar, S.A. de C.V., Hermosillo, Sonora, Mexico*

Abstract. The vegetative portion of the seawater irrigated halophyte, *Salicornia bigelovii* Torr., was fed either washed (to reduce salt content) or unwashed at various inclusion rates to sheep and goats in four trials in the United Arab Emirates and Kuwait during 1987–1990. The weight gain of the small-ruminants was not significantly different than that in diets composed of traditional forages up to a *Salicornia* inclusion rate in the forage component of the diets to 50% for sheep and 100% for goats as long as the total dietary crude protein content was at an acceptable level.

Key words: Dietary protein, Feed-conversion ratios, Feed intake, Goats, Halophytes, Kuwait, Saudi Arabia, Sheep, United Arab Emirates

Background

Meat is an important component in the diets of the people on the Arabian peninsula. Although there is an increase in the consumption of imported frozen beef, goats and sheep are the primary sources of fresh meat. It is estimated that there are 23 million goats and sheep on the Arabian peninsula (Pillai 1989). In addition there is a significant importation of live sheep from Australia to the countries bordering on the Persian (Arabian) Gulf. Alfalfa (*Medicago sativa*) and Rhodes grass (*Chloris gayana*) are the primary forage crops for feeding these and other ruminants. These require large amounts of high quality irrigation water as soils are generally sandy, temperatures high and solar radiation intense. The quantities of fresh water on the Peninsula are limited and the need for conservation is recognized

An introduced halophyte, *Salicornia bigelovii* Torr., has demonstrated good agronomic potential in the United Arab Emirates, Kuwait and Saudi Arabia when irrigated with seawater or highly saline water with the same salinity as seawater (Mota 1987; Riley & Abdal 1992; Koch 1992). *Salicornia* is a potential source of oilseed and forage (Glenn et al. 1991). It produces a biomass of 15–20 mt/ha in a growing period of 8–10 months. About 10% of the biomass is seed, which is low in salt, high in oil (28.2%) and rich in protein (31.2%) (Glenn et al. 1991). The remaining vegetative material is high in salt content (31.5%) and low in protein (6.2%) (Glenn et al. 1992).

Four small-ruminant feeding trials in which the vegetative portion of *Salicornia* was substituted for or combined with various amounts of traditional forages were conducted in the United Arab Emirates and Kuwait from 1987 to 1990. Other trials are planned for Saudi Arabia in the coming year (Koch pers. comm.). The trial design and results are summarized below. The term 'hay' is used to indicate dried vegetative material containing immature seeds. The term 'straw' is used to describe dried vegetative material from which the seeds have been removed.

Description of trials

A 95–103 day feeding trial with 55 Damascus kid goats was initiated in the spring of 1987 in Kalba, Sharjah, United Arab Emirates (Glenn et al.

V.R. Squires & A.T. Ayoub (eds.), *Halophytes as a resource for livestock and for rehabilitation of degraded lands*, 273–276.

1992). The animals were divided approximately equally between four treatments which differed only in the composition of the forage component of their diets. The forage in treatment A was chopped Rhodes grass. In treatment B Rhodes grass was mixed with *Salicornia* straw on an equal basis by weight. The forage component of treatment C consisted of only *Salicornia* straw. However, in C the *Salicornia* straw was washed in seawater to reduce its salt content. The forage in treatments A-C were fed ad lib. The composition of the forage in treatment D was the same as in treatment A (i.e. pure Rhodes grass). The amount of D offered was restricted to equal the amount consumed by the goats in treatment B. A high protein concentrate (16.8% protein) was provided equally to all treatments at a rate of 250 g/head/day.

A 60 day feeding trial with 54 Australian wether lambs was initiated in Kuwait in the spring of 1989 (Abdal et al. 1992). Washed *Salicornia* straw and alfalfa were fed at various rates in 6 treatments. In treatments 1 and 6 alfalfa and *Salicornia* straw, respectively, were fed as sole sources of nutrition. In all other treatments 50% of the diet was composed of a high protein concentrate (13.8% protein). The other 50%, the forage component, was composed of mixtures of *Salicornia* straw and alfalfa. The *Salicornia* straw component in the total diet in these treatments was treatment 2, 0%; treatment 3, 12.5%; treatment 4, 25%; and treatment 5, 50%.

A similar trial with local species of sheep was initiated in the late spring of 1990 in Kuwait. Five diets with inclusion rates of unwashed *Salicornia* hay (9.5% protein, 39% ash) ranging from 0–50% of the total diet were fed to 30 local sheep. There were 6 animals per treatment (i.e., 4 Naeami and 2 Erbei). The total protein content of all the diets was maintained within the range of 16–17% by adjusting the composition of the concentrate (Riley & Abdal 1992).

A 97 day feeding trial with 12 goats and 12 sheep (unspecified breeds) was initiated in December 1988 in Kalba, Sharjah, United Arab Emirates, using the same facilities as the 1987 goat trial (Planetary Design Corporation 1989). Similar to the previously described trials, washed *Salicornia* hay was substituted for Rhodes grass. In addition, differing amounts of *Salicornia* seed

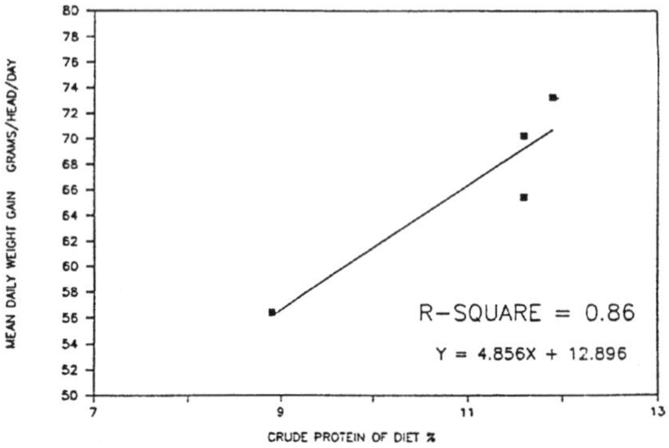

Figure 1. Relationship between mean daily weight gain of Damascus kid goats and crude protein of diet, Kalba, Sharjah, UAE. 1987.

were added to the diets. In treatment 1, the diet was composed of 50% *Salicornia* hay and 25% *Salicornia* seed; in treatment 2, *Salicornia* hay component equalled 60% and 20% *Salicornia* seed was added; in treatment 3, 70% of the diet was made up of *Salicornia* hay and 15% by *Salicornia* seed; and treatment 4, the control diet, was composed of 50% Rhodes grass. In all the diets various amounts of yellow corn, dates and wheat bran were added to balance the constituents.

Results and discussion

In the 1987 Kalba goat trial the only difference in weight gain among the treatments was that male goats gained more weight than the female goats. The mean daily gain per head was: males, 77.8 g/head/day and females, 54.8 g/head/day. Although the weight gain differences between treatments were not significant, there was a tendency for gains to be greater in diets with a higher protein content (Figure 1). As *Salicornia* straw (5.8% protein) was substituted for Rhodes grass (9.5% protein), the greater the amount of *Salicornia* straw in the diet the lower the total dietary protein content. The goats tended to compensate by consuming more of the lower protein content diet resulting in an increase in the food conversion ratio, FCR (Table 1).

In the 1989 Kuwait sheep trial, the animal weight gain was highly correlated with the dietary

Table 1. Inclusion rate of *Salicornia* straw in the forage component of the diet, dietary crude protein content (as consumed), and feed conversion ratio, FCR, for Damascus kid goats, Kalba, Sharjah, UAE, 1987

Treatment	*Salicornia* forage (%)	Dietary protein (%)	FCR
A & D	0	11.8	7.2 a
B	50	11.6	8.3 a
C	100	8.9	11.4 b

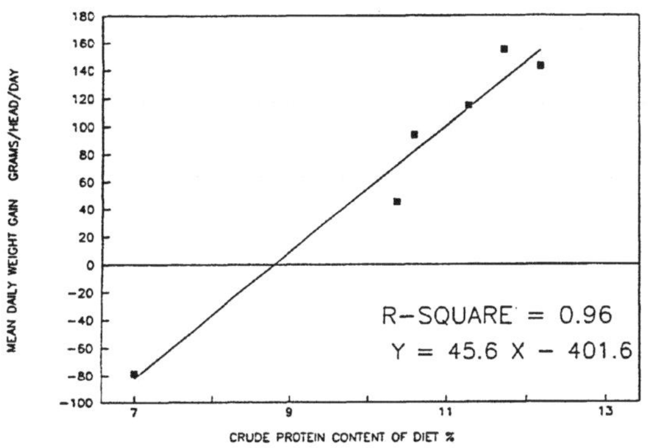

Figure 2. Relationship between mean daily weight gain of Australian wether lambs and crude protein of diet, Kuwait, 1989. Source: Riley & Abdal 1992.

protein content (Figure 2). The regression equation indicates that wether lambs fed a diet with a protein content below 8.8% would have no weight gain. In the four diets to which a high protein concentrate was added, the FCR increased with the substitution rate of washed *Salicornia* straw (7.0% protein) for alfalfa (10.6% protein). See Table 2. There were no significant differences between the weight gains of sheep fed diets with *Salicornia* straw inclusion rates from 0–25%. However, when the *Salicornia* straw inclusion rate increased to 50%, both the daily rate of gain and consumption decreased. The gain decreased more than the consumption which accounts for the higher FCR in this treatment (#5). The addition of concentrate to diets containing only *Salicornia* straw increased the protein content from 7.0–10.4% and the mean daily weight gain of the sheep from −78.5 g/head/day (treatment 6) to 45.4 g/head/day (treatment 5).

The initial observations of the performance of

Table 2. Amount of *Salicornia* straw in the forage component of the diet, dietary protein content (as formulated), and feed conversion ratio, FCR, for Australian wether lambs, Kuwait, 1989

Treatment	*Salicornia* forage (%)	Dietary protein (%)	FCR
2	0	12.2	7.3
3	25.0	11.7	7.9
4	50.0	11.3	9.4
5	100.0	10.4	18.2

the local sheep fed unwashed *Salicornia* hay in Kuwait in 1990 indicated that animals were all growing well except those fed a diet composed of 50% unwashed *Salicornia* hay (*Salicornia* hay composed 100% of the forage component), which lost weight in the initial observation period. These trials were interrupted by the Kuwait crisis and no further data were collected (Riley & Abdal 1992).

Two of the goats in the 1988–89 feeding trial in Kalba died of rinderpest. Another died after giving birth. Since there were only 3 animals per treatment, this made statistical differences between treatments non-significant. The average daily weight gain for all animals was 12.9 g/head/day with a range of −12.9–54.2 g/head/day.

In the 1988–89 sheep feeding trial in Kalba there were no significant differences in weight gain. The mean weight gain for all treatments was 70.8 g/head/day, which compares favorably with the mean daily weight gain of the sheep in the Kuwait trials (84.3 g/head/day). It is worth noting that no acceptability or growth problems were observed in diets containing up to 25% of *Salicornia* seed in the diet of the goats and sheep.

Conclusions

Small ruminants maintain good productivity when *Salicornia* forage replaces local forages up to a level of 100% of the total forage component for goats and up to 50% for sheep. Maintaining adequate protein levels in diets may enable the *Salicornia* straw inclusion level in sheep diets to be increased and the FCR for both small ruminants to be lowered. Based on the results of the Kalba

88–89 trials, it may be possible to use *Salicornia* seed (or meal) as a principal constituent of the high protein concentrate in small ruminant diets.

Acknowledgements

We are grateful to UNEP for a travel grant that allowed one of us, Edward P. Glenn, to attend the international workshop in Nairobi on Halophytes as a Resource for Livestock Husbandry, for which this paper was prepared.

References

Abdal, M.S., Husseini, M.D, Riley, J.J., Diab, M.F. & Awadi, A A. (1992). Feeding value of the halophyte *Salicornia bigelovii* (SOS-7) for wether lambs. Manuscript prepared for submittal for publication by the Kuwait Institute for Scientific Research.

Glenn, E P., O'Leary, J.W., Watson, M.C., Thompson, T.L. & Kuehl, R.O. (1991). *Salicornia bigelovii* Torr.: An oilseed halophyte crop for seawater irrigation. Science 251: 1065–1067.

Glenn, E.P., Coates, W.E., Riley, J.J., Kuehl, R.O. & Swingle, R.S. (1992). *Salicornia bigelovii* Torr.: A seawater-irrigated forage crop for goats. Animal Feed Sci. and Technology 40: 21–30.

Mota, C. (1987). Kalba Field Report. Unpublished report to Planetary Design Corporation, c/o Environmental Research laboratory, Tucson, AZ.

Pillai, P.K. (ed.) (1989). Arab Agriculture 1989. Falcon Publishing WWL, Manama, Bahrain, 272 pp.

Planetary Design Corporation (1989). Kalba farms feed trials – evaluation of SOS-7 threshed seed and seawater washed hay as a feed meal for goats and sheep. Unpublished report.

Riley, J.J. & Abdal, M.S. (1992). Preliminary evaluation of *Salicornia* production and utilization in Kuwait. In: H. Lieth & A. Al-Massoum (eds.), Towards rational use of high salinity tolerant plants. Vol. 1. Deliberations about high salinity tolerant plants and ecosystems. Kluwer Acad. Publ., Dordrecht, The Netherlands, pp. 319–329.

27. Effect of feeding *Atriplex amnicola* on growth and carcass quality of dwarf goats

S. NAWAZ,[1] S.H. HANJRA[1] & R.H. QURESHI[2]
Departments of [1]Livestock Management and [2]Soil Science, University of Agriculture, Faisalabad, Pakistan

Abstract. *Atriplex* species are becoming strong candidates for introduction into salt-affected lands. A study was conducted to see the effect of *Atriplex amnicola*, grown on salt-affected land, alone and mixed with Sudex in varying proportions, on the performance of dwarf goats and the blood picture along with carcass and meat quality. The results of the study reveal that dwarf goats can be maintained on *Atriplex* alone and for growth, supplementation with good quality local fodders or concentrates is required. The study also reveals that carcass quality along with organoleptic value remains unchanged due to *Atriplex amnicola* feeding.

Key words: Blood composition, Halophytes, Irrigation, Nutritive value, Organoleptic sensory evaluation, Pakistan, Rangeland

Introduction

Pakistan with approximately 6 m ha of salt-affected land faces serious shortage of animal products (NCA 1988). The area already under fodders represents about 16% of the cultivated area, the 2nd largest after wheat, and cannot be increased further due to competition with other more urgently required crops. However, the experience in Australia (Malcolm 1986) indicated that highly saline lands can be utilized to increase fodder production using highly salt tolerant *Maireana* and *Atriplex* species while Qureshi (Person. comm 1992) has reported that *Atriplex amnicola* and *Atriplex lentiformis* were the most successful species under a variety of environmental and edaphic situations in Pakistan. We have studied feeding value of *Atriplex amnicola* for dwarf (teddy) goats, a very versatile small ruminant used for meat production in Pakistan. Earlier, Rehman et al. (1988, 1989, 1990) had reported good grazing value of salt bushes in southern Pakistan (Baluchistan province).

Experimental techniques

Twenty five healthy female dwarf goats were divided into 5 groups of 5 animals each in such a way that total weight in all the groups remained almost the same. Five feeding regimes having 0, 25, 50, 75 and 100% *Atriplex amnicola* and/or Sudex (Sorghum × Sudan grass hybrid), were used in the trial which lasted for 90 days (Table 1).

The feeding regimes were randomly allotted to the experimental groups maintained in separate pens of loose housing. Both experimental feeds and water were offered ad libitum and refuse recorded daily. A digestibility trial was conducted at the end of the experimental period.

Jugular blood samples were collected from each animal at 4 week interval for studying blood picture. Haematological studies (see results) were conducted according to Benjamin (1978) and the biochemical analysis was done using Automatic Chemistry Analyzer Model FP 901 (M/s Lab. Systems, Finland).

V.R. Squires & A.T. Ayoub (eds.), *Halophytes as a resource for livestock and for rehabilitation of degraded lands*, 277–280.

Table 1. Composition of feeding regimes

	Feeding regimes				
	I	II	III	IV	V
Sudex (%)	100	75	50	25	0
Atriplex (%)	0	25	50	75	100
Moisture (%)	75.44	68.31	61.66	61.03	58.93
Dry matter (%)	24.56	31.69	38.34	38.97	41.01
Ether extract (%)	1.95	2.67	1.84	2.20	2.97
Crude protein (%)	12.69	9.19	10.76	9.63	9.63
Crude fibre (%)	28.71	30.26	29.70	32.15	30.21
Total ash (%)	8.64	11.85	16.55	16.30	16.15
Nitrogen free (%) extract	48.01	46.03	41.15	39.72	42.04

Results and discussion

Growth and feed efficiency

Significant differences were observed in weight gain, feed and water intake and the feed efficiency for goats maintained on different experimental regimes (Table 2). Maximum weight gain (4.50 kg) and feed intake (3.3 kg) were obtained in feeding regime I and a slightly negative growth and the minimum feed intake were in the group fed on 100% *Atriplex*. Non-significant differences with respect to weight gain were observed in feeding regimes having 0 to 50% *Atriplex*. Feeding regimes having 50 and 75% *Atriplex* level were also not significantly different from each other. Average daily water intake increased with increase in quantity/proportion of *Atriplex* in various regimes and the differences were significant ($p < 0.05$) when *Atriplex* percentage increased to 50% or more.

Best feed efficiency was observed in feeding regime II having 25% *Atriplex* based feed followed by Sudex based regime I and regime III (50% *Atriplex*). The differences among these were non-significant. Significantly the lowest feed efficiency was with 100% *Atriplex*.

Fibre digestibility

The maximum digestibility of crude fibre was observed to be 77.51% with regime II (25% *Atriplex* + 75% Sudex). Increase in *Atriplex* percentage to 75% in the ration reduced digestibility to 60.78% while the lowest digestibility of 38.34 was with 100% *Atriplex*. In the case of purely Sudex based regime digestibility of crude fibre was 69.31%. These differences were significant ($p < 0.01$). Thus with low level feeding of *Atriplex*, digestibility of fibre is improved to some extent which supports previous findings of Rehman et al. (1990).

Carcass quality and organoleptic sensory evaluation

Differences in all the feeding regimes for various carcass parameters were non-significant (Table 3). Similarly, the differences in organoleptic evaluation by a taste panel of the meat from goats raised on various feeding regimes and cooked by 3 different methods, i.e. oven cooking, steam roast and conventional cooking were also non-significant (see Table 3, only Pakistani conventional cooking method is given).

Blood picture

Results on haematological and biochemical values are reported in Table 4. Both haematological as well as biochemical values remained within the normal ranges and the differences in these values were also non-significant except for cholesterol and triglycrides – cholesterol was significantly de-

Table 2. Weight gain, feed and water intake and feed efficiency in dwarf goats

Feeding regime	Wt. gain (kg)	Feed intake (kg)	Feed efficiency (%)	Water intake (litres)	Crude fibre digestibility (%)
I	4.50 a	3.3 a	66 a	0.56 a	69.31 c[a]
II	4.14 a	2.96 ab	64 a	0.72 a	77.51 a
III	2.84 ab	2.92 ab	91 ab	1.73 bc	71.20 b
IV	1.50 bc	2.45 b	144 ab	1.93 c	60.78 d
V	−0.30 c	1.87 c	187 c	2.22 c	38.34 e

[a]Figures followed by different letters differ significantly at $p < 0.05$.

Table 3. Carcass and sensory evaluations of dwarf goat meat fed on different feeding regimes (sudex + *Atriplex*)

Parameters	Feeding regimes				
	I	II	III	IV	V
Sudex (%)	100	75	50	25	0
Atriplex (%)	0	25	50	75	100
CARCASS					
Carcass wt. (%)	39.28	40.44	39.21	39.48	37.07^{NS}
Skin wt. (%)	5.18	5.89	5.72	6.00	5.75
Liver wt. (%)	1.41	1.30	1.10	1.48	1.34
Heart wt. (%)	0.36	0.36	0.38	0.37	0.39
Lungs wt. (%)	0.97	0.92	0.91	1.07	0.99
Kidney wt. (%)	0.26	0.27	0.27	0.33	0.35
Head wt. (%)	5.86	6.55	6.59	6.57	6.97
Legs wt. (%)	1.82	2.01	2.12	2.34	2.74
Stomach wt. (%)	3.32	3.20	3.24	3.28	3.27
MEAT (organoleptic sensory evaluation)					
Colour/general appearance	14.44	12.56	14.89	12.33	13.22^{NS}
Taste	13.22	11.89	12.89	13.22	12.00
Flavour	13.78	12.33	13.44	14.33	11.78
Juiciness	14.11	12.11	14.00	13.00	11.11
Chewability	14.33	11.89	13.78	13.33	10.56
Tenderness	14.44	12.33	13.67	12.67	10.78

NS = Differences between means were non-significant.

creased while triglycrides increased with *Atriplex* based regimes. However, various values remained within the standard ranges for goats.

Conclusions

The study reveals that *Atriplex amnicola* can be used with advantage to partially replace the conventional green fodders like Sudex for feeding dwarf goats on a long term basis. Up to 25% replacement of Sudex with *Atriplex amnicola* was at par in terms of weight gain and was superior to Sudex alone in terms of feed intake and feed efficiency and crude fibre digestibility. There was no detrimental effect on carcass, meat acceptability, blood picture and haematology of the animals fed on fodder mixtures with up to 50% *Atriplex amnicola*.

Acknowledgement

A travel grant was provided by UNEP to Prof. R.H. Qureshi to attend the workshop in Nairobi.

Table 4. Effect of feeding *Atriplex* and Sudex, at varying levels on the blood picture of dwarf goats

Parameters	Units	Feeding regimes				
		I	II	III	IV	V
HAEMATOLOGY						
Haemoglobin	%	66.17	73.51	70.47	67.85	70.55
RBC counts	106 m^{-3}	13.12	14.21	14.40	14.50	16.85
WBC counts	103 m^{-3}	12.73	16.83	12.85	13.41	14.16
Packed cell	%	25.67	25.83	26.60	26.40	27.37
Mautrophils	%	36.11	40.02	38.73	41.25	40.83
Lymphocytes	%	53.85	52.01	54.87	51.66	51.43
Monocytes	%	6.93	6.54	5.23	4.75	5.89
Eosinophils	%	3.10	1.03	1.18	1.34	1.19
Basophils	%	0.00	0.00	0.00	0.00	0.00
BIOCHEMISTRY						
Glucose	mg 100 mL^{-1}	31.26	29.06	29.53	26.31	26.91
Cholesterol	mg 10 mL^{-1}	74.87 a	56.20 b	49.9 b	58.53 b	54.89 b
Triglycerides	mg 100 mL^{-1}	30.88 bc	27.76 bc	25.86 c	35.51 ab	39.79 a
Total bilirubin	mg 100 mL^{-1}	0.22	0.22	0.22	0.24	0.21
Indirect bill	mg 100 mL^{-1}	0.06	0.08	0.11	0.13	0.07
Direct bill	mg 100 mL^{-1}	0.06	0.12	0.11	0.12	0.14
Total protein	g L^{-1}	81.83	81.12	77.91	80.73	77.94
Albumin	g L^{-1}	41.52	41.04	40.45	41.57	42.71
Globulin (TP-Alb)	g L^{-1}	40.31	40.08	37.46	39.19	35.24
Albu/glob ratio	ratio	1.05	1.03	1.09	1.09	1.23

*Figures in rows followed by different letters differ significantly at $p < 0.05$.

References

Benjamin, M.M. (1978). Outline of Veterinary Clinical Pathology, 3rd Ed. Iowa State Univ. Press, Ames. IA, USA.

Malcolm, C.V. (1986). Grazing and management of salt land shrubs. J. Agri. West. Australia. 27: 59–63.

National Commission on Agriculture (1988). Report of the National Commission on Agri., Govt. of Pakistan, Ministry of Food, Agriculture & Cooperatives, Islamabad.

Rehman, A.U., Khan, K.N.M., Asghar, M. & Sultani, M.I. (1988). Four wing salt bush forage compared with conventional feeds for yearling sheep. MART/AZRI Research Report 16, ICARDA, Quetta, Pakistan.

Rehman, A.U., Rafique, S. & Aro, R.S. (1989). Fourwing salt bush as a winter maintenance forage for sheep in upland Baluchistan. MART/AZRI Research Report 37, ICARDA, Quetta, Pakistan.

Rehman, A.U., Rafique, S., Ali, A. & Munir, M. (1990). Nutritive value of fourwing salt bush in growth and digestibility trials with Harnai lambs in upland Baluchistan. Aust. J. Anim. Sci. 3(4): 299–303.

28. Halophytes as animal feeds in Egyptian deserts

H.M. EL SHAER[1] & E.A. GIHAD[2]
[1]*Desert Research Centre, Mataria, Cairo, Egypt;*
[2]*Faculty of Agriculture, Cairo University, Giza, Cairo, Egypt*

Abstract. The Egyptian flora comprises about 2300 species of which 80 are halophytic, belonging to 32 genera and 17 families. Halophytic plants in Egyptian deserts vary greatly in their chemical composition, digestibility (*in vitro*) and palatability. They are widely distributed in high density in the north-western coast (NWC) and Sinai. They contain a moderate level of CP (10.0%) and high levels of CF and ash (21.7 and 27.0%, respectively). Bedouins move their livestock flocks from area to area to take advantage of grazing and/or browsing the most palatable forages. The factors affecting their nutritive value are: plant species, stage of growth, stocking rate and other environmental factors. These are discussed. The process of ageing and maturation were associated with a decline in digestibility, CP and consequently nutritive value. Palatability is affected by the chemical and physical factors. The constraints to the utilisation of halophytic plants are described. Feeding halophytes is a feasible solution to minimise the problem of feed shortage in Egypt.

Key words: Chemical composition, Nutritive value, Palatability, Sinai

Introduction

Animal production practice in Egyptian deserts is mainly based on natural vegetation as animal feeds. The Egyptian flora comprises about 2300 species out of these there are almost 80 halophytic species, belonging to 32 genera and 17 families (Batanouny & Abo Sitta 1977). Though the number of halophytic species is low, they constitute the vegetation of wide areas in Egypt. Most of halophytic plants exhibit a great potential for the rehabilitation of degraded arid lands, as well as a good source for livestock feed. Traditional feeding system disregarded the necessity of proper range and grazing management. This system is the main obstacle to develop animal production on such native ranges. Developed programs to improve the utilization efficiency of native ranges should be practised. This study will focus on the present situation of these rangelands and their potentialities as fodders for livestock.

The native range

The native natural range is described, generally, as an open shrub vegetation, characterized by a sparse cover of semi-shrubs. They vary greatly in their production from year to year due to environmental changes especially the amount of rainfall and its distribution. The annual variability of rainfall (30–200 mm y^{-1}) affects their quantity and quality from area to area. Several halophytes were recognized in the Egyptian deserts, particularly in the north western coast of Egypt (NWC) and Sinai. Halophytes vary greatly in their palatability and are widely distributed in high density and resistant to salinity (Agricultural and Water Investigation of Sinai 1981).

The most dominant halophytes in the Egyptian deserts, particularly in the NWC of Egypt and Sinai peninsula, are listed in Table 1 along with their chemical composition, in vitro digestibility and palatability. On overall average basis, it can

V.R. Squires & A.T. Ayoub (eds.), *Halophytes as a resource for livestock and for rehabilitation of degraded lands*, 281–284.

Table 1. Proximate analysis (%), digestibility (%) on DM basis, and palatability rate (PR) of the dominant halophytic ranges in Egypt

Plant	PR[b]	DM	CP	EE	CF	Ash	NFE	DMD[c]
Alhagi maurorum	S, G	44.0	9.45	4.42	29.5	25.9	30.73	46.4
Arthrocnemon glaucum	C	22.9	3.38	1.26	12.1	51.9	31.36	50.4
Atriplex halimus	A	34.2	12.6	2.28	25.4	22.7	37.02	66.7
Atriplex leucoclada	S, G	25.6	15.1	2.69	27.4	31.7	23.11	52.2
Atriplex nummularia	A	21.7	13.3	5.09	24.2	26.7	30.71	58.8
Halocnemon strobilaceum	C	29.7	6.69	2.22	7.04	40.3	43.75	63.0
Haloxylon salicornicum	Nil	42.4	14.8	6.11	24.1	15.9	39.09	46.5
Juncus acutus	A	35.0	7.11	2.35	28.5	12.3	49.94	34.4
Nitraria retusa	A	37.6	10.2	2.46	32.6	33.0	21.74	61.8
Salicornia fructicosa	C	37.6	13.5	1.89	18.9	14.3	51.41	70.5
Salsola tetrandra	A	37.1	6.32	2.37	36.1	35.9	19.31	68.0
Suaeda fruticosa	A	25.0	10.0	5.00	33.2	16.1	35.7	70.4
Limoniastrum monopetalum	A	48.6	11.5	3.49	14.6	23.6	46.81	68.2
Tamarix aphylla	G, C	34.9	12.9	3.99	13.6	20.1	49.41	48.7
Tamarix mannifera	A	40.0	8.19	3.57	11.6	24.9	51.74	59.6
Zygophyllum album	Nil	24.7	7.76	2.46	11.2	34.2	44.38	76.3
Zygophyllum simplex	C	40.5	11.1	2.12	16.7	29.8	40.28	56.9
Zygophyllum decumbens	C, G	37.7	9.37	1.80	24.1	26.9	37.83	49.6
Overall average		34.3	10.1	3.08	21.7	27.0	38.12	58.2

[a]Data cited from (1,2,5,6).
[b]S: sheep; G: goats, C: camels; A: all these animal species.
[c]DMD%: *In vitro* dry matter digestibility % (15).

be noticed, from these data, that such halophytic forages contain a moderate level of CP (10.0%) and high levels of CF and ash (21.7 and 27%, respectively). They are deficient in energy content in the form of NFE. Several of these halophytes are palatable for animals raised in the Egyptian desert, i.e. camels, sheep and goats, some forages are less palatable. Chemical composition, nutritive value and palatability vary greatly from area to area due to the effect of location factor (Abd El Aziz 1982; El Bassosy 1983). Therefore, Bedouins move their livestock flocks from area to area to have the advantage of grazing and/or browsing the most palatable forages.

Factors affecting the nutritive value

Plant species

Halophytic plant species vary considerably in their chemical composition; the nutritive value and palatability will be consequently different (Table 1). *Limoniastrum monopetalum*, *Suaeda fruticosa*, *Nitraria retusa*, *Atriplex nummularia*, *Atriplex halimus*; *Salsola tetrandra* and *Zygophyllum simplex* are considered the most important halophytic plants as animal feeds due to their higher content of CP and higher palatability. The protection of such forage from over-grazing is necessary to insure available green nutritious fodders for animals around the year.

Stage of growth

The process of ageing and maturation of the studied ranges was found to be associated with a decline in digestibility, CP and consequently nutritive value (El Bassosy 1983; El Shaer 1981). Through two successive years the variations in organic and inorganic constituents of the natural forages at Wadi Sudr, Southern Sinai, were studied (El Shaer et al. 1984). Halophytic plants were dominant in this area, it was concluded that, on overall, the percent of CP, NFE and EE contents and DMD% decreased, while DM, ash, NDF, ADL, cellulose contents increased with advancing maturity from April (spring season) to September (autumn season). The performance of grazing animals; DM intake and nutrients utilization of these forages, estimated by feeding sheep and goats, increased during the wet season (El Shaer 1981) as such halophytic shrubs were nutritious and palatable during the wet season.

Palatability

Halophytic species differ greatly in their palatability (Table 1) which is also affected by other chemical and physical factors (Gihad & El Shaer 1994; Wilson 1994). Using oesophageal fistulated sheep to determine the palatability of halophytic plants at Wadi Sudr, Southern Sinai, during four maturity stages, El Shaer et al. (1984) showed that *Nitraria retusa* was the most palatable followed by *Suaeda fruticosa* then *Salsola tetranda* and *Atriplex halimus*.

Constraints of halophytic plant utilization

Silification of exposed tissues is a major physical form of defense as it reduces feed intake and digestibilities of plant tissues; it contributes to the development of oesophageal cancer and it may cause fatal silica urolithiasis due to formation of calculi (Bayoumi & El Shaer 1994). The native ranges in Sinai showed higher levels of Silica, particularly during the dry season (El Shaer 1984), which eliminated their intake and digestibilities by sheep and goats. Recently, it was found also, that the intake of *Tamarix mannifera* was less as it contained higher levels of silica as compared with other plants (Kandil 1991).

The salt contents in both halophytic plants and the underground water, as the main source of drinking water in the Egyptian deserts, considerably limit the utilization of such forages and affect the performance of animal (Bayoumi et al. 1990; El Bassosy 1983).

Moreover, the high levels of NDF, ADL and hemicellulose in halohpytes appeared to limit their utilization by sheep and goats (Kandil & El Shaer 1990). Such forages were poor during dry seasons due to high contents of lignin and hemicellulose. Chemical defenses may be the most important plant characteristics influencing feed choice by many livestock. Most of halophytes contain secondary metabolites, i.e. alkaloids, tannins, oxalates, glucosides and nitrate (Rizk 1986). Unfortunately, few studies have dealt with the impact of such constituents of grazing livestock in the Egyptian deserts. Particular attention should be given to the role of secondary metabolites because of their harmful impact on the performance of livestock (Bayoumi & El Shaer 1994).

Practical feeding of halophytes

Palatable and nutritious halophytic plants during their early stage of growth could be the sole feed for native livestock. On the other hand, when they are scarce or of poor nutritive value, supplementary feeding should be practiced (El Shaer 1981). This requires a good knowledge of livestock needs as well as availability of nutrients from these plants. Supplementary feeding sometimes constitutes an economic burden upon the production enterprize. The moderate protein content in halophytes might fullfil the livestock needs, but their energy content is deficient. Type, quantity and quality of supplements for sheep and goats on the rangelands of Wadi Sudr (Sinai) were intensively studied (El Shaer 1981; El Shaer et al. 1986). Conventional energy concentrate supplements improved DM intake, growth rate and milk production. Bedouin in such areas grow rain-fed barley as well as olive and date-palm plantations which produce large quantities of by-products. These were used successfully in the unconventional supplementary feed of livestock grazing the native rangelands in Sinai as it considerably reduced the feeding costs in comparison with the conventional supplements (El Shaer et al. 1986; Gihad 1990).

Processing of unpalatable or less-palatable halophytic plants to produce improved feeds and upgrade the halophytes palatability and feeding value were carried out in early nineties (El Shaer et al. 1990, 1991). Ensiling *Halocnemum strobilaceum* with broiler litter and molasses increased animal acceptability and intake of goats by 13% over the control (El Shaer 1991). Judged from the nutritive value and consumption of such silages, they would be of great value in alleviating feed shortage in Egyptian deserts.

Conclusion

Feeding halophytes is a feasible solution to minimize the problem of feed shortage in Egypt, where desert represents 96% of its total area. Therefore, proper range management and utilization of halophytes as an animal feed should have the priority of the development plans.

References

Abd El Aziz, D.M. (1982). A study of the nutritive value of some range plants in the North Western Coastal Desert. PhD Thesis, Fac. Agric., Ain Shams Univ., Egypt.

Agricultural and Water Investigation of Sinai. (1981). Part VI. Animal Production Desert Research Institute, Cairo, Egypt.

Batanouny, K.H. & Abo Sitta, Y.M. (1977). Eco-physiological studies on halophytes in arid and semi-arid zones. 1. Autecology of the salt-secreting halophyte *Limmiastrum monopetalum* (L.) Boiss. Acta Botanice Academiae Scientiarum Hungaricae, Tomus 23–31.

Bayoumi, M.T., El Shaer, H.M. & Fawzia Assad. (1990). Survival of sheep and goats fed salt marsh plants. J. Arid Environments 18: 75–78.

Bayoumi, M.T. & El Shaer, H.M. (1994). Impacts of halophytes on animal health and nutrition (this volume, Ch. 25).

El Bassosy, A.A. (1983). A study of the nutritive value of some range plants from Saloom to Mersa Mattroh. PhD Thesis, Fac. Agric., Ain Shams Univ. Egypt.

El-Shaer, H.M. (1981). A comparative nutrition study on sheep and goats grazing southern Sinai desert range with supplements. PhD Thesis, Fac., Agric., Ain Shams Univ., Egypt.

El Shaer, H.M., Hassan, N.I., El Ashry, M.A. & El Serafy, A.M. (1984). Seasonal variation in the nutritive value of the Southern Sinai native range pastures. Proc. 1st Egyptian-British Conf. on Animal Production. Sept. 11–13, Zagazig, Egypt.

El Shaer, H.M., Kandil, H.M. & Farid, M.F.A. (1986). The use of agro-industrial by-products in supplementary feeding of pregnant and lactating sheep and goats. Proc. 2nd Egyptian-British Conf. on Animal & Poultry Production. August 26–28, Bangor, Gwynedd, UK.

El Shaer, H.M., Salem, O.A., Khamis, H.S., Shalaby, A.S. & Farid, M.F.A. (1990). Nutritional comparison studies of goats and sheep fed broiler litter ensiled with desert shrubs in Sinai. Proc. Inter. Goat Production Symp. Oct. 22–26. Tallahassee, FL, USA, pp. 70–74.

El Shaer, H.M., Kandil, H.M. & Khamis, H.S. (1991). Salt marsh plants ensiled with dried broiler litter as a feedstuff for sheep and goats. J. Agric. Sci. Mansoura Univ. 16: 1524–1534.

Gihad, E.A. (1990). Utilization of high salinity tolerant plants and saline water by desert animals. In: H. Lieth & A. Al Massoom (eds.), Towards national use of high salinity tolerant plants, Vol. 1. Kluwer Academic Publishers, Dordrecht, pp. 443–447.

Gihad, E.A. & El Shaer, H.M. (1994). Utilization of halophytes by livestock on rangelands: Problems and prospects (this volume, Ch. 6).

Kandil, H.M. & El Shaer, H.M. (1990). Comparison between goats and sheep in utilization of high fibrous shrubs with energy feeds. Proc. Inter. Goat Production Symp., Oct. 22–26, 1990. Tallahassee, FL, USA.

Kandil, H.M., El Shaer, H.M., Khamis, H.S. & Ahmed, A.M. (1991). Nutritional value of hyperarid forage species for sheep in Upper Egypt. J. Agric. Sci. Mansoura Univ. 16: 518–525.

Rizk, A.M. (1986). The phytochemistry of the flora of Qatar. Scientific and Applied Research Center, Univ. of Qatar, pp. 33.

Tilley, J.M.A. & Terry, R.A. (1963). A two-stage technique for the in vitro digestion of forage crops. J. Br. Grassland Soc. 18: 104–111.

Wilson, A.D. (1994). Halophytic shrubs in semi-arid regions of Australia: Value for grazing and land stabilization (this volume, Ch. 8).

29. *Alhagi sparsifolia* Schap.: A potentially utilizable forage in saline soils

QIHONG JIN
Institute of Botany, Academia Sinica, Beijing, PR China

Abstract. *Alhagi sparsifolia*, a semi-shrub in the family Leguminosae is widely distributed on saline meadows. It has a highly developed root system. Seeds can survive at a salinity level of 100 m Mole sodium chloride. It has potential in the land rehabilitation. Research on *Alhagi sparsifolia* has been conducted in low-land saline meadows in Xinjiang, western China since 1990. The present paper reviews the ecology and biology of *Alhagi sparsifolia* and reports some data on nutritive value of the plant.

Key words: Deserts, Germination, Land rehabilitation, Nutritive value

Introduction

Xinjiang, a vast region of 1.65 million km^2 located at the upper northwest corner of China is typical of Inner Asia (Table 1). This is a vast region of deserts and mountains. Natural salines and secondary (induced) salinity are common problems (Fan 1992; Zhang et al. 1994).

Within this region, a dry climate prevails because of the impact of the rain shadow caused by high mountains to the west and south and the location deep in the heart of Asia. Desert, including sand desert and gravel (gobi) dominates the landscape at lower altitudes (Figure 1).

The water resources come mainly from melting snow and ice in the high mountains (Tian Shan, Altai and Kunlun). Many ephemeral rivers, reservoirs and underground channels (karez) are supplemented mainly from melting snows in the summer season. The oases* (Figure 1) as the land first exploited for agriculture, have been cropped for thousands of years. They are deteriorating to a greater or lesser extent (Fen 1992; Squires 1989). One of the main adverse changes of soil is salinization, which is particularly serious in low lying areas. The salt content of the upper layers of the soil (0–30 cm) is now as high as 1–4%.

To maintain animal production, which is one of the supporting industries in the region (Squires & Bayan 1992 and Table 2), and to avoid destruction of the fragile environmental system, care must be taken in exploiting such a natural resource and of course rehabilitation is desirable.

Research objective

My research could be regarded as one of the efforts to benefit land use in two ways: to obtain a higher economic value from it and to assist in land rehabilitation.

Alhagi – ecology and biology

Among many species of native plants which have wide ranges of salt tolerance, *Alhagi sparsifolia* Schap. a salt tolerant, drought resistant plant was chosen. *A. sparsifolia* is a sub-shrub in the family Papilionatae, a sub family of Leguminosae (Yiang 1989). It is the only species in this genus recorded so far in China (Liu 1987). According to Wang

*In the Chinese context, an oasis may be man made. It usually refers to irrigated areas in the desert.

V.R. Squires & A.T. Ayoub (eds.), *Halophytes as a resource for livestock and for rehabilitation of degraded lands*, 285–288.

Table 1. Geographic location and some climate attributes of Urumqi, Xinjiang

Longitude	Latitude	Altitude (M)	Annual Precip. (mm)	Annual sunshine (hr)	Max. Ave. monthly Temp. (°C)	Frost free period (days)
87°37′ E	43°47′ N	912	273	2744	23.7	161

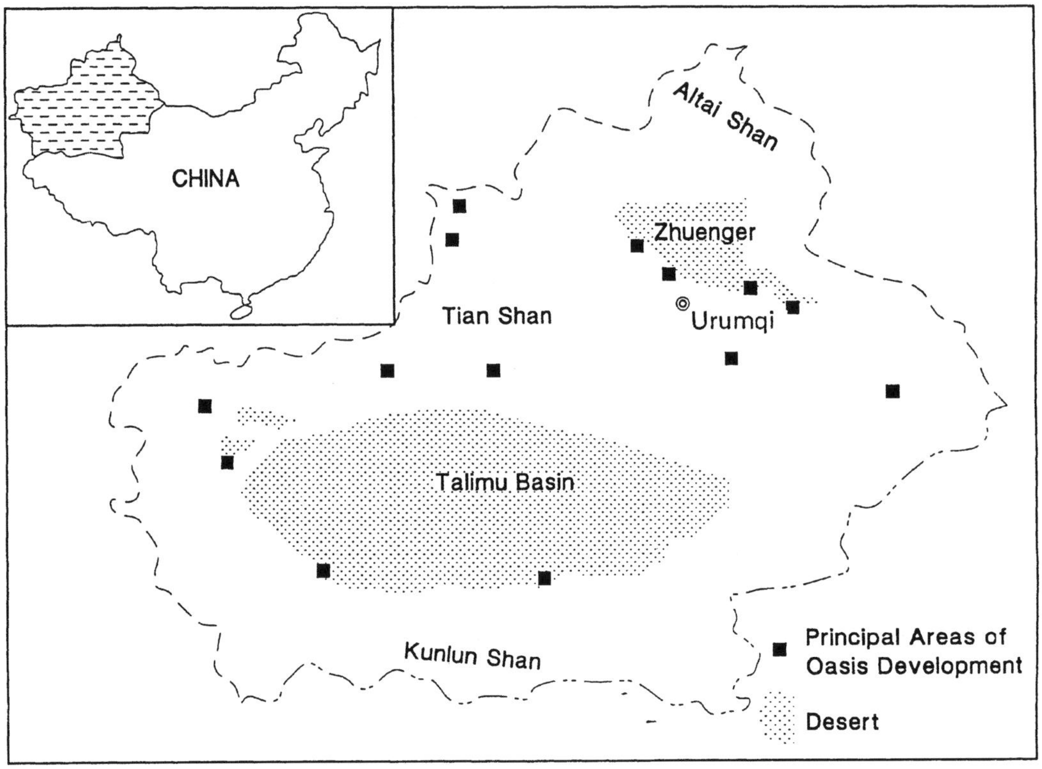

Figure 1. Sketch map showing the geographic location of Xinjiang and the location of oases in relation to the principal mountains, deserts and rivers.

1964; Ma 1989; Liu 1987; and Li 1983 it is distributed in NW China (Xinjiang, Gansu, Ninxia and Inner Mongolia). More or less similar species can be found in Mongolia, Pakistan, Afghanistan, Iran, India and some localities in the Middle East and central Asia (Zohary 1962; Western 1989).

Its long growing season, from spring to autumn, alternative reproduction modes (vegetative or via seeds, depending on the ambient conditions) and the extensive root system allows it

Table 2. Xinjiang and area of land utilizable by livestock in Xinjiang

Land type (km^2)			
Grazing lands	Meadows	Saline & abandoned	Oases
4.8×10^5	5.4×10^4	5.2×10^4	7.0×10^4

Table 3. Chemical composition of *Alhagi sparsifolia* (data from Yiang 1962)

Water (%)	Percentage in dry matter						
	Ash	Protein	Fat	Fibre	Soluble carbohydrate	Calcium	Phosphorus
6.87	6.08	17.67	1.99	20.97	46.43	4.0	0.11

to compete and survive in the adverse soil environments so characteristic of the region. *A. sparsifolia* was chosen as material in my experiment for developing potential fodder in that region, because:

1. it is regarded as a good grazing plant by shepherds and camel-drivers, and also a good food in spring for cattle;
2. it has high nutrient content with over 50% digestible organic matter (Table 3), and even higher digestible protein than some species of *Medicago* (Yiang 1989; Hou 1982).
3. it has a wide distribution in a comparatively wide range of habitats from abandoned cropping lands, saline meadows and sand-dunes. It is not always a dominant species or subdominant species (Hou 1982, 1988; Wang 1964). With these advantages, this species is getting more and more attention recently.

Table 4 Effects of salinity on germination (%) of *A. sparsifolia* at constant temperature of 20 °C on day 7

Salt concentration mM	0	25	50	100
NaCl treatment		42	33	29
Na_2SO_4 treatment	35	43	34	34
$NaHCO_3$ treatment		42	42	42
Na_2CO_3 treatment		55	46	9

Research objectives

My interest focuses upon both its potential for establishing as a sown pasture and on the mechanism for survival in saline habitats. As part of the first approach, some effort was made to develop it from seeds at different salt conditions. Germination of seeds was assessed in the presence of salts such as sodium chloride, sodium sulphate, sodium bicarbonate and sodium carbonate. A 100 m mole concentration, which is compatible with its distributed soil salt conditions, was chosen. The results of seed germination experiment (Table 4) showed a strong resistance to salt stress.

The next step is to sow seeds in the field on a small scale and to try to find a suitable way for growing it on a large scale. If successful, the main role for *A. sparsifolia* could be as supplementary fodder for livestock.

References

Fan, Zili (1992). The evolution trends of Xinjiang's ecological environment and its control. Arid Zone Research 91: 26 (in Chinese).

Fen, Huilan (1992). The land use type and its characteristics in Xinjiang. Arid Land Geography, Vol. 15-1 (in Chinese).

Hou, Xueyu (1982). Vegetation geography of China and the chemical composition of dominant plant species. Science Press, Beijing (in Chinese).

Hou, Xueyu (1988). Plant geography. In: Natural geography of China. Science Press, Beijing (in Chinese).

Li, Shiying (1983). A review of plant ecology in arid zone of China. Acta Ecologica Sinica 34: 357–365 (in Chinese).

Liu, Yengxin, ed. (1987). Flora in desertis republicae populorum sinarum, Vol. 2, Science Press, Beijing (in Chinese).

Ma, Yu-chuan, ed. (1989). Flora intramongolica, Vol. 3. Innermongolia Renmin Press, Huhhot (in Chinese).

Squires, V.R. (1989). Farming on the desert fringe: A case study from Xinjiang, China. J. Arid Envir. 16: 229–233.

Squires, V.R. & Bayan, Onerhan (1992). Biology and ecology of sheep raising on the desert fringe: A case study from Xinjiang, China. In: G.K. Perrier & C.W. Gay (eds.), Current natural resources development activities in developing nations. Utah State University, Logan Utah, pp. 118–125.

Wang, Heshen (1964). The distribution of halophyte communities in Xinjiang and the relation with soil and underground water. Acta Phytoecologica Et Geobotanica Sinica, Vol. 2-1: 57–69 (in Chinese).

Western, A.R. (1989). The flora of the United Arab Emirates – An introduction, UAE University.

Xinjiang Integrated Investigation Group (1978). The topographical features of Xinjiang. Science Press (in Chinese).

Yiang, Zenzhong (1989). Alhagi sparsifolia Schap. In: Flora of Chinese Fodder. Agriculture Press, Beijing (in Chinese).
Zhang, Liyun., Xia, Yang. & Zou Yun (1994). Halophytes and halophytic communities in Inner-Asia (this volume, Ch. 9).
Zohary, M. (1962). Plant life of Palestine. The Ronald Press Company.

30. Massive propagation of halophytes (*Distichlis spicata* and *Tamarix* spp.) on the highly saline-alkaline soils in the ex-Lake Texcoco, Mexico

F.A. LLERENA V.
National Water Commission, S.A.R.H. Mexico City, Mexico

Abstract. The Lake Texcoco, a former salt lake in Mexico City, covers 12,000 ha. It has serious drainage problems in the wet season and is a major source of dust in the dry season. The soils are highly saline (>120 mmhos cm^{-1} in the saturation extract). The exchangeable sodium is near 100% and the pH is >9.5. A massive re-vegetation scheme was implemented in 1971 using the locally-adapted native grass *Distichlis spicata* (L. Greene) and various *Tamarix* species. Vegetative propagation has been favoured because of lack of suitable viable seed source. By 1992 over 5,000 ha have planted. A methodology for the establishment of wind barriers has been developed and over 28 km of double wind breaks have been planted using *T. parviflora*, *T. articulata* and *T. plumosa*. These procedures are explained and an assessment given of progress to date.

Key words: Exchangeable sodium, pH, Vegetative propagation, Windbreaks

Introduction

The metropolitan area of Mexico City ranks among those with the highest population in the world. It is located in the central part of the country, inside a basin known as the Valley of Mexico. Five hundred years ago, the enclosed basin had a system of six lakes which covered approximately 200,000 ha.

Of these, Lake of Texcoco, located in the east, was special because it was the oldest and the only salt water lake. About 400 years ago, changes in human activity within the valley started a gradual drying up of lakes. In the present century the rate of desiccation has speeded up.

The drying up of the newer lakes permitted the growth of vegetation on the newly exposed surfaces. These areas were then incorporated into the urban areas of the valley. However, in the case of the ex-Lake of Texcoco, because of drainage problems and the high salinity levels of its soils, the natural propagation of plants was impossible.

During the dry season, large areas of exposed soils created a great quantity of dust on the surface of the dried lakebed. Because waste water and detritus from the metropolitan drainage system passes across the same area, highly polluting dust storms were a frequent occurrence.

In an attempt to solve these problems, The Lake of Texcoco Project was created in 1971. It covers over 12,000 ha and has the following objectives:

- A solution to the problems of the dust storm.
- Management and use of the hydraulic resources.
- Development of the area.

Special characteristics of the soils

The soils of the ex-Lake of Texcoco are characterized by five extreme conditions:

1. High salt content, greater than 120 mmhos cm^{-1} (or greater than 8%) in the saturation extract.

V.R. Squires & A.T. Ayoub (eds.), *Halophytes as a resource for livestock and for rehabilitation of degraded lands*, 289–292.

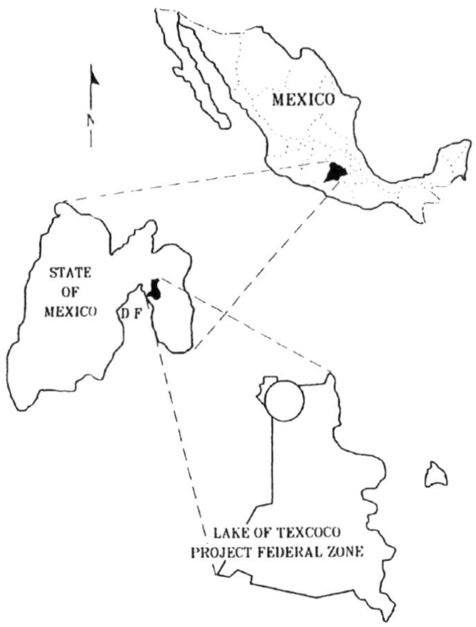

Figure 1. Location of the federal zone of the Lake of Texcoco project.

2. pH values over 9.5 and reaching a maximum 11.5
3. Content of exchangeable sodium (ESP) near 100%.
4. Presence of highly saline, shallow water table.
5. Presence of an alluvial soil stratum formed of volcanic ash (called 'ALOFANO'). It has a high water content and low permeability which makes soil management difficult.

Dust storm control actions

Establishment of vegetation cover

From the beginning, the most economically and technically viable option was the establishment of a vegetative cover. Only after testing several halophytes, both native and those which were brought from other areas, was it possible to determine the most suitable species. The salt grass, *Distichlis spicata* (L. Greene) a native of this zone offered a number of advantages:
- it is perennial grass;
- it is tolerant of highly saline soils;
- it is resistance to frost, drought and flooding.

Since 1973 programs for massive grass propagation have been carried out. The vegetative propagation method was used because the grass did not produce seed.

The method followed to cultivate this grass is to first take a cube of soil containing grass with roots and leaf blades. The cube is then taken to a previously ploughed target area and placed in the deepest part of the furrow. Finally, it is irrigated with low salinity waste water, which is the only plentiful water source in the zone. It must be irrigated for at least two years, after which the grass is fully rooted and must depend on pluvial precipitation and/or stored soil moisture.

Today, about 5,000 ha are planted with this grass. 60–80% of it has a good cover over it. There are also 1,000 ha on which plantings have been made, of these, about 600 ha are still bare and the remainder has a cover of between 1 and 40%.

Establishment of natural wind barriers

In 1980 a methodology for the establishment of wind barriers was developed. It consists of planting trees along the irrigation canals. A double row barrier (over 28 km long) has been established using *Tamarix parviflora*, *T. plumosa* and *T. articulata* trees. However, some problems have arisen such as a short life-span and a high proportion of sickly plants. This has been found to be a result of poor and superficial root development due to extreme soils conditions.

Support infrastructure

In order to carry out the Project objectives, several infrastructure projects have been constructed. These include two artificial lakes which cover about 5000 ha and store rain and treated waste water for further reuse. Another two reservoirs were constructed for the regulation and storage of the waste water.

Other projects, such as re-afforestation and soil and water conservation are in progress in the mountains adjacent to the basin.

Figure 2. A surface showing grass cover.

Figure 3. A panoramic view showing the natural wind barriers.

Evaluation

What are the results after years of intensive work?

Since 1985 the dust storm has been eliminated. The *Tamarix* spp. wind barriers are now serving two main functions:
- to decrease the effect of the wind,
- to stop the dust particles generated in other areas, and
- they also contribute to the air quality improvement.

* Extensive production of the beef cattle, sheep, and horses has been successfully practiced, supported by the great extension of 'Salt rangelands' where the salt grass has demonstrated its relatively high performance as a forage.
* The artificial lakes have served as an environment for the proliferation of several species of migratory birds which yearly arrive in hundred of thousands from October to April:
* In the mountains adjacent to the basin about 16,000 ha have been re-afforested. This is about the 25% of the area affected by erosion.

Figure 4. Beef cattle grazing on the salt grass rangeland.

* Several plant and animal species have gradually appeared in the new ecosystem. Likewise the climatic parameters tend to be mild and regular.
* The experiences generated on halophytes propagation specially of *Tamarix* spp. and *Distichlis spicata* may be applied in other areas with problems of salinity and/or drainage.

Although The Lake of Texcoco Project is now considered as a model to follow, it has only been possible thanks to the continuity of the conducted efforts and the protection of the zone as a Federal Ecological Reserve, which has prevented the intrusion of the urban areas.

References

Cruickshank, G.G. (1986). Avances del Proyecto Lago de Texcoco. Reporte Interno. Sarh. Mexico.

Llerena, V., Llerena, F.A. & Tarin, V.M. (1978). Establecimiento de Pasto Salado (*Distichlis spicata*), Como Cubierta Vegetal en Suelos Extremadamente Salino-Sodicos del Ex-Lago de Texcoco. Memorias del XI Congreso Nacional de la SMCS. Villahermosa, Tabasco, Mexico.

Llerena, V. & Llerena, F.A. (1982). Desarrollo de una Metodologia Economica Para el Establecimiento de Barreras Rompevientos de *Tamarix* spp. en el Lago de Texcoco, Mexico. Reporte Interno CLT-Sarh. Mexico.

Llerena, V., Llerena, F.A. & Tarin M.V. (1980). Una Alternativa para Integrar a la Productividad Algunas de las Areas Altamente Salino-Sodicas del Pais. Memorias de la Segunda Conferencia Interamericana Sobre Tecnologias de la Salinidad y Manejo del Agua. Cuidad Juarez, Chihuahua, Mexico.

Morales, R.J.A. (1980). Evaluacion del Pasto Salado *Distichlis spicata*) Como Fuente de Forraje Para Rumiantes. Tesis Profesional Uach., Chapingo, Mexico.

31. Vetiver grass, its potential in the stabilisation and rehabilitation of degraded saline land

P.N.V. TRUONG
Natural Resource Management Unit, Department of Primary Industries, Indooroopilly, Queensland, Australia

Abstract. Vetiver grass (*Vetiveria zizanioides*) is a perennial grass well adapted to humid and arid, tropical and temperate climates. When planted as contour hedges the Vetiver system can stop sheet erosion and improve soil moisture. It can be equally effective in stabilising steep slopes, eroded gullies and waterways. High soil salinity is having a major impact on Australian agriculture, at least 1.3 million hectares of agricultural land are affected by high soil salinity. Vetiver grass was found to be moderately salt tolerant and it has a saline threshold value of 8 mS cm^{-1} which is comparable to some of the most salt tolerant crop and pasture species grown in Australia. Soil EC_{se} at 17.5 mS cm^{-1} reduces growth by 50%. The degree of tolerance was even higher in mature plants. Vetiver plants could survive soil salinity level at 47.88 mS cm^{-1} and established on a highly alkaline (pH = 9.65) and sodic coal mine spoil. The findings of this work potentially opens up tens of thousands of hectares of degrading saline land worldwide within the climatic limits of 30° latitude and average annual rainfall of 500 mm.

Key words: Gully stabilisation, Salt tolerance, Scald reclamation, Vetiver grass

Introduction

The World Bank since 1986 and recently FAO have promoted the use of Vetiver system in soil and water conservation worldwide particularly in developing countries. Recently 85 delegates from nine countries attended the first Vetiver Workshop which was held in Malaysia to review its application to date and to assess its potential uses in other environments including saline lands.

Vetiver (*Vetiveria zizanioides* L.) is a perennial grass reported to be fire-resistant, well adapted to both humid and arid, tropical and temperate climates and it can also survive under water for a long period of time. Vetiver has been used in India for centuries, when planted as contour hedges it can stop sheet erosion and improve soil moisture content. It does not compete with crops, forms a barrier against creeping weeds and produces material for thatch roofs, fodder for livestock, and roots for scented mats (World Bank 1990).

An alternative to engineering structures

In a review of existing soil conservation technologies, Grimshaw (1988) concluded that vegetative systems are more effective both for soil and moisture conservation than constructed conservation systems which require accurate engineering, design and maintenance. These systems often reduce soil loss but do not lower runoff significantly resulting in nutrients and soil moisture losses. Vetiver hedges, on the other hand when planted on contour, form a protective barrier across the slope, slowing the runoff water, depositing the silt, and water oozes and spreads through the hedges across the slope thus conserving water and nutrient *in situ*. Vetiver hedges are found to be

V.R. Squires & A.T. Ayoub (eds.), *Halophytes as a resource for livestock and for rehabilitation of degraded lands*, 293–296.

equally effective in stabilising waterways, gullies and steep slopes.

Nature and extent of saltland in Australia

High soil salinity is having a major impact on all states of Australia. Statistics from 1988–1989 surveys showed that salt affected land in Western Australia was 444,000 ha, Victoria 385,000 ha, New South Wales 250,000 ha, South Australia 221,000 ha (Barrett-Lennard et al. 1990) and Queensland 15,000 ha. These figures do not include saline scalds which accounts for at least 950,000 ha in New South Wales and more than 600,000 ha in Queensland. In Western Australia over 2 million ha have the potential to become salt affected. Given these statistics, the prognosis for salinity in Australia is not promising. It is likely that the area of saltland will continue to substantially increase, with many millions of hectares finally becoming affected.

Extent and distribution of soil salinity in Queensland

Although only 15,000 ha (0.01% of the State) of land are salt affected in Queensland, 66% of this land occurs in the most productive regions of the State and area affected on individual farms can exceed 50%. Two thirds of the affected area occurs in grazing lands. The land affected by saline scalding is presently assessed at 600,000 ha which represents only 1.5% of the area of saline and/or sodic soils with potential for developing scalding. These sodic soils often contain high levels of exchangeable sodium and magnesium and low levels of exchangeable calcium (Isbell 1957). These soils are highly erodible in semi-arid regions of Queensland, drought and overgrazing often accelerate the erosion processes. Both sheet and particularly gully erosion are prominent features of the semi arid landscape. Gully stabilisation and pasture rehabilitation on these soils by conventional engineering methods are very difficult and often not economical. A low cost, low maintenance method which requires minimal technical input is needed on the saline/potentially saline grazing lands of Queensland. The Vetiver hedge system provides the most promising solution to this problem in Queensland so far.

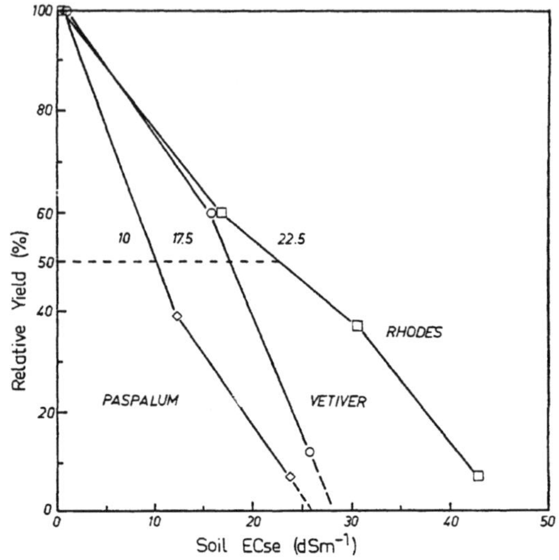

Figure 1. Comparative salt tolerance between Rhodes, Vetiver and Paspalum grasses.

Salt tolerance of Vetiver grass

To be effective in the rehabilitation of salt affected lands in Queensland, Vetiver needs to be moderately salt tolerant. In a series of glasshouse and field trials, Vetiver was found to have a moderate to high level of salt tolerance, comparable to Rhodes grass (*Chloris gayana*), the most salt tolerant of the commonly used pasture species in sub-tropical Australia. Figure 1 shows that soil EC_{se} at 17.5 dS m^{-1} reduces Vetiver yield by 50%, soil salinity at 16 dS m^{-1} or higher are considered to be highly saline by the US Salinity Laboratory (1954). With the saline threshold at 8 dS m^{-1} (Figure 2) Vetiver is among the group of highly salt tolerant crop and pasture species grown in Australia (Table 1).

Results from a column experiment (60 cm deep) indicate that the high level of salt tolerance in Vetiver was partly due to its deep rooting characteristics. The majority of Vetiver roots could escape from the high salt accumulation in the surface soil, a characteristics of dryland salt-

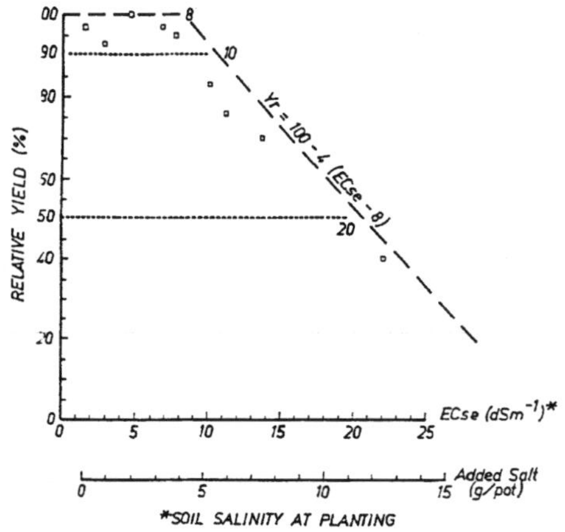

Figure 2. Salinity threshold of Vetiver grass.

Table 1. Salt tolerance levels of Vetiver grass as compared with some crop and pasture species grown in Australia

Species	Soil EC_e ($dS\ m^{-1}$)	
	Saline threshold	50% Yield reduction
Bermuda Grass (*Cynodon dactylon*)	6.9	14.7
Rhodes Grass (*C.V. Pioneer*) (*Chloris gayana*)	7.0	22.5
Tall Wheat Grass (*Thynopyron elongatum*)	7.5	19.4
Cotton (*Gossypium hirsutum*)	7.7	17.3
Barley (*Hordeum vulgare*)	8.0	18.0
Vetiver (*Vetiveria zizanioides*)	8.0	17.5

Table 2. Soil saline levels corresponding to different species establishment following rain

Species	Soil EC_{se} ($mS\ cm^{-1}$)		
	0–5 cm	10–20 cm	40–50 cm
Chloris gayana	4.83	9.59	12.53
Paspalum vaginatum	9.73	11.51	15.35
Vetiveria zizanioides	18.27	18.06	5.88
Bareground	49.98	23.94	4.2

ing. Under field conditions, Vetiver growth correlated best with the average saline level of soil between 0–50 cm depth. It was also shown that when saline treatments were imposed on mature Vetiver plants (8–10 weeks) saline levels which adversely affected establishing plants, did not affect mature Vetiver. This was supported by the field trials where Vetiver plants could survive very high soil saline level and would resume vigorous growth following good rains.

A very active gully head caused by saline seepage, was successfully stabilised by two Vetiver rows. The seepage water with EC ranged from 6 to 19 $dS\ m^{-1}$ (mostly NaCl and $MgCl_2$) increased the average soil EC_{se} (0–50 cm depth) from 1.96 to 9.04 $dS\ m^{-1}$. This gully head was not affected by a 150 mm storm with a flow depth of at least 60 cm.

In an attempt to revegetate a highly saline area (caused by shallow saline groundwater), Rhodes grass, salt water couch (*Paspalum vaginatum*) and Vetiver were planted on a slope completely bare of vegetation, with EC_{se} values for the surface soil (0–5 cm) ranged from 3.85 to 47.88 $dS\ m^{-1}$ (top and bottom of the slope respectively). Due to dry weather little establishment or growth occurred over the next two months. All the plants remain in that state through the winter but following good rain in early summer (9 months after planting) vigorous growth of the three species occurred. Vetiver survived and resumed growth under very highly saline conditions (Table 2), reaching the height of 60 mm in eight weeks. In another trial, Vetiver was able to establish and thrive (with N and P applications) on a very alkaline (pH: 9.65) and sodic (Na at 26,400 ppm) coal mine spoil in Central Queensland.

Potential use of Vetiver in saline land rehabilitation in Australia and elsewhere

The above results have clearly shown that Vetiver grass, at least the ecotype used in Queensland, tolerates considerably high levels of soil salinity. Prior to this work, the World Bank Newsletter (1991) stated that indication from literature had led to the unchallenged belief that Vetiver grass was highly sensitive to saline conditions, the results of this work potentially opens up tens of thousands of hectares of degrading saline lands worldwide on sites where contour vegetative barrier of Vetiver grass may be used to halt and reverse land degradation.

Although Vetiver can survive extreme climatic conditions, it thrives best in areas having at least

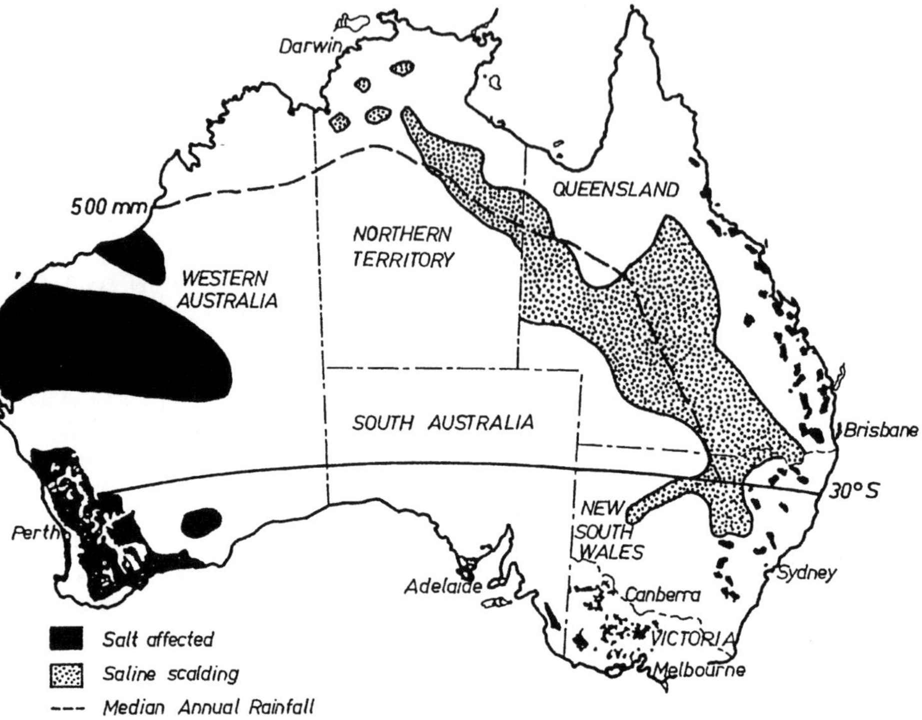

Figure 3. Map showing salt affected areas of mainland Australia.

500 mm of annual rainfall and between the latitude of 30° north and south (Grimshaw pers. comm.). Within these climatic limits, exists most of the highly productive agricultural land in Australia and potentially the Vetiver system could provide an invaluable means to combat land degradation in general and saline lands in particular (Figure 3). In Queensland the Vetiver system can be applied in all the salt affected lands mentioned above. In the saline scalding areas of Queensland and other states, particularly in the semi-arid regions within the above climatic limits, Vetiver hedges when established as water conservation measures, would promote revegetation by leaching out the high salt concentration of the surface soil (Jones 1969). Similarly this application together with its gully stabilising effect could be used on saline lands in India, Thailand and Pakistan to rehabilitate degrading saline lands.

References

Barrett-Lennard, E.G., Warren, B.E. & Malcom, C.V. (1990). Revegetation of saline lands. Vic. Dept. Agri. and Rural Affairs.

Grimshaw, R.G. (1988). A review of existing soil conservation technologies. ASTAG, World Bank. Washington DC, USA.

Isbell, R.F. (1957). The soils of the Inglewood-Talwood-Tara-Glenmorgan region, Queensland. QBC Tech. Bul. No. 5.

Jones, R.M. (1969). J. Soil Cons. NSW 25: 104–120.

US Salinity Laboratories Staff (1954). Agricultural Handbook No. 60. USDA.

Vetiver Newsletter (1991). Special Issue, No. 6. ASTAG. World Bank, Washington DC, USA.

World Bank (1990). Vetiver grass, the hedge against erosion (3rd ed.). ASTAG. World Bank, Washington DC, USA.

32. Some features of salt tolerance in Senna (*Cassia acutifolia*), in Sudan

ALI T. AYOUB
UNEP, Nairobi, Kenya

Abstract. Senna plants, native to the arid parts of the northern Sudan, were studied for mineral cation contents and salt tolerance. Plants tended to accumulate salts in tops, particularly when growing in saline soils. Highest concentrations of salts were in the leaves with a gradient from bottom to top leaves. Under high salinity and soil moisture stress in the field, plant growth was reduced without injury symptoms, but with shedding of some lower leaves. Lower leaf shedding could, therefore, be a physiological measure for reducing plant salt level and transpiration. Studies on the germination of senna seeds under controlled conditions indicated that they were highly tolerant of salinities up to 16.0 mmhos cm^{-1}, and 50% reduction in germination occurred at about 20.0 mmhos cm^{-1}. Air temperature significantly effected senna germination at substrate salinities between 10.0 and 25.0 mmhos cm^{-1}. Studies in open air indicated that plants were more sensitive to salinity and alkalinity at seedling stage than at germination. Young plant survival and total pod yield were significantly reduced in soil salinities higher than 11.0 mmhos cm^{-1}, being more pronounced at shorter than at longer irrigation intervals with saline waters.

Key words: Cations, Drought tolerance, Germination, Mineral composition, Pharmaceutical, Sudan

Introduction

Senna (*Cassia acutifolia* c.v. Delile) is a wild plant native to the arid and semi-arid parts of the Sudan and survives as a perennial crop with occasional irrigation (Tothill 1954). The crop has an important export market in the pharmaceutical trade and the Sudan senna is one of the best known.

The plant seems to be resistant to drought and salinity and grows on lands where usually Na is the principal problem and requires abundance of sunlight and high temperatures. No work has been reported on its tolerance to salinity or alkali or even on its cationic composition. Information of this nature would assist in understanding the environmental requirements of this promising crop, and in assessing the potential use of saline soils.

The purpose of this study, was therefore, to examine the cationic composition of this plant with samples taken from specimens grown in saline and non-saline natural habitats. Studies were also undertaken on the effects of salt level, salt types, and soil alkalinity on senna germination, vegetative growth, yield, and mineral composition. Some environmental factors which may effect salt tolerance by senna, such as soil moisture and air temperature were also included in these studies.

1. Materials and methods

Plant and soil samples from natural habitats

Samples of naturally growing senna were collected from localities along the eastern side of the Nile in northern Sudan at a distance of about 150

V.R. Squires & A.T. Ayoub (eds.), *Halophytes as a resource for livestock and for rehabilitation of degraded lands*, 297–301.

km in a typical arid zone with an average annual precipitation of about 50–100 mm and potential evaporation of more than 2,000 mm. Sampling was carried out during the dry season and each sample consisted of five plants which were analyzed individually. Plant samples were separated into leaves, stems, roots, flowers, and pods, and some samples were further separated into top, middle, and bottom leaves and stems. Soils are mostly saline/sodic and medium in texture, but most of the samples were collected from watercourses which are usually leached sandy soils. Soil samples were also collected from under and between plant samples and were separated into profile depths 0–30 and 30–60 cm. Na, K, Ca, Mg and Cl were determined in plant and soil samples.

Germination studies in the laboratory

Germination experiments were carried out in 10-cm-diameter Petri dishes in a single randomized design with 11 levels of NaCl solutions (0–40.0 mmhos cm^{-1}) replicated four times. The Petri dishes which contained 10 seeds of senna were covered and placed in the dark in an incubator maintained at 30 ± 2 °C. The germination count was taken every 24 hr. In another germination test of interaction of temperature and salinity four replicates of salt levels of 0.5, 8.3, 16.3, 24.5 and 33.6 mmhos cm^{-1} NaCl were incubated at 12–15, 28–30, 32–35, and 40–45 °C and germination counts were made on the 7th day.

Studies in the greenhouse

Exchangeable sodium tolerance studies at levels 5, 15, 24, 45 and 55% were done in pots containing 5.5 kg of sodium treated soils. Effects of salt types and levels on plant growth were also studied in pots. Equal concentrations of NaCl, $CaCl_2$, Na_2SO_4, and an equal mixture of NaCl and $CaCl_2$ were compared at the levels shown in Table 2.

Studies in lysimeters

Frequency of irrigation with NaCl on senna growth was studied in metallic lysimeters with drainage holes at the bottom and containing about 130 kg of river silt. The experimental design was a factorial combination of five concentrations of NaCl (0.2, 2.5, 5.0, 10.0, and 20.0 mmhos cm^{-1}) and two intervals of watering (every 7 or 14 d with the salt solution). These lysimeters, with three replicates, were arranged randomly in the open air where daily temperatures ranged between 25 and 39 °C. Observations concerning plant survival and development were recorded. Samples of dead plants were also taken throughout the duration of the experiment. Pod yields were recorded periodically and at the termination of the experiment yield components and some plant characters were determined. Leaf samples were also taken for Na and Cl analysis.

Table 1. Cationic composition of senna plants growing in non-saline soil

Plant parts	mEq per 100 g dry wt.				
	Na	K	Ca	Mg	Cation sum
Flowers	1.3	51.7	28.0	86.9	167.9
Pods	1.2	44.0	18.0	67.2	130.4
Top leaves	1.5	44.8	37.5	103.3	187.1
Middle leaves	2.5	54.5	81.5	47.5	186.0
Bottom leaves	3.5	51.2	112.5	59.0	226.2
Top stem	2.2	41.4	45.0	99.2	148.7
Bottom stem	1.4	18.4	15.0	65.6	100.4
Roots	1.7	14.4	33.8	57.9	107.9
s.e. mean	0.8	5.40	11.93	7.27	15.26

Adapted from Ayoub (1975).

Results

Soil analysis

Major differences among sites from which plant samples were taken are not readily apparent, though there were some saline spots. Soils were non-saline up to 30 cm depth, but at lower depths salinity varied from non to about 15 mmhos cm^{-1}. Variation of minerals between sites included Na, Ca, and Mg, especially in the lower horizon where part of the upper active roots were found.

Cationic composition of plant parts

The cationic composition of senna plant parts growing in normal soils is shown in Table 1. The Na contents of the plant parts ranged from 1–

Table 2. Concentrations of cations in roots, stems, and leaves of *C. acutifolia* on saline and non-saline soils

Plant parts	mEq per 100 g dry wt.				
	Na	K	Ca	Mg	Cation sum
Saline soil					
Roots	3.7	11.0	21.5	41.8	78.0
Stems	10.9	21.7	39.5	69.7	141.8
Leaves	18.5	24.0	200.5	70.5	313.5
s.e. mean	4.27	4.00	56.91	9.43	70.32
Non-saline soil					
Roots	1.9	17.7	65.0	45.1	129.7
Stems	1.9	22.0	59.0	80.3	163.2
Leaves	2.5	51.9	101.5	111.5	267.4
s.e. mean	0.20	10.76	13.28	19.18	41.48

Adapted from Ayoub (1975).

$4\,\text{mEq}\ 100\,\text{g}^{-1}$. The leaves retained the highest level of Na and a decreasing gradient existed from leaves to stems to roots, with the older leaves retaining most. Potassium content of the plant was generally low but fairly evenly distributed in the upper parts. The lower parts of the stem and the root contained the least proportion (less than $20\,\text{mEq}\ 100\,\text{g}^{-1}$). Highest concentrations of Ca were in the leaves, the bottom leaves retaining most. However, Ca in the stems showed a steep gradient from top to bottom. Magnesium contents were generally high throughout the plant. Top leaves and top stem contained the highest proportion (103 and $99\,\text{mEq}\ 100\,\text{g}^{-1}$, respectively). The cation mEq sum was in the order leaves > stems > roots. The decrease in cation mEq sum seemed to be associated with decrease in Ca content, though the decrease in Ca was partially compensated for by increased Mg.

From Table 2 we see that plants growing on normal soils contained high amounts of K and Mg, especially in their leaves. In plants growing in saline soils excessive amounts of Na and Ca were accumulated in the leaves, while the amounts of K and Mg tended to decrease in the leaves without their contents being much affected in the stems and roots. In contrast to K and Mg, Na and Ca accumulated in the leaves at the expense of their contents in the stems and roots. Comparing mineral contents in the leaves with their contents in the roots, Na increased about five-fold and Ca about ten-fold in the leaves when the plants were growing on saline soils. The ability of this plant to accumulate salts in the leaves is noteworthy. The cation mEq sum also increased in the leaves at the expense of the stems and roots on saline soil.

Table 3. Effect of NaCl levels in the soil on plant growth and Na and Cl contents of senna tops

Salt level (mmhos cm^{-1})	Relative growth	mEq per 100 g dry wt.	
		Na	Cl
0.9	100.0	2.7	18.7
2.4	110.4	3.4	14.6
4.8	81.6	7.3	14.6
9.2	72.2	10.9	19.9
13.0	65.9	12.3	30.8
17.9	49.1	12.0	38.3
s.e. mean	9.21	1.75	3.93
		$r = -0.9367$	$r = -0.8226$

Adapted from Ayoub (1975).

Effect of NaCl levels on senna growth

To examine the effect of seasonal fluctuation in soil salinity on senna growth and mineral content, naturally growing plants were transferred to pots and irrigated with water containing different levels of NaCl and their growth and mineral content measured (Table 3). Each value is a mean of three replicates of five plants each.

The growth rate of senna was reduced when the NaCl concentration was increased above $9\,\text{mmhos cm}^{-1}$, and up to 50% reduction in growth was reached at salt concentration above $17\,\text{mmhos cm}^{-1}$. However, even at this level, no injury symptoms were noted except for some leaf shedding. NaCl levels above $9\,\text{mmhos cm}^{-1}$ did not result in appreciable increase in the Na content of the tissue, but that of Cl was doubled when the substrate NaCl was increased from 9.2 to $17.9\,\text{mmhos cm}^{-1}$. This may suggest cessation of active Na^+ uptake at higher salinity levels in mature senna plants.

Effects of salt levels, temperature, ESP, salt types, and watering intervals on seed germination (from Ayoub 1977)

Seed germination was satisfactory up to a salt level of about $16.0\,\text{mmhos cm}^{-1}$ NaCl. Between salt levels of 20.0 and $24.0\,\text{mmhos cm}^{-1}$ germination was reduced 50% and beyond the salt level of 30.0 mmhos cm^{-1} germination was almost inhibited.

Substrate salinity, air temperature, and their interaction highly significantly affected senna germination. Up to a salinity of 6.0 mmhos cm^{-1} there was no difference between temperatures from 28 to 45 °C in germination percentage. Air temperature had its greatest effect on germination at salinities between 10.0 and 25.0 mmhos cm^{-1}. Fifty per cent reduction in germination occurred at salinities of about 10.0, 17.0, and 22.0 mmhos cm^{-1} when air temperatures were 40–45, 32–35, and 28–30 °C, respectively. Senna did not germinate at temperatures below 15 °C.

Seed germination was satisfactory up to an ESP level of 35. Beyond that level some seeds germinated, but most of the seedlings died shortly after that. Seedlings growing in ESP levels of 25 or more were stunted and had bleached-yellowish (chlorotic) secondary leaves. The effect of ESP was, therefore, more pronounced at the seedling growth stage than during seed germination.

Plants receiving $CaCl_2$ were affected more markedly than those receiving the corresponding levels of NaCl while plants grown in Na_2SO_4 media showed less changes in growth than those grown in NaCl.

Watering senna plants every 7 d with lower-salt waters resulted in appreciably higher yields than watering every 14 d. Contrary to this, frequent irrigation with high-salt waters injured plants and resulted in lower pod yields than irrigation every 14 d with the same water. This might indicate that the effect of NaCl salinity on senna was not solely an osmotic phenomenon.

Discussion

Senna in the area surveyed seemed to be subjected to a wide range of salinity and great seasonal fluctuations, but plant survival was noted to be highly dependent on soil moisture availability. Great variability was observed between plant growth in different localities. Samples varied between healthy vigorous plants and stunted defoliated ones. Our experience with the crop shows that senna is a hardy plant that grows normally in the poorest types of soil as long as adequate soil moisture is available, but in heavy clays the crop may suffer from water logging especially at germination and seedling stages. In summer most of the plants displayed visual symptoms of water stress, like defoliation, retardation of growth, and die-back in severe cases.

As can be seen from Tables 1 and 2 the plant accumulates salt in the tops, and the older leaves retain most of it. Lower leaf shedding is therefore a physiological measure of reducing plant salt level, but may also reduce productivity. Rickard (1965) postulated that early summer leaf abscission of hopsage (*Grayia spinosa*) functioned to reduce transpiration and might also be an effective means of removing annual accumulation of ions from the plant system, but in greasewood he found that leaves abscissed when Na content was high irrespective of soil moisture. He concluded that salt desert communities do not require Na for growth but merely to tolerate alkaline conditions. Gale et al. (1970), working with *A. halimus*, a halophyte, found that relatively small concentrations of NaCl were required to alleviate the adverse effect of moisture stress on the plant.

Data in Table 3 show that naturally growing senna when transfered to saline media showed some tolerance of salts in the sense that no injury symptoms were noticed even at high salt levels (17.9 mmhos cm^{-1}), but it virtually made little growth at that salinity level (about 50 per cent reduction in growth). This habit is of great significance in a desert ecosystem where large seasonal fluctuations in salinity exist due to rainfall or other leaching action. The fact that senna accumulates relatively large amounts of ions in the tops shows that some degree of salt tolerance may be associated with its drought resistance.

Senna showed to be fairly tolerant of salinity at the germination stage with 50% reduction reached at salt levels between 20.0 and 24.0 mmhos cm^{-1}. The relatively higher tolerance of senna to salinity during germination is not surprising since it is a crop originating from regions with a dry climate and where salinity problems usually exist. The optimum temperature at which senna could germinate satisfactorily at high salinity levels was found to be between 28 and 30 °C. The significant interaction of soil salinity and air temperature reported in this study might be important in determining optimum sowing dates on saline soils.

Senna seeds germinated in soils high in exchangeable sodium (ESP of 35), but most of the seedlings died shortly after. The chlorotic symptoms appearing in seedlings growing in high ESP levels could be due to the toxic effects of bicarbonate ions resulting in high pH values. Salt type effects were considered important in senna since Na_2SO_4 effects were generally less detrimental to the plant growth than were those of $CaCl_2$ or NaCl of the same salinity. With chloride salinity, $CaCl_2$ was more toxic than NaCl. It would appear that the high chloride concentration in the plant tissue produces the drastic decrease in plant growth. It is possible that the high sodium levels in plants found in some treatments could be due to chloride injury to membranes giving rise to passive sodium uptake. In its sensitivity to chloride injury, senna behaves in a like manner to avocado, citrus, and grapes but contrary to most plant species which are more sensitive to sulphate salts (Mengel 1973).

For good pod yields of senna, NaCl salinity should not exceed about 8.0 mmhos cm^{-1} in the soil, but satisfactory yields may be obtained up to 15.0 mmhos cm^{-1}, similar to safflower and wheat (Bernstein 1970). More frequent irrigation with saline water induced earlier and more severe injury than less frequent irrigation. This could be due to greater osmotic shock (Bernstein & Francois 1975).

In conclusion, one can say that senna is moderately tolerant to salinity. It is more salt tolerant at germination and later growth stages, the seedling stage being the most sensitive. Avoiding adverse environmental conditions like high air temperature, low relative humidity, and bad cultural practices at that critical stage may improve the salinity tolerance by senna and its survival in the field.

References

Ayoub, Ali T. (1975). Sodium and chloride accumulation by senna (*Cassia acutifolia*). J. Exp. Bot. 26: 891–896.

Ayoub, Ali T. (1977). Some primary features of salt tolerance in senna (*Cassia acutifolia*). J. Exp. Bot. 28: 484–492.

Bernstein, L. (1970). Salt tolerance of plants. Agric. Inform. Bull. No. 283, US Dept. Agric.

Bernstein, L. & Francois, L.E. (1975). Agron. J. 67: 185–90.

Gale, J., Naaman, R. & Poljakof-Mayber, A. (1970). Growth of *Atriplex halimus* L. in sodium chloride salinated culture solutions as affected by the relative humidity of the air. Aust. J. Biol. Sci. 23: 947–952.

Mengel, K. (1973). First FAO/SIDA Seminar for plant scientists from Africa and Near East, Cairo, Egypt, 1–20 September 1973.

Rickard, W.H. (1965). Bot. Gaz. 126, 116–119.

Tothill, J.D., ed. (1954). Agriculture in the Sudan. London, Oxford University Press.

33. Characterization of progeny clones belonging to natural Sicilian *Atriplex halimus* populations

L. STRINGI, D. GIAMBALVO, A. ACCARDO & G. AMATO
Istituto di Agronomia generale e Coltivazioni erbacee, Universita degli Studi de Palermo, Italy

Abstract. The aim was to assess the forage potential and protective function of Sicilian *Atriplex halimus* populations. The objective was to characterize 34 progeny clones of superior phenotypes selected among four Sicilian populations. A plant breeding plan was implemented to achieve an improved and stable population in the short term (5 years) and to constitute synthetic varieties in the medium term (9 years). Sixty progeny clones (20 plants replicated three times with a 1.5×2.0 m layout) obtained by cuttings were directly planted in the field in January 1991. Altogether there were 2,040 plants. In August 1992 all of the plants of two replications (1,360 plants) were cut 50 cm above ground level and the total dry matter, edible fraction were determined. Regrowth volume 50 days after cutting was also observed. The progeny clones of the third replication (680 plants) were grazed by sheep at a stocking rate of 70 head ha^{-1} d^{-1} during August. Palatability was evaluated using visual estimation criteria (0 = no grazing; 10 = max. grazing). There was considerable variability in total dry matter production and the proportion of edible fraction. The progeny clones which were generally characterized by a lower total dry matter production showed a higher ratio of edible dry matter to total dry matter. A preliminary evaluation of palatability showed that variability exists within each progeny group and that some clones, in each of the groups seemed to be highly palatable. A cluster analysis on all of the observed characters of the clones of different origin was carried out. Regrowth capacity was significantly correlated with total dry matter production of the plants.

Key words: Cluster analysis, Dry matter production, Palatability

Introduction

Genetic research on shrubs has not received the same attention as cultivated species. There are several reasons for this;

- The breeding systems, the genetic control mechanisms, and the heritability of characters are not yet known for many of the shrub species.
- The marked variability of the breeding systems.
- The present low economic value of shrubs and the fact that modest results are obtained in plant breeding programmmes while costs are high.

Despite this, the known high potential of many shrubby species, especially *Atriplex*, and the increasing interest in exploiting them has induced some researchers to carry out shrub improvement programmes.

The genus *Atriplex* has 417 species which are very diversified for:

- tolerance to salinity, drought, cold, and water-logging;
- the capacity to adapt to quite degraded soils;
- protein content;
- palatability;
- total and edible biomass production;
- regrowth capacity;
- content of antinutritional substances.

Many of these factors affect potential use as forage and for rehabilitation of degraded land. Salt

V.R. Squires & A.T. Ayoub (eds.), *Halophytes as a resource for livestock and for rehabilitation of degraded lands*, 303–310.

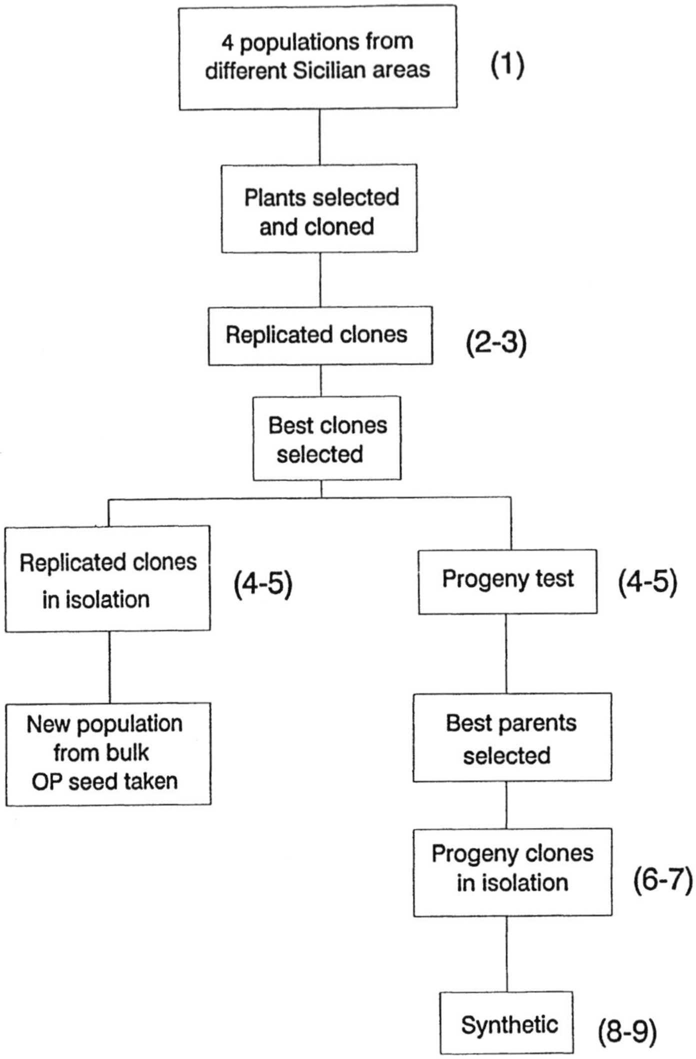

Figure 1. A flow chart of the *Atriplex halimus* selection and breeding plan.

tolerance is also related to the option of using saline water for irrigation. Studies in the USA (O'Leary 1985, 1986; Watson et al. 1987; Watson 1990) and in Israel (Forti 1986) show that brackish water irrigation is feasible and that halophytes have potential as agricultural field crops.

In the semi-arid environment of southern Italy social, economic and environmental demands are focussing attention on:

- the problems of improvement of eroded soils (frequently quite saline) which are in an advanced state of degradation;
- the protection of hilly areas already with a low productive potential from meteoric water erosion;
- the supply of forage for sheep during the drought season (summer and early autumn).

Shrubs have potential role in addressing these

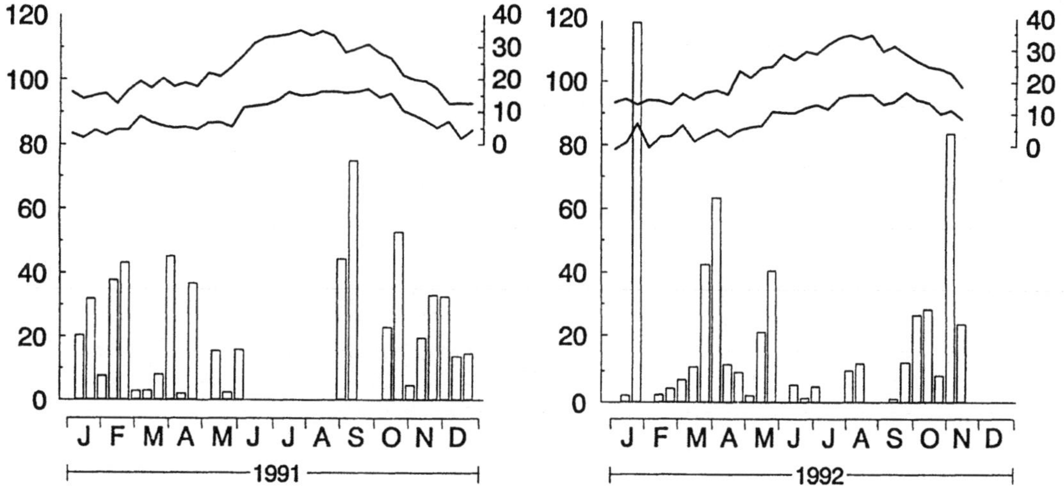

Figure 2. Climatic data (10-day averages of rainfall and temperature) for the study site.

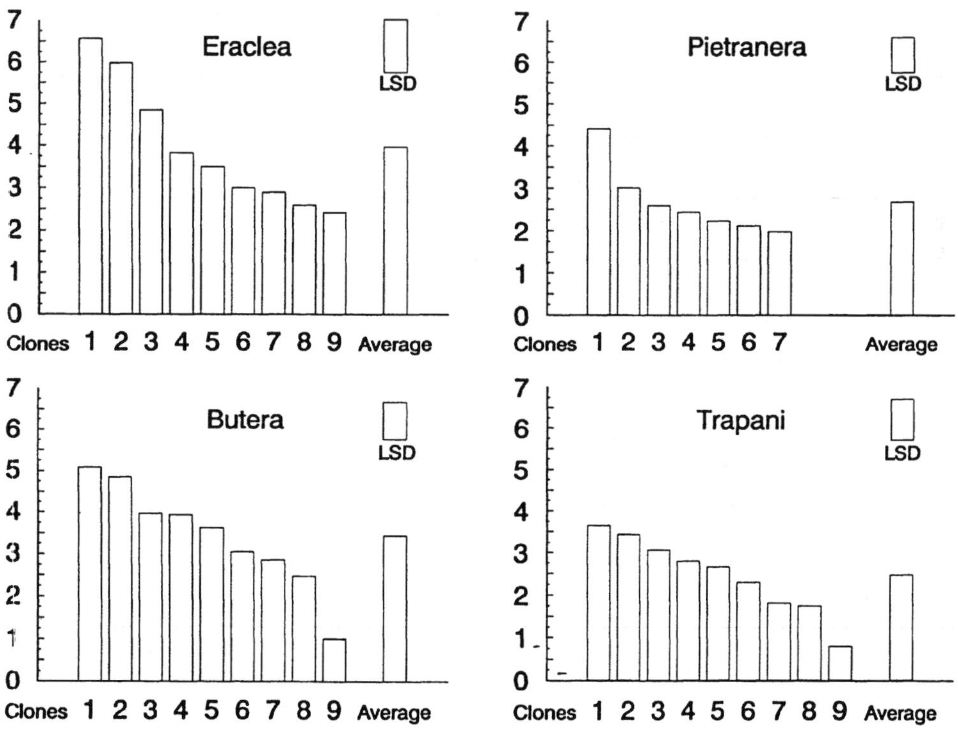

Figure 3. Total dry matter/plant (kg) at cutting.

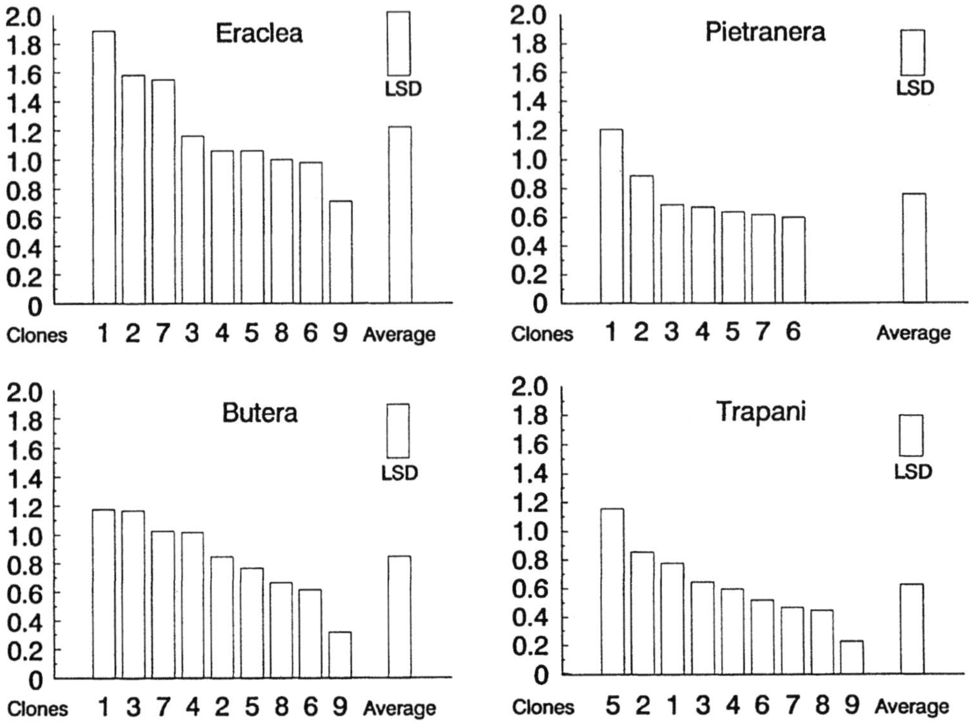

Figure 4. Edible dry matter/plant (kg) at cutting.

problems. The present study is aimed at assessing the forage potential and the protective function of Sicilian *Atriplex* populations through:

- the exploitation of natural variability of *Atriplex* populations of different origin;
- the development of synthetic populations with high palatability, high production of edible biomass and rapid regrowth capacity, good adaptability to the limiting factors of the semiarid mediterranean environment (summer drought, degraded soils and high salinity).

A breeding plan (Figure 1) was adopted. It should make it possible to obtain an improved and stable population in the short term (5 years) and a synthetic variety in the medium term (9 years).

Materials and methods

Selections were made from *Atriplex h.* populations in four different areas of Sicily. Two sites were coastal (Eraclea and Trapani) and two were inland (Butera and Pietranera). A total of 34 phenotypes, which appeared to be superior for parameters such as height, crown diameter, leafiness (visually estimated), were selected within natural populations. From each selected phenotype 60 clones were obtained. The number of progenies for each population was 9 for the Eraclea, Trapani and Butera progeny groups and 7 for the Pietranera one.

By the end of December 1991, 60 cuttings of each phenotype were planted directly into field plots with a 1.5×2.0 m layout. Each replicate contained 20 cuttings. A randomized block design with three replications was used. Climatic data for the study site are summarized in Figure 2.

In August 1992 all of the plants of two replications were cut 50 cm above ground level and the following observations were made on 1360 plants:

- total fresh and dry matter per plant;
- edible fresh and dry matter per plant (on the basis of one plant per replicate);
- assessment of plant volume 50 days after cut-

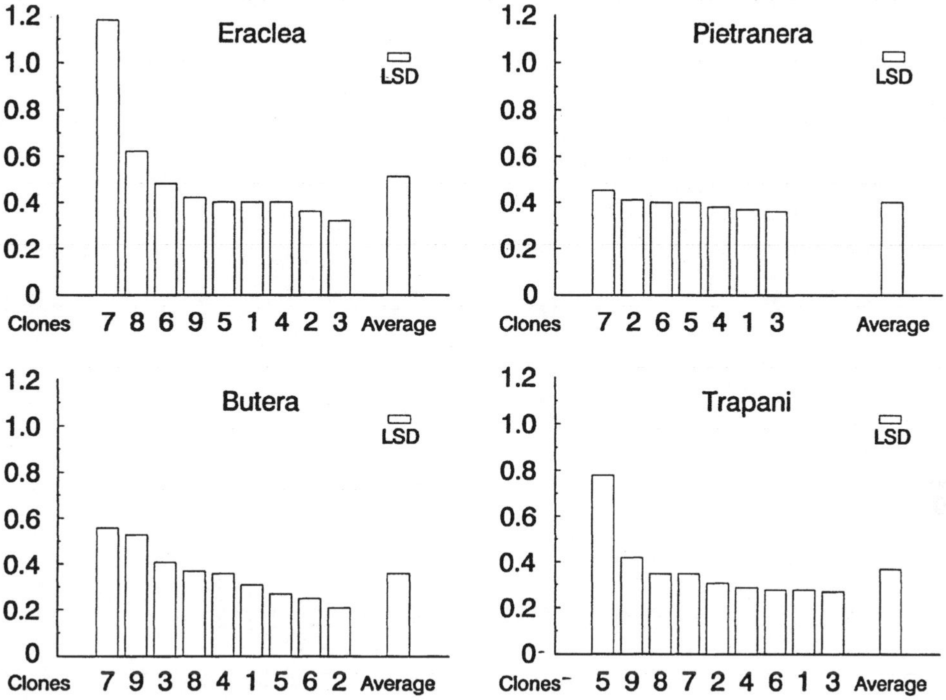

Figure 5. Edible dry matter/total dry matter ratio.

ting, based on measurements of height and crown diameter.

On the remaining replication, a month-long sheep grazing trial was carried out in August 1992 using 70 head ha^{-1} day^{-1}. A preliminary assessment of palatability was estimated on the basis of visual criteria of the utilization intensity (0 = no grazing; 10 = maximum).

Results and discussion

The mean production of total dry matter/plant of the progeny clones (Figure 3) ranged from 2.5 kg in the Trapani group to 4.0 kg in the Eraclea one. The Butera and Pietranera progeny clones (inland populations) were intermediate with mean values ranging from 2.7–3.4 kg $plant^{-1}$. Variability for this character was high for all of the progeny groups. The widest range was expressed by the Butera progenies with values ranging from 1–5 kg $plant^{-1}$ and the lowest by the Pietranera (2.1–4.4 kg $plant^{-1}$). Edible dry matter (Figure 4) was also quite variable; there was a good correlation between total dry matter production and production of edible dry matter. The edible dry matter to total dry matter ratio (Figure 5) was often higher in the least productive clones; e.g. clone 7 of the Eraclea group and clone 5 of Trapani.

The preliminary evaluation of palatability to sheep (Figure 6) showed high variability. On average, the Trapani progeny clones were the least palatable, whereas Eraclea was most acceptable to the sheep. Four clones from Eraclea (4, 6, 2 and 7) had mean values over 8 on a scale of 10. Highly palatable genotypes were found in this preliminary phase of the evaluation in each group (progeny 8 and 4 of Trapani, progeny 8 of Butera and progeny 4 of Pietranera).

Plant volume measured 50 days after cutting (Figure 7) averaged 0.24 m^3 in the Eraclea progenies and slightly lower in the Pietranera and Butera progenies. Within each progeny group differ-

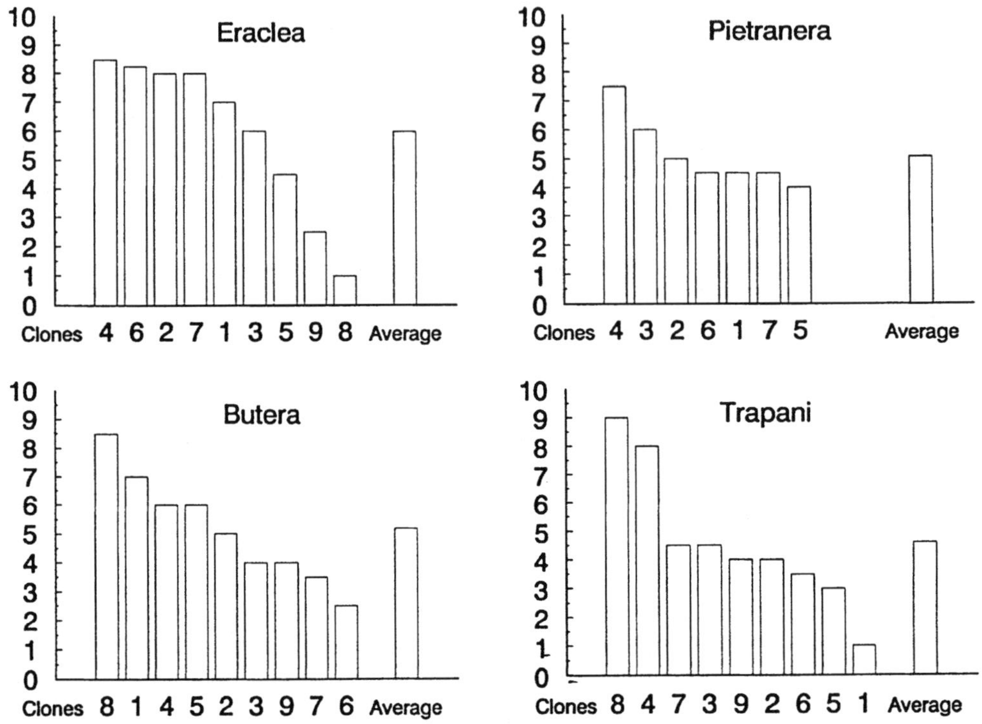

Figure 6. Palatabilty ratings (0 = no grazing; 10 = maximum grazing).

ences were small among the clones, although variability in Eraclea and Trapani was slightly greater. There was a significant correlation (r = 0.637) between regrowth capacity (expressed as plant volume) and total dry matter of the plants prior to cutting in August (Figure 8). The Trapani progenies had the lowest regrowth capacity (0.17 m^3) after the summer cutting.

The mean values of the observed characters of the progenies of different origin were treated with a global statistical analysis using the single linkage method cluster analysis. Figure 9 summarizes the results. There are four fundamental groups, mostly diversified on the basis of palatability to the sheep. There are 10 progeny clones (all characterized by high palatability) in the first cluster group, five of them from the coastal population Eraclea. These might form a useful basis for the breeding work.

As concerns the edible dry matter and regrowth capacity expressed by this first cluster group progeny 1 Eraclea and progeny 1 Butera ranked well.

The possibility of using cuttings directly planted in the field was confirmed. This supports the work of Franclet & Le Houérou (1971) and Amato et al. (1990).

The preliminary evaluation of palatability needs to be extended to consider only the best clones evaluated under wider range of conditions.

The study showed that there is high variability among the local populations of *Atriplex* as evidenced by the performance of progeny clones from different origins. This variability might be exploited in a breeding programme.

Acknowledgments

This study was supported by a grant from the Italian Ministry of Agriculture and Forestry within the project 'Research on forage crops'.

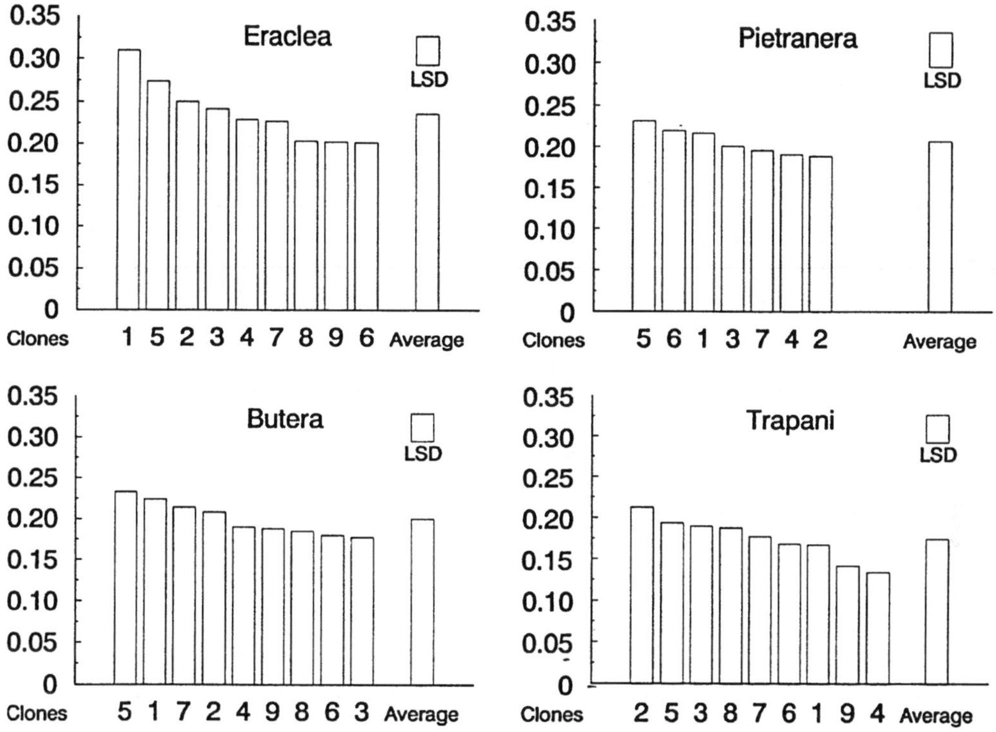

Figure 7. Plant volume (m^3) 50 days after cuttings.

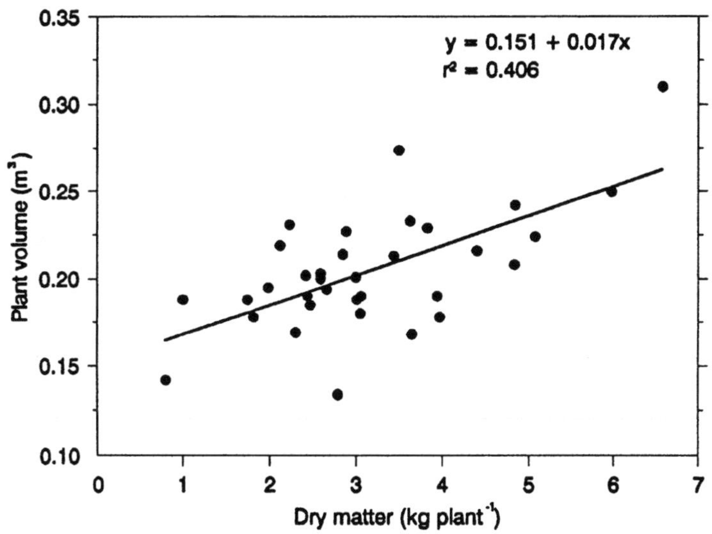

Figure 8. Relation between plant volume (m^3) of various *Atriplex halimus* selections 50 days after cutting and total dry matter/plant (kg) at time of harvest.

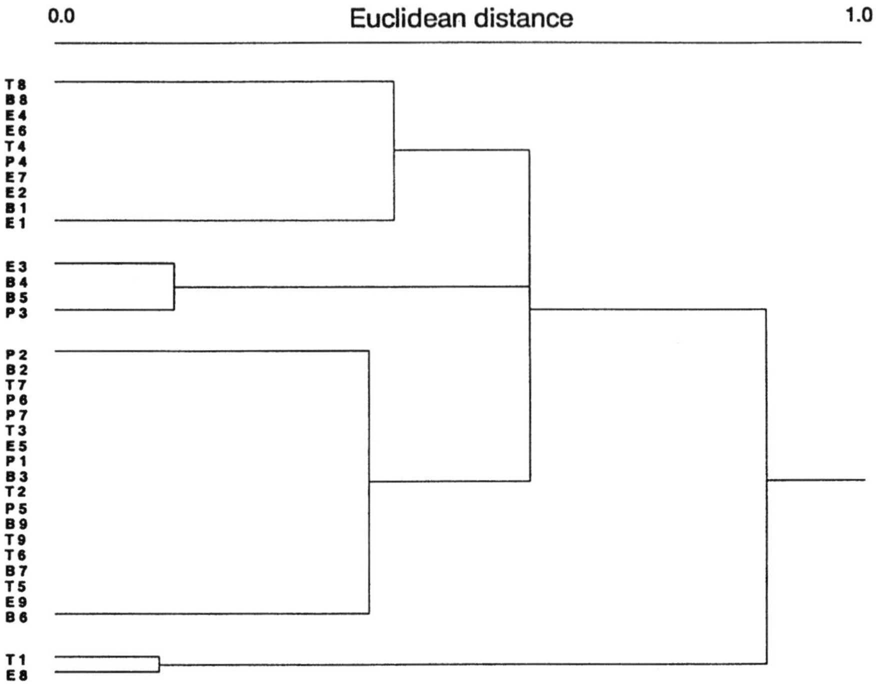

Figure 9. Cluster analysis (single linkage method) of the progeny clones of *Atriplex halimus*. Populations: E = Eraclea; P = Pietranera; B = Butera; T = Trapani.

The authors contributed equally to the paper. We are grateful to Dr V. Squires for his help in editing this paper.

References

Amato, G., Stringi, L., Gristina, L., Meli, R. & Accardo, A. (1990). Propagation by cuttings of *Atriplex halimus* L. as affected by various rooting factors. FAO, Bari, 17–19 October, pp. 135–137.

Forti, M. (1986). Salt tolerant halophytic plants in Israel. Reclam. Reveg. Res. 5: 83–96.

Franclet, A. & Le Houérou, H.N. (1971). Les *Atriplex* en Tunisie et en Afrique du Nord. FO: SF/TUN. Rapport technique 7 PNUD/FAO, Rome.

O'Leary, J.W. (1986). A critical analysis of the use of *Atriplex* species as crop plants for irrigation with highly saline water. Proc. US-Pakistan Biosaline Res. Workshop, Botany Dept. University of Karachi, Univ. of Karachi, pp. 415–432.

O'Leary, J.W., Glenn, E.P. & Watson, M.C. (1985). Agricultural production of halophytes irrigated with seawater. Plant & Soil 89: 311–321.

Watson, C.M. (1990). *Atriplex* species as irrigated forage crops. Agric. Ecosys. & Envir. 32: 107–118

Watson, M.C., O'Leary, J.W. & Glenn, E.P. (1987). Evaluation of *Atriplex lentiformis* (Tarr.) S. Wats. and *Atriplex nummularia* Lindl as irrigated forage crops. J.Arid Envir. 13: 293–303.

Key words index

Author index

Tasks for vegetation science

1. E.O. Box: *Macroclimate and plant forms.* An introduction to predictive modelling in phytogeography. 1981 ISBN 90-6193-941-0
2. D.N. Sen and K.S. Rajpurohit (eds.): *Contributions to the ecology of halophytes.* 1982 ISBN 90-6193-942-9
3. J. Ross: *The radiation regime and architecture of plant stands.* 1981 ISBN 90-6193-607-1
4. N.S. Margaris and H.A. Mooney (eds.): *Components of productivity of Mediterranean-climate regions.* Basic and applied aspects. 1981 ISBN 90-6193-944-5
5. M.J. Müller: *Selected climatic data for a global set of standard stations for vegetation science.* 1982 ISBN 90-6193-945-3
6. I. Roth: *Stratification of tropical forests as seen in leaf structure* [Part 1]. 1984 ISBN 90-6193-946-1
   For Part 2, see Volume 21
7. L. Steubing and H.-J. Jäger (eds.): *Monitoring of air pollutants by plants.* Methods and problems. 1982 ISBN 90-6193-947-X
8. H.J. Teas (ed.): *Biology and ecology of mangroves.* 1983 ISBN 90-6193-948-8
9. H.J. Teas (ed.): *Physiology and management of mangroves.* 1984 ISBN 90-6193-949-6
10. E. Feoli, M. Lagonegro and L. Orláci: *Information analysis of vegetation data.* 1984 ISBN 90-6193-950-X
11. Z. Šesták (ed.): *Photosynthesis during leaf development.* 1985 ISBN 90-6193-951-8
12. E. Medina, H.A. Mooney and C. Vázquez-Yánes (eds.): *Physiological ecology of plants of the wet tropics.* 1984 ISBN 90-6193-952-6
13. N.S. Margaris, M. Arianoustou-Faraggitaki and W.C. Oechel (eds.): *Being alive on land.* 1984 ISBN 90-6193-953-4
14. D.O. Hall, N. Myers and N.S. Margaris (eds.): *Economics of ecosystem management.* 1985 ISBN 90-6193-505-9
15. A. Estrada and Th.H. Fleming (eds.): *Frugivores and seed dispersal.* 1986 ISBN 90-6193-543-1
16. B. Dell, A.J.M. Hopkins and B.B. Lamont (eds.): *Resilience in Mediterranean-type ecosystems.* 1986 ISBN 90-6193-579-2
17. I. Roth: *Stratification of a tropical forest as seen in dispersal types.* 1987 ISBN 90-6193-613-6
18. H.-G. Dässler and S. Börtitz (eds.): *Air pollution and its influence on vegetation.* Causes, Effects, Prophylaxis and Therapy. 1988 ISBN 90-6193-619-5
19. R.L. Specht (ed.): *Mediterranean-type ecosystems.* A data source book. 1988 ISBN 90-6193-652-7
20. L.F. Huenneke and H.A. Mooney (eds.): *Grassland structure and function.* California annual grassland. 1989 ISBN 90-6193-659-4
21. B. Rollet, Ch. Högermann and I. Roth: *Stratification of tropical forests as seen in leaf structure*, Part 2. 1990 ISBN 0-7923-0397-0
22. J. Rozema and J.A.C. Verkleij (eds.): *Ecological responses to environmental stresses.* 1991 ISBN 0-7923-0762-3
23. S.C. Pandeya and H. Lieth: *Ecology of Cenchrus grass complex.* Environmental conditions and population differences in Western India. 1993 ISBN 0-7923-0768-2
24. P.L. Nimis and T.J. Crovello (eds.): *Quantitative approaches to phytogeography.* 1991 ISBN 0-7923-0795-X
25. D.F. Whigham, R.E. Good and K. Kvet (eds.): *Wetland ecology and management.* Case studies. 1990 ISBN 0-7923-0893-X

Tasks for vegetation science

26. K. Falinska: *Plant demography in vegetation succession.* 1991 ISBN 0-7923-1060-8
27. H. Lieth and A.A. Al Masoom (eds.): *Towards the rational use of high salinity tolerant plants,* Vol. 1: Deliberations about high salinity tolerant plants and ecosystems. 1993 ISBN 0-7923-1865-X
28. H. Lieth and A.A. Al Masoom (eds.): *Towards the rational use of high salinity tolerant plants,* Vol. 2: Agriculture and forestry under marginal soil water conditions. 1993 ISBN 0-7923-1866-8
29. J.G. Boonman: *East Africa's grasses and fodders.* Their ecology and husbandry. 1993 ISBN 0-7923-1867-6
30. H. Lieth and M. Lohmann (eds.): *Restoration of tropical forest ecosystems.* 1993 ISBN 0-7923-1945-1
31. M. Arianoutsou and R.H. Groves (eds.): *Plant-animal interactions in Mediterranean-type ecosystems.* 1994 ISBN 0-7923-2470-6
32. V.R. Squires and A.T. Ayoub (eds.): *Halophytes as a resource for livestock and for rehabilitation of degraded lands.* 1994 ISBN 0-7923-2664-4

GPSR Compliance

The European Union's (EU) General Product Safety Regulation (GPSR) is a set of rules that requires consumer products to be safe and our obligations to ensure this.

If you have any concerns about our products, you can contact us on ProductSafety@springernature.com

In case Publisher is established outside the EU, the EU authorized representative is:

Springer Nature Customer Service Center GmbH
Europaplatz 3
69115 Heidelberg, Germany

FSC
www.fsc.org
MIX
Papier | Fördert
gute Waldnutzung
FSC® C083411

Kolibri 360 GmbH
Ferdinand-Jühlke-Straße 7
99095 Erfurt, Deutschland
produktsicherheit@kolibri360.de